1 DEBYE $= 3.33 \times 10^{-30}$ C·m (ELECTRIC DIPOLE MOMENT)

Conversions (continued)

1 foot	$= 0.3048 \ldots \text{m}$
1 foot·pound$_f$	$= 1.355 \ldots \text{J}$
1 gallon (U S. liq.)	$= 3.78 \ldots \times 10^{-3} \text{ m}^3$
1 gram	$= 0.602 \ldots \times 10^{24} \text{ amu}$
	$= 2.20 \ldots \times 10^{-3} \text{ lb}_m$
1 gram/cm^3	$= 62.4 \ldots \text{ lb}_m/\text{ft}^3$
	$= 1000 \text{ kg/m}^3$
	$= 1 \text{ Mg/m}^3$
1 inch	$= 0.0254 \ldots \text{m}$
1 joule	$= 0.947 \ldots \times 10^{-3} \text{ Btu}$
	$= 0.239 \ldots \text{ cal, gram}$
	$= 6.24 \ldots \times 10^{18} \text{ eV}$
	$= 0.737 \ldots \text{ ft·lb}_f$
	$= 1 \text{ watt/sec}$
1 joule/meter^2	$= 8.80 \ldots \times 10^{-5} \text{ Btu/ft}^2$
1 [joule/(m^2·s)]/[°C/m]	$= 1.92 \ldots \times 10^{-3} \text{ [Btu/(ft}^2\text{·s)]/[°F/in.]}$
1 kilogram	$= 2.20 \ldots \text{ lb}_m$
1 megagram/meter^3	$= 1 \text{ g/cm}^3$
	$= 10^6 \text{ g/m}^3$
	$= 1000 \text{ kg/m}^3$
1 meter	$= 10^{10} \text{ Å}$
	$= 10^9 \text{ nm}$
	$= 3.28 \ldots \text{ ft}$
	$= 39.37 \ldots \text{ in.}$
1 micrometer	$= 10^{-6} \text{ m}$
1 nanometer	$= 10^{-9} \text{ m}$
1 newton	$= 0.224 \ldots \text{ lb}_f$
1 ohm·inch	$= 0.0254 \ldots \Omega \cdot \text{m}$
1 ohm·meter	$= 39.37 \ldots \Omega \cdot \text{in.}$
1 pascal	$= 0.145 \ldots \times 10^{-3} \text{ lb}_f/\text{in.}^2$
1 poise	$= 0.1 \text{ Pa·s}$
1 pound (force)	$= 4.44 \ldots \text{ newtons}$
1 pound (mass)	$= 0.453 \ldots \text{ kg}$
1 pound/foot^3	$= 16.0 \ldots \text{ kg/m}^3$
1 pound/inch^2	$= 6.89 \ldots \times 10^{-3} \text{ MPa}$
1 watt	$= 1 \text{ J/s}$
1 (watt/m^2)/(°C/m)	$\dot{=} 1.92 \ldots \times 10^{-3} \text{ [Btu/(ft}^2\text{·s)]/[°F/in.]}$

SI prefixes

giga	G	10^9
mega	M	10^6
kilo	k	10^3
milli	m	10^{-3}
micro	μ	10^{-6}
nano	n	10^{-9}

FOURTH EDITION # Elements of Materials Science and Engineering

LAWRENCE H. VAN VLACK
THE UNIVERSITY OF MICHIGAN
ANN ARBOR, MICHIGAN

ADDISON-WESLEY PUBLISHING COMPANY

READING, MASSACHUSETTS • MENLO PARK, CALIFORNIA
LONDON • AMSTERDAM • DON MILLS, ONTARIO • SYDNEY

This book is in the
Addison-Wesley series in metallurgy and materials

Consulting Editor
Morris Cohen

Sponsoring Editor: Thomas Robbins
Production Editor: Marion E. Howe
Designer: Marshall Henrichs
Illustrator: Richard L. Morton
Cover Design: Ann Scrimgeour

Library of Congress Cataloging in Publication Data

Van Vlack, Lawrence H
    Elements of materials science and engineering.

    Includes bibliographical references and index.
    1. Materials. I. Title.
TA403.V35   1980   620.1 '1   79–19352
ISBN 0–201–08090–7

Second printing, June 1980

ISBN 0-201-08090-7
ABCDEFGHIJK-HA-89876543210

Preface

Materials Science and Engineering (MSE) has come into its own as a field of endeavor during the past 25 years. There has been a central theme in this development. This is the concept that the properties and behavior of a material are closely related to the internal structure of that material. As a result, in order to modify properties, appropriate changes must be made in the internal structure. Also, if processing or service conditions alter the structure, the characteristics of the material are altered.

Within this developmental period, there have been noticeable changes in the procedures for introducing the science and engineering of materials to the undergraduate technical student. Current introductory courses build upon the basic college chemistry and physics courses, rather than being an abridgment of the more sophisticated solid-state physics and chemistry, as was the pattern 15 to 25 years ago. The second (1964) and third (1975) editions of this book contributed to that trend.

Concurrently, there is a second trend in the introductory teaching of materials. That is the incorporation of study aids for the student, so that the introductory text is more than a presentation of organized concepts and useful relationships. The previous edition of this text addressed this trend by incorporating chapter previews, greater emphasis on calculations as a learning procedure, discussion topics, and review and summary sections at the end of each chapter. The National Science Foundation–sponsored development of Educational Modules for the teaching of Materials Science and Engineering (EMMSE) represents a concentrated country-wide effort to produce study-aids for a full scope of materials topics.

This new edition of Elements of Materials Science and Engineering is designed to capitalize on the merits of the third edition, by extending those features that have facilitated self-study. At the same time, it seeks to retain the numerous advantages inherent in the textbook format, such as course integration, portability, individual availability, and an introductory reference source. The feature of the previous edition most favored by students was the sets of example problems. Their numbers have been nearly doubled in this edition to 200. In addition, study objectives have been added to the preview of each chapter for study orientations. The requests of various instructors have led to additional home study problems, now more than 600. Furthermore, these study problems are separated into two categories. The first is a trial category (identified with italic numbers, *1–6.1*). These study problems parallel example problems closely or relate directly to equations. Their purpose is to allow the students to "get their feet wet," before proceeding to the problems that require more analysis and/or greater integration with previous topics. (These are identified with boldface numbers, **1–6.2.**) Finally, the discussion topics for classroom use have been revised and increased in number, where that mode of instruction is possible as a supplement to lectures.

The general format of previous editions has been maintained, whereby initial attention is given to the engineering properties of interest, and pertinent chemical bonding principles that are encountered in general chemistry are reviewed (Chapter 2). Structural characteristics are considered on the atomic and electronic levels in Chapters 3 through 5 to provide the basis for the properties of single-phase (a) metals in Chapter 6, (b) polymers in Chapter 7, and (c) ceramics in Chapter 8. Multiphase materials are presented in Chapters 9 through 11, with sequential emphasis on phase diagrams, microstructure, and methods of developing the desired properties through microstructural control. The major importance of corrosion warrants a separate Chapter 12. Chapter 13 has undergone a significant change. It now contains specific attention to the four materials that command major usage: cast irons, concrete, composites, and wood. Although complex, each illustrates the relationship between structure and properties. This chapter is written so that the instructor may use none, one, or more of these sections, depending upon available time and the local curricula needs.

Other significant changes include a reorganization of the effects of grain size upon properties in Chapter 6, and an introduction of recrystallization kinetics with annealing and hot working. Processing receives added attention at various points throughout this edition, not to make this a textbook on manufacturing processes but rather to show the student that the structure–property concept extends to processing as well as to service behavior. These added "touch points" with processing also provides a number of opportunities to build that processing interface with examples out of the instructor's experience.

This edition utilizes SI units as the primary dimension throughout. However, it still retains a parenthetical relationship with selected English units, because many engineers must be "bilingual" in this respect for a number of years as they communicate with less technical workers who possess a wealth of practical experience that should not be ignored.

As in previous editions, those sections and subsections are identified that may be deleted by the instructor in deference to time or to supplementary course content. Those topics marked with a bullet (•) contain subject matter that is not a primary prerequisite for later (unmarked) sections.

I regret it is impossible to give individual thanks to the innumerable students at The University of Michigan who have contributed in their way to this new edition. So that students may know that their contributions are greatly appreciated. I dedicate this edition *To Today's Students, who will be Tomorrow's Engineers.*

Likewise, each of my professorial colleagues in Ann Arbor and many elsewhere deserve special thanks. Among these, however, Professor Morris Cohen of MIT and Professors W. C. Bigelow, W. F. Hosford, E. E. Hucke, and T. Y. Tien of Ann Arbor deserve more than anonymous recognition. The job of revision was greatly facilitated by the help of Mrs. Ardis Vukas in my office, and of Dick Morton and Marion Howe at Addison-Wesley. Last, but far from least, Fran's patience and encouragement were indispensable.

Ann Arbor, Michigan   L. H. VV.
September 1979

To Today's Students, who will be Tomorrow's Engineers

Contents

Foreword

Materials In Human Affairs

MORRIS COHEN

Materials are all about us; they are engrained in our culture and thinking as well as in our very existence. In fact, materials have been so intimately related to the emergence and ascent of man that they have given names to the Stone, Bronze, and Iron Ages of civilization. Naturally-occurring and man-made materials have become such an integral part of our lives that we often take them for granted, and yet materials rank with food, living space, energy, and information as basic resources of mankind. Materials are indeed the working substance of our society; they play a crucial role not only in our way of life but also in the well-being and security of nations.

But what are materials? How do we understand, manipulate, and use them? Materials are, of course, a part of the matter in the universe, but more specifically *they are substances whose properties make them useful in structures, machines, devices, or products.* For example, these categories include metals, ceramics, semiconductors, superconductors, polymers (plastics), glasses, dielectrics, fibers, wood, sand, stone, and many composites. The production and processing of these materials into finished goods account for about one fifth of the jobs and gross national product in the United States.

Since the human body might be regarded as a structure or machine or device, we could also embrace foods, drugs, biomatter, fertilizers, etc., among the classes of materials, but it is presently customary to leave these materials to the life and agricultural sciences. For similar reasons, even though fossil fuels, water, and air likewise fall within the broad definition of materials, they are usually dealt with in other fields.

The materials of mankind can be visualized to flow in a vast *materials cycle*—a global cradle-to-grave system. Raw materials are taken from the earth by mining, drilling, excavating, or harvesting; then converted into bulk materials like metal ingots, crushed stone, petrochemicals, and lumber; and subsequently fabricated into engineering materials, like electric wire, structure steel, concrete, plastics, and plywood, for meeting end-product requirements in society. Eventually, after due performance in the service of man, these materials find their way back to earth as scrap, or preferably re-enter the cycle for reprocessing and further use before their ultimate disposal.

An important aspect of the materials–cycle concept is that it reveals many strong interactions among materials, energy, and the environment, and that all three must be taken into account in national planning and technological assessment. These considerations are becoming especially critical because of mounting shortages in energy and materials just at a time when the inhabitants of this planet are manifesting deeper concern for the quality of their living space. As a case in point, if scrap aluminum can be effectively recycled, it will require only about one twentieth the energy needed for an equivalent tonnage of primary aluminum from the ore, and the earth would be that much less scarred by the associated removal operations.

Consequently, the materials cycle is a system that intertwines natural resources and human needs. In an all-embracing way, materials form a global connective web that ties nations and economies not only to one another on this planet, but also to the very substance of nature.

Clearly, then, in the development of human knowledge, one is not surprised to find a *science and engineering of materials* taking its place among all the other bodies of inquiry and endeavor that extend the reach of man. Simply stated, *materials science and engineering* (MSE) *is concerned with the generation and application of knowledge relating the composition, structure, and processing of materials to their properties and uses.* As suggested in Fig. I, there is a linkage that interrelates the structure, properties, processing, function, and performance of materials. MSE operates as a knowledge-conduction band that stretches from basic science and fundamental research (on the left) to societal needs and experience (on the right). The countercurrent flow of scientific understanding in one direction and empirical information in the other direction intermix very synergistically in MSE.

If we wish to highlight the *materials science* part of this spectrum, we focus on understanding the nature of materials, leading to theories or descriptions that explain how structure relates to composition, properties, and behavior. On the other hand, the *materials engineering* part of the spectrum deals with the synthesis and use

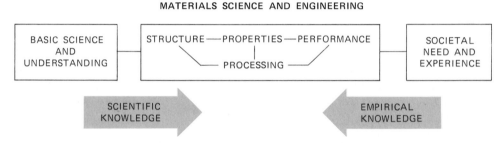

MATERIALS SCIENCE AND ENGINEERING

Fig. I   A representation of the central elements of MSE, in relation to the countercurrent flows of scientific and empirical knowledge.

BOILING WATER REACTOR

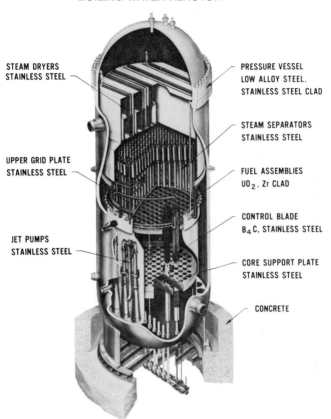

STEAM DRYERS
STAINLESS STEEL

PRESSURE VESSEL
LOW ALLOY STEEL,
STAINLESS STEEL CLAD

STEAM SEPARATORS
STAINLESS STEEL

UPPER GRID PLATE
STAINLESS STEEL

FUEL ASSEMBLIES
UO_2, Zr CLAD

CONTROL BLADE
B_4C, STAINLESS STEEL

JET PUMPS
STAINLESS STEEL

CORE SUPPORT PLATE
STAINLESS STEEL

CONCRETE

Fig. II   Materials system for energy conversion. Any integrated design such as a boiling-water nuclear reactor requires specific contributions from a wide variety of materials. The role, properties, and limitations of each material must be understood by the scientist or engineer who wishes to design a workable, safe, cost-effective, and environmentally reliable system. (Courtesy General Electric Company.)

of both fundamental and empirical knowledge in order to develop, prepare, modify, and apply materials to meet specified needs. It is evident that the distinction between materials science and materials engineering is primarily one of viewpoint or center of emphasis; there is no line of demarcation between the two domains, and we find increasing logic in adopting the combined name of *materials science and engineering.* Actually, this textbook joins the two, both in its title and coverage, by using the term *materials science* generically to include many aspects of *materials engineering.*

Figures II and III illustrate the selection and function of materials in two materials applications, each requiring numerous professional judgments concerning performance, reliability, maintainability, economic trade-offs, and environmental considerations. In the nuclear reactor shown, the fuel consists of uranium oxide pellets encased in a zirconium alloy that is heat resistant and does not waste neutrons. The neutron flux, and hence the operating temperature of the fuel, is controlled by a boron carbide device that can absorb neutrons as desired but has to be corrosion-protected by stainless steel. The generated steam is contained in an outer higher-pressure vessel built of low-alloy steel for providing strength at reasonable cost but clad with more expensive stainless steel for corrosion protection. This nuclear-energy conversion involves a complex materials system in which the selected materials must perform *inter*dependently, over and above their specific functions. This is a striking application of materials for producing energy. Conversely, about one half of the energy consumed by all manufacturing industries in the United States goes into the production and forming of materials.

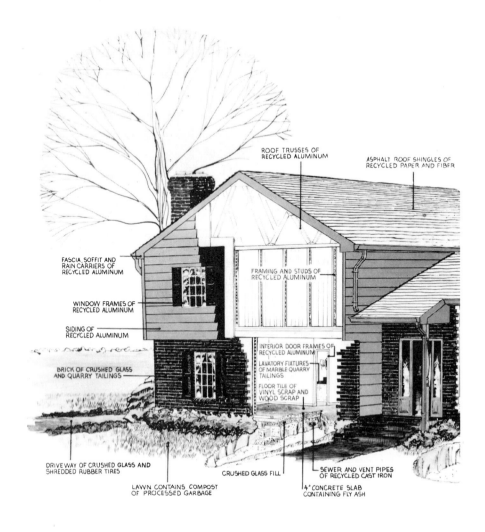

Another materials system, showing how materials enter into providing shelter, is indicated in Fig. III. Although housing may seem like a less sophisticated case of materials science and engineering than is the nuclear reactor, the example given here is, nevertheless, one of challenge and opportunity; for this house has been built almost entirely of *recycled* materials, thereby conserving the earth's resources and energy while reducing damage to the environment. In order to manufacture the recycled metals, glass, concrete, roofing shingles, sheathing paper, insulation, and wood products, it is necessary to use a high level of materials development and technology, to compete with conventional materials in cost as well as in quality.

So we see that MSE is a purposeful enterprise, reaching down into the microworld of atoms and electrons and tying the condensed state of matter to the macroworld

Fig. III Reusable materials. Recycling is very desirable and is receiving increased national emphasis. However, it cannot be an effective answer to our resource and environmental problems unless the recycled materials are comparable to virgin materials on the basis of appearance, properties, and service qualities—and without the penalty of added cost. These requirements present many technological challenges for the engineer. (Courtesy Reynolds Metals Company.)

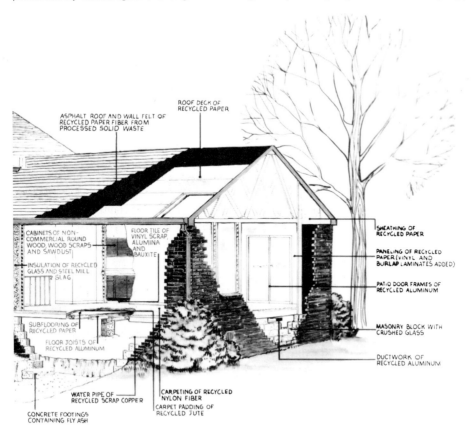

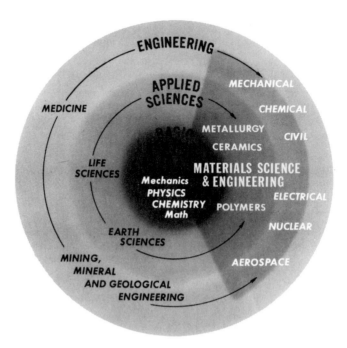

Fig. IV Materials science and engineering. This text draws from introductory science principles of physics and chemistry to relate *composition and structure* of materials to the *properties and service behavior* that are important to engineers. More than one fourth of our national technical effort involves the development and technology of materials. This effort draws from the physical sciences and all branches of engineering. (Courtesy National Academy of Science.)

world of material function and service to meet societal problems. The circular chart in Fig. IV portrays a large section of human knowledge, extending from basic sciences at the core, through applied sciences in the middle ring, to various engineering fields in the outer rim. In the center, we show physics and chemistry flanked by mathematics and mechanics; and on moving out radially, we pass through various applications-oriented disciplines. The part of this map to be visualized as materials science and engineering is the shaded sector at the right, which may be compared to other sectors designating the life sciences and the earth sciences. In its broad sense, MSE is a multidiscipline that embraces (but does not replace!) some disciplines (e.g., metallurgy and ceramics) and some subdisciplines (e.g., solid-state physics and polymer chemistry), and also overlaps several engineering disciplines.

There are, of course, many scientists and engineers who are materials specialists —metallurgists, ceramists, and polymer chemists—and who are wholly involved in materials science and engineering. Equally important, government data, when analyzed, reveal that one of every six hours of professional work done by *all other engineers* directly involves materials and their utilization. The time fractions are even higher for chemists and physicists. As a result, the equivalent of one-half million of the nearly two million scientists and engineers in this country contribute to this major segment of our national product and our national well-being.

Thus, materials science and engineering constitutes a framework in which professionals in many disciplines work creatively to probe nature's processes and, at the same time, advance knowledge in response to the pull of human needs.

Cambridge, Massachusetts   Morris Cohen
September 1979

CHAPTER **1**

Introduction to Materials: Selected Characteristics

PREVIEW

The engineer adapts materials and energy into useful products. In doing so, the engineer endeavors to select materials with optimum properties. This chapter introduces a number of common properties that must be considered. We will encounter others later.

The theme of this text is that *the properties and behavior of materials depend on their internal structure.* The mechanical, thermal, and electrical properties and characteristics that are introduced in Chapter 1 will serve as a basis for structure–property discussion in later chapters. A brief consideration is also given in Chapter 1 to data presentation, their variations, and the accuracy that is warranted in calculations.

CONTENTS

1−1 Structure ↔ Properties ↔ Processing.

1−2 Mechanical Behavior:
deformation, strength (and hardness), toughness, and example calculations.

1−3 Thermal Characteristics:
heat capacity, thermal expansion, thermal conductivity, and example calculations.

1−4 Responses to Electric Fields:
conductivity (and resistivity), dielectric behavior, and example calculations.

1−5 Presentations of Properties:
qualitative information, quantitative data, and variance of data.

This chapter, like all the chapters in this text, is concluded with a Review and Study section that includes a summary, a list of key terms and concepts, discussion topics, and study problems.

STUDY OBJECTIVES

1 To become alerted to the concept that materials possess an internal structure. (The various details of internal structure will be the focus of later chapters.)

2 To become familiar with selected terms and concepts that you will encounter (a) later in this text, and (b) during communication with other engineers in your future career.

3 To learn the relationships among various mechanical properties of materials, particularly those cited in Table 1−2.1.

4 To review your previous Physics course with respect to the more common thermal and electrical properties, and to develop the concept of conductivity in terms of charge carriers.

5 To demonstrate your knowledge of properties through simple calculations. (The Study Problems at the end of each chapter are for this purpose.)

6 To review the concepts of mean, median, and standard deviation, and to present your answers with an appropriate number of significant figures.

7 To become "bilingual" with respect to dimensional units (if you are a U.S. student).

1-1 STRUCTURE ↔ PROPERTIES ↔ PROCESSING

Every applied scientist and engineer—mechanical, civil, electrical, or other—is vitally concerned with the materials available for use. Whether the product is a bridge, computer, space vehicle, heart pacemaker, nuclear reactor, or an automobile exhaust system, the engineer must have an intimate knowledge of the properties and behavioral characteristics of the materials he or she proposes to use. Consider, for a moment, the variety of materials used in the manufacture of an automobile: iron, steel, glass, plastics, rubber—to name a few (Fig. 1–1.1). And for steel alone there are as many as 2000 varieties or modifications. On what basis is the selection of the material for a particular part to be made?

In making a choice, the designer must take into account such properties as strength, electrical and/or thermal conductivity, density, and others. Further, one

Fig. 1–1.1 There are several hundred varieties of materials in this automobile engine. (The reader is asked to identify those that are apparent.) The design engineer selects each material on the basis of fabricability, properties, service behavior, cost, and availability. Improvements in materials will be required during the coming decade as weight limitations gain importance, as the engine is redesigned for more efficient energy conversion, and as the supply of raw materials tightens. (Courtesy of the Ford Motor Company.)

must consider the behavior of the material during processing and use (where form-ability, machinability, electrical stability, chemical durability, and radiation be-havior are important), as well as cost and availability. For example, the steel for transmission gears must machine easily in production but must then be toughened enough to withstand hard usage. Fenders must be made of a metal that is easily shaped but that will resist deformation by impact. Electrical wiring must be able to withstand extremes of temperature, and semiconductors must have constant amperage/voltage characteristics over long periods of time.

Many improved designs depend on the development of completely new materials. For example, the transistor could not have been built with the materials available only a few years ago; the development of the lasers required new kinds of crystals and glasses; and although engineering designs for gas-turbine engines are far advanced, there still is a need for an inexpensive material that will resist ever higher temperatures, for the turbine blades.

Internal structure and properties   Since it is obviously impossible for the engineer or scientist to have detailed knowledge of the many thousands of materials already available, as well as to keep abreast of new developments, he or she must have a firm grasp of the underlying principles that govern the properties of *all* materials. The principle that is of most value to engineers and scientists is *properties of a material originate from the internal structures of that material.* This is analogous to saying that the operation of a TV set or other electronic product (Fig. 1–1.2) depends upon the components, devices, and circuits within that product. Anyone can twirl knobs, but electronic technicians must understand the internal circuits if they are going to repair a TV set efficiently; and the electrical engineer and the physicist must know the characteristics of each circuit element if they are to design or improve the performance of the final product.

The *internal structures* of materials involve atoms and the way atoms are asso-ciated with their neighbors in crystals, molecules, and microstructures. In the following chapters we shall devote much attention to these structures, because technical persons must understand them if they are going to produce and use materials, just as mechanical engineers must understand the operation of an internal-combustion engine if they are going to design or improve a car for the demands of the next decade.

Processing and properties   Materials must be processed to meet the specifications that the engineer requires for the product being designed. The most familiar pro-cessing steps simply change the shape of the material, e.g., machining, or forging. Of course, properties are important for easy processing. Extremely hard materials immediately destroy the edge of a cutting tool, and soft materials like lead can "gum-up" saw blades, grinding wheels, and other tools. Likewise, strong materials are not amenable to plastic deformation, particularly if they are also nonductile, i.e., brittle. For example, it would be prohibitively expensive to produce sheet metal for most car fenders with anything other than the softest of steels.

Processing commonly involves more than simply changing the shape of the material by machining or by plastic deformation. Not uncommonly, the manu-

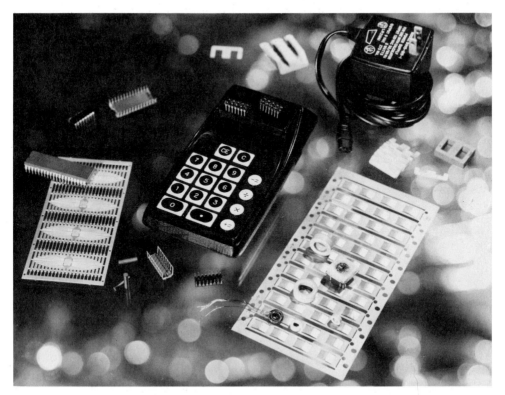

Fig. 1–1.2 Electronic calculator. The performance of a calculator depends on the arrangement of the components of its internal circuit. Likewise, the behavior of a material depends upon the structure of its internal components. We shall see that these arrangements involve electron structures around atoms, the coordination of atoms, the structure of crystals, and the microstructure of adjacent crystals. The engineer and scientist can select and modify these internal structures to meet design needs, just as the circuit designer modifies electrical components. To do this intelligently, we must know the relationships between structure and properties of materials. (Courtesy of the Arnold Engineering Company.)

facturing process changes the properties of a material. For example, a wire is strengthened and hardened if it is drawn through a die to decrease its diameter. Typically, this hardening is not desired in a copper wire to be used as an electrical conductor; conversely, the engineer depends upon this strengthening that develops during processing the steel wire used in a steel-belted, radial tire. Whether desired or not, the properties should be expected to be modified whenever the manufacturing process changes the internal structure of the material. The internal structure of a material is altered when it is deformed; hence, there is a change in properties.

Thermal processing may also affect the internal structure of a material. Such processing includes annealing, quenching from elevated temperatures, and a number of other heat treatments. Our purpose will be to understand the nature of the structural changes, so that we, as engineers, may specify appropriate processing steps.

Service behavior   A material in the completed product possesses a set of properties—strength, hardness, conductivity, density, color, etc.—chosen to meet the design requirements. It will retain these properties indefinitely, *provided* there is no change in the internal structure of the material. However, if the product encounters a service condition that alters the internal structure, we must expect the properties and behavior of the material to change accordingly. This explains why rubber gradually hardens when exposed to light and the air; why aluminum can not be used in many locations of a supersonic plane; why a metal can fatigue under cyclic loading; why a drill of ordinary steel cannot cut as fast as a drill of high-speed steel; why a magnet loses its polarity in an *rf* field and why a semiconductor can be damaged by nuclear radiation. The list is endless. The consequence to the engineer is that consideration must be given not only to the initial demand but also to those service conditions that will alter the internal structure and hence the properties of a material.

The pattern of this course   The "structure" designed into this text assigns the first chapter to discuss some property terms and measurements. Then in Chapter 2, we will review those science fundamentals from introductory chemistry and physics that bear directly upon structure–property relationships. Starting with Chapter 3, we will consider, in turn, the various structural features (from atoms to composites) on which properties depend. The roles of processing and service factors will be cited often since their consideration must always be present in design decisions.

Example problems and study problems are used (1) to illustrate structure and/or property principles, and (2) to extend the problem-solving capabilities that are so important to engineers and applied scientists.*

1-2 MECHANICAL BEHAVIOR

Deformation occurs when forces are applied to a material. *Strain, e,* is the amount of deformation per unit length, and *stress, s,* the force per unit area. Energy is absorbed by a material during deformation because a force has acted along the deformation distance. *Strength* is a measure of the level of the *stress* required to make a material fail. *Ductility* identifies the amount of *permanent strain* prior to fracture, while *toughness* refers to the amount of *energy* absorbed by a material during failure (Table 1–2.1).

The design engineer makes various specifications for these mechanical properties. Commonly one prescribes high strength, e.g., in a pipeline steel. One may also want high ductility to enhance toughness. Since the two (strength and ductility) tend to be incompatible, the engineer often must make trade-offs between the two to optimize the specifications. Also, there are various ways to define strength and ductility. For

---

*   This text will utilize SI units throughout. However, certain English and non-SI metric units are included parenthetically where there may be a need for the engineer to be "bilingual." Not only does the engineer solve problems which are facilitated by SI units, the engineer will find it necessary and desirable to communicate for some period of time with individuals who are not immediately fluent with SI units, but who possess an accumulation of valuable technical experience.

Table 1–2.1
Mechanical properties of materials

Property, or characteristic	Symbol	Definition (or comments)	Common units SI	Common units English
Stress	s σ	Force/unit area (F/A)	pascal* ($N^\dagger/m^2$)	psi* $lb_f/in.^2$
Strain	e ϵ	Fractional deformation ($\Delta L/L$)	—	—
Elastic modulus	E	Stress/elastic strain	pascal	psi
Strength		Stress at failure		
Yield	S_y	Resistance to initial plastic deformation	pascal	psi
Tensile	S_t	Maximum strength (based on original dimensions)	pascal	psi
Ductility		Plastic strain at failure		
Elongation	e_f	$(L_f - L_o)/L_o$	§	§
Reduction of area	R of A	$(A_o - A_f)/A_o$	§	§
Toughness		Energy for failure by fracture	joules	ft-lb
Hardness‡		Resistance to plastic indentation	Empirical units	

* 1 pascal (Pa) = 1 newton/m^2 = 0.145 × 10^{-3} psi; 1000 psi = 6.894 MPa.
† A load of 1 kg mass produces a force F of 9.8 newtons (N) by gravity.
‡ Three different procedures are commonly used to determine hardness values:
    Brinell (BHN):   A large indenter is used. The hardness is related to the diameter (1 to 4 mm) of the indentation.
    Rockwell (R):   A small indenter is used. The hardness is related to the penetration depth. Several different scales are available, based on the indenter size and the applied load.
    Vickers (DPH): A diamond pyramid is used. A very light load may be used to measure the hardness in a microscopic area.
§ Dimensionless. Based on original (o) and final or fractured (f) measurements (usually expressed as percent).

example, has a steel rod failed when it bends, or must it actually *fracture* to have failed? The answer, of course, depends on the requirements of the engineering design; but the contrast shows us the advantage of identifying at least two strengths—one for *initial yielding*, and one for *maximum load* a material may support. To do this, we will consider the stress–strain (*s-e*) diagrams of Fig. 1–2.1, giving attention in turn to deformation, to strength and hardness, and to toughness.

Deformation   The initial strain is essentially proportional to the stress; furthermore, it is reversible. After the stress is removed, the strain disappears. We call this lineal, reversible strain, *elastic strain*. The *modulus of elasticity* (*Young's modulus*) is the ratio between the stress s and this reversible strain e:

$$E = s/e. \qquad (1\text{–}2.1)$$

The metric units of Young's modulus E are *pascals* (or more commonly megapascals, MPa) as shown in Table 1–2.1. The corresponding English units are psi, pounds/

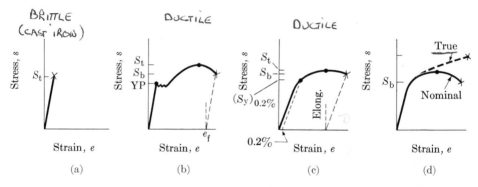

Fig. 1–2.1 Stress–strain diagrams. (a) Nonductile material with no plastic deformation (example: cast iron). (b) Ductile material with yield point (example: low-carbon steel). (c) Ductile material without marked yield point (example: aluminum). (d) True stress–strain curve versus nominal stress–strain curve. S_b = breaking strength; S_t = tensile strength; S_y = yield strength; e_f = elongation (strain before fracture); X = fracture; YP = yield point.

square inch. The values of Young's moduli for selected materials are included in Appendix C. We shall see, in Chapter 2, that the *elastic modulus* (Young's modulus) is a measure of the interatomic bonding forces. The engineer must be fully aware of this property, since it relates directly to the rigidity of the resulting engineering product.

At higher stresses, permanent displacement can occur among the atoms within a material in addition to the elastic strain. This permanent strain is not reversible when the applied stresses are removed; we call this *plastic strain*. This kind of strain is necessary during the processing of materials (e.g., during the rolling of aluminum plate, first to a thinner sheet, and then to a very thin foil). In product applications, we commonly design to *avoid* plastic deformation and therefore must design at stresses within the elastic (proportional) range of Fig. 1–2.1(b) and (c).

The elastic strain, which was the sole mode of deformation below the yield strength, continues to increase in response to the higher stresses that introduce plastic strain. The elastic, but not the plastic strain, recovers when the material fractures (or is simply unloaded).

Ductility, the plastic strain required for fracture, e_f, may be expressed as *percent elongation*. Like all strain, it is dimensionless, $(L_f - L_o)/L_o$ or $\Delta L/L_o$. From Fig. 1–2.2, however, note that, since plastic deformation is commonly localized in the necked area, the percent of elongation depends on the *gage length*. Whenever reporting ductility, one must be specific about the gage length.

A second measure of ductility is the *reduction in area*, $(A_o - A_f)/A_o$, at the point of fracture. Highly ductile materials are greatly reduced in cross section before breaking. Elongation is a measure of plastic "stretching," whereas reduction in area is a measure of plastic "contraction." The reduction in area is preferred by some engineers as a measure of ductility because it does not require a gage length, and because it can be used to determine the *true strain* at the point of fracture. (See the comment following Example 1–2.4.) One cannot establish an exact correlation

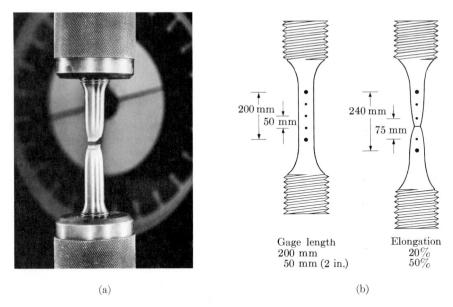

Gage length
200 mm
50 mm (2 in.)

Elongation
20%
50%

(a)   (b)

Fig. 1–2.2 Tensile tests. (a) Completed test on a round specimen. (Courtesy of the U.S. Steel Corp.)
(b) Elongation versus gage length. Since final deformation is localized, an elongation value is
meaningless unless the gage length is indicated. For routine testing, a 50-mm (2-in.) gage length
is common. (In inches, 2.0 in. → 3.0 in., and 7.9 in. → 9.5 in., for 50% and 20% elongation,
respectively.)

between elongation and reduction in area, since the plastic deformation may be
highly localized. Of course, a very ductile material will have high values of each, and
a nonductile material has near-zero values of each.

Strength (and hardness)   The ability of a material to resist plastic deformation is
called the *yield strength*, S_y, and is computed by dividing the force initiating the yield
by the cross-sectional area. In materials such as some of the softer steels, the yield
strength is marked by a definite *yield point* (Fig. 1–2.1b). In other materials, where the
proportional limit is less obvious, it is common to define the yield strength as that
stress required to give 0.2% (or some other value specified by the design engineer)
plastic offset (Fig. 1–2.1c).

The *tensile strength* (S_t) of a material is calculated by dividing the maximum
force by the *original* cross-sectional area. This strength, like all other strengths, is
expressed in the same units as stress. Note particularly that tensile strength is based
on the original cross-sectional area. This is important, inasmuch as a ductile material
will have its cross-sectional area somewhat reduced when the maximum load is
exceeded.

In Table 1–2.1, *strength* is described as the *stress to cause failure*. However,
observe that failure is a matter of definition. The steel of an angle iron in a radio
tower has failed if it is bent in service, because it could permit the tower to collapse.

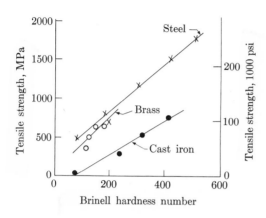

Fig. 1–2.3 Tensile strength versus Brinell Hardness Number (BHN). Examples: Steels, brasses, and cast irons.

Since bending involves plastic deformation, the engineer designs the structural members of the tower to hold the stresses below the yield strength S_y of the angle iron. However, in making the guy wire for the radio tower, it was necessary to plastically deform the wire by drawing it through a die. That plastic deformation did not constitute failure even though the stresses were in excess of the yield strength. It is the maximum load that the wire can support in service without breaking that enters the design calculations. By dividing the tensile strength S_t into the maximum expected force, the engineer obtains the cross-sectional dimension that is required to avoid failure of the wire. These dimensions must be the original design dimensions prior to any possible plastic deformation.

When desired, the engineer can calculate the *true stress, σ*, which is the force divided by the *actual* area. Of course, this will vary along the length of the test sample that has undergone plastic deformation and will be the greatest at the necked region (Fig. 1–2.2). This means that the true stress σ_f for *fracture* will always be greater than the *breaking strength* S_b that is based on the original area. (Cf. Figs. 1–2.1c and 1–2.1d.) The true stress permits us to analyze the actual forces during deformation and fracture; however, the *nominal stress* (based on the original area) is more useful to engineers, who must, of course, make their designs on the basis of initial dimensions.

Hardness is defined as the resistance of a material to penetration of its surface. As might be expected, the hardness and the strength of a material are closely related, as indicated in Fig. 1–2.3. *The Brinell hardness number* (BHN) is a hardness index calculated from the area of penetration by a large spherical indenter. The indentation is made by a very hard steel or tungsten-carbide ball under a standardized load. The *Rockwell hardness* (R), another of several common indexes of hardness used by engineers, is related to BHN but is measured by the depth of penetration by a small standardized indenter. Several different Rockwell scales for materials of different hardness ranges have been established by selecting various indenter shapes and loads.

Toughness   This is a measure of the *energy* required to break a material (Fig. 1–2.4). This is in contrast to *strength*, which is a measure of the *stress* required to deform or

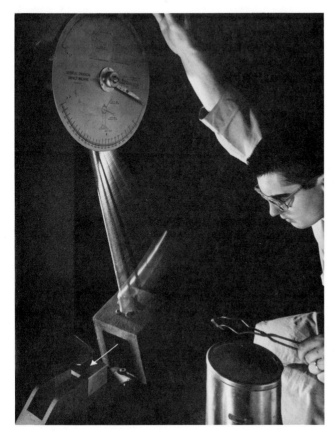

Fig. 1–2.4 Toughness test. The test specimen (arrow) is broken by the impact of the swinging pendulum. The amount of absorbed energy can be calculated from the arc of the follow-through swing. (Courtesy of the U.S. Steel Corp.)

break a material. Energy, the product of force times distance, is expressed in joules, or foot-pounds; it is closely related to the area under the stress–strain curve. A ductile material with the same strength as a nonductile material will require more energy for breaking and be tougher. Standardized *Charpy* and *Izod* tests are two of several procedures used to measure toughness. They differ in the shape of the test piece and in the method of applying the energy. Since toughness also depends on the geometry of stress concentrations, one must be careful to identify the test procedure that was used.

Study aids (stress–strain curves)   The reader who has questions about the stress–strain curve or the related terms is referred to the paperback *Study Aids for Introductory Materials Courses.** The stress–strain curve is detailed in terms of load, stress, and strain. In addition, the topics of true stress and strain are presented on an optional basis. The latter is germane to the detailed attention that is given in later courses to strain-hardening calculations.

---

* *Study Aids for Introductory Materials Courses*, Reading, Mass.: Addison-Wesley Publishing Co., 1977.

Example 1–2.1 Which part has the greater stress: (a) an aluminum bar of 24.6 mm × 30.7 mm (0.97 in. × 1.21 in.) cross section, under a load of 7640 kg, and therefore a force of 75,000 N (16,800 lb$_f$), or (b) a steel bar whose cross-sectional diameter is 12.8 mm (0.505 in.), under a 5000-kg (11,000-lb) load?

Solution

Units: $\dfrac{\text{Newtons}}{(\text{m})(\text{m})} = \text{Pascals.}$ Units: $\dfrac{\text{Pounds}}{(\text{in.})(\text{in.})} = \text{psi.}$

Calculation:

a) $\dfrac{(7640)(9.8)}{(0.0246)(0.0307)} = 100 \text{ MPa};$ a) $\dfrac{16,800}{(0.97)(1.21)} = 14,300 \text{ psi};$

b) $\dfrac{(5000)(9.8)}{(\pi/4)(0.0128)^2} = 380 \text{ MPa.}$ b) $\dfrac{11,000}{(\pi/4)(0.505)^2} = 55,000 \text{ psi.}$

Comment. Recall that newtons = $ma = (\text{kg})(9.8 \text{ m/s}^2)$. ◀

Example 1–2.2 A 50-mm (1.97 in.) gage length is marked on a copper rod. The rod is strained so that the gage marks are 59 mm (2.32 in.) apart. Calculate the strain.

Solution

Units: $\dfrac{(\text{mm} - \text{mm})}{\text{mm}} = \dfrac{\text{mm}}{\text{mm}} = \dfrac{\text{percent}}{100}.$

Calculation: $\dfrac{59. - 50.}{50.} = 0.18 \text{ mm/mm}$ (or 18%).

Comment. Measurements in inches give the same answer. ◀

Example 1–2.3 If the average modulus of elasticity of the steel used is 205,000 MPa (30,000,000 psi), how much will a wire 2.5 mm (0.1 in.) in diameter and 3 meters (10 ft) long be extended when it supports a load of 500 kg (= 1100 lb$_f$ and 4900 N)?

Solution: Modulus of elasticity = stress/strain, or,

$$\text{Strain} = \frac{s}{E}.$$

Units: $\text{m/m} = \dfrac{\text{N/m}^2}{\text{pascals}}.$ $\text{in./in.} = \dfrac{\text{lb/in.}^2}{\text{psi}}.$

Calculation:

$\text{Strain} = \dfrac{4900/(\pi/4)(0.0025)^2}{205,000 \times 10^6}$ $\text{Strain} = \dfrac{1100/(\pi/4)(0.1)^2}{30,000,000}$

$= 0.005 \text{ m/m.}$ $= 0.005 \text{ in./in.}$

$\text{Extension} = (0.005 \text{ m/m})(3 \text{ m})$ $\text{Extension} = (0.005 \text{ in./in.})(120 \text{ in.})$

$= 15 \text{ mm.}$ $= 0.6 \text{ in.}$ ◀

•**Example 1–2.4** A copper wire had a nominal breaking strength of 300 MPa (43,000 psi). Its ductility was 77% reduction of area. Calculate the true stress σ_f for fracture.

Solution: Based on the original area, A_o,

$$\frac{F}{A_o} = 300 \text{ MPa}; \qquad F = (300 \times 10^6 \text{ N/m}^2)\, A_o.$$

$$\sigma_f = \frac{F}{A_{tr}} = \frac{F}{(1 - 0.77)A_o} = \frac{(300 \times 10^6 \text{ N/m}^2)\, A_o}{0.23 A_o} = 1300 \text{ MPa};$$

or

$$\frac{F}{A_{tr}} = \frac{(43,000 \text{ psi})\, A_o}{0.23 A_o} = 187,000 \text{ psi}.$$

Comment. The *true strain* ϵ may be determined from the cross-sectional dimensions. If we define the true strain ϵ as

$$\epsilon \equiv \int_{l_o}^{l} \frac{dl}{l} = \ln\left(\frac{l}{l_o}\right),$$

and assume constant volume, $Al = A_o l_o$, then

$$\epsilon = \ln\left(\frac{l}{l_o}\right) = \ln\left(\frac{A_o}{A}\right). \qquad\qquad •(1-2.2)$$

This gives a definition of true strain that holds for all strains and is independent of gage length. ◀

1–3 THERMAL CHARACTERISTICS

Heat capacity   The distinction between temperature and heat content of a material is an important one for the engineer and scientist. *Temperature* is a level of thermal activity, whereas the *heat content* is thermal energy. The two are related through heat capacity.

In the absence of any volume change, the *heat capacity* c is the change in heat content per °C. Not uncommonly in technical tables, specific heat is used in place of heat capacity. The *specific heat* of a material is defined as the ratio of the heat capacity of the material to that of water. Thus, with the heat capacity of water equalling 1 cal/g·°C ($=4.184$ joules/g·°C $= 1$ Btu/lb·°F), one can make thermal calculations in the units of his or her choice.

Various heats of transformation are of importance in materials. The better known of these are the *heat of fusion* and the *heat of vaporization,* which are the heats required to produce melting and gasification, respectively. Each involves a change within the material from one atomic or molecular structure to another. We shall learn later that there are various structural changes possible *within solids,* and that these changes also require a change in heat or thermal-energy content of the material.

● Examples preceded by a bullet, ●, may be omitted at the discretion of the instructor. (See Preface.)

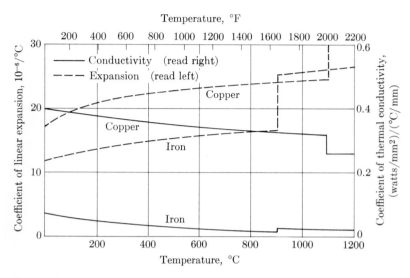

Fig. 1–3.1   Thermal properties versus temperature. The discontinuity for copper at 1084.5°C (1984°F) is a result of melting. Iron has a discontinuity because there is a rearrangement of the atoms at 912°C (1673°F). See Chapter 3.

Thermal expansion   The expansion that normally occurs during the heating of a material arises from the more intense thermal vibrations of the atoms. To a first approximation, the increase in length, $\Delta L/L$, is proportional to the change in temperature, ΔT:

$$\Delta L/L = \alpha_L \Delta T. \; \approx \sigma/E \qquad (1\text{–}3.1)$$

Closer examination shows that the *linear expansion coefficient*, α_L, generally increases slightly with temperature (Fig. 1–3.1). The thermal expansion data listed in Appendix C are for 20°C (68°F).

The *volume expansion coefficient*, α_V bears the same relationship between the *volume change*, $\Delta V/V$, and temperature change, ΔT, as shown previously in Eq. (1–3.1). For all intents and purposes, α_V is three times the value of α_L. (See Study Problem 1–3.3b.)*

---

* Materials such as graphite and mica are anisotropic; i.e., their properties vary with crystal direction. Therefore, a more general statement would be that

$$\alpha_V = \alpha_x + \alpha_y + \alpha_z, \qquad (1\text{–}3.2a)$$

where the x, y, and z subscripts refer to the linear expansion coefficient in the three coordinate directions. When these are equal, as is the case in many engineering materials,

$$\alpha_V = 3\alpha_L. \qquad (1\text{–}3.2b)$$

The discontinuities in the expansion coefficients of Fig. 1–3.1 occur with changes in the atomic arrangements within the material. Specifically, copper melts at 1084.5°C (1984°F), and the iron atoms rearrange themselves from having eight neighbors below 912°C (1673°F) to having twelve neighbors above that temperature. Since structural changes such as these affect many engineering properties, we will consider atomic arrangements in later chapters.

Thermal conductivity   Heat transfer through solids most commonly occurs by conduction. Thermal conductivity k is the proportionality constant relating the *heat flux* $\vec{Q}$ and the thermal gradient, $\Delta T/\Delta x$:

$$\vec{Q} = k \left(\frac{T_2 - T_1}{x_2 - x_1} \right). \tag{1–3.3}$$

The preceding coefficient of thermal conductivity is also temperature-sensitive. However, unlike the coefficient of thermal expansion, this coefficient *decreases* as the temperature is raised above room temperature. (The reasons for this behavior will be discussed later.) The changes of atomic packing, which accompany melting and other atomic rearrangements arising from temperature variations, produce discontinuities in the thermal-conductivity values (Fig. 1–3.1).

The relationships for Eq. (1–3.3) are

$$\frac{\text{energy}}{\text{area} \cdot \text{time}} = k \left[\frac{\text{temp. diff.}}{\text{thickness}} \right].$$

Thus, thermal conductivity k is most simply written as $J/(mm^2 \cdot s)/(°C/mm)$, or $(W/mm^2)/(°C/mm)$.*

Example 1–3.1 An aluminum wire is stressed 34.5 MPa (5000 psi) in tension. What temperature increase is required to change its length by the same amount?

Solution: Obtain the elastic and expansion data from Appendix C.

$$\Delta L/L = s/E = (34.5 \times 10^6 \text{ Pa})/(70{,}000 \times 10^6 \text{ Pa})$$

$$= \alpha \, \Delta T = (22.5 \times 10^{-6}/°C)(\Delta T);$$

$$\Delta T = 22°C \, (= 40°F).$$

Comment. The temperature *difference*, ΔT, may also be expressed as 22 K, since the two scales have the same intervals. ◀

Example 1–3.2 A steel guy wire is 2.8 mm (0.11 in.) in diameter and 7.2 m (23.6 ft) long. After tightening to 31 MPa (4500 psi) the temperature rises 10°C. (a) What was the force on the wire before the temperature rise? (b) After the temperature rise? (No change in total length was permitted. $\alpha_{st} = 11.7 \times 10^{-6}/°C$.)

---

* Unfortunately there is not always consistency when dimensions for conductivity data are presented in tables. The dimensions listed in the footnotes of Appendix C for conversion between various English and metric values are consistent with the above dimensions.

Solution

a) $F_1 = (31 \times 10^6 \text{ N/m}^2)(\pi/4)(0.0028 \text{ m})^2 = 190 \text{ N};$

   or

$$F_1 = (4500 \text{ psi})(\pi/4)(0.11 \text{ in.})^2 = 42.8 \text{ lb}_f.$$

b)   Write your own equation. $\alpha \Delta T + (e_2 - e_1) = 0 = \alpha \Delta T + \Delta F/AE;$

$$-(F_2 - F_1) = \alpha \Delta T A E.$$

   Obtain data from Appendix C.

$$F_1 - F_2 = (11.7 \times 10^{-6}/°\text{C})(+10°\text{C})(\pi/4)(0.0028 \text{ m})^2(205 \times 10^9 \text{ N/m}^2)$$
$$= 148 \text{ N}.$$
$$F_2 = 42 \text{ N} \quad \text{(or 9.4 lb}_f\text{)}. \quad \blacktriangleleft$$

Example 1–3.3 A stainless steel plate 0.4 cm thick has circulating hot water on one side and a rapid flow of air on the other side, so that the two metal surfaces are 90°C and 20°C, respectively. How many joules are conducted through the plate per minute?

Solution: Obtain data from Appendix C.

$$\vec{Q} = [0.015 \text{ (J/mm}^2\cdot\text{s)}/(°\text{C/mm})](70°\text{C}/4 \text{ mm})(60 \text{ s/min})$$
$$= 15.75 \text{ joules/mm}^2\cdot\text{min}.$$

Comment. We shall learn later that stainless steel has lower thermal (and electrical) conductivity than other metals because it contains major amounts of alloying elements (Chapter 5). ◀

1–4   RESPONSES TO ELECTRIC FIELDS*

Conductivity (and resistivity)   Metals and semiconductors will conduct electrical charges when they are placed in an electrical field. The conductivity σ depends on the number of carriers n, the charge q carried by each, and the mobility μ of the charge carrier. The conductivity is the reciprocal of the *resistivity* ρ:

$$\frac{1}{\rho} = \sigma = nq\mu. \tag{1–4.1a}$$

The units are

$$\text{ohm}^{-1}\cdot\text{m}^{-1} = \left(\frac{\text{carriers}}{\text{m}^3}\right)\left(\frac{\text{coul}}{\text{carrier}}\right)\left(\frac{\text{m/sec}}{\text{volt/m}}\right). \tag{1–4.1b}$$

In metals and in those semiconductors in which electrons are the charge carriers, the charge per carrier is 0.16×10^{-18} coul, or 0.16×10^{-18} amp·sec. The *mobility* may be considered as a net, or *drift velocity* $\bar{v}$ of the carrier, which arises from the *electric field* $\mathscr{E}$, where the units are (m/sec) and (volts/m), respectively:

$$\mu = \frac{\bar{v}}{\mathscr{E}}. \tag{1–4.2}$$

---

* The instructor may choose to defer the topic of conductivity until Chapter 5 is assigned and the topic of dielectric constant until either Section 7–6 or Section 8–6 is assigned.

Mobility is commonly expressed as $m^2/\text{volt}\cdot\text{sec}$. The relationships of Eqs. (1–4.1) and (1–4.2) will be particularly useful in Chapter 5 when we consider semiconductors.

Resistivity ρ is a property of a material, and therefore independent of shape. With a uniform geometry, resistivity may be converted to resistance R:

$$R = \rho L/A, \qquad (1\text{–}4.3)$$

where L is length and A is cross-sectional area. With R, the engineer may use the basic equations of physics for current I in amperes and power P in watts; that is, $I = E/R$, and $P = EI = I^2 R = E^2/R = \text{J/s}$.

Dielectric behavior   Electrical insulators do not, of course, transport electric charges. However, they are not inert to an electric field. We can show this by separating two electrode plates by a distance d and applying a voltage E between them (Fig. 1–4.1 a). The *electric field* $\mathscr{E}$ is the voltage gradient:

$$\mathscr{E} = \frac{E}{d}. \qquad (1\text{–}4.4)$$

Under these conditions, when there is nothing between the plates, the *charge density* $\mathscr{D}_0$ on each plate is proportional to the field $\mathscr{E}$. For each volt/m of field, there are 8.85×10^{-12} coulombs per square meter of electrode.

$$\mathscr{D}_0 = (8.85 \times 10^{-12} \text{ C/V}\cdot\text{m})\mathscr{E}. \qquad (1\text{–}4.5)$$

This charge density requires 55×10^6 electrons/m² for each volt/m, since the *electron charge* is 0.16×10^{-18} coulombs.

If a material m is placed between the electrodes in Fig. 1–4.1, the charge density can be increased to $\mathscr{D}_m$ from the $\mathscr{D}_0$ just described. The ratio, $\mathscr{D}_m/\mathscr{D}_0$, is called the relative *dielectric constant* κ of the material that is used as the dielectric spacer between the electrodes:

$$\kappa = \frac{\mathscr{D}_m}{\mathscr{D}_0}. \qquad (1\text{–}4.6)$$

The relative dielectric constant will be important to us in Chapters 7 and 8, in which we look at the *dielectric* properties of the plastics and ceramics that are used in capacitors.* The relative dielectric constant is always greater than 1.0 because both electrons and the positive and negative ions are displaced within the material when it is inside an electric field.

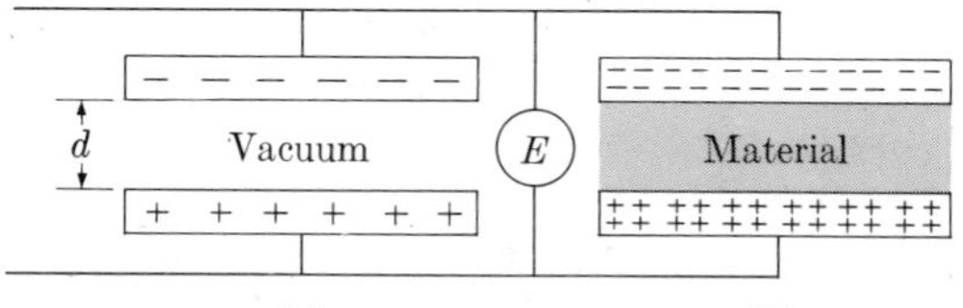

Fig. 1–4.1   Charge density $\mathscr{D}$ versus relative dielectric constant, κ. The presence of a material increases the charge density, $\mathscr{D}_m$, held by the capacitor plates in proportion to the relative dielectric constant, κ:

$$\mathscr{D}_m = \kappa\mathscr{D}_0.$$

(a)   (b)

---

*   For a parallel-plate capacitor, the capacitance C is determined by

$$C = \kappa(8.85 \times 10^{-12} \text{ C/V}\cdot\text{m})(A/d), \qquad (1\text{–}4.7)$$

where A is the plate area and d is the spacing between plates.

Example 1-4.1 A copper wire has a diameter of 0.9 mm. (a) What is its resistance per foot (0.305 m)? (b) How many watts are expended if 1.5 volts are applied to 30 m of this wire?

Solution: From Appendix C, $\rho = 17$ ohm·nm.

a)
$$R = (17 \times 10^{-9} \text{ ohm·m})(0.305 \text{ m})/(\pi/4)(9 \times 10^{-4} \text{ m})^2$$
$$= 0.008 \text{ ohm.}$$

b)
$$P = \frac{E^2}{R}$$

$$= \frac{E^2 A}{\rho L} = \frac{(1.5 \text{ V})^2(\pi/4)(9 \times 10^{-4} \text{ m})^2}{(17 \times 10^{-9} \text{ ohm·m})(30 \text{ m})} = 2.8 \text{ watts.} \quad \blacktriangleleft$$

Example 1-4.2 Some impure silicon has a resistivity of 0.03 ohm·m and a charge mobility of 0.19 $m^2/V \cdot s$. (a) How many charge carriers are there per m^3? (b) What is the drift velocity of the charge carriers when 5 mV are applied across a 0.4-mm silicon chip? With zero voltage?

Solution: From Eqs. (1-4.1) and (1-4.2):

a)
$$= 1/(0.16 \times 10^{-18} \text{ amp·sec})(0.19 \text{ m}^2/\text{V·sec})(0.03 \text{ ohm·m})$$
$$= 1.1 \times 10^{21}/\text{m}^3.$$

b)   With 5 mV, $\bar{v} = (0.19 \text{ m}^2/\text{V·s})(0.005 \text{ V}/4 \times 10^{-4} \text{ m}) = 2.4$ m/sec. With zero, $\bar{v} = 0$ m/sec.

Comments. Actual electron movements are essentially identical in both situations. At zero voltage, the movements are random, and no net transfer results. With increased fields, the net, or drift velocity, increases. ◀

Example 1-4.3 A 6-V light uses 16 watts of dc current. How many electrons move through the filament per minute?

Solution: Since $P = EI$,

$$\text{Current} = 16 \text{ watts}/6\text{V} = 2.7 \text{ amp;}$$

$$(2.7 \text{ amp}) (60 \text{ sec}) = 160 \text{ amp·sec.}$$

$$\text{Electrons} = 160 \text{ amp·sec}/(0.16 \times 10^{-18} \text{ amp·sec/el})$$
$$= 10^{21} \text{ electrons.} \quad \blacktriangleleft$$

Example 1-4.4 Two capacitor plates (30 mm × 20 mm each) are parallel and 2.2 mm apart, with nothing between them. What voltage is required to produce a charge of 0.24×10^{-10} coul on the electrodes?

Solution: Rearranging Eqs. (1-4.4) and (1-4.5) and using meters, we get

$$\text{Voltage} = \frac{(0.24 \times 10^{-10} \text{ C})/(0.03 \text{ m})(0.02 \text{ m})}{(8.85 \times 10^{-12} \text{ C/V·m})/(0.0022 \text{ m})} = 10 \text{ volts.} \quad \blacktriangleleft$$

$\varepsilon = S/E$

1–5   PRESENTATIONS OF PROPERTIES   Skim

Qualitative information   Schematic diagrams that show the effect of one variable on a dependent property are indispensable aids in translating complicated empirical relationships into qualitative terms. Figure 1–5.1, for example, illustrates the change in the strength of concrete in relation to the amount of water added. Thus, concrete is strongest when a minimum amount of water is used, although there must be sufficient water to make the concrete workable.

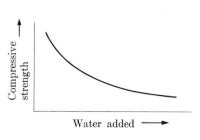

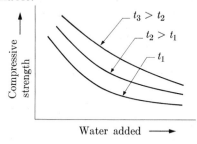

Fig. 1–5.1   Schematic representation of two variables. Strength of concrete versus water content. The water content is the independent variable.

Fig. 1–5.2   Schematic representation of three variables. Strength of concrete as related to time t and water content.

Other variables may be shown schematically by the use of additional parameters. Figure 1–5.2 adds the parameter of time, t, to the relationship previously given in Fig. 1–5.1. Figure 1–5.2 tells us that (1) for any given addition of water to cement, the strength increases as the period of time increases; (2) for any given period of time, the strength is less if excess water was used; and (3) a given strength may be attained in less time if less water is used.

Schematic representations help the engineer to determine in advance what variables can be controlled to obtain the desired result. With such information engineers can anticipate possible modifications of their materials in production or in service.

Quantitative data   It is commonly important to secure quantitative data concerning the properties of materials. Thus, from Fig. 1–5.3, the design engineer observes that concrete may have a compressive strength of 33 MPa (4800 psi) if the water-to-portland cement ratio is 0.5. However, to make the information complete, the parameters of time as well as data on particle size and temperature should be included, since each of these influences the quantitative relationships.

Variance of data   All laboratory and industrial data contain a spread in values. Figure 1–5.4 shows the scatter obtained in the toughness testing of fifty samples of steel at 21°C (70°F). This occurred even though the samples and testing procedures were identical, so far as could be determined. The variation in toughness can arise from several sources: (1) undetected differences within the steel which was sampled; (2) subtle differences in the preparation of the samples; (3) slight differences in the testing procedure.

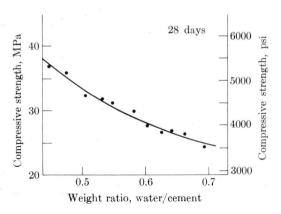

Fig. 1–5.3 Quantitative values. Strength of concrete versus water content. (ASTM Testing Standards No. C 39 was followed.)

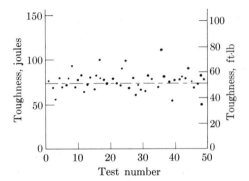

Fig. 1–5.4 Variance of data. Variations encountered with Charpy impact testing of SAE 1040 steel (21°C). All tests were identical.

Because of this range of data, results are commonly reported as the *mean* (average) value, or *median* (middle) test value. In Example 1–5.1, which cites 21 strength values for a plastic, the mean and median strengths are 20.7 MPa (3010 psi) and 20.9 MPa (3030 psi), respectively. Although the mean $\bar{X}$ and median $\bar{M}$ values are close to each other, they are not identical. The difference arises in this case because individual results are more dispersed at the lower strengths.

Reporting either the mean or median does not fully satisfy the design engineer and the applied scientist, because they do not know the range of strengths; they therefore ask for the *standard deviation* (SD). The standard deviation is a statistical measure of *scatter*. It is calculated as follows:

$$SD = \sqrt{\sum(X_i - \bar{X})^2/(n-1)}, \qquad \bullet \,(1\text{–}5.1)^*$$

---

* See Preface for the bullet, ●, notation.

where $\bar{X}$ is the mean and X_i are the individual values (up to n). In practice, the standard deviation has significance in that approximately two thirds of the results lie within ± 1 SD, and 5% of statistical data lie outside ± 2 SD, when there are sufficient data to follow normal distribution laws.

Accuracy of calculations   It is commonly desirable to calculate properties as closely as the available data warrant. In this text, the available data include two or three, and occasionally four, *significant figures*. This is typical of data that are generally provided for commercial materials. With a hand calculator, answer displays commonly contain nonsignificant figures. The practice of reporting nonsignificant figures should be avoided. The recommended practice is to round off the *final* answer (not intermediate values) to the number of significant figures possessed by the least accurate *noninteger* datum (+ one).*

Example 1–5.1 The test data for the breaking stresses of 21 samples of a plastic have been arranged in descending order in Fig. 1–5.5.

a)   Determine the two measures of central tendency: median $\bar{M}$ and mean $\bar{X}$.
• b)   Calculate the scatter or standard deviation, SD.

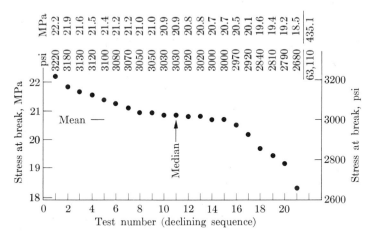

Fig. 1–5.5   Variance in test data. (See Example 1–5.1) The mean is the average value; the median is the middle value.

Solution: We may use either MPa or psi.

a)   The 11th (median) value is 20.9 MPa (3030 psi). The mean value is

$$\bar{X} = \frac{\sum X}{21} = \frac{435.1 \text{ MPa}}{21} = 20.7 \text{ MPa},$$

or

$$\frac{63{,}110 \text{ psi}}{21} = 3010 \text{ psi}.$$

• b) $$\text{SD} = \sqrt{\sum (X - \bar{X})^2 / 20} = \sqrt{16.79/20} = 0.9 \text{ MPa}$$

or

$$\sqrt{355{,}900/20} = 130 \text{ psi}.$$

Comments. The data of Fig. 1–5.5 have a typical statistical variation, since 14 of the 21 values lie within 20.7 ± 0.9 MPa (3010 ± 130 psi). The median is higher than the mean because there is more dispersion among the low test data. ◀

REVIEW AND STUDY

SUMMARY

If technology is to meet the needs and desires of society, there must be an intelligent use of materials. This involves selecting the materials with the optimum characteristics, cost, and availability, and incorporating them into a design that is safe, reliable, and compatible with everyone's well-being. In order to provide the technology, we must select materials to meet specific requirements, such as strength, thermal and electrical conductivity, and fabricability. Therefore, we are interested in the various properties that give us a basis for our designs and products.

The properties and behavior of materials arise from their internal structures. If a specific set of properties is required, the materials must be chosen appropriately so that they have suitable structures of atoms, crystals, and other internal arrangements. Should the internal structure of a material be changed during processing or service, there will be corresponding changes in properties.

In this chapter, we have cited some—but not all—of the properties that require attention. The ones we have encountered, however, will enable us to discuss in later chapters the features of internal structures that (a) provide resistance to mechanical alteration by applied stresses; (b) account for several common characteristics of a material when it is thermally agitated at higher temperatures; and (c) relate to the behavior of a material within an electric field. These features are important because knowledge of them enables the designer to specify, and the technologist to produce, materials with optimum structures and characteristics for present and future products.

Solid materials undergo elastic strain that is proportional to the applied stress at all stress levels. This strain is recoverable after the stress is removed. A permanent, plastic strain develops when a sufficiently high stress is applied to a ductile material. Strong materials that undergo significant plastic strain are tough, in that they require considerable energy before failure.

The input of thermal energy to raise the temperature increases the thermal agitation within the material. This extra energy expands the structure. Thermal energy is transferred to cooler parts of the material by atom vibrations and (in metals) by electron conduction.

The diffusion of electrons accounts for the electrical conductivity of most conducting materials. The charges within an insulator do not migrate in electric fields, but are displaced, so that, in addition to simply being "isolators," dielectrics have useful functional electrical properties.

Finally, attention was given to the significance and accuracy of property data. Measures of central tendency (mean and median) and of scatter (standard deviation) should become familiar to the reader. Furthermore, the reader is alerted against the use of nonsignificant figures in presenting an answer from a hand calculator.

KEY TERMS AND CONCEPTS

Brinell hardness number, BHN

Brittle

Charge density, $\mathscr{D}$

Conductivity

   electrical, σ

   thermal, k

Dielectric constant, κ

Drift velocity, $\bar{v}$

Ductility

Electric field, $\mathscr{E}$

Electron charge, q

Elongation, e_f

Gage length

Hardness

Heat capacity, c

Internal structure

Mean, $\bar{X}$

Median, $\bar{M}$

Mobility, μ

Modulus of elasticity, E

Nominal stress

Plastic strain

Properties

Reduction of area, % R of A

Resistivity, ρ

Rockwell hardness, R

Significant figures

Specific heat

• Standard deviation, SD

Strain, e

   elastic

   plastic

Strength

   breaking, S_b

   tensile, S_t

   yield, S_y

Stress

   nominal, s

   true, σ

Système International, SI

Thermal expansion coefficient

   linear, α_L

   volume, α_V

Toughness

• True strain, ϵ

Yield point

Young's modulus, E

FOR CLASS DISCUSSION

A_1 Name three varieties of steel and indicate the characteristic property of each.

B_1 A car fender could be made out of armor plate so that it would not crumple in a collision. Give at least three reasons why it is not designed that way.

C_1 Examine a piece of wood. Describe some of its structural characteristics and indicate how they affect the properties.

D_1 Before World War II, one company made one car body out of stainless steel. To date it shows negligible signs of rusting. What complications, if any, would exist if this design had been used for all cars?

E_1 Indicate three critical materials requirements that are encountered in your engineering field (i.e., M.E., C.E., E.E., etc.).

F_1 Compare and contrast: strength, ductility, and toughness.

G_1 Why is the true stress always higher than the nominal stress? Why does the engineer use the nominal stress when he or she designs a bridge?

H_1 Refer to Fig. 1–2.1(c). Why is yield strength commonly defined on the basis of 0.2% permanent strain?

I_1 Loading is removed when the top point of the curve in Fig. 1–2.1(b) is reached. Will the unloading curve be vertical or sloping? Why?

J_1 Refer to I_1. After unloading, gage marks are placed on the test piece. It is now reloaded and carried to failure. In what way will the calculated elongation be affected?

K_1 The percent reduction in area cannot be calculated from percent elongation (nor vice versa). Why is this true, when both are measures of ductility?

L_1 The nominal breaking strength (based on original area) is commonly less than the tensile strength. However, the true stress for breaking is higher than the tensile strength. Explain.

M_1 Toughness could be defined as the area under the stress–strain curve. Is this consistent with the concept that toughness is the energy required for fracture?

N_1 The terms *elastic limit* and *yield strength* are sometimes used as synonyms. Why is this not always strictly true?

O_1 "The tensile strength is the stress where the strain switches from elastic to plastic." Comment on the validity of this statement.

P_1 Cite a design situation where we would multiply kilograms by a number other than 9.8 to get newtons.

Q_1 European practice, which has also been widely adopted in English-speaking countries, indexes hardness values as the number of kg/mm^2 (i.e., kg load and indentation area). What problems, if any, will this produce in engineering calculations?

R_1 Examine the prongs of an extension cord for a power saw. Why should there be specifications for (a) yield strength, (b) modulus of elasticity, (c) resistivity, and (d) corrosion resistance?

S₁   The values for the thermal expansion and thermal conductivity coefficients are identical whether we use °C or the absolute temperature, K. In later chapters we will have problems where the choice of °C or K does make a difference. Why does it not matter here?

T₁   Consider the data for iron and copper in Appendix C. Under what service conditions will these have to be altered if they are to be used in engineering calculations?

U₁   Provide proof for Eq. (1–3.2 b).

V₁   Iron has discontinuities in its thermal conductivity coefficient slightly above 900°C (Fig. 1–3.1). It also changes volume at that temperature. Suggest whether this volume change is an expansion or contraction. Why?

W₁   List the kinds of carriers that can transport electrical charge through a liquid.

X₁   Is it possible to have a relative dielectric constant of less than one. Why?

Y₁   Why is it better practice to delete excess (insignificant) figures in the final answer rather than in intermediate steps?

Z₁   Refer to the metals of Appendix C. Identify two sets of data that can be correlated. Draw a schematic curve of the correlation. Discuss a possible basis for the correlations.

STUDY PROBLEMS*

1–1.1   Take the cord of a household appliance, such as a toaster or coffee-maker. List the materials used and the probable reason for their selection.

➤ **1–1.2**   Examine an incandescent light bulb closely. How many different types of materials can you name? What thermal and electrical characteristics are required of each?

1–2.1   A rod of copper should not be stressed to more than 70 MPa (10,200 psi) in tension. What diameter is required if it is to carry a load of 2000 kg (4400 lbs)?

Answer: 19 mm (or 0.74 in.)

1–2.2   What is the elastic strain in a copper rod that is stressed 70 MPa (10,200 psi)? (Moduli of elasticity and other properties of common metals are given in Appendix C.)

1–2.3   (a) A steel bar 12.7 mm in diameter supports a load of 7000 kg (and therefore a force of 68,600 N). What is the stress placed on the bar? (b) If the bar of part (a) has a modulus of elasticity of 205,000 MPa, how much will the bar be strained with this load?

Answer: (a) 540 MPa   (b) 0.0026

---

*   Study Problems that parallel examples of the text or that closely relate to proceeding topics are identified by italicized numbers. They may be used as "starters." The remaining problems, like engineering calculations, are not simply variations of a previously available problem. Rather, they require the engineer to establish the required relationships. These are identified by bold-face numbers. The student should be able to handle a significant number of these comprehensive-type problems.

   Those Study Problems relating to topics that may be assigned at the option of the instructor are marked with a bullet, ●.

1–2.4 The bar in Study Problem 1–2.3 supports a maximum load of 11,800 kg (26,000 lb) without plastic deformation. What is its _____ strength?

1–2.5 The bar of Study Problem 1–2.3 breaks with a load of 11,400 kg (and therefore a force of 111,000 N). Its final diameter is 7.87 mm. (a) What is its true breaking strength? (b) What is its nominal breaking strength? (Based on original area.)

Answer: (a) 2280 MPa   (b) 875 MPa

• *1–2.6* Bronze has a modulus of elasticity of 110,000 MPa (16×10^6 psi) and a yield strength of 158 MPa (22,900 psi), a nominal breaking strength of 238 MPa (34,500 psi), and a reduction in area of 34%. (a) What is the true strain just before yielding starts? (b) What is the true stress at the point of fracture? (Cf. Example 1–2.4.)

1–2.7 A copper alloy has a modulus of elasticity of 110,000 MPa (16,000,000 psi), a yield strength of 330 MPa (48,000 psi), and a tensile strength of 350 MPa (51,000 psi). (a) How much stress would be required to stretch a 3-m (118-in.) bar of this alloy 1.5 mm? (b) What size round bar would be required to support a force of 22,000 N (5000 lb$_f$) without yielding?

Answer: (a) 55 MPa (8,000 psi)   (b) 9.2 mm (0.36 in.)

1–2.8 A wire of a magnesium alloy is 1.05 mm (0.04 in.) in diameter. According to Appendix C, its modulus of elasticity is 45,000 MPa (6.5×10^6 psi). Plastic deformation starts with a load of 10.5 kg (which is _____ N), or 23 lb$_f$. The total strain is 0.0081 after loading to 12.1 kg (26.6 lb$_f$). (a) How much permanent strain has occurred with a load of 12.1 kg (26.6 lb$_f$)? • (b) Rework this problem with English units.

1–2.9 Aluminum (6151 alloy) has a modulus of elasticity of 70,000 MPa (10^7 psi) and a yield strength of 275 MPa (40,000 psi). (a) How much load can be supported by a 2.75-mm (0.108-in.) wire of this alloy without yielding? (b) If a load of 44 kg (97 lb) is supported by a 30.5-m (100-ft) wire of this size, what is the total extension?

Answer: (a) 1630 N, or 167 kg (366 lb$_f$)   (b) 32 mm (1.27 in.)

1–2.10 Monel metal (70Ni–30Cu) has a modulus of elasticity of 180,000 MPa (26×10^6 psi) and a yield strength of 450 MPa (65,300 psi). (a) How much load could be supported by a rod 18 mm (0.71 in.) in diameter without yielding? (b) If a maximum total stretching of 2.5 mm (0.1 in.) is permissible in a 2.1-m (6.9-ft) bar, how large a load could be applied to this rod?

• **1–2.11** Refer to Example 1–2.4 and its comments. What was the true strain at the point of fracture?

Answer: 1.47 (or 147%)

1–2.12 A 6.4-mm (0.25-in.) diameter 1020 steel bar 1.83 m (6 ft) long supports a weight of 4450 N (1000 lb$_f$). What is the difference in length if the bar is changed to a 70–30 Monel? (See Problem 1–2.10.)

1–2.13 A long-used rule of thumb says that in English units a *steel* has a tensile strength of approximately 500 times its Brinell hardness number, that is, $(S_t)_{psi} \approx 500$ (BHN). (a) Check this against Fig. 1–2.3. (b) Establish a similar constant relating $(S_t)_{MPa}$ to BHN.

Answer: $(S_t)_{MPa} \approx 3.5$ (BHN)

• **1–2.14** Refer to Study Problem 1–2.6. What is the true strain at the point of fracture?

1–2.15 A 1040 steel wire has a diameter of 0.89 mm (0.035 in.). Its yield strength is 980 MPa (142,000 psi) and its tensile strength is 1130 MPa (164,000 psi). An aluminum alloy is also available with a yield strength of 255 MPa (37,000 psi), and a tensile strength of 400 MPa (58,000 psi). (a) How much (%) heavier, or lighter, will a steel wire be than an aluminum wire to support a 40-kg (88-lb) load with the same elastic deformation as the steel wire? (b) How much (%) heavier, or lighter, must an aluminum wire be to support the same maximum load without deforming? (c) Without breaking? (*Hint:* establish a ratio of masses without calculating the actual areas.)

Answer: (a) $+0.7\%$ (b) $+32\%$ (c) -3%

1–3.1 The temperature of a 2-meter brass rod is raised 80°C (144°F). Based on the data of Appendix C, what stress would be required to change the length by the same amount?

Answer: 176 MPa (25,600 psi)

1–3.2 An iron wire is fastened between two buildings and stressed 20 MPa (2900 psi). The air temperature suddenly drops from 25°C to 19°C. What is the new stress?

1–3.3 The average linear coefficient of thermal expansion of a special steel rod is 12×10^{-6} °C^{-1}. (a) How much temperature change is required to provide the same linear change as a stress of 620 MPa (90,000 psi)? (b) What volume change does this temperature change produce?

Answer: (a) $\Delta T = 250$°C (450°F)   (b) 0.9%

1–3.4 A welded steel rail is laid in place at 35°C (95°F) and anchored so that contraction cannot occur. When the temperature drops to 0°C, how much stress develops in the rail?

1–3.5 (a) Calculate the heat transfer through a plate of copper that is substituted for the stainless steel in Example 1–3.3. (b) What is the wattage per mm^2? (Joules are volt·amp·sec, or watt·sec.)

Answer: (a) 420 joules/mm^2·min   (b) 7 watts/mm^2

1–3.6 A hot liquid is pumped through a 50-mm diameter copper pipe that has a wall 1.0 mm thick. The metal is 140°C on the inside and 25°C on the outside. Estimate the heat transferred per hour per meter of pipe length.

1–3.7 Repeat Study Problem 1–3.6 for iron where the inside and outside temperatures are 840°C and 725°C, respectively. (The data from Appendix C may not be appropriate.)

Answer: ~1500 MJ/hr

1–3.8 (a) Using the data from Appendix C, determine the ratios of thermal to electrical conductivities, k/σ, for each metal. (b) Based on your calculations, make a generalization about the relationship between electrical and thermal conductivities.

1–4.1 There are 10^{20} electrons per m^3 in a semiconductor that has an electron mobility of 0.19 m^2/V·sec. (a) What is the resistivity? (b) The conductivity?

Answer: (a) 0.33 Ω·m   (b) 3 ohm^{-1}·m^{-1}

1–4.2 A semiconductor has 10^{21} electrons that serve as charge carriers per m^3, and a resistivity of 0.1 ohm·m. (a) What is the mobility μ of the electrons? (b) What is the drift velocity $\bar{v}$ when the voltage gradient is 0.016 volts/mm?

1–4.3 (a) If a pure copper wire (resistivity = 17 ohm·nm) that is 1 mm in diameter is used for an electrical circuit carrying 10 amp, how many watts of heat will be lost per foot (0.305 m)? (b) How many more watts would be lost if the copper wire were replaced by a brass wire of the same size (resistivity = 32 ohm·nm)?

Answer: (a) 0.66 W   (b) $\Delta = 0.6$ W

1–4.4 Refer to Study Problem 1–2.8. Before loading, the wire length was 3.07 m. (a) Its resistance is _____. (b) With 1.5 volts across this wire length, there will be _____ electrons entering the wire per millisecond.

1–4.5 Measurements indicate that the drift velocity of electrons in a semiconductor that has 10^{20} carriers/m³ is 200 mm/sec when 30 millivolts are placed across 57 mm. What is the expected resistivity?

Answer: 0.164 ohm·m

1–4.6 What is the electron density on the electrodes of Example 1–4.4?

1–4.7 The space between the plates of Example 1–4.4 is filled with a 2.2-mm thick sheet of polystyrene. With 10 volts, the charge is increased from 0.24×10^{-10} coul to 0.6×10^{-10} coul. What is the relative dielectric constant of the polystyrene?

Answer: 2.5

1–4.8 The dielectric constant of a glass ribbon is 6.1. Would a capacitor using such a glass ribbon 0.25 mm thick have greater or less capacitance than another capacitor having the same area, but using a 100-μm plastic with a dielectric constant of 2.1?

1–4.9 What new relative dielectric constant will be required for a capacitor to retain the same charge density if the spacing between the foil electrodes is reduced from 0.10 mm to 0.06 mm without changing the applied voltage? The present insulation has a relative dielectric constant of 3.3.

Answer: 2.0

1–4.10 The steel wire of Study Problem 1–2.15 is replaced by a copper wire with the same dimensions. How much more ($+\%$), or less ($-\%$), resistance will the copper wire have?

1–4.11 A copper wire (1.1-mm dia.) was chosen to meet resistance specifications. How much heavier ($+\%$), or lighter ($-\%$), would a pure aluminum wire be which has the same resistance? (*Hint:* Establish a ratio of masses without calculating actual areas.)

Answer: -48%

1–5.1 There are 6.1×10^{21} carriers/m³ in a semiconductor. Their mobility is 0.21 m²/V·sec. The hand calculator gives an answer of 0.00487900781 ohm·m as the resistivity. Express this answer with an appropriate number of significant figures.

Answer: 0.00488 (or 0.0048_8)

1–5.2 The "4 horsemen" of the backfield weigh 215, 197, 211, and 181 lbs. What is their mean weight expressed in an appropriate number of significant figures?

1–5.3 The points (out of 145) scored by 15 students in a Materials examination were 89, 137, 95, 98, 57, 77, 121, 69, 101, 108, 51, 88, 117, 80, 109. (a) What is the median score? (b) The mean score? (c) The quartiles?

Answer: (a) 95   (b) 93.1   (c) 77, 109

• **1–5.4** Obtain the scores for last year's basketball games. Determine the mean, median and standard deviation of the scores of your team (a) at home, (b) away.

1–5.5 What are the (a) mean and (b) median of the following density data for a certain type of foam rubber: 0.26, 0.30, 0.23, 0.24, 0.21, 0.26, 0.27, 0.29, 0.22, 0.28 g/cm³ (or Mg/m³)?
• (c) What is the standard deviation of the data?

Answer: (a) 0.256 Mg/m³   (b) 0.26 Mg/m³   (c) 0.030 Mg/m³ ($=$0.030 g/cm³)

2

Introduction to Materials: Review of Chemical Bonding

PREVIEW

In all solids, atoms are held together by bonds. They provide strength and related electrical and thermal properties to the solids. For example, strong bonds lead to high melting temperatures, high moduli of elasticity, shorter interatomic distances, lower thermal expansion coefficients.

In this chapter, we shall draw upon the concepts introduced in general chemistry. We will need to examine the role of the valence electrons in the primary bonds—*ionic*, *covalent*, and *metallic*—in sufficient detail so we may anticipate their effect on interatomic distances and atomic coordination. These bonds also provide a basis for categorizing materials as metals, polymers (plastics), and ceramics.

CONTENTS

2–1 Individual Atoms and Ions:
atomic mass units, periodic table, electrons.

2–2 Strong Bonding Forces (Primary Bonds):
ionic bonds, covalent bonds, metallic bonds.

2–3 Molecules:
bond lengths and energies, bond angles, isomers.

•**2–4 Secondary Bonding Forces:**
induced dipoles, polar molecules, hydrogen bridge.

2–5 Interatomic distances:
coulombic forces, electronic repulsion forces, bonding energy, atomic and ionic radii.

2–6 Coordination Numbers.

2–7 Types of Materials:
metals, polymers (plastics), ceramics.

The Review and Study section of this chapter includes *generalizations regarding properties* in addition to a list of key terms and concepts, discussion topics, and study problems.

STUDY OBJECTIVES

1 To review those parts of your general chemistry that pertain to bonds between atoms.

2 To understand key terms and concepts, so that you may follow later topics more readily.

3 To know the characteristics of the three major types of primary bonds (but to appreciate that actual bonds usually involve contributions from more than one of these bond types).

4 To develop a qualitative picture of how the forces and energies of bonds vary with interatomic distances (as presented in Fig. 2–5.2) to give us a basis for interatomic distance, modulus of elasticity, thermal expansion, etc., in later chapters.

5 To understand various factors that affect atom size. You should be able to rationalize size differences in Table 2–5.1.

6 To know the characteristics of the three major types of materials (but to appreciate that many materials will have intermediate characteristics).

7 To develop some generalizations regarding properties as they relate to bond type.

2–1   INDIVIDUAL ATOMS AND IONS

The atom is the basic unit of internal structure for our studies of materials. The initial concepts involving individual atoms are familiar to most of the readers. They include *atomic number, atomic mass*, and the relationships in the *periodic table*. We shall also give attention to the energy levels that are established by the electrons of the atoms.

Atomic mass unit   Since atoms are extremely small in comparison to our day-to-day concepts of mass, it is convenient to use the *atomic mass unit* as the basis for many calculations. The amu is defined as one twelfth of the atomic mass of carbon-12, the most common isotope of carbon. There are $0.6022 \ldots \times 10^{24}$ amu per g. We will use this conversion factor (called *Avogadro's number N*) in various ways. Since natural carbon contains approximately one percent C^{13} along with 98.9% C^{12}, the average atomic mass of a carbon atom is $12.011 \ldots$ amu. This is the value presented in the *periodic table* (Fig. 2–1.1) and in tables of selected elements (Appendix B). Those atomic masses encountered most commonly by the reader include

> H $1.0079 \ldots$ amu (or 1.0079 g/$(0.602 \times 10^{24}$ atoms))
> C $12.011 \ldots$
> O $15.9994 \ldots$
> Cl $35.453 \ldots$
> Fe $55.847 \ldots$

These values may be rounded off to 1, 12, 16, 35.5, and 55.8 amu, respectively, for all but the most precise of our calculations.

The *atomic number* indicates the number of electrons associated with each neutral atom (and the number of protons in the nucleus). Each element is unique with respect to its atomic number. Appendix B lists selected elements, from hydrogen, with an atomic number of one, to uranium (92). It is the electrons, particularly the outermost ones, that affect most of the properties of engineering interest: (1) they determine the chemical properties; (2) they establish the nature of the interatomic bonding, and therefore the mechanical and strength characteristics; (3) they control the size of the atom and affect the electrical conductivity of materials; and (4) they influence the optical characteristics. Consequently, we shall pay specific attention to the distribution and energy levels of the electrons around the nucleus of the atom.

Periodic table   The *periodicity of elements* is emphasized in chemistry courses. We shall not repeat those characteristics here except to observe that the periodic table (Fig. 2–1.1) arranges the atoms of sequentially higher atomic numbers so that the vertical columns, called *groups*, possess atoms of similar chemical and electronic characteristics. In brief, those elements at the far left of the periodic table are readily ionized to give positive ions, *cations*. Those in the upper right corner of the periodic table more readily share or accept electrons. They are *electronegative*.

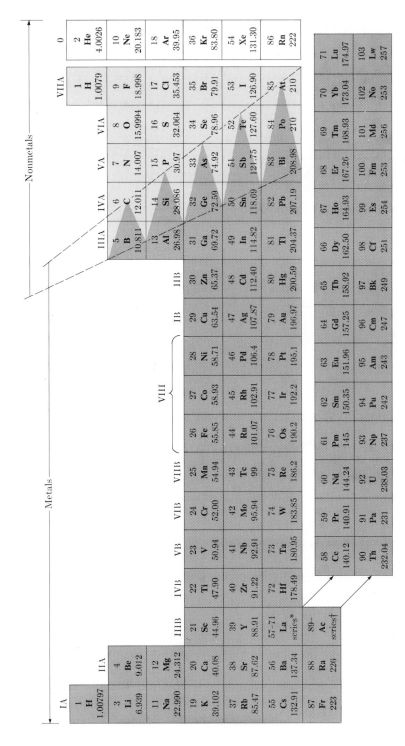

Fig. 2–1.1 Periodic table of the elements, showing the atomic number and atomic mass (in amu). There are 0.6×10^{24} amu per gram; therefore the atomic masses are grams per 0.6×10^{24} atoms. Metals readily release their outermost electrons. Nonmetals readily accept or share additional electrons.

Electrons   Since electrons are components of all atoms, their negative electrical charge is commonly regarded as unity. In physical units, this charge is equal to 0.16×10^{-18} A·s per electron (or 0.16×10^{-18} coul/electron).

The electrons that accompany an atom are subject to very rigorous rules of behavior because they have the characteristics of standing waves during their movements in the neighborhood of the atomic nuclei. Again the reader is referred to introductory chemistry texts; however, let us summarize several features here. With individual atoms, electrons have specific energy states—*orbitals*. As shown in Fig. 2–1.2, the available electron energy states around a hydrogen atom can be very definitely identified.* To us, the important consequence is that there are large ranges of intermediate energies *not* available for the electrons. They are forbidden because the corresponding frequencies do not permit standing waves. Unless excited by external means, the one electron of a hydrogen atom will occupy the lowest orbital of Fig. 2–1.2.

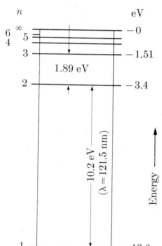

Fig. 2–1.2   Energy levels for electrons (hydrogen). The electron of hydrogen normally resides in the lowest energy level. (At this level, it would take 13.6 eV, or 2.2×10^{-18} J, to separate the electron from the nucleus.) Electrons can be given additional energy, but only at specific levels. Gaps exist between these levels. These are forbidden energies.

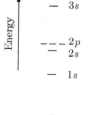

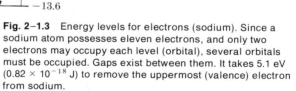

Fig. 2–1.3   Energy levels for electrons (sodium). Since a sodium atom possesses eleven electrons, and only two electrons may occupy each level (orbital), several orbitals must be occupied. Gaps exist between them. It takes 5.1 eV (0.82×10^{-18} J) to remove the uppermost (valence) electron from sodium.

Single atom

Figure 2–1.3 shows schematically the energies of the lowest orbitals for sodium. Each orbital can contain no more than two electrons. These must be of opposite spins. Again, there are forbidden energy gaps between the orbitals that are unavailable for electron occupancy.

---

*   This is done by spectrographic experiments.

In our considerations, the topmost occupied orbital will have special significance since it contains the *valence electrons*. These electrons may be removed by a relatively small electric field, to give us the positive *cations* mentioned a few paragraphs ago. The energy requirements are called the *ionization energies*. In the next section we will see that these outermost or valence electrons are *delocalized* in metallic solids and free to move throughout the metal rather than remaining bound to individual atoms. This provides the basis for electrical and thermal conductivity.

When the valence orbitals are not filled, the atom may accept a limited number of extra electrons within these unfilled energy states, to become a negative ion, *anion*. These electronegative atoms with unfilled valence orbitals may also share electrons. This becomes important in covalent bonding and will be reviewed in the next section.

Example 2–1.1 Sterling silver contains approximately 7.5 w/o* copper and 92.5 w/o silver. What are the a/o* copper and a/o silver?

Solution: Basis: 10,000 amu alloy = 9250 amu Ag + 750 amu Cu.

Ag: 9250 amu Ag/(107.87 amu Ag/atom) = 85.75 atoms = 88 a/o

Cu: 750 amu Cu/(63.54 amu Cu/atom) = 11.80 atoms = 12 a/o. ◀

$$\overline{97.55}$$

Example 2–1.2 The mass of small diamond is 3.1 mg. (a) How many C^{13} atoms are present if carbon contains 1.1 a/o of that isotope? (b) What is the weight percent of that isotope?

Solution

a)
$$\frac{0.0031 \text{ g}}{(12.011 \text{ g}/0.6022 \times 10^{24} \text{ atoms})} = 1.55 \times 10^{20} \text{ C atoms};$$

$$(1.55 \times 10^{20})(0.011) = 1.7 \times 10^{18} \text{ C}^{13} \text{ atoms.}$$

b)   Basis: 3.1 mg.

$$\frac{\text{Mass}_{13}}{\text{Mass}_{\text{total}}} = \frac{(1.7 \times 10^{18})(13 \text{ amu})}{(1.55 \times 10^{20})(12.011 \text{ amu})} = 1.2 \text{ w/o.}$$

Comment. Since the mass of the C^{13} is greater than the average atom, the weight percent will be greater than the atom percent. ◀

Example 2–1.3 From Appendix B, indicate the orbital arrangements for a single iron atom, and for Fe^{2+} and Fe^{3+} ions.

Solution: The argon core is $1s^2 2s^2 2p^6 3s^2 3p^6$; therefore,

$$\text{Fe: } 1s^2 2s^2 2p^6 3s^2 3p^6 3d^6 4s^2. \tag{2–1.1}$$

---

* Weight percent, w/o; atom percent, a/o; linear percent, l/o; volume percent, v/o; mole percent, m/o; etc. In condensed phases (*solids and liquids*) weight percent is implied unless specifically stated otherwise. In *gases*, v/o or m/o are implied unless specifically stated otherwise.

During ionization, the $4s$ electrons are removed first; therefore:

$$Fe^{2+}: 1s^2 2s^2 2p^6 3s^2 3p^6 3d^6; \tag{2-1.2}$$

$$Fe^{3+}: 1s^2 2s^2 2p^6 3s^2 3p^6 3d^5. \blacktriangleleft \tag{2-1.3}$$

Example 2-1.4 Ten grams of nickel are to be electroplated on a steel surface with an area of 0.8953 m². The electrolyte contains Ni^{2+} ions. (a) How thick will the nickel plate be? (b) What amperage is required if this is to be accomplished in 50 minutes?

Solution

$$M_{Ni}/A_{ST}(\text{Density } Ni) =$$

a) $10 \text{ g}/(8.9 \times 10^6 \text{g/m}^3)(0.8953 \text{ m}^2) = 1.25 \times 10^{-6} \text{ m (or 1.25 } \mu m).$

b)

$$\left[\frac{10 \text{ g Ni}}{58.71 \text{ g Ni}/0.6 \times 10^{24} \text{ atoms}}\right]\left[\frac{(2 \text{ el/atom})(0.16 \times 10^{-18} \text{ A·s/el})}{(3000 \text{ s})}\right] = 10.9 \text{ amperes}.$$

Comment. By now, the student should be alert to the data available in the Appendices. $\blacktriangleleft$

2-2 STRONG BONDING FORCES (PRIMARY BONDS)

Since most products are designed with solid materials, it is desirable to understand the attractions that hold the atoms together. The importance of these attractions may be illustrated with a piece of copper wire, in which each gram contains $(0.602 \times 10^{24}/63.54)$ atoms. Based on the density of copper, each cubic centimeter contains 8.9 times this number, i.e., 8.4×10^{22} atoms/cm³ (or 8.4×10^{28}/m³). Under these conditions, the forces of attraction that bond the atoms together are strong. If this were not true, they would easily separate, the metal would deform under small loads, and atomic vibrations associated with thermal energy would gasify the atoms at low temperatures. As in the case of this wire, the engineering properties of any material depend on the interatomic forces which are present.

Interatomic attractions are caused by the electronic structure of atoms. The noble (inert or chemically inactive) gases, such as He, Ne, Ar, have only limited interactions with other atoms because they have a very stable arrangement of eight electrons (2 for He) in their outer, or valence, electron orbitals. Furthermore, they have no net charge as a result of an unbalanced number of protons and electrons. Most other elements, unlike the noble gases, must achieve the relatively stable configuration of having eight electrons available for their outer orbitals through one of the following procedures: (1) receiving extra electrons, (2) releasing electrons, or (3) sharing electrons. The first two of these processes produce ions with a net negative or positive charge and thus provide the ions with coulombic attractions to other ions of unlike charge. The third process obviously requires an intimate association between atoms in order for the sharing of electrons to be operative. Where applicable, the above three processes produce strong or primary bonds. Energies approximating 500 kJ/mole (i.e., 500,000 joules per 0.602×10^{24} bonds) are required to rupture these bonds. Other weaker or secondary bonds (less than 40 kJ/mole) are also always present, but gain importance when they are the only forces present (Section 2-4).

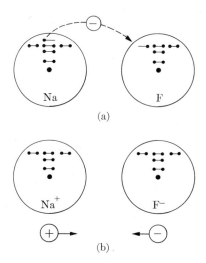

(a)

(b)

Fig. 2–2.1 Ionization. (a) Electron transfer from the outer orbital of sodium to fluorine. (b) The resulting positive and negative ions are mutually attracted, by coulombic forces, to form an ionic bond.

Ionic bonds   The interatomic bond that is easiest to describe is the ionic bond, which results from the mutual attraction of positive and negative charges. Atoms of elements such as sodium and calcium, with one and two electrons in their valence orbitals, respectively, easily release these outer electrons and become positively charged ions. Likewise, chlorine and oxygen atoms readily add to their valence orbitals until they have eight electrons by accepting one or two electrons and thus becoming negatively charged ions. Since there is always a *coulombic attraction* between negatively and positively charged materials, a bond is developed between neighboring ions of unlike charges as shown schematically in Fig. 2–2.1.

A negative charge possesses an attraction for *all* positively charged particles and a positive charge for *all* negatively charged particles. Consequently, sodium ions surround themselves with as many negative chlorine ions as possible, and chlorine ions surround themselves with the maximum number of positive sodium ions, the attraction being equal in all directions (Fig. 2–2.2). The major requirement in an ionically bonded material is that the number of positive charges equals the number of negative charges. Thus, sodium chloride has a composition of NaCl. Magnesium chloride has a composition of $MgCl_2$, because the magnesium atom can supply two electrons from its valence shell but each chlorine atom can accept only one.

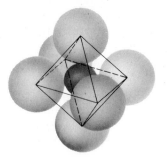

Fig. 2–2.2 Three-dimensional structure of sodium chloride. The positive sodium ion is coordinated with and has equal attraction for all six neighboring negative chlorine ions. (Compare with Fig. 3–1.1 where the structure reveals that Na^+ ions also surround Cl^- ions.)

Since these coulombic attractions involve all neighbors, ionically bonded materials may be very stable, particularly if multivalent ions are involved. As an example, when magnesium and oxygen combine to form MgO, 570 kJ/mole are released, i.e., 570,000 joules (or 136,000 calories) per 0.6×10^{24} Mg^{2+} ions and 0.6×10^{24} O^{2-} ions in the product. Thus, MgO must be raised to approximately $2800°C$ ($\sim 5000°F$) before it overcomes this energy and melts.

Covalent bonds Another primary force of strong attraction is the *covalent* bond in which electrons are shared. Figure 2–2.3 shows two representations of this sharing for two fluorine atoms in F_2. Commonly the first representation (electron dots, or a "bond line") will suffice for our purposes, e.g., Fig. 2–2.4(a) for carbon. However, the reader should be aware that electrons cannot be precisely located without a

Fig. 2–2.3 Covalent bonding (fluorine). (a) Either the "electron dot" or the "bond line" is commonly used for simplicity. (b) Orbital energy levels (schematic). It takes 160,000 joules (38,000 calories) to break a mole (0.6×10^{24}) of these bonds. (Only the $2p$ electrons are shown in part (b).)

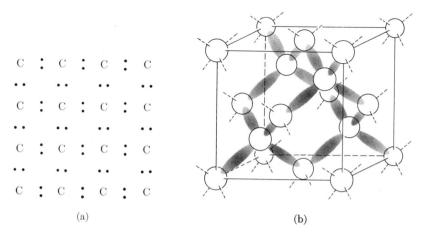

Fig. 2–2.4 Diamond structure. The strength of the covalent bonds is what accounts for the great hardness of diamond. (a) Two-dimensional representation. (b) Three-dimensional representation with the bond shown as the region of high electron probability (shaded).

degree of uncertainty. Therefore, this presentation will not always be satisfactory. An alternative is the schematic representation in Fig. 2–2.4(b) where a shaded region is used to indicate the location of high probability for the pair of shared electrons.

Figure 2–2.3(b) focuses on the energies of the valence electrons (2*p orbitals*) of fluorine as they combine into molecular orbitals. Note that the *average* energy of the outer or valence electrons drops when the molecule is formed from the two individual atoms. Therefore, we may consider covalent bonds in terms of energy, because energy would be required to reverse the reaction sketched in Fig. 2–2.3(b).

That covalent bonds provide strong attractive forces between atoms is evidenced in diamond, which is the hardest material found in nature, and which is wholly carbon. Each carbon has four valence electrons. These are shared with adjacent atoms to form a three-dimensional lattice entirely bonded by covalent pairs (Fig. 2–2.4). The strength of the covalent bond in carbon is demonstrated not only by the great hardness of diamond but also by the extremely high temperature ($>3000°C$) to which it can be heated before the structure is disrupted (melted) by thermal energy.*

Unlike coulombic attractions that bring as many unlike ions into neighboring positions as space will allow, covalent bonds are formed between specific atoms. In the diamond of Fig. 2–2.4(b), the number of neighbors is limited by the number of bonds and not by the available space. In Fig. 2–2.3, the two fluorines are held together with a covalent connection of $160 \text{ kJ}/0.6 \times 10^{24}$ bonds. However, neither of these two atoms develops strong attractions to other fluorine atoms (or molecules) that may approach them. As evidence, F_2 vaporizes to a gas at 85 K ($-188°C$ or $-306°F$) with only $\sim 3 \text{ kJ/mole}$. When the bond involves a given pair of atoms, we apply the term *stereospecific*, and they are therefore directional.

An exception to the above stereospecificity of covalent bonds occurs in compounds with a benzene ring, which is discussed in chemistry texts (Fig. 2–2.5). One electron per carbon atom (for a total of six) is *delocalized*. These six electrons have equal probability of being found anywhere around the ring.[†] They can respond to alternating electric fields by moving from one side of the molecule to another but cannot leave the molecule (except under unusually catastrophic conditions). There are as many wave patterns for these delocalized electrons *as there are atoms in the ring.*

Metallic bonds   In addition to ionic and covalent bonds, a third type of primary interatomic attractive mechanism is the metallic bond. The model of metallic bonding is not as simple to construct as the other two. However, we can adapt the concept of delocalized electrons from the previous paragraph to serve our purpose. First consider graphite (Fig. 2–2.6); the layers of carbon atoms in graphite possess delocalized electrons (as well as electron pairs between specific atoms). The delocalized electrons can respond to electric fields, moving within the graphite sheet in a wavelike pattern.[‡] In fact, conductivity becomes possible if a positive electrode is present to

---

* It would take about 750 kJ, or 180 kcal, to break all the bonds in a mole (0.6×10^{24}) of carbon atoms.
† They reside in π bonds which are perpendicular to the plane of the molecule.
‡ As with the benzene ring, there are as many wave patterns possible *as there are carbon atoms in the layer.* Also note that while the delocalized electrons can readily move within the layer, there is not a comparable mechanism to move from one layer to another. (See discussion question Z_2 at the end of the chapter.)

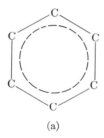

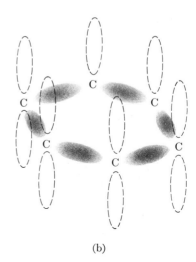

(a)

Fig. 2–2.5 Delocalized electrons. (a) Benzene ring.
(b) The orbitals between the carbon atoms are
stereospecific. Electrons in the other orbitals can move
from one side of the molecule to the other in response to
internal and external electrical fields; they are
delocalized. (The hydrogen atoms, which lie in the
plane of the carbon atoms, have been omitted for
clarity.)

(b)

Fig. 2–2.6 Delocalized electrons in
graphite layers. Each layer contains
"multiple benzene rings" (Fig. 2–2.5).
The conductivity is more than 100 times
greater in the parallel direction than in the
perpendicular direction.

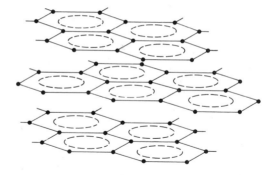

remove electrons from one end of the layer, and a negative electrode is available to
supply electrons to the other end.

Typical metals have delocalized electrons that can move in three dimensions.
It is thus common to speak of an electron "cloud" or "gas" because the outer, least
strongly bonded electrons are able to move throughout the metal structure. The orbitals for metals are sketched schematically in Fig. 2–2.7 for sodium. Just as the
molecular orbitals of F_2 on the right side of Fig. 2–2.3(b) are modified from the
atomic orbitals, the energy levels in multiatomic sodium differ from that of the single-
atom orbitals that were first shown in Fig. 2–1.3. The prime change between Fig.
2–2.7(a) and 2–2.7(b) is that the upper valence orbital has split into as many levels
as there are atoms in the system. Note that the average energy of the electrons in the
valence band of Fig. 2–2.7(b) is below the energy of the 3s orbital for the individual
atom. This accounts for the bonding in metals; in brief, energy would have to be
supplied to overcome the metallic bond and separate the atoms from one another
and to reestablish the individual atomic orbitals. Qualitatively speaking, we find very
strong bonds holding tungsten atoms together. Its melting and boiling temperatures

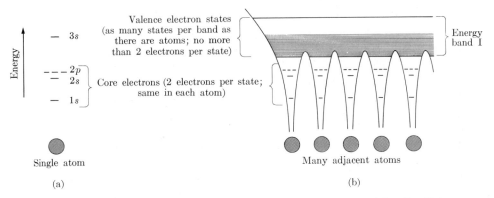

Fig. 2–2.7 Valence electrons in metal (sodium). The valence electrons are delocalized into an energy band. These electrons are able to move throughout the metal. The valence electrons fill only the bottom half of the band. Their average energy is lower than that of the 3s electrons with individual atoms. This energy difference provides the metallic bond.

are very high (~ 3400 and $5900°C$, respectively); also, it has an extremely high modulus of elasticity (50,000,000 psi, or 345,000 MPa; see Appendix C for comparisons). In contrast, the bonds in sodium are low as evidenced by its melting point ($97.8°C$) and soft behavior. Both have delocalized electrons for electrical and thermal conduction.

Before concluding this section on primary bonds, or strong attractive forces, we should observe that while one bond type may be prevalent in a material, other bond types can be present, too; so *mixed-bond types* are widespread.

Example 2–2.1 The covalent bond between two carbon atoms, C—C, is 370 kJ/mole (88 kcal/0.6×10^{24} bonds). The energy of light is

$$E = h\nu \tag{2–2.1}$$

where h is Planck's constant (0.66×10^{-33} joule·sec) and ν is the frequency of light. What wavelength λ is required to break a C—C bond?

Solution

$$\nu = c/\lambda$$

$$370{,}000 \text{ J}/0.6 \times 10^{24} = (0.66 \times 10^{-33} \text{ J·s})\nu,$$

$$\nu = 9.34 \times 10^{14}/\text{s} = c/\lambda,$$

where c is the velocity of light.

$$\lambda = (0.299 \times 10^9 \text{ m/s})/(9.34 \times 10^{14}/\text{s})$$

$$= 0.320 \times 10^{-6} \text{ m } (=320 \text{ nm}).$$

Comments. This is in the ultraviolet range. It is for this reason that ultraviolet light can cause deterioration in plastics that contain C—C covalent bonds. ◀

2–3 MOLECULES

A *molecule* may be defined as a group of atoms that are strongly bonded together, but whose bonds to other, similar groups of atoms are relatively weak. Our prototype for a molecule may be fluorine, F_2, which was discussed in the previous section. Recall that 160 kJ/mole (~ 1.65 eV/bond) would have to be present to break the covalent bond joining the two atoms (Fig. 2–2.3). In contrast, only 3 kJ/mole (0.03 eV/bond) provided the thermal agitation that is required to separate the molecules into a gas by boiling.

The more common examples of molecules include compounds such as H_2O, CO_2, CCl_4, O_2, N_2, and HNO_3. Other small molecules are shown in Fig. 2–3.1. Within each of these molecules, the atoms are held together by strong attractive forces that usually have covalent bonds, although ionic bonds are not uncommon. Unlike the forces that hold atoms together, the bonds between molecules are weak and consequently each molecule is free to act more or less independently. These observations are borne out by the following facts: (1) Each of these molecular compounds has a low melting and a low boiling temperature compared with other materials. (2) The molecular solids are soft because the molecules can slide past each other with small stress applications. (3) The molecules may remain intact in the liquid and gaseous forms.

The molecules listed above are comparatively small; other molecules have large numbers of atoms. For example, pentatriacontane (shown in Fig. 2–3.2c) has over 100 atoms, and some molecules contain many thousand. Whether the molecule is small like CH_4 or much larger than that shown in Fig. 2–3.2(c), the distinction between the strong *intra*molecular and the weaker *inter*molecular bonds still holds.

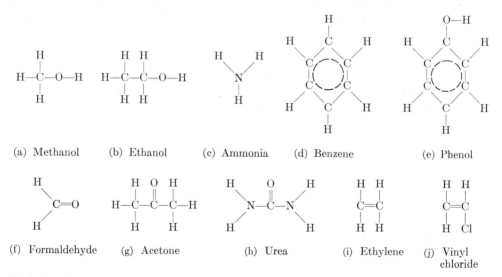

(a) Methanol (b) Ethanol (c) Ammonia (d) Benzene (e) Phenol

(f) Formaldehyde (g) Acetone (h) Urea (i) Ethylene (j) Vinyl chloride

Fig. 2–3.1 Small organic molecules. Each carbon is surrounded by four bonds, each nitrogen by three, each oxygen by two, and each hydrogen and chlorine by one.

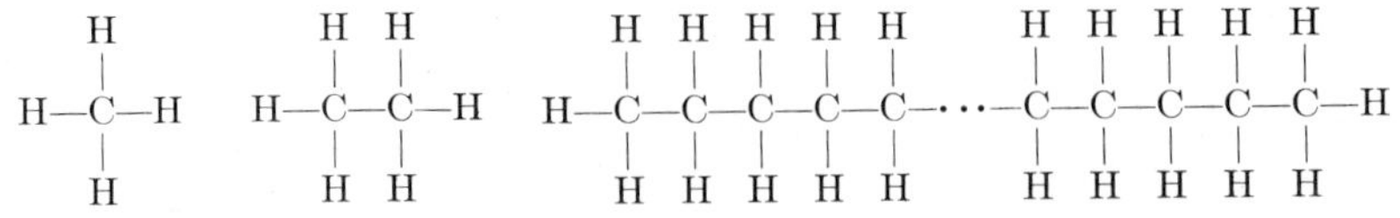

```
    H                H  H              H  H  H  H  H          H  H  H  H  H
    |                |  |              |  |  |  |  |          |  |  |  |  |
H — C — H        H — C — C — H     H — C— C— C— C— C—· · ·— C— C— C— C— C—H
    |                |  |              |  |  |  |  |          |  |  |  |  |
    H                H  H              H  H  H  H  H          H  H  H  H  H

(a) Methane      (b) Ethane         (c) C₃₅H₇₂, pentatriacontane (i.e., 35-ane)
```

(a) Methane   (b) Ethane   (c) $C_{35}H_{72}$, pentatriacontane (i.e., 35-ane)

Fig. 2–3.2 Examples of molecules. Molecules are discrete groups of atoms. Primary bonds hold the atoms together within the molecule. Weaker, secondary forces attract molecules to each other.

Other materials such as metals, MgO, SiO_2, and phenol-formaldehyde plastics have continuing three-dimensional structures of primary bonds. The difference between the structures of molecular materials and those with primary bonds continuing in all three dimensions produces major differences in properties. These differences will be considered in subsequent chapters.

Bond lengths and energies   The strength of bonds between atoms in a molecule, of course, depends on the kind of atoms and the other neighboring bonds. Table 2–3.1 is a compilation of bond lengths and energies for those atom couples most frequently encountered in molecular structures. The energy reported is the amount required to break one mole (Avogadro's number) of bonds. For example, 370,000 joules of energy are required to break 0.602×10^{24} C—C bonds, or $370,000/(0.602 \times 10^{24})$ joules per bond. Likewise, this same amount of energy is released $(-0.61 \times 10^{-18} \text{ J})$ if one of these C—C bonds is formed. Only the sign is changed.

Bond angles   The chemist recognizes hybrid orbitals in certain covalent compounds, where the s and p orbitals are amalgamated. The most important hybrid for us to review is the sp^3 orbital. We have already sketched it in Fig. 2–2.4 for the four bonds of carbon in diamond. Four equal orbitals are formed instead of having distinct $2s$ and $2p$ orbitals that occur in individual atoms (e.g., the individual sodium and fluorine atoms of Fig. 2–2.1a). Methane (CH_4) and carbon tetrachloride (CCl_4), like diamond, have sp^3 orbitals that connect four identical atoms to the central carbon. Therefore, we find them equally spaced around the central carbon at 109.5° from each other.* Geometrically, this is equivalent to placing the carbon at a cube center and pointing the orbitals toward four of the eight corners (Fig. 2–3.3a). However, if the orbitals do not bond identical atoms to the central carbon, these time-averaged angles are distorted slightly, as shown for CH_3Cl (Fig. 2–3.3b).

Greater distortions occur in hybrid orbitals when some of the electrons occur as *lone pairs* rather than in the covalent bond. This is particularly evident in NH_3 and H_2O (Fig. 2–3.4) where 107.3° and 104.5° are the time-averaged values for H—N—H and H—O—H, respectively.

---

* The angle 109.5° is a time-averaged value. Any particular H—C—H angle in CH_4 will vary rapidly as a result of thermal vibrations.

Table 2–3.1
Bond energies and lengths

Bond	Bond energy*		Bond length, nm
	kcal/mole	kJ/mole	
C—C	88†	370†	0.154
C=C	162	680	0.13
C≡C	213	890	0.12
C—H	104	435	0.11
C—N	73	305	0.15
C—O	86	360	0.14
C=O	128	535	0.12
C—F	108	450	0.14
C—Cl	81	340	0.18
O—H	119	500	0.10
O—O	52	220	0.15
O—Si	90	375	0.16
N—H	103	430	0.10
N—O	60	250	0.12
F—F	38	160	0.14
H—H	104	435	0.074

* Approximate. The values vary with the type of neighboring bonds. For example, methane (CH_4) has the above value for its C—H bond; however the C—H bond energy is about 5% less in CH_3Cl, and 15% less in $CHCl_3$.

† All values are negative for forming bonds (energy is released), and positive for breaking bonds (energy is required).

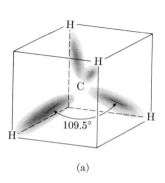

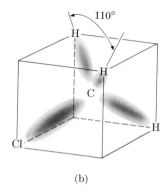

(a)   (b)

Fig. 2–3.3 Bond angles. (a) Methane, CH_4, is symmetrical with each of the six angles equal to 109.5°. (b) Chloromethane, CH_3Cl, is distorted.

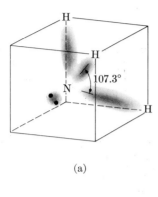

(a)

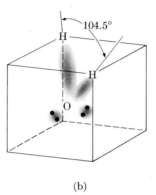

Fig. 2–3.4   Bond angles. (a) Ammonia, NH$_3$, and (b) water have angles between the 109.5° of Fig. 2–3.3(a) and 90°. Ammonia has one lone-pair of electrons; water, two.

(b)

One of the bond angles most frequently encountered in the study of materials is the C—C—C angle of the hydrocarbon chains (Fig. 2–3.5). While this will differ slightly, depending upon whether hydrogen or some other side radical is present, we may assume for our purposes that the C—C—C angle is close to 109.5°.

Fig. 2–3.5   Bond angles (butane). Although we commonly draw straight chains (Figs. 2–3.2c and 2–3.7b), there is a C—C—C bond angle of about 109°.

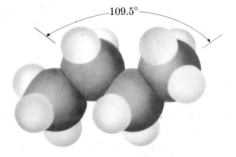

Isomers   In molecules of the same composition, more than one atomic arrangement is usually possible. This is illustrated in Fig. 2–3.6 for propyl and isopropyl alcohol. Variations in the structure of molecules with the same composition are called *isomers*. Differences in structure affect the properties of molecules because of the resulting change in molecular polarity (Section 2–4). For example, the melting and boiling temperatures for propyl alcohol are − 127°C and 97.2°C, respectively, whereas for isopropyl alcohol the corresponding temperatures are − 89°C and 82.3°C.

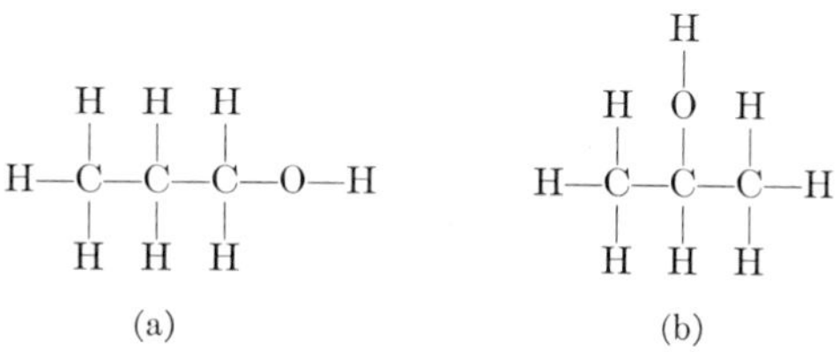

Fig. 2–3.6 Isomers of propanol. (a) Normal propyl alcohol. (b) Isopropyl alcohol. The molecules have the same composition but different structures. Consequently, the properties are different. Compare with polymorphism of crystalline materials (Section 3–4).

$$
\begin{array}{ccc}
& \text{H} \quad \text{H} \quad \text{H} & \\
& |\quad\ |\quad\ | & \\
\text{H} & -\text{C}-\text{C}-\text{C} & -\text{O}-\text{H} \\
& |\quad\ |\quad\ | & \\
& \text{H} \quad \text{H} \quad \text{H} &
\end{array}
$$

(a)

$$
\begin{array}{ccc}
& \quad\quad \text{H} & \\
& \quad\quad | & \\
& \text{H} \quad \text{O} \quad \text{H} & \\
& |\quad\ |\quad\ | & \\
\text{H} & -\text{C}-\text{C}-\text{C} & -\text{H} \\
& |\quad\ |\quad\ | & \\
& \text{H} \quad \text{H} \quad \text{H} &
\end{array}
$$

(b)

Example 2–3.1 How much energy is given off when 70 g of ethylene (Fig. 2–3.7a) react to give polyethylene (Fig. 2–3.7b)?

Solution: Each added C_2H_4 molecule breaks one $C{=}C$ bond and forms two $C{-}C$ bonds. From Table 2–3.1:

$$\frac{+680{,}000 \text{ J}}{0.602 \times 10^{24} \text{ molecules}} - \frac{2(370{,}000 \text{ J})}{0.602 \times 10^{24} \text{ molecules}} = -9.96 \times 10^{-20} \text{ J}/C_2H_4,$$

$$70 \text{ g } (0.602 \times 10^{24} \text{ amu/g})/(28 \text{ amu}/C_2H_4) = 1.5 \times 10^{24} \ C_2H_4,$$

$$(-9.96 \times 10^{-20} \text{ J}/C_2H_2)(1.5 \times 10^{24} \ C_2H_4 = -150{,}000 \text{ J},$$

or

$$-150{,}000 \text{ J } (0.239 \text{ cal/J}) = -36 \text{ kcal.}$$

Comment. Convention treats required energy as $(+)$ and released energy as $(-)$. The reaction of Fig. 2–3.7 is the basic reaction for making large vinyl-type molecules that are used in plastics (Chapter 7). ◀

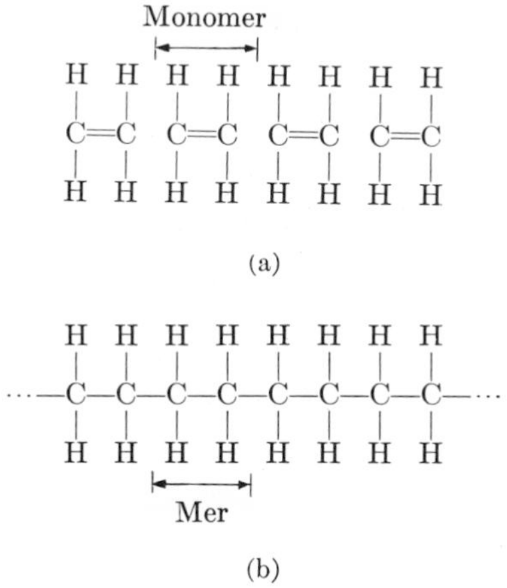

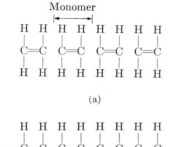

Monomer

(a)

Mer

(b)

Fig. 2–3.7 Addition polymerization of ethylene. (a) Monomers of ethylene. (b) Polymer containing many C_2H_4 mers, or units. The original double bond of the ethylene monomer is broken to form two single bonds and thus connect adjacent mers.

Example 2–3.2 Show sketches (carbon atoms only) of the various isomers of heptane, C_7H_{16}.

Sketches

a) C—C—C—C—C—C—C

b)
```
          C
          |
   C—C—C—C—C—C
```

c)
```
     . C .
       |
   C—C—C—C—C—C
```

d)
```
     C  C
     |  |
   C—C—C—C—C
```

e)
```
     C        C
     |        |
   C—C—C—C—C
```

f)
```
       C
       |
   C—C—C—C—C
       |
       C
```

g)
```
         C
         |
   C—C—C—C—C
         |
         C
```

h)
```
     C  C
     |  |
   C—C—C—C
       |
       C
```

Comment. Other arrangements, for example,
```
       C
       |
       C
       |
   C—C—C—C,
       |
       C
```
duplicate one of the above alternatives (in this case, sketch *g*). ◄

•2–4 SECONDARY BONDING FORCES*

The three types of bonds considered in Section 2–2 are all relatively strong primary bonds that hold atoms together. Weaker, secondary bonds, which supply interatomic attraction, are grouped here as *van der Waals forces*, although there are actually several different mechanisms involved. Were it not for the fact that sometimes they are the only forces that operate, van der Waals bonding might be overlooked.

In a noble gas like helium, the initial orbital, with its two electrons, is complete; and other noble gases, such as neon and argon, have a full complement of eight electrons in their valence orbitals. In these stable situations none of the primary bonds can be effective, since covalent, ionic, and metallic bonds all require adjustments in the valence electrons. As a result, atoms of these noble gases have little attraction for one another, and with rare exceptions they remain monatomic at ordinary temperatures. Only at extremely low temperatures, when thermal vibrations have been greatly reduced, do these elemental gases condense (Table 2–4.1A). It is this condensation that makes it evident that there are weak interatomic attractions that pull the atoms together.

Similar evidence for these weak attractions is found in the molecules listed in Table 2–4.1(B). As pointed out earlier, these gases have satisfied their valence requirements by covalent bonding within the molecule. The condensation of these simple molecules occurs only when thermal vibrations are sufficiently reduced in energy to permit the weak van der Waals forces to become noticeable.

---

* See Preface for the bullet, •, notation.

Table 2–4.1

Melting and boiling temperatures of gases (absolute temperature)

| | A. Noble gases | | | B. Simple molecules | | |
Gas	Melting temperature, K	Boiling temperature, K		Molecule	Melting temperature, K	Boiling temperature, K
				Symmetric		
He	0.96*	4.25		H_2	14.02	21
Ne	24.5	27		N_2	63	78
Ar	84	87.5		O_2	55	90
Kr	116	120		CH_4	88	145
Xe	161	166		CCl_4	250	349
Rn	202	211		C_4H_{10}	135	274
				Polar		
				NH_3	195	240
				CH_3Cl	113	259
				H_2O	273	373

* Melting point with 26 atmospheres of pressure. At one atmosphere pressure, helium remains as a liquid as 0 K ($-273.16°C$) is approached.

Induced dipoles All but the last three of the gases and molecules of Table 2–4.1 are symmetric; i.e., over any extended period of time, the center of positive charges from the protons in the nuclei, and the center of negative charges from the electrons, are at the center of each molecule (or noble gas atom). Continually, however, the electron motions and the atom vibrations disrupt this electrical symmetry. When this happens, a small electrical *dipole* is established. In each small fraction of a second,* the centers of positive and negative charges are not coincident, so an electrical dipole is established giving the molecule a positive end and a negative end. In turn, this induces a dipole into the adjacent molecules by displacing their electrons in response to this minute electric field. Therefore, attractive forces are established, though admittedly weak. This is demonstrated by the data of Table 2–4.1, which shows that these gases do condense; however, not until low temperatures are reached.

Polar molecules Asymmetric molecules, such as NH_3, CH_3Cl, and H_2O, always have a noncoincidence of their positive and negative charges. This *polarity* is best illustrated in Fig. 2–3.4 for ammonia (NH_3). Three hydrogen nuclei, which are no more than bare protons (+), are exposed to the upper right. The lone pair of electrons on the other side of the molecule makes that position the negative end of the molecule. The *inter*molecular bonding forces of symmetrical CH_4 and asymmetric NH_3 may be compared through their melting and boiling temperatures, since each has about the same mass (16 amu and 17 amu, respectively). Table 2–4.1 shows that NH_3 must be raised to 240 K ($-33°C$) before thermal agitation breaks the *inter*molecular bonds

* Between 10^{-16} and 10^{-12} sec.

to form a gas. The more weakly bonded, symmetrical CH_4 is vaporized at 145 K ($-128°C$). In further contrast, CH_3Cl (Fig. 2–3.3b) does not vaporize until 259 K ($-14°C$) because the chlorine possesses a large number of electrons that are located relatively far from the center of the molecule and therefore give strong attractions to neighboring molecular dipoles.

Hydrogen bridge   This third type of van der Waals bonding force is actually a special case of a polar molecule. However, it is by far the strongest of these secondary bonding forces and is widely encountered. It therefore warrants special attention and a special name.

The exposed proton at the end of a C—H, O—H, or N—H bond is not screened by electrons. Therefore, this positive charge can be attracted to valence electrons of adjacent molecules. A coulombic-type bond is developed, called a *hydrogen bridge*. Our most common example of this is found in water, where the proton of the hydrogen in one molecule is attracted to the lone pairs of electrons on the oxygen in an adjacent molecule (Fig. 2–4.1). The maximum energy of this bond is about 30 kJ/mole (7 kcal/mole). This is in contrast to (1) a maximum of 5 kJ/mole (and usually $\ll$ 1 kJ/mole) for the other types of van der Waals bonds, and (2) several hundred kilojoules per 0.6×10^{24} bonds for primary bonds (Section 2–2).

Chemists point out that it would be difficult to overemphasize the importance of the hydrogen bridge. For example, H_2O, with a molecular weight of only 18 amu, has the highest boiling temperature of any molecule with a molecular weight of less than 100 amu. If it had a boiling point comparable to other 3- or 4-atom molecules, our oceans would be nonexistent and all biological and geological conditions would be completely altered. In our study of materials, the hydrogen bridge affects the properties and behavior of plastics (Chapter 7) and certain ceramics (Chapter 8).

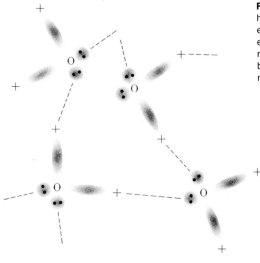

Fig. 2–4.1 Hydrogen bridge (in water). The hydrogen at the end of the orbital is an exposed proton ($+$). It is attracted to the electron lone-pairs of the adjacent water molecules. (Cf. Fig. 2–3.4b.) The hydrogen bridge makes water the highest boiling of any material with a low molecular weight (18 amu).

Example 2–4.1 Each OH arm of a water molecule has an electric dipole of moment 5×10^{-30} coul·m. What is the dipole moment of the whole molecule?

Solution: Solve for the resultant of the two dipoles at $104.5°$ (Fig. 2–3.4).

$$p = 2(5 \times 10^{-30} \text{ coul·m}) \cos (104.5°/2) \qquad P_{TOTAL} = 2 \left(P_{ARM}\right) \cos\left(\frac{\theta}{2}\right)$$
$$= 6 \times 10^{-30} \text{ C·m}.$$

Comment. The *dipole moment* is the product of the charge and the distance between the centers of positive and negative charges. ◀

*

Example 2–4.2 The H_2O molecule has one proton in each hydrogen atom and eight in the oxygen atom. (a) How far apart are the centers of positive and negative charges?

The H_2O molecule may also be considered to have a positive end from the two exposed protons (H^+), and a negative end from the oxygen atom with its eight protons and complement of ten electrons (O^{2-}). (b) How far apart are the ends of the dipole defined in this manner?

Solution: From Example 2–4.1, the dipole moment $(p = Qd)$ of the H_2O molecule is 6×10^{-30} coul·m.

a)
$$Qd = 10(0.16 \times 10^{-18} \text{ C})d = 6 \times 10^{-30} \text{ C·m};$$
$$d = (6 \times 10^{-30} \text{ C·m})/(1.6 \times 10^{-18} \text{ C})$$
$$= 3.8 \times 10^{-12} \text{ m} \qquad (=0.004 \text{ nm}).$$

b)
$$Qd = 2(0.16 \times 10^{-18} \text{ C})d = 6 \times 10^{-30} \text{ C·m};$$
$$d = 19 \times 10^{-12} \text{ m} \qquad (=0.02 \text{ nm}).$$

Comments. Either concept is satisfactory for pointing out that the centers of positive and negative charges are not coincident. The first considers the elementary charged particles (electrons and protons); the latter assigns the charges to the ionized atoms. ◀

2–5 INTERATOMIC DISTANCES

Although in the case of diatomic molecules there is bonding and coordination of only two atoms, most materials involve a coordination of many atoms into an integrated structure. Two main factors, interatomic distances and spatial arrangements, are of importance. Let us therefore consider them in some detail.

The forces of attraction between atoms, which we considered in the preceding sections, pull the atoms together; but what keeps the atoms from being drawn still closer together? It should be apparent from the preceding figures and the discussion that there is much vacant "space" in the volume surrounding the nucleus of an atom. The existence of this space is evidenced by the fact that neutrons can move through the fuel and the other materials of a nuclear reactor, traveling among many atoms before they are finally stopped (see Fig. 6–9.1).

The space between atoms is caused by interatomic repulsive forces, which exist in addition to the interatomic attractive forces described in Sections 2–2 and 2–4. Mutual repulsion results primarily because the close proximity of two atoms places

Fig. 2–5.1   Balance of forces (ceramic ring magnets). Downward force on upper ring is caused by gravity. Upward force is caused by magnetic repulsion. Space remains between the two magnets at the equilibrium position. (Of course the forces in this analogy are not identical to those between atoms; however, the principle of opposing forces is comparable.) (Courtesy of the North American Philips Co.)

too many electrons into interacting locations. The equilibrium distance is that distance at which the repulsive and the attractive forces are equal. An analogy may be made between the interatomic distances among atoms and the spacing between the two ring magnets of Fig. 2–5.1. (In this example, the magnets are aligned to give repulsion rather than attraction.) Of course the forces in this analogy are not identical to those between atoms; however, the principle of force balances is comparable. The top ring magnet is moved by a force (gravity) toward the lower ring magnet (which in this case is fixed by the container). Since the force of gravity is essentially constant over the distance considered here, the top magnet falls to the point where it is repulsed by an equal magnetic force, of opposite direction. Because the repulsive force increases as an inverse function of the distance, equilibrium distance is achieved. Note that the magnets remain separated by space. (A non-magnetic material may move through this space, just as a neutron (neutrons have no charge) may move among atoms in a solid.)

Coulombic forces   The ionic bond will be used to illustrate the balance between attractive and repulsive forces in materials. The coulombic force F_C developed between two point charges is related to the quantity of the two charges Z_1q and Z_2q, and their separation distance a_{1-2} as follows:

$$F_C = -k_0(Z_1q)(Z_2q)/a_{1-2}^2, \qquad (2\text{–}5.1)$$

where Z is the valence ($+$ or $-$) and q is 0.16×10^{-18} coulomb. The proportionality constant k_0 depends on the units used when we are considering adjacent ions.*

Electronic repulsive forces   The repelling force F_R between the electronic fields of two atoms or ions is also an inverse function with distance, but to a higher power:

$$F_R = -bn/a_{1-2}^{n+1}. \qquad \bullet(2\text{–}5.2)$$

---

* With SI units, k_0 is 9×10^9 V·m/C, since $k_0 = 1/4\pi\epsilon_0$.

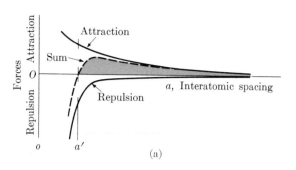

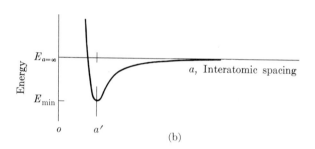

Fig. 2–5.2 Interatomic distances.
(a) The equilibrium spacing o–a' is
the distance at which the attractive
forces equal the repulsive forces.
(b) The lowest potential energy
occurs when o–a' is the interatomic
distance. Since $E = \int F\,da$, the
shaded area of (a) equals the depth
of the energy trough in (b).

Both b and n are empirical constants, with n equal to approximately 9 in ionic
solids. Comparing, $F_C \propto a^{-2}$, and $F_R \propto a^{-10}$. Thus, the attractive forces predominate
at greater distances of atomic separation, and the repulsive forces predominate at
closer interatomic spacings (Fig. 2–5.2a). The equilibrium spacing, o–a', is a natural
result when

$$F_C + F_R = 0. \tag{2–5.3}$$

A tension force is required to overcome the predominant forces of attraction if the
spacing is to be increased. Conversely, a compressive force has to be applied to push
the atoms closer together against the rapidly increasing electronic repulsion.

   The equilibrium spacing is a very specific distance for a given pair of atoms, or
ions. It can be measured to five significant figures by x-ray diffraction (Chapter 3),
if temperature and other factors are controlled. It takes a large force to stretch or
compress that distance as much as one percent. (Based on Young's modulus, a stress
of 2000 MPa (300,000 psi) is required for iron.) It is for this reason that the *hard ball*
provides a usable model for atoms for many purposes where strength or atom
arrangements are considered.*

---

* The hard-ball model is not suitable for all explanations of atomic behavior. For example, a neutron
(which doesn't have a charge) can travel through the space among the atoms without being affected by
the electronic repulsive forces just described. Likewise, atomic nuclei can be vibrated vigorously by increased thermal energy, with only a small expansion of the *average* interatomic spacing. Finally, by a
momentary distortion of their electrical fields, atoms can move past one another in a crowded solid. (See
diffusion in Chapter 4.)

Bonding energy   The sum of the above two forces provides us with a basis for bonding energies (Fig. 2–5.2b). Since the product of force and distance is energy,

$$E = \int_{\infty}^{a} (F_C + F_R)\, da. \tag{2–5.4}$$

We will use infinite atomic separation as our energy reference, $E_{a=\infty} = 0$. As the atoms come together, energy is *released* in an amount equal to the shaded area of Fig. 2–5.2(a). The amount of energy released is shown in Fig. 2–5.2(b). Note, however, that at o–a', where $F = 0 = dE/da$, there is a minimum of energy because energy would have to be supplied to force the atoms still closer together. The depth of this *energy well*, $E_{a=\infty} - E_{min}$, represents the bonding energy, because that much energy would be released $(-)$ as two atoms are brought together (at 0 K). The data of Table 2–3.1 list such values for covalent bonds.*

The schematic representations of Fig. 2–5.2 will be useful to us on various occasions in subsequent sections, when we pay attention to elastic moduli, thermal expansion, theoretical strengths, melting and vaporization temperatures, etc.

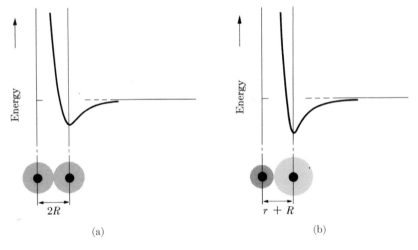

(a)   (b)

Fig. 2–5.3   Bond lengths. The distance of minimum energy between two adjacent atoms is the bond length. It is equal to the sum of the two radii. (a) In a pure metal, all atoms have the same radius. (b) In an ionic solid, the radii are different because the two adjacent ions are never identical.

Atomic and ionic radii   The equilibrium distance between the centers of two neighboring atoms may be considered to be the sum of their radii (Fig. 2–5.3). In metallic iron, for example, the mean distance between the centers of the atoms is 0.2482 nm (or 2.482 Å) at room temperature. Since both atoms are the same, the radius of the iron atom is 0.1241 nm.

---

* The attraction forces of covalent bonds are more complicated than those of the coulombic forces which were presented in Eq. (2–5.1). However, a comparable energy well exists.

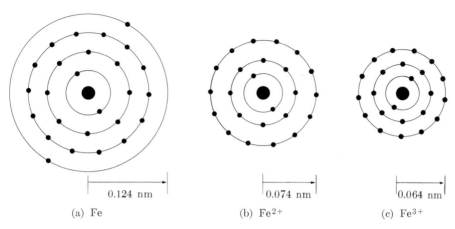

(a) Fe   (b) Fe^{2+} (c) Fe^{3+}

Fig. 2–5.4 Atom and ion sizes (schematic). (a) Both iron atoms and iron ions have the same number of protons (26). (b) If two electrons are removed, the remaining 24 electrons and adjacent negative ions are pulled closer to the 26-proton nucleus. (c) A ferric ion holds its 23 electrons still closer to the nucleus.

Several factors can change this distance between atom centers. The first is temperature. Any increase in energy above the minimum point shown in Fig. 2–5.2(b) will increase the mean distance because the energy trough is asymmetric. This increase in the mean spacing between atoms accounts for the thermal expansion of materials.

Ionic valence also influences interatomic spacing. The ferrous iron ion (Fe^{2+}) has a radius of 0.074 nm, which is smaller than that of the metallic iron atom (Table 2–5.1 and Appendix B*). Since the two outer valence electrons of the iron ion have been removed (Fig. 2–5.4), the remaining 24 electrons are pulled in closer to the nucleus, which still maintains a positive charge of 26. A further reduction in interatomic spacing is observed when another electron is removed to produce the ferric ion (Fe^{3+}). The radius of this ion is 0.064 nm or only about one half that of metallic iron.

A negative ion is larger than its corresponding atom. Since there are more electrons surrounding the nucleus than there are protons in the nucleus, the added electrons are not as closely attracted to the nucleus as were the original electrons.

A third factor affecting the size of the atom or ion is the number of adjacent atoms. An iron atom has a radius of 0.1241 nm when it is in contact with eight adjacent iron atoms, which is the normal arrangement at room temperature. If the atoms are rearranged to place this one iron atom in contact with twelve other iron atoms, the radius of each atom is increased slightly, to ~0.127 nm. With a large number of adjacent atoms, there is more electronic repulsion from neighboring atoms, and consequently the interatomic distances are increased. (Table 2–5.1).

* The metallic radii used in this book are from the ASM *Metals Handbook*. The ionic radii are patterned after Ahrens.

Table 2–5.1
Selected atomic radii

Element	Metallic atoms		Ions			Covalent bonds	
	CN*	Radius, nm	Valence	CN*	Radius, nm[†]	(Bond distance)/2, nm	
Carbon						Single	0.077
						Double	0.065
						Triple	0.06
Silicon			4+	6	0.042	Single	0.117
			4+	4	0.038		
Oxygen			2−	8	0.144	Single	0.075
			2−	6	0.140	Double	0.065
			2−	4	0.127		
			2−	2	~0.114		
Chlorine			1−	8	0.187	Single	0.099
			1−	6	0.181		
Sodium	8	0.1857	1+	6	0.097		
Magnesium	12	0.161	2+	6	0.066		
Aluminum	12	0.1431	3+	6	0.051		
			3+	4	0.046		
Iron	8	0.1241	2+	6	0.074		
	12	~0.127	3+	6	0.064		
Copper	12	0.1278	1+	6	0.096		

* CN = coordination number, i.e., the number of immediate neighbors. For ions, $1.1 R_{CN=4} \approx R_{CN=6} \approx 0.97 R_{CN=8}$.
† These values vary slightly with the system used. Patterned after Ahrens.

We generally do not speak of atomic radii in covalently bonded materials because the electron distributions may be far from spherical (Fig. 2–3.3a). Furthermore, with stereospecific bonds (Section 2–2), the limiting factor in atomic coordination is not the atom size, but rather the number of electron pairs available. Even so, we may make some comparisons of interatomic distances when we look at Table 2–5.1. In ethane with a single C—C bond, this nucleus-to-nucleus distance is 0.154 nm as compared with 0.13 nm for a C=C bond and 0.12 nm for the C≡C bond. This change is to be expected, since the bonding energies are greater with the multiple bonds (Table 2–3.1).

• **Example 2–5.1** MgO and NaCl are comparable, except that Mg^{2+} and O^{2-} ions are divalent and Na^+ and Cl^- ions are monovalent. Therefore, the Mg—O interatomic distance is 0.21 nm while the latter is 0.28 nm. Compare the coulombic attractive forces ($\rightarrow\leftarrow$) that are developed at these two distances for the two pairs of ions.

Solution: From Eq. (2–5.1) and the adjacent footnote,

$$F_{Mg \rightarrow \leftarrow O} = -(9 \times 10^9 \text{ V·m/C})\left(\frac{(+2)(-2)(0.16 \times 10^{-18} \text{ C})^2}{(0.21 \times 10^{-9} \text{ m})^2}\right)$$

$$= 20.9 \times 10^{-9} \text{ J/m}.$$

Similarly,

$$F_{\text{Na}\rightarrow\leftarrow\text{Cl}} = 2.9 \times 10^{-9} \text{ J/m}.$$

Comments. The opposing electronic repulsive forces (Eq. 2–5.2) will be -20.9 nJ/m and -2.9 nJ/m at these equilibrium distances. Thus, from Eq. (2–5.2), the empirical constant b is $\sim 0.4 \times 10^{-105}$ Jm^9 for MgO, and $\sim 10^{-105}$ Jm^9 for NaCl (assuming $n = 9$). ◀

• **Example 2–5.2** Compare the energy of the $\text{Mg}^{2+} \rightarrow\leftarrow \text{O}^{2-}$ bond with the energy of the $\text{Na}^{+} \rightarrow\leftarrow$ Cl^{-} bond. By combining Eqs. (2–5.1) and (2–5.2) into Eq. (2–5.4), and integrating from ∞ to a, we obtain

$$E = k_0 Z_1 Z_2 q^2 / a + b/a^n. \tag{2-5.5}$$

Solution: Using data from Example 2–5.1,

$$E_{\text{Mg}-\text{O}} = \frac{(9 \times 10^9 \text{ V·m/C})(-4)(0.16 \times 10^{-18} \text{ C})^2}{0.21 \times 10^{-9} \text{ m}} + \frac{0.4 \times 10^{-105} \text{ Jm}^9}{(0.21 \times 10^{-9} \text{ m})^9}$$

$$= -4.4 \times 10^{-18} \text{ J} + 0.5 \times 10^{-18} \text{ J}$$

$$= -3.9 \times 10^{-18} \text{ J};$$

$$E_{\text{Na}-\text{Cl}} = -0.8 \times 10^{-18} \text{ J} + 0.1 \times 10^{-18} \text{ J}$$

$$= -0.7 \times 10^{-18} \text{ J}.$$

Comments. We integrated from ∞ to a, rather than from a to ∞, since our reference energy is at infinite separation. The negative values indicate that energy is given off as the ions approach each other from a distance. In this energy range, it is common to use electron volts (1 joule = 6.24×10^{18} eV); therefore, the two calculated energies are -24 eV and -4.4 eV, respectively. ◀

2-6 COORDINATION NUMBER

Much of our discussion has been about diatomic combinations that involve only two atoms. However, since most engineering materials have coordinated groups of many atoms, attention must be given to polyatomic groups. Therefore, when we are analyzing the bonding of atoms within materials, we speak of a *coordination number*. The coordination number, CN, simply refers to the number of first neighbors that an atom has. Thus, in Fig. 2–3.3, the coordination number for carbon is four. In contrast, the hydrogens have only one immediate neighbor, so that their coordination numbers are only one. In Fig. 2–6.1, the Mg^{2+} ion has CN $= 6$.

Two factors control the coordination number of an atom. The first is covalency. Specifically, the number of covalent bonds around an atom is dependent on the number of its valence electrons. Thus the halides, which are in Group VII of the periodic table (Fig. 2–1.1), form only one bond and thus have a coordination number of one when bonded covalently. The members of the oxygen family in Group VI are held in a molecule with two bonds and normally have a maximum coordination number of two. (Of course, oxygen may be coordinated with only one other atom through a double bond.) The nitrogen elements have a maximum coordination number of three since they are in Group V. Finally, carbon and silicon, in Group IV, have four bonds with other atoms, and a maximum coordination number of four (Fig. 2–2.4b).

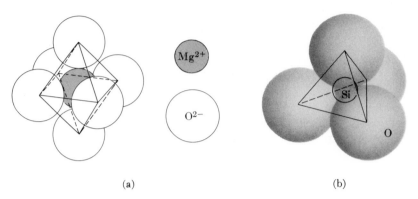

Fig. 2–6.1 Coordination numbers for ionic bonding. (a) A maximum of six oxygen ions (O^{2-}) can surround each magnesium ion (Mg^{2+}). (b) The coordination number of Si^{4+} among O^{2-} is only four because the ion-size ratio is less than 0.41 (Table 2–6.1).

The second factor affecting the coordination number is efficient atomic packing. Since energy is released as ions of unlike charges approach each other, ionic compounds generally have high *coordination numbers*, i.e., as many neighbors as possible without introducing the strong mutual repulsion forces between ions of like charges. This was illustrated in Fig. 2–2.2 with NaCl and is shown again in Fig. 2–6.1(a) with Mg^{2+} ions surrounded by O^{2-} ions. The Mg^{2+} ion has a radius r of 0.066 nm (Table 2–5.1 and Appendix B). This is large enough to permit six O^{2-} ions ($R = 0.140$ nm) to surround it without direct "contact" of negative ions with one another. The minimum radius ratio (r/R), possible for six neighbors without interference, is 0.41 (Table 2–6.1). A coordination number of six (CN = 6) is encountered widely in ionic compounds.

Table 2–6.1

COORDINATION NUMBERS VERSUS MINIMUM
RADII RATIOS

Coordination number	Radii ratios, r/R*	Coordination geometry
3-fold	≥ 0.155	△
4	≥ 0.225	△
6	≥ 0.414	◇
8	≥ 0.732	▢
12	1.0	—

* r—smaller radius; R—larger radius.

Later, in Chapter 8, we will observe that silicon of SiO_2 has $CN = 4$ because an Si^{4+} ion is too small to have six coordinating oxygen ions. Since r/R for Si/O is approximately 0.3, this is consistent with the prediction of Table 2–6.1 and is shown schematically in Fig. 2–6.1(b). Another factor also favors $CN = 4$ for silicon among oxygens. There is considerable electron sharing between the two atoms. (Recall that the last short paragraph of Section 2–2 indicated that mixed bonding is widespread.) As with carbon, four is the maximum number of covalent bonds for silicon under normal conditions. Thus, with sharing, the probability of a $CN = 4$ is increased over what it would be on the basis of radius alone.

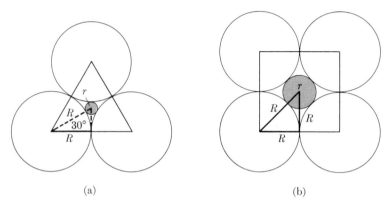

(a)   (b)

Fig. 2–6.2   Coordination calculations. (a) Three-fold coordination. (b) Six-fold coordination. (Compare with Example problems and Fig. 2–6.1a.)

Example 2–6.1   Show the significance of 0.15 as the minimum ratio for a coordination number of three (Table 2–6.1).

Solution: The minimum ratio of sizes possible to permit a coordination number of three is shown in Fig. 2–6.2(a). In this relationship,

$$\cos 30° = \frac{R}{R + r} = 0.866, \qquad \frac{r}{R} = \frac{1 - 0.866}{0.866} = 0.15. \quad \blacktriangleleft$$

Example 2–6.2   Show the significance of 0.41 as the minimum ratio for a coordination number of six.

Solution: The minimum ratio of sizes possible to permit a coordination number of six is shown in Fig. 2–6.2(b). In this relationship,

$$(r + R)^2 = R^2 + R^2, \qquad r = \sqrt{2}\,R - R, \qquad \text{and} \quad \frac{r}{R} = 0.41.$$

Comment. From Fig. 2–6.1(a), note that the fifth and sixth ions sit above and below the center ion of Fig. 2–6.2(b). ◀

2–7 TYPES OF MATERIALS

It is convenient to group materials into three main types: *metals*, *polymers* (or *plastics*), and *ceramics*. In reality, these are three idealized categories, since most materials have intermediate characteristics.

Metals   These materials are characterized by their high thermal and high electrical conductivity. They are opaque and usually may be polished to a high luster (Fig. 2–7.1). Commonly, but not always, they are relatively heavy and deformable.

What accounts for the above characteristics? The simplest answer is that metals owe their behavior to the fact that some of the electrons are delocalized and can leave their "parent" atoms (Fig. 2–2.7b). (Conversely, electrons are not free to roam to the same extent in polymers and ceramics.) Since some of the electrons are delocalized in metals, they can quickly transfer an electric charge and thermal energy. The opacity and reflectivity of a metal arises from the response of these delocalized electrons to electromagnetic vibrations at light frequencies. It is another result of the partial independence of some of the electrons from their parent atoms.

Fig. 2–7.1   Metallic materials (stainless steel utensils for commercial kitchens). Metals possess ductility for the required processing. The luster and conductivity of metals arise from the electronic characteristics of the interatomic bonds. (Courtesy of the Vollrath Company.)

Polymers (commonly called plastics)   Plastics are noted for their low density and their use as insulators, both thermal and electrical. They are poor reflectors of light, tending to be transparent or translucent (at least in thin sections). Finally, some of them are flexible and subject to deformation. This latter characteristic is used in manufacturing (Fig. 2–7.2).

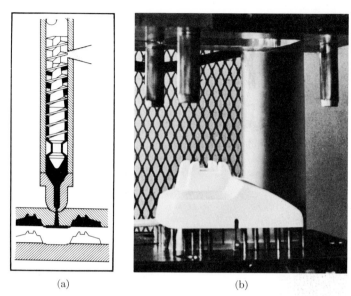

(a)   (b)

Fig. 2–7.2 Plastic products (injection-molded telephone receiver). (a) The soft heated
plastic was forced through the sprues into the die cavity. (b) The cooled product, which is
rigid, required only one manufacturing step. (Courtesy of the Western Electric Company.)

Unlike metals, which have some migrant electrons, the *nonmetallic elements* of
the upper right corner of the periodic table (Fig. 2–1.1) have an *affinity* to *attract* or
share additional electrons. Each electron becomes associated with a specific atom (or
pair of atoms). Thus, in plastics, we find only limited electrical and thermal conductivity because all the thermal energy must be transferred from hot to cold regions by
atomic vibrations, a much slower process than the electronic transport of energy
that takes place in metals. Furthermore the less mobile electrons in plastics are more
able to adjust their vibrations to those of light and therefore do not absorb light rays.

Materials that contain *only* nonmetallic elements share electrons to build up
large molecules. These are often called *macromolecules*. We shall see in Chapter 7
that these large molecules contain many repeating units, or *mers*, from which we get
the word *polymers*.

Ceramics Simply stated, ceramics are *compounds that contain metallic and nonmetallic elements*. There are many examples of ceramic materials, ranging from the
cement of concrete (and even the rocks themselves) to glass, to spark-plug insulators
(Fig. 2–7.3), and to oxide nuclear fuel elements of UO_2, to name but a few.

Each of these materials is relatively hard and brittle. Indeed, hardness and
brittleness are general attributes of ceramics, along with the fact that they tend to be
more resistant than either metals or polymers to high temperatures and to severe
environments. The basis for these characteristics is again the electronic behavior of

Fig. 2–7.3   Ceramic insulator (in a spark plug). The insulator is primarily Al_2O_3, a compound of metal and nonmetallic elements. (Courtesy of the Champion Spark Plug Company.)

the constituent atoms. Consistent with their natural tendencies, the metallic elements release their outermost electrons and give them to the nonmetallic atoms, which retain them. The result is that these electrons are immobilized, so that the typical ceramic material is a good insulator, both electrically and thermally.

Equally important, the positive metallic ions (atoms that have lost electrons) and the negative nonmetallic ions (atoms that have gained electrons) develop strong attractions for each other. Each *cation* (positive) surrounds itself with *anions* (negative). Considerable energy (and therefore considerable force) is usually required to separate them. It is not surprising that ceramic materials tend to be hard (mechanically resistant), refractory (thermally resistant), and inert (chemically resistant).

REVIEW AND STUDY

GENERALIZATIONS REGARDING PROPERTIES

Several of the engineering properties of Chapter 1 may be related *qualitatively* to the atomic bonding characteristics described in this chapter.

1.   Density is controlled by atomic weight, atomic radius, and coordination number. The latter is a significant factor because it controls the packing factor.

2.   Melting and boiling temperatures can be correlated with the depth of the energy trough shown in Fig. 2–5.2(b). Atoms have minimum energy (at the bottom of the trough) at a temperature of absolute zero. Increased temperatures raise the energy until the atoms are able to separate themselves one from another.

3.   Strength is also correlatable with the height of the total force or sum curve of Fig. 2–5.2(a). That force, when related to the cross-sectional area, gives the stress required to separate atoms. (As we shall see in Section 6–4, materials can deform through a process other than direct separation of the atoms. However, the amount of stress required to deform them is still governed by the interatomic forces.) Also, since larger interatomic forces of attraction imply deeper energy troughs, we observe that materials with high melting points are often the harder materials: e.g., diamond, Al_2O_3, TiC. In contrast, in materials with weaker bonds there is a correlation between softness and low melting point: e.g., lead, plastics, ice, and grease. Apparent exceptions to these generalizations can arise when more than one type of bond is present, as in graphite and clay.

4.   The modulus of elasticity can be calculated from the slope of the sum curve of Fig. 2–5.2(a) because at the equilibrium distance, where the net force is zero, dF/da relates stress to strain. As long as the strain or change in interatomic distance is less than one percent, the modulus of elasticity remains essentially constant. Extreme compression or extreme tension respectively raise or lower the modulus of elasticity.

5.   Thermal expansions of materials with comparable atomic packing factors vary inversely with their melting temperatures. This indirect relationship exists because the higher-melting-point materials have deeper and therefore more symmetrical energy troughs. Thus their mean interatomic distances increase less with a given change in thermal energy. Examples of several metals include Hg, melting temperature equals $-39°C$, coefficient of linear expansion equals 40×10^{-6} m/m·°C; Pb, 327°C, 29×10^{-6} m/m·°C; Al, 660°C, 22×10^{-6} m/m·°C; Cu, 1084°C, 17×10^{-6} m/m·°C; Fe, 1538°C, 12×10^{-6} m/m·°C; W, 3387°C, 4.2×10^{-6} m/m·°C.

6.   Electrical conductivity is very dependent on the nature of the atomic bonds. Both ionically and covalently bonded materials are extremely poor conductors, because electrons are not free to leave their host atoms. On the other hand, the delocalized electrons of metals easily move along a potential gradient. Semiconductors will be considered in Chapter 5; however, we can note here that their conductivity is controlled by the freedom of movement of their electrons.

7.   Thermal conductivity is high in materials with metallic bonds, because delocalized electrons are very efficient carriers of thermal as well as electrical energy.

8.   The influence of the structure of atoms on chemical properties has not been elaborated on here, since the chemical differences between elements depend primarily on the number of valence electrons. Furthermore, all chemical reactions involve the

formation and the disruption of bonds. So far as engineering materials are concerned, the *corrosion* reaction (Chapter 12) is probably the most obvious chemical reaction. In corrosion, the separation of a metallic ion from the metal proper involves the removal of valence electrons from the outer shell of the atom and its *ionization potential*.

In subsequent chapters, we shall develop the principles that relate the properties of engineering materials to the existing structures.

KEY TERMS AND CONCEPTS

Atomic mass

Atomic mass unit, amu

Atomic number

Atomic radii

Avogadro's number, N

Bond angle

Bond energy

Bond length

Ceramics

Coordination number, CN

Coulombic forces

Covalent bonds

Delocalization

Dipole moment

Electronic repulsion

Energy well

Hydrogen bridge

Induced dipole

Ionic bond

Ionic radii

Isomer

Lone pair

Metallic bond

Metals

Molecules

  polar

Orbital

Periodic table

Polymer

Primary bond

Secondary bond

Stereospecific

Valence electrons

van der Waals forces

FOR CLASS DISCUSSION

A_2 Distinguish between atomic number and atomic mass.

B_2 Explain why electrons can have only selected energy levels within an atom, and all other levels are forbidden.

C_2 Refer to Fig. 2–1.3. Based on your knowledge of chemistry, make a similar sketch for potassium.

D_2 Refer to Fig. 2–2.1. Make a comparable sketch for KCl.

E_2 The "hard-ball" model is widely used for metal atoms and ions but is less applicable to molecules. Why?

F_2 A small diamond has 10^{20} atoms. How many covalent bonds?

G_2 Silicon tetrafluoride (SiF_4) is a stable compound but has a low melting temperature ($-77°C$). Account for these facts by predicting the nature of its bonds. (Use a sketch if necessary.)

H_2 Sulfur dichloride has a molecular weight of 103 amu and a boiling point of 59°C. Draw a diagram showing the valence electron structure of this compound by using the "dot" representation (Fig. 2–2.3a).

I_2 Using the "dot" representation, sketch the electron structure for ClO_4^-, SO_4^{2-}, PO_4^{3-}, SiO_4^{4-} ions.

J_2 Figure 2–2.3(b) is drawn schematically with energy plotted vertically. Indicate the average energy of the valence electrons for the two separate fluorine atoms (left of arrow); for the F_2 molecule (right of arrow). Why does fluorine form molecules?

K_2 The electric dipole moment decreases from HF to HCl to HBr to HI. Suggest a reason.

L_2 HBr boils at a higher temperature ($-67°C$) than HCl ($-85°C$) even though it has a shorter electric dipole. Explain.

M_2 Sulfur can form ring molecules with approximately eight atoms. Use the "electron dot" representation to show bonds. What type of bond does the molecule have?

N_2 Hydrogen peroxide (H_2O_2) is covalently bonded. It is unstable and readily forms 2 OH rather than H and OOH. Suggest a reason why. The resulting OH's are very reactive. Use an "electron dot" representation to explain why.

O_2 The energy of the $C\equiv N$ bond is not listed in Table 2–3.1. However, its value is probably 915 kJ/mole; greater than that; less than that. Explain your choice.

P_2 Why can a neutron move through materials, even though the atoms "touch"?

Q_2 Barium is not listed in Appendix B. On the basis of the periodic table (Fig. 2–1.1) and other data from Appendix B, predict the radius of the Ba^{2+} ion; Ba atom; Ag^{2+} ion.

R_2 Why is there a lower limit, but not an upper limit, on the radii ratios for CN = 6?

S_2 List the factors that affect the coordination number for ionic compounds; covalent compounds.

T_2 Categorize the following into the three types of materials of Section 2–7: bronze, portland cement, "rubber" cement, wood, window glass, bakelite, rust, ethylene glycol (antifreeze).

U_2 Cite materials that lie intermediate between metals and ceramics; between polymers and ceramics.

V_2 List various metals that are used to make a kitchen stove; various ceramics; various polymers.

W_2 Plot α_L vs. T_m data of paragraph (5) in Review and Study. Add other metals to this list from data in the handbook. Explain your results in your own words.

X$_2$ Obtain the melting points for the pure metals (99+%) listed in Appendix C. Plot these values against their moduli of elasticity. Discuss the patterns of your results.

Y$_2$ Gold and nickel are among the elements listed in Appendix B but are not among the materials in Appendix C. Provide an estimate of the thermal expansion coefficient and elastic modulus of each.

Z$_2$ Graphite is very anisotropic, i.e., its properties vary markedly with direction. Explain why the conductivity of graphite is more than 100 times greater in the horizontal direction of Fig. 2–2.6 than in the vertical direction.

STUDY PROBLEMS

2–1.1 (a) What is the mass of an aluminum atom? (b) The density of aluminum is 2.70 Mg/m^3 (2.70 g/cm^3); how many atoms per mm^3?

Answer: (a) 4.48×10^{-23} g/atom   (b) 6.02×10^{19} atoms/mm^3

2–1.2 A copper wire weighs 1.312 g, is 2.15 mm in diameter and 40.5 mm long. (a) How many atoms are present per mm^3? (b) Calculate its density.

2–1.3 A solder contains 60 w/o tin and 40 w/o lead. What are the atom percents of each element?

Answer: 72 a/o Sn; 28 a/o Pb   w/o   % by weight

2–1.4 (a) How many iron atoms are there per gram? (b) What is the volume of a grain of metal containing 10^{20} iron atoms?

2–1.5 (a) Use data from the appendices to identify the mass of a single silver atom. (b) How many atoms are there per mm^3 of silver? (c) Based on its density, what is the volume of a grain of silver that contains 10^{21} atoms? (d) Assume the silver atoms are spheres ($R_{Ag} = 0.1444$ nm), and ignore the space among them. What is the volume occupied by the 10^{21} atoms? (e) What volume percent of the space is occupied?

Answer: (a) 1.79×10^{-22} g/Ag (b) 5.86×10^{19} Ag/mm^3
(c) 17 mm^3 (d) 12.6 mm^3 (e) 74%

2–1.6 Change Study Problem 2–1.5 to nickel and answer the same questions.

2–1.7 (a) Al$_2$O$_3$ has a density of 3.8 Mg/m^3 (3.8 g/cm^3). How many atoms are present per mm^3? (b) Per gram?

Answer: (a) 1.12×10^{20} atoms/mm^3 (b) 2.95×10^{22} atoms/g

2–1.8 A cubic volume of MgO that is 0.42 nm along each edge contains 4 Mg^{2+} ions and 4 O^{2-} ions. What is the density of MgO?

2–1.9 A compound contains 33 a/o Cu and 67 a/o Al. What is the weight percent of each?

Answer: 54 w/o Cu; 46 w/o Al

2–1.10 Silver was plated onto a brass surface (1610 mm^2) until it was 7.5 μm thick. (a) How much silver (Ag$^+$) is required? (b) How many amperes were required to do this in 5 minutes?

2–1.11 An electroplated surface of sterling silver (92.5 w/o Ag; 7.5 w/o Cu) is to be plated on some silverware. How many amperes are required to plate 1 milligram per second? (Ag^+ and Cu^{2+}.)

Answer: 1.05 amp.

• 2–1.12 Indicate the orbital arrangements for a single atom (a) of chlorine; (b) of potassium.

• 2–1.13 Give the notation for the electronic structure (a) of zirconium atoms, (b) of Zr^{4+} ions.

Answer: (a) $1s^2 2s^2 2p^6 3s^2 3p^6 3d^{10} 4s^2 4p^6 4d^2 5s^2$ (b) $1s^2 2s^2 2p^6 3s^2 3p^6 3d^{10} 4s^2 4p^6$

• 2–1.14 Indicate the number of $3d$ electrons in each of the following:

a) Ti^{2+} b) Ti^{4+} c) Cr^{3+} d) Fe^{3+}

e) Fe^{2+} f) Mn^{2+} g) Mn^{4+} h) Ni^{2+}

i) Co^{2+} j) Cu^+ k) Cu^{2+}

2–2.1 It takes approximately 5×10^{-19} joule to break the covalent bond between carbon and nitrogen. What wavelength would be required of a photon to supply this energy? (See Appendix A for constants.)

Answer: 400 nm

2–2.2 An electron absorbs all the energy from a photon of ultraviolet light ($\lambda = 276.8$ nm). How many eV are absorbed?

2–3.1 Determine the molecular weight for each of the molecules of Fig. 2–3.1.

Answer: (a) 32 (b) 46 (c) 17 (d) 78 (e) 94 (f) 30 (g) 58 (h) 60 (i) 28 (j) 62.5

• 2–3.2 An organic compound contains 62.1 w/o carbon, 10.3 w/o hydrogen, and 27.6 w/o oxygen. Name a possible compound.

• 2–3.3 Refer to Fig. 2–3.3. Methyliodide (CH_3I) has a similar structure; however, the H—C—H angles equal 111.4°. What is the H—C—I angle in CH_3I? (This is a problem in trigonometry; but it will illustrate the distortion in a polar molecule.)

Answer: 107.5°

2–3.4 Sketch three of the four possible isomers of butanol (C_4H_9OH).

2–3.5 Sketch the structure of the various possible isomers for octane, C_8H_{18}.

• 2–4.1 Calculate the electric dipole moment of each H—S arm of H_2S if the electric dipole of the molecule is 3.1×10^{-30} C·m and the H—S—H bond angle is 93°.

Answer: 2.3×10^{-30} C·m

• 2–4.2 The H—F distance in HF is 0.1 nm, and there are a total of 10 electrons present. The electric dipole moment of an HF molecule is 6.4×10^{-30} C·m. (a) How far from the center of positive charge (our reference point) is the center of negative charge? (b) What fraction is this of the total bond length?

• 2–4.3 Consider the HCl molecule to be H^+Cl^-. How far apart are the centers of the two ions on the basis that the dipole moment is 3.5×10^{-30} C·m?

Answer: 0.02 nm

• 2–5.1 Refer to Example 2–5.1 and its comments. What are the attractive and repulsive forces between Mg^{2+} and O^{2-} ions (a) at $a = 0.2$ nm? (b) At 0.22 nm?

Answer: (a) 23 nJ/m, -35 nJ/m (b) 19 nJ/m, -13 nJ/m

• 2–5.2 Plot the net force $(F_C + F_R)$ versus interatomic distance for $Mg^{2+} \rightarrow \leftarrow O^{2-}$ from 0.19 nm to 0.23 nm, that is, $a_{Mg-O} \pm 0.02$ nm. (Use data from Example 2–5.1 and Study Problem 2–5.1 as needed.)

2–5.3 Continue the plot of Study Problem 2–5.2 to determine the maximum net force between the Mg^{2+} and O^{2-} ions. What is the distance of separation from maximum force?

Answer: At 0.24 nm, 10.3 nJ/m; at 0.25 nm, 11.0 nJ/m; at 0.26 nm, 11.08 nJ/m; at 0.27 nm, 10.9 nJ/m (By setting $d(F_C + F_R)/da = 0$, the maximum is at 0.258 nm, with $(F_C + F_R) = 11.1$ nJ/m.)

• 2–5.4 Paralleling the calculation of Example 2–5.2, determine the energy of the $Mg^{2+} \rightarrow \leftarrow O^{2-}$ bond at several selected distances between 0.19 and 0.23 nm, that is, $a_{Mg-O} \pm 0.02$ nm.

• **2–5.5** Using Eq. (2–5.5), plot the energy versus separation distance for Na^+ and Cl^- from 0.24 to 0.32 nm, that is, $a_{Na-Cl} \pm 0.04$ nm. (From Example 2–5.1, b for $NaCl = \sim 10^{-105}$ Jm^9.)

Answer: at 0.24 nm, -0.58×10^{-18} J   at 0.22 nm, -0.21×10^{-18} J
at 0.26 nm, -0.70×10^{-18} J   at 0.40 nm, -0.57×10^{-18} J
at 0.28 nm, -0.73×10^{-18} J
at 0.30 nm, -0.72×10^{-18} J
at 0.32 nm, -0.69×10^{-18} J

(*Hint:* Your calculator will not overflow if you divide the numerator and the denominator by 10^{-90}.)

2–5.6 Based on the data of Table 2–5.1, compare the sphere volumes of Fe^{2+} and Fe^{3+}.

2–5.7 Compare the volumes of the iron atoms when they have 8 and 12 neighbors on the basis of their radii (Table 2–5.1).

Answer: (as spheres) $V_{CN=8} = 0.008$ nm^3, $V_{CN=12} \sim 0.0086$ nm^3

2–5.8 The ionic radii listed in Appendix B apply when six neighbors are present $(CN = 6)$. (a) What radii do the halide ions have when they have eight neighbors? (b) What is the radius of Zn^{2+} with only four neighbors?

2–6.1 Show the origin of 0.73 in Table 2–6.1.

Answer: $2(r + R) = \sqrt{3}(2R)$

2–6.2 (a) What is the radius of the smallest cation that can have a six-fold coordination with O^{2-} ions without distortion? (b) Eight-fold coordination?

2–6.3 Show the origin of 0.22 in Table 2–6.1. (*Hint:* Place the four large ions of Fig. 2–6.1(b) at four corners of a cube and the small ion at the cube center.) (See Fig. 2–3.3a.)

Answer: $2(r + R)/\sqrt{3} = 2R/\sqrt{2}$

2–6.4 Look ahead to Fig. 3–1.1. (a) What is the coordination number for each Na^+ ion? (b) For each Cl^- ion? (Assume that the structure continues with the same pattern beyond the present sketch.)

2–6.5 Six O^{2-} ions surround an Mg^{2+} ion. Consider the ions to be hard balls with the radii shown in Table 2–5.1. What is the distance between the surfaces of the O^{2-} ions?

Answer: gap $= 0.01$ nm

2–6.6 Six Cl^- ions surround an Na^+ ion. What is the distance between the surfaces of the Cl^- spheres based on the data of Table 2–5.1?

2–6.7 (a) From Appendix B, cite three divalent cations that can have CN $= 6$ with S^{2-}, but not CN $= 8$. (b) Cite two divalent ions that can have CN $= 8$ with F^-.

Answer: (b) Ca, Hg, Pb

CHAPTER **3**

Atomic Order
in Solids

PREVIEW

After considering atom-to-atom bonding, our next step
along the structural scale is to look at the *long-range
patterns of atomic order.* This is made relatively simple
in crystalline solids because unit cells are formed that
repeat in each of the three dimensions. Each unit cell has
all of the geometric characteristics of the total crystal.

It will be the purpose of this chapter to look
specifically at the atomic arrangements in a few of the
more simple structures (bcc, fcc, and hcp) and establish
a credibility for their existence through density
calculations.

The reader should become familiar with the notations
for unit cell locations, crystal directions, and crystal
planes, because they will be used subsequently to relate
the crystal structure to the properties and behavior of
materials.

CONTENTS

STUDY OBJECTIVES

1 To develop the concept of the unit cell so that you may use the unit cell to visualize (a) the atomic arrangements that exist, (b) the long-range order that is present along various directions and planes, and (c) the packing densities in 1-, 2-, and 3-dimensions.

2 To identify the ordering patterns (called lattices) that are encountered in some of the simpler metals, specifically bcc and fcc, and also other patterns such as hcp and bct.

3 To be able (a) to make calculations relating the atomic radius in metals (i) to unit cell size, (ii) to atomic packing factors, and (iii) to densities, and (b) to modify the calculations appropriately for simple compounds such as NaCl.

4 To visualize crystal directions and planes from their indices, so that we may discuss anisotropic properties later without redescribing the geometric relationships each time.

5 To know that many solids can change crystal structure as a result of temperature and/or pressure changes; and that such changes require bond breaking, atom movements, and new bond formations, all of which take time.

• **6** To review the physics of wave motions so that you can understand the basis of x-ray diffraction, which is the very precise method that can be used to obtain interplanar spacings, atomic radii, etc.

3–1 CRYSTALLINITY

Essentially all metals, a significant fraction of ceramics, and certain polymers crystallize when they solidify. By this, we mean that the atoms arrange themselves into an ordered, repeating, 3-dimensional pattern. Such structures are called *crystals* (Fig. 3–1.1).

This ordered pattern over a *long range* of many atomic distances arises from the atomic coordinations (Section 2–6) within the material; in addition, the pattern sometimes controls the external shape of the crystal; the six-pointed outline of snowflakes is probably the most familiar example of this. The planar surfaces of gems, quartz (SiO_2) crystals, and even ordinary table salt (NaCl) are all external manifestations of internal crystalline arrangements. In each case the internal atomic arrangement persists even though the external surfaces are altered. For example, the internal structure of a quartz crystal is not altered when the surfaces of the crystal are abraded to produce round silica beach sand. Likewise, there is a hexagonal arrangement of water molecules in chunks of crushed ice as well as in snowflakes.

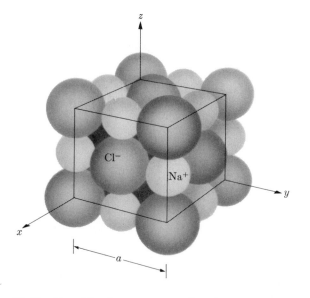

Fig. 3–1.1 Crystal structure. The cubic faces of table salt are the crystal faces of the NaCl structure. MgO has the same structure.

Unit cells The long-range order that is characteristic of crystals may be observed in Fig. 3–1.2. This model shows one of several patterns, or *lattices*, that may develop when only one kind of atom is present. Since the atomic pattern is repeated indefinitely, it is convenient to subdivide the crystal lattice into *unit cells*. These are small volumes, each one having all of the characteristics found in the total crystal.

The repeat distance, called the *lattice constant*, in the long-range pattern of a crystal dictates the size of the unit cell. Thus, *a*, the repeatable dimension of Fig. 3–1.2, is also the dimension of the unit cell edge. Since the crystal pattern of Fig. 3–1.2 is identical in the three perpendicular coordinate directions, this unit cell is *cubic*, and *a* is the lattice constant for all three dimensions. In noncubic crystals, the lattice constant differs in the three coordinate directions.

Fig. 3–1.2   Unit cell. The small repeating volume within a crystal is called a unit cell. The lattice constant *a* is the same in the three coordinate directions when the crystal is cubic.

The corner of the unit cell can be placed *anywhere* within the crystal. Thus, the corner may be at the center of an atom, elsewhere within the atom, or at any position between atoms, e.g., point ⊠ of Fig. 3–1.3. Wherever it is, that small volume is duplicated by an identical volume next door (provided the cell has the same orientation as the crystal pattern). Each cell has all of the geometric features found in the total crystal.

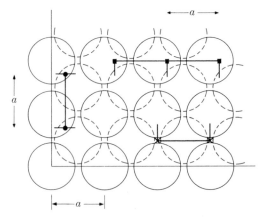

Fig. 3–1.3   Lattice constant. The lattice constant *a* is the repeat distance between equivalent positions in the crystal. It is parallel to the crystal axes. The unit cell is commonly, *but not necessarily*, positioned with an atom at the corner.

Crystal systems   Cubic crystals have identical patterns along the three perpendicular directions: $a_1 = a_2 = a_3$. The majority of metals, and a significant number of ceramic materials, are cubic. (Very few molecular crystals are cubic.)

Noncubic crystals arise when the repeating pattern is not the same in the three coordinate directions, or the angles between the three *crystal axes* are not 90°. There are seven possible *crystal systems*. These are listed in Table 3–1.1, together with their geometric characteristics. Most of our attention in this introductory materials course will be focused on the simpler, more symmetric cubic crystals. However, we must also be familiar with the hexagonal system. In addition, tetragonal and orthorhombic crystals, which have the unit cell characteristics shown in Fig. 3–1.4, will be encountered occasionally in this text.

Table 3–1.1
Crystal systems

System	Axes	Axial angles
Cubic	$a_1 = a_2 = a_3$	All angles = 90°
Tetragonal	$a_1 = a_2 \neq c$	All angles = 90°
Orthorhombic	$a \neq b \neq c$	All angles = 90°
Monoclinic	$a \neq b \neq c$	2 angles = 90°; 1 angle ≠ 90°
Triclinic	$a \neq b \neq c$	All angles different; none equal 90°
Hexagonal	$a_1 = a_2 = a_3 \neq c$	Angles = 90° and 120°
Rhombohedral	$a_1 = a_2 = a_3$	All angles equal, but not 90°

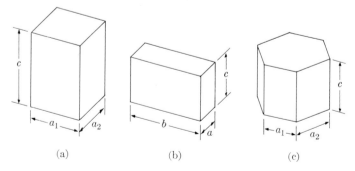

(a)   (b)   (c)

Fig. 3–1.4   Noncubic crystals. (a) Tetragonal: $a_1 = a_2 \neq c$; angles = 90°. (b) Orthorhombic: $a \neq b \neq c$; angles = 90°. (c) Hexagonal: $a_1 = a_2 \neq c$; angles = 90° and 120°.

Example 3–1.1 The unit cell of chromium is cubic and contains two atoms. Use the data of Appendix B to determine the lattice constant for chromium.

Solution: The density of chromium is 7.20 Mg/m^3.

$$\text{Mass per unit cell} = 2 \text{ Cr}(52.0 \text{ g})/(0.602 \times 10^{24} \text{ Cr})$$
$$= 172.76 \times 10^{-24} \text{ g}.$$
$$\text{Volume} = a^3 = (172.76 \times 10^{-24} \text{ g})/(7.20 \times 10^6 \text{ g/m}^3)$$
$$= 23.994 \times 10^{-30} \text{ m}^3.$$
$$a = 0.2884 \times 10^{-9} \text{ m} \qquad \text{(or 0.2884 nm)}.$$

Comment. This same value is obtained experimentally by x-ray diffraction (Section 3–8). ◀

3–2   CUBIC LATTICES

Cubic crystals possess one of three lattices. These are *simple cubic, body-centered cubic,* and *face-centered cubic.* A lattice is the 3-dimensional repeating pattern that is developed within the crystal. A significant majority of our metals possess either a body-centered cubic (bcc) lattice or a face-centered cubic (fcc) lattice.

Bcc   **Body-centered cubic metals**   Iron has a cubic structure. At room temperature, the unit cell of iron has an atom at each corner of the cube, and another atom at the body-center of the cube (Fig. 3–2.1). Iron is the most common metal with this body-centered cubic structure, but not the only one. Chromium and tungsten, among others listed in Appendix B, also have body-centered cubic arrangements.

Each iron atom in this body-centered cubic (bcc) structure is surrounded by eight adjacent iron atoms, whether the atom is located at a corner or at the center of

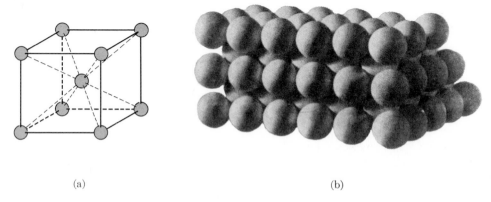

(a)   (b)

Fig. 3–2.1   Body-centered cubic structure of a metal. Part (a) is a schematic view showing the location of atom centers. (b) Model made from hard balls. (G. R. Fitterer. Reproduced by permission from B. Rogers. *The Nature of Metals,* 2d ed., American Society for Metals, and the Iowa State University Press, Chapter 3.)

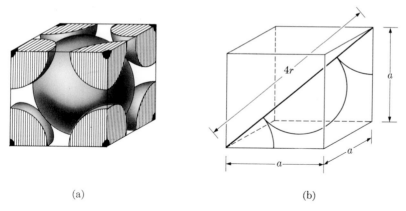

(a)   (b)

Fig. 3–2.2 Body-centered cubic unit cell (metal). The structure of a bcc metal has two atoms per unit cell and an atomic packing factor of 0.68. The lattice constant a is related to the atomic radius as shown by Eq. (3–2.1).

the unit cell. Therefore each atom has the same geometric environment (Fig. 3–2.1a). There are two atoms per unit cell in a bcc metal. One atom is at the center of the cube, and eight octants are located at the eight corners (Fig. 3–2.2). In a metal, the lattice constant a is related to the atomic radius R by

$$(a_{\text{bcc}})_{\text{metal}} = 4R/\sqrt{3}. \tag{3–2.1}$$

We can apply the concept of the atomic *packing factor* (PF) to a bcc metal. This assumes spherical atoms (hard-ball model) and is the volume fraction of the unit cell that is occupied by those spheres:

$$\text{Packing factor} = \frac{\text{Volume of atoms}}{\text{Volume of unit cell}}. \tag{3–2.2}$$

Since there are two atoms per unit cell in a bcc metal,

$$\text{PF} = \frac{2[4\pi R^3/3]}{a^3}$$

$$= \frac{2[4\pi R^3/3]}{[4R/\sqrt{3}]^3} = 0.68.$$

Fcc **Face-centered cubic metals**   The atom arrangement in copper (Fig. 3–2.3) is not the same as that in iron, although it is cubic. In addition to an atom at the corner of each unit cell of copper, there is one at the center of each face, but none at the center of the cube.

This face-centered cubic (fcc) structure is somewhat more common among metals than the body-centered cubic structure is. Aluminium, copper, lead, silver, and nickel possess this atomic arrangement (as does iron at elevated temperatures).

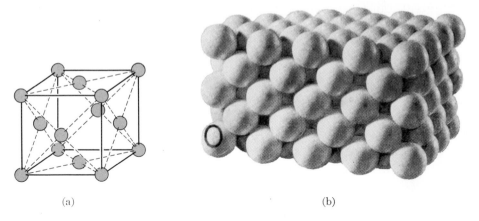

(a)   (b)

Fig. 3–2.3 Face-centered cubic structure of a metal. Part (a) is a schematic view showing location of atom centers. (b) Model made from hard balls. (G. R. Fitterer. Reproduced by permission from B. Rogers, *The Nature of Metals*, 2d ed., American Society for Metals, and the Iowa State University Press, Chapter 3.)

A metal with an fcc lattice has four times as many atoms as it has unit cells. The eight corner octants contribute a total of one atom, and the six face-centered atoms contribute a total of three atoms per unit cell (Fig. 3–2.4). In a metal, the lattice constant a is related to the atomic radius R by

$$(a_{fcc})_{metal} = 4R/\sqrt{2}. \tag{3–2.3}$$

As shown in Example 3–2.1, the packing factor for an fcc metal is 0.74, which is greater than the 0.68 for a bcc metal. This is to be expected since each atom in a bcc metal has only eight neighbors. Each atom in an fcc metal has 12 neighbors. This is verified in Fig. 3–2.4 where one observes that the front face-centered atom has four adjacent neighbors, four neighbors in contact on the backside, and four comparable neighbors sitting in front.

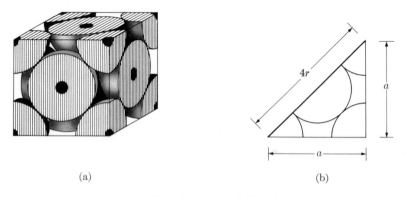

(a)   (b)

Fig. 3–2.4 Face-centered cubic unit cell (metal). The structure of an fcc metal has four atoms per unit cell and an atomic packing factor of 0.74. In metals, the lattice constant a is related to the atomic radius as shown by Eq. (3–2.3).

Fcc compounds   Compounds may also have face-centered cubic lattice, as revealed in Fig. 3–1.1 for NaCl. The center of each face is identical in every respect with the corners. In the compound NaCl, where unlike atoms touch, the dimension of the fcc unit cell is obtained from the sum of the two radii:

$$(a_{fcc})_{NaCl} = 2(r_{Na^+} + R_{Cl^-}).\tag{3-2.4}$$

From Eqs. (3–2.1), (3–2.3), or (3–2.4), one can determine the volume of the unit cell. Also the number of atoms per unit cell permit a calculation of its mass. Together, they let us calculate the density (Example 3–2.2). Highly accurate answers are possible.

Example 3–2.1   Calculate (a) the atomic packing factor of an fcc metal (Fig. 3–2.4); (b) the ionic packing factor of fcc NaCl (Fig. 3–1.1).

Solution: (a)   From Eqs. (3–2.2) and (3–2.3),

$$PF = \frac{4(4\pi R^3/3)}{a^3} = \frac{16\pi R^3(2\sqrt{2})}{(3)(64R^3)} = 0.74.$$

b)   From Eqs. (3–2.2) and (3–2.4), Fig. 3–1.1, and radii from Appendix B,

$$PF = \frac{4(4\pi r^3/3) + 4(4\pi R^3/3)}{(2r + 2R)^3} = \frac{16\pi(0.097^3 + 0.181^3)}{3(8)(0.097 + 0.181)^3} = 0.67.$$

Comments.   It is apparent, from this example, that the packing factor is independent of atom size, if only one size is present. In contrast, the relative sizes do affect the packing factor when more than one type of atom is present. The face-centered cubic structure has the highest packing factor (0.74) that is possible for a pure metal, and thus this structure could also be called a *cubic close-packed* (ccp) structure. As we might expect, many metals have this structure, although we shall see in a moment that a hexagonal close-packed structure also has a packing factor of 0.74.   ◄

Example 3–2.2   Copper has an fcc structure and an atom radius of 0.1278 nm. Calculate its density and check this value with the density listed in Appendix B.

Solution: From Eq. (3–2.3),

$$a = \frac{4}{\sqrt{2}}(0.1278 \text{ nm}) = 0.3615 \text{ nm}.$$

From Figure 3–2.4,

$$\frac{\text{Atoms}}{\text{Unit cell}} = \frac{8}{8} + \frac{6}{2} = 4,$$

$$\text{Density} = \frac{\text{Mass/Unit cell}}{\text{Volume/Unit cell}}\tag{3-2.5a}$$

$$= \frac{(\text{Atoms/Unit cell})(\text{g/atom})}{(\text{Lattice constant})^3}.\tag{3-2.5b}$$

$$\text{Density} = \frac{4[63.5/(0.602 \times 10^{24})]}{(0.3615 \times 10^{-9} \text{ m})^3} = 8.93 \text{ Mg/m}^3 \, (= 8.93 \text{ g/cm}^3).$$

The experimental value listed in Appendix B is 8.92 Mg/m³.   ◄

Example 3–2.3 Calculate the volume of the unit cell of LiF, which has the same structure as NaCl (Fig. 3–1.1).

Solution: Although LiF is fcc, we cannot use the geometry of Fig. 3–2.4(b), since the fluoride ions do not touch, as did the metal atoms. Rather, a is twice the sum of r_{Li} and R_{F^-}. (*Check* Fig. 3–1.1 again.) From Appendix B,

$$a = 2(0.068 + 0.133) \text{ nm},$$

$$a^3 = 0.065 \text{ nm}^3 \qquad (\text{or } 65 \times 10^{-30} \text{ m}^3).$$

Comment. We use ionic radii since LiF is an ionic compound. Furthermore, each ion has six neighbors; therefore, the radii of Appendix B do not have to be modified. ◀

3–3 HEXAGONAL CRYSTALS

The lattices of Figs. 3–3.1(a) and 3–3.1(b) are two representations of *hexagonal* unit cells. The angles within the base are 120° (and 60°). These cells have no internal positions that are equivalent to the corner positions. Although the volume of the cell is three times as great in Fig. 3–3.1(a) as in Fig. 3–3.1(b), there are three times as many atoms (3 versus 1); therefore the number of atoms per unit volume remains the same. Metals do not crystallize with atoms arranged according to Fig. 3–3.1 because the packing factor is too low.

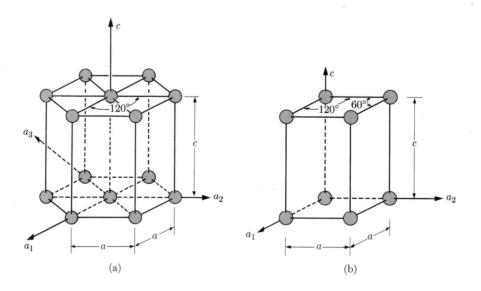

(a)   (b)

Fig. 3–3.1 Simple hexagonal unit cells. (a) Hexagonal representation. (b) Rhombic representation. The two are equivalent, with $a \neq c$, a basal angle of 120°, and vertical angle of 90°.

HCP **Hexagonal close-packed**   The specific hexagonal structure formed by magnesium is shown in Fig. 3–3.2. Such a structure, which is more densely packed than that rep-

FoRMED By 3
CLOSE PACKED RHOMBIC
2 ATOMS/UNIT CELL

(a)   (b)

Fig. 3–3.2 Hexagonal close-packed structure. (a) Schematic view showing the location of atom centers. (b) Model made from hard balls.

resented by Fig. 3–3.1, is called a *hexagonal close-packed* (hcp) structure. It is characterized by the fact that each atom in one layer is located directly above or below interstices among three atoms in the adjacent layers. Consequently, each atom touches three atoms in the layer below its plane, six atoms in its own plane, and three atoms in the layer above for CN = 12. There is an average of six atoms per unit cell in the hcp structure of Fig. 3–3.2 (or two per unit cell if we use the related rhombic representation).

The atomic packing factor for an hcp metal may be calculated and is found to equal 0.74. This is identical to the packing factor of an fcc metal, which is predictable because each has a coordination number of 12.

Example 3–3.1 The atomic packing factor of magnesium, like all hcp metals, is 0.74. What is the volume of its unit cell, which is shown in Fig. 3–3.2(a)?

Solution: From Appendix B, magnesium has $\rho = 1.74$ Mg/m^3 (or 1.74 g/cm^3), and its atomic mass = 24.31 amu. From Fig. 3–3.2(a), 12/6 + 2/2 + 3 = 6 atoms/unit cell.

Basis: 1 m^3. $(1.74 \times 10^6 \text{ g/m}^3)/(24.31 \text{ g}/0.602 \times 10^{24} \text{ atoms}) = 4.31 \times 10^{28}$ atoms./m^3

$$\text{Vol}_{\text{uc}} = 6/(4.3 \times 10^{28} \text{ atoms/m}^3)$$
$$= 1.4 \times 10^{-28} \text{ m}^3 = 0.14 \text{ nm}^3.$$

Alternate solution: From Appendix B, $R_{\text{Mg}} = 0.161$ nm.

$$6 \text{ atoms } (4\pi/3)(0.161 \text{ nm})^3/0.74 = 0.14 \text{ nm}^3.$$

Comments. The average radius of the magnesium atom is 0.161 nm. However, x-ray diffraction data show that the Mg atoms are compressed almost 1 percent to become oblate spheroids. (See Comments with Example 3–3.2.) ◀

● **Example 3–3.2** Assume spherical atoms. What is the c/a ratio of an hcp metal?

Solution: Refer to Fig. 3–3.2(a) and consider the three central atoms plus the one at the center of the top. This is an equilateral tetrahedron with edges $a = 2R$. From geometry,

$$h = a\sqrt{2/3}.$$

$$c = 2h = 2a\sqrt{2/3} = 1.63a. \tag{3–3.1}$$

Comments. The c/a ratios of hcp metals depart somewhat from this figure: Mg, 1.62; Ti, 1.59; Zn, 1.85. This means we must envision magnesium and titanium atoms as slightly compressed spheres, and zinc atoms as prolate spheriods. ◀

● **Example 3–3.3** The unit-cell volume of hcp titanium at 20°C is 0.106 nm³, as sketched in Fig. 3–3.2(a). The c/a ratio is 1.59. (a) What are the values of c and a? (b) What is the radius of the Ti in a direction that lies in the base of the unit cell?

Solution

a) From geometry,

$$\text{area of base} = 6(1/2)(a)(a \sin 60°) = 2.60a^2.$$

$$\text{Volume} = (1.59a)(2.60a^2) = 4.13a^3 = 0.106 \text{ nm}^3.$$

$$a^3 = 0.02566 \text{ nm}^3; \quad \text{and} \quad a = 0.2950 \text{ nm}.$$

$$c = 1.59(0.295 \text{ nm}) = 0.469 \text{ nm}.$$

b) $\qquad\qquad\qquad a = 2R_{base}; \quad \therefore R = 0.1475 \text{ nm}.$

Comments. The average radius is 0.146 nm (Appendix B). The atoms of titanium (and the unit cell) are slightly compressed in the c-direction. (See Comments with Example 3–3.2.) ◀

3–4 POLYMORPHISM

We recall from Section 2–3 that two molecules may possess different structures even though their compositions are identical. We called those molecules *isomers*. An analogous situation that will be extremely important to us occurs in crystalline solids. *Polymorphs* are two or more distinct types of crystals that have the same composition.* The most familiar example is the dual existence of graphite and diamond as two polymorphs of carbon.

The prime example of polymorphism in metals will be iron, since our whole ability to heat-treat steel and modify its properties stems from the fact that, as iron is heated, it changes from the bcc to an fcc lattice. Furthermore, the change is reversible as iron cools. At room temperature bcc iron has a coordination number of 8, an atomic packing factor of 0.68, and an atomic radius of 0.1241 nm. Pure iron changes to fcc at 912°C, at which point its coordination number is 12, its atomic packing factor is 0.74, and its atomic radius is 0.129 nm. [At 912°C (1673°F) the atomic radius of bcc iron, due to thermal expansion, is 0.126 nm.]

---

* When this is found in elemental solids such as metals, the term *allotropes* is sometimes used.

Many other compositions have two or more polymorphic forms. In fact some, such as SiC, have as many as 20 crystalline modifications; however, this is unusual. Invariably, polymorphs have differences in density and other properties. In succeeding chapters we shall be interested in the property variations and in the time required to change from one crystal modification (phase) to another.

Example 3–4.1 Iron changes from bcc to fcc at 912°C (1673°F). At this temperature the atomic radii of the iron atoms in the two structures are 0.126 nm and 0.129 nm, respectively.
a)   What is the percent of volume change, v/o, as the structure changes?
b)   Of linear change, l/o?
(*Note*: As indicated in Section 2–5 and Table 2–5.1, the higher the coordination number the larger the radius.)

Solution: Basis: 4 iron atoms, or *two* unit cells of bcc iron, and *one* unit cell of fcc iron.
a)   In bcc, Eq. (3–2.1):

$$\text{Volume} = 2a_{bcc}^3 = 2\left[\frac{4(0.126)}{\sqrt{3}}\right]^3 = 0.0493 \text{ nm}^3.$$

In fcc, Eq. (3–2.3):

$$\text{Volume} = a_{fcc}^3 = \left[\frac{4(0.129)}{\sqrt{2}}\right]^3 = 0.0486 \text{ nm}^3;$$

$$\frac{\Delta V}{V} = \frac{0.0486 - 0.0493}{0.0493} = -0.014 \quad (\text{or} -1.4 \text{ v/o change})$$

b) $(1 + \Delta L/L)^3 = 1 + \Delta V/V,$

$$\Delta L/L = \sqrt[3]{1 - 0.014} - 1 = -0.0047 \quad (\text{or} -0.47 \text{ l/o change}).$$

Comment. Iron expands by thermal expansion until it reaches 912°C, where there is an abrupt shrinkage; further heating continues the expansion (Fig. 10–1.5a). ◀

Example 3–4.2 The densities of ice and water at 0°C are 0.915 and 1.0005 Mg/m³ (or g/cm³), respectively. What is the percent volume expansion during the freezing of water?

Solution: Basis: 1 Mg.

$$\text{Volume ice} = 1.093 \text{ m}^3;$$

$$\text{Volume liquid} = 0.9995 \text{ m}^3.$$

$$\Delta V/V = (1.093 \text{ m}^3 - 0.9995 \text{ m}^3)/0.9995 \text{ m}^3$$

$$= +0.0935 \quad (\text{or } 9.35 \text{ v/o}).$$

Comments. We are familiar with the major changes that occur during freezing and thawing. In principle, polymorphic changes are similar. That is, there is a change in structure within the solid. This brings about changes in volume, density, and almost every other physical property. ◀

3–5   UNIT CELL GEOMETRY

The general convention that we will follow is to orient a crystal so its x-axis points toward us, the y-axis points to our right as we face it, and the z-axis upward. Thus, by convention the origin is at the left, lower, rear corner of the unit cell.* The opposing directions are negative.

Unit cell locations   Every point within a unit cell can be identified in terms of the coefficients along the three coordinate axes. Thus the origin is 0,0,0. Since the center of the unit cell is at $\frac{a}{2},\frac{b}{2},\frac{c}{2}$, the index of that location is $\frac{1}{2},\frac{1}{2},\frac{1}{2}$. The *coefficients of locations* are always expressed in unit cell dimensions. Thus, the far corner of the unit cell is 1,1,1 regardless of the crystal system—cubic, tetragonal, orthorhombic, etc.

• **Translations**   A translation from any selected point within a unit cell by an integer multiple of lattice constants (a, b, and/or c) leads to an *equivalent site* in another unit cell. Thus, in the two-dimensional lattice of Fig. 3–5.1, the two points labeled ∗ are separated by translations of 3b (parallel to y) and 2c (parallel to z). This example is obviously noncubic (or nonsquare). However, integer multiples apply to the duplication of equivalent sites in all crystal systems.

omit ⟋

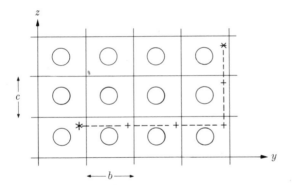

Fig. 3–5.1   Unit translations. A translation of an integer number of unit dimensions, i.e., lattice constants, leads to an equivalent site that is identical in every respect to the original site.

Additional translations exist in a number of space lattices. For example, in a body-centered lattice, *any* translation of $\pm\frac{1}{2},\pm\frac{1}{2},\pm\frac{1}{2}$ leads to another equivalent site. The most obvious is a translation from a corner to the cell center of $\pm\frac{1}{2},\pm\frac{1}{2},\pm\frac{1}{2}$. Note, however, that the two sites in Fig. 3–5.2 that are marked with ∗'s are also related by $\frac{1}{2},\frac{1}{2},-\frac{1}{2}$. One is at the center of the upper back edge, the other at the center of the right-hand face. Each location is midway between two adjacent atoms; each is central to four atoms in 45° directions. Likewise, a point of contact between two atoms is duplicated with any of the variety of $\pm\frac{1}{2},\pm\frac{1}{2},\pm\frac{1}{2}$ translations. This leads to a formal description of a *body-centered* lattice. A body-centered lattice has *equivalent*

---

*   We are not locked into this convention if there is reason to change. However, we will assume this convention applies in our discussions *unless* informed otherwise. Conversely, your answers should comply with the convention *unless* you advise your reader of the reorientation that you have used.

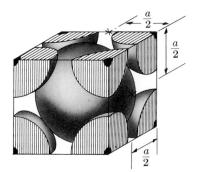

Fig. 3–5.2. Translations in bcc. A translation from any location of

$$\pm\frac{a}{2},\ \pm\frac{a}{2},\ \pm\frac{a}{2}$$

leads to an equivalent site, for example, ∗ to ∗.

sites that are related by $\pm\frac{a}{2}, \pm\frac{b}{2}, \pm\frac{c}{2}$ *translations.* This gives us *two* equivalent sites per unit cell.

A variety of translations exist in a *face-centered* lattice. For example, Fig. 3–5.3 represents an extended plane parallel to the cubic surface of NaCl (Fig. 3–1.1). A translation of $\pm\frac{a}{2}, \pm\frac{a}{2}$ in *two* of the three dimensions leads to a second point that is equivalent to the first (∗ to ∗). Again the initial point may be selected at random. Were we to examine the planes parallel to other faces in the crystal in Fig. 3–1.1, we would need to expand our formal description of the *face-centered* lattice. A face-centered lattice has *equivalent sites that are related by* $\pm\frac{a}{2}, \pm\frac{b}{2}, 0$; *by* $\pm\frac{a}{2}, 0, \pm\frac{c}{2}$; *and by* $0, \pm\frac{b}{2}, \pm\frac{c}{2}$. This gives us *four* equivalent sites per unit cell—the initial site and the three additional sites.

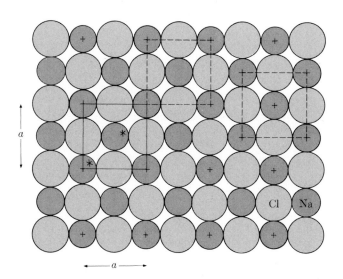

Fig. 3–5.3 Translations in fcc. A translation from *any* location of a/2 in two of the three coordinate directions leads to an equivalent site, for example ∗ to ∗.

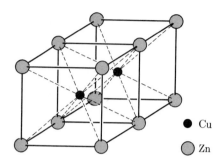

Fig. 3–5.4 Simple cubic compound (β' brass).
Each copper atom is coordinated with eight
zinc atoms; and each zinc atom is coordinated
with eight copper atoms. The lattice is simple
cubic, because the center site is not equivalent
to the corner sites. The prototype for this
structure is CsCl (Fig. 8–2.1a).

● Cu

◯ Zn

Simple cubic lattice Although we mentioned the simple cubic lattice in the second
sentence of Section 3–2, we have not described it. Unlike the bcc and fcc lattices that
have two and four equivalent sites, respectively, per unit cell, a *simple lattice has only
one equivalent site per unit cell*. This does not necessarily say that such crystals are
"simple." In fact, they always have more than one kind of atom, which in itself adds
more complexity. Rather, their translations are limited to integer vectors. Figure 3–5.4
shows the CsCl structure, a simple cubic compound. While this reminds one of a bcc
structure, the $\frac{a}{2},\frac{a}{2},\frac{a}{2}$ translation from the corner to the cell center does *not* connect two
identical points. One is Cs^+; the other is Cl^-. Thus, it is *not* bcc.

Example 3–5.1 (a) Sketch a noncubic unit cell and show the locations which have the following
coefficients:

$$0,0,0; \qquad 0,0,\frac{1}{2}; \qquad \frac{1}{2},\frac{1}{2},\frac{1}{2}; \qquad \frac{1}{2},\frac{1}{2},0; \qquad 1,1,0; \qquad 1,1,1; \qquad 1,1,2.$$

b)   Assume the cell is orthorhombic (all axial angles $= 90°$), and $a = 0.270$ nm, $b = 0.403$ nm,
and $c = 0.363$ nm. What is the translation distance t between locations $0,-1,0$ and $1,1,2$?

Solution: (a) See Fig. 3–6.1(a).
b)   Since the axial angles are $90°$,

$$t = \sqrt{[(1-0)(0.270 \text{ nm})]^2 + [(1-(-1))(0.403 \text{ nm})]^2 + [(2-0)(0.363 \text{ nm})]^2}$$
$$= 1.118 \text{ nm}.$$

Comment. This procedure for identifying crystal locations does not limit us to the reference unit
cell. ◀

Example 3–5.2 Consider the following four points within a unit cell:

$$\frac{1}{4},\frac{1}{4},\frac{3}{4}; \qquad \frac{7}{8},\frac{3}{4},\frac{1}{5};$$

$$\frac{1}{8},\frac{1}{8},\frac{1}{8}; \qquad 0.53, 0.25, 0.17.$$

a)   For a body-centered lattice, identify an equivalent site *within* the reference unit cell for each
of the four points.

b)   For a face-centered lattice, identify three equivalent sites *within* the reference unit cell for each of the four points.

Answers: (a) In a body-centered lattice, the translations between equivalent points are $\pm0.5a$, $\pm0.5a$, $\pm0.5a$. Therefore,

For $\frac{1}{4},\frac{1}{4},\frac{3}{4}$: $(0.25 + 0.5),\ (0.25 + 0.5),\ (0.75 - 0.5) = \frac{3}{4},\frac{3}{4},\frac{1}{4}$.

For $\frac{1}{8},\frac{1}{8},\frac{1}{8}$: $(0.125 + 0.5), (0.125 + 0.5), (0.125 + 0.5) = \frac{5}{8},\frac{5}{8},\frac{5}{8}$.

For $\frac{7}{8},\frac{3}{4},\frac{1}{5}$: $(0.875 - 0.5),\ (0.75 - 0.5),\ (0.20 + 0.5) = \frac{3}{8},\frac{1}{4},0.7$.

For $0.53, 0.25, 0.17$: $(0.53 - 0.5),\ (0.25 + 0.5),\ (0.17 + 0.5) = 0.03, 0.75, 0.67$.

b)   The translations in face-centered lattices are permutations of $\pm0.5a$, $\pm0.5a$, and 0.

For $\frac{1}{4},\frac{1}{4},\frac{3}{4}$: $\frac{3}{4},\frac{3}{4},\frac{3}{4}$ $\frac{3}{4},\frac{1}{4},\frac{1}{4}$ $\frac{1}{4},\frac{3}{4},\frac{1}{4}$.

For $\frac{1}{8},\frac{1}{8},\frac{1}{8}$: $\frac{5}{8},\frac{5}{8},\frac{1}{8}$ $\frac{5}{8},\frac{1}{8},\frac{5}{8}$ $\frac{1}{8},\frac{5}{8},\frac{5}{8}$.

For $\frac{7}{8},\frac{3}{4},\frac{1}{5}$: $\frac{3}{8},\frac{1}{4},\frac{1}{5}$ $\frac{3}{8},\frac{3}{4},0.7$ $\frac{7}{8},\frac{1}{4},0.7$.

For $0.53, 0.25, 0.17$: $0.03, 0.75, 0.17$ $0.03, 0.25, 0.67$ $0.53, 0.75, 0.67$.

Comment. These translations within body-centered and face-centered lattices are independent of whether the system is cubic, orthorhombic (or tetragonal). ◀

Example 3–5.3 Calculate the distances in NaCl between the center of a sodium ion and the center of
a)   its nearest neighbor;
b)   its nearest positive ion;
c)   its second nearest Cl^- ion;
d)   its third nearest Cl^- ion;
e)   its nearest equivalent site.

Solution: Refer to Fig. 3–1.1. From Appendix B, the radii of Na^+ and Cl^- are 0.097 nm and 0.181 nm, respectively (since each has CN = 6).

$$a = 2(0.097 + 0.181 \text{ nm}) = 0.556 \text{ nm}.$$

a)   Distance $= a/2 = 0.278$ nm.
b)   Distance $= \sqrt{(a/2)^2 + (a/2)^2} = 0.393$ nm.
c)   Distance $= \sqrt{(a/2)^2 + (a/2)^2 + (a/2)^2} = 0.482$ nm.
d)   Distance $= \sqrt{(a/2)^2 + a^2} = 0.622$ nm.
e)   Same as (b), since the nearest Na^+ ions are at equivalent sites, with a translation of $\frac{a}{2},\frac{a}{2},0$. ◀

Example 3–5.4 Cesium iodide (CsI) has the structure shown in Fig. 3–5.4. What is its packing factor if the radii are 0.172 nm and 0.227 nm, respectively?

Solution: The body diagonal of the unit cell is equal to $(2r + 2R)$. Therefore,

$$a = [2(0.172 + 0.227)]/\sqrt{3} = 0.461 \text{ nm}.$$

$$\text{Packing factor} = (4\pi/3)(0.172^3 + 0.227^3)/(0.461)^3 = 0.72.$$

Comments. The packing factor is greater than that obtained in the calculation following Eq. (3–2.2) because a bcc metal does not contain a second (different) atom in the center.

These radii differ from the radii of Appendix B by a factor of 0.97, since the data in the appendix are for CN = 6. Here, CN = 8. ◀

3–6 CRYSTAL DIRECTIONS

When we correlate various properties with crystal structures in subsequent chapters, it will be necessary to identify specific crystal directions, because many properties are directional. For example, the elastic modulus of bcc iron is greater parallel to the body diagonal than to the cube edge. Conversely, the magnetic permeability of iron is greatest in a direction parallel to the edge of the unit cell.

Directions *Crystal directions* are indexed simply as a *ray* extending from the origin through the locations with the lowest integer index (Fig. 3–6.1). Thus the $[111]$ direction passes from 0,0,0 through 1,1,1. Note, however, that this direction also passes through $\frac{1}{2},\frac{1}{2},\frac{1}{2}$ (and 2,2,2). Likewise the $[112]$ passes through $\frac{1}{2},\frac{1}{2},1$; but for simplicity's

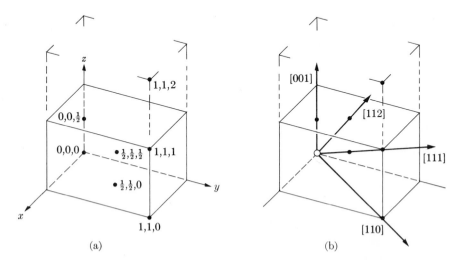

Fig. 3–6.1 Orthorhombic unit cell locations (a), and crystal directions (b). The location of the origin is commonly, but not necessarily, at the lower left rear corner. By convention we use square brackets $[uvw]$ to indicate specific directions, and $\langle uvw \rangle$ to indicate a family of directions. This is to avoid confusion with indices for planes (hkl). Locations are indexed without closures: x, y, z.

sake, we use the integer notation. Observe that we enclose the direction indices in square brackets $[uvw]$, and use the letters u, v, and w for the indices arising from the three principal directions, x, y, and z, respectively. Parallel directions always have the same indices. Finally, note that we may have negative coefficients, which we designate with an overbar; a $[11\bar{1}]$ direction will have a component in the minus z direction.

[handwritten: angle θ between (x_1, y_1, z_1) & (x_2, y_2, z_2) is ——→]

Angles between directions   In certain calculations (e.g., resolved shear stresses), it will be necessary to calculate the angle between two different crystal directions. For most of the calculations we will encounter, this may be performed by simple inspection. Thus, in Fig. 3–6.1, the angle between $[110]$ and $[112]$ directions (i.e., $[110] \nmid [112]$) is $\arctan 2c/\sqrt{a^2 + b^2}$. If that unit cell had been cubic rather than orthorhombic, so that $a = b = c$, the angle would have been $\arctan 2a/a\sqrt{2}$, or $\arccos a\sqrt{2}/a\sqrt{6}$. In fact, with cubic crystals (*only*), we can determine $\cos [uvw] \nmid [u'v'w']$ by the *dot product*. This latter procedure will be useful to us since most of our calculations in this introductory text will involve these symmetric cubic crystals.

[handwritten, right margin: $\cos \theta = \dfrac{(x_1 x_2 + y_1 y_2 + z_1 z_2)}{(x_1^2 + y_1^2 + z_1^2)^{\frac{1}{2}}(x_2^2 + y_2^2 + z_2^2)^{\frac{1}{2}}} = [\ ]$ ↑ TAKE cos⁻¹ FOR θ $= \cos^{-1}$]

- **Linear density**   The repeating distance between equivalent sites differs from direction to direction and from lattice to lattice. For example, in the $[111]$ direction of a bcc metal, an equivalent lattice site is repeated every $2R$, or $a\sqrt{3}/2$. The repeating distance is $a\sqrt{2}$ in the $[110]$ direction of bcc, but $a/\sqrt{2}$ in fcc. These can be checked out in Figs. 3–2.1 and 3–2.3.

  Conversely, the reciprocals of these distances are the *linear densities* of equivalent sites. Thus, in the $[110]$ direction of aluminum, which is fcc with a equal to 0.405 nm, the linear density is 1 per $a/\sqrt{2}$, or $\sqrt{2}/(0.405 \times 10^{-6}$ mm), that is, 3.5×10^6/mm.

$$\text{Linear density} = \frac{\text{Number}}{\text{Unit length}}. \qquad (3\text{–}6.1)$$

We will observe in Chapter 6 that deformation occurs most readily in those directions with the greatest linear density of equivalent sites.

  In most metals, there is one atom per lattice site; therefore, the linear density of atoms equals the linear density of equivalent sites. However, Example 3–6.4 shows that the two may differ in some crystals.

Families of directions   In the cubic crystal, the following directions are identical except for our arbitrary choice of the x-, y-, and z-labels on the axes:

$$[111] \quad [11\bar{1}] \quad [1\bar{1}1] \quad [\bar{1}11]$$
$$[\bar{1}\bar{1}\bar{1}] \quad [\bar{1}\bar{1}1] \quad [\bar{1}1\bar{1}] \quad [1\bar{1}\bar{1}]$$

Any directional property* will be identical in these eight directions. Therefore, it is convenient to identify this *family of directions* as $\langle 111 \rangle$ rather than writing the eight individual directions. Note that the closure symbols are angle brackets $\langle \ \rangle$.

---

* For example, Young's modulus, magnetic permeability, index of refraction.

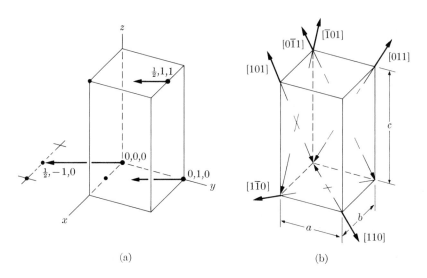

Fig. 3–6.2   Crystal directions. (a) [1$\bar{2}$0] (see Example 3–6.1). (b) $\langle 101 \rangle$ (see Example 3–6.2b).

Example 3–6.1   All parallel directions possess the same directional indices. Sketch rays in the [1$\bar{2}$0] direction that pass through locations (a) 0,0,0; (b) 0,1,0; and (c) $\frac{1}{2}$,1,1.

Solution: See Fig. 3–6.2(a). ◀

Example 3–6.2   The $\langle 101 \rangle$ family of directions include what individual directions (a) in a cubic crystal? ● (b) in a tetragonal crystal?

Solution: (a) In cubic, $a = a = a$. Therefore,

$$[110], [101], [011], [1\bar{1}0], [10\bar{1}], [01\bar{1}].$$

b)   In tetragonal cells $a = a \neq c$ (Fig. 3–6.2b). Therefore, only the u and v indices of $\langle uvw \rangle$ are interchangeable; the w index is not:

$$[101], [011], [\bar{1}01], [0\bar{1}1].$$

Comments.   The [$\bar{1}$01] and the [10$\bar{1}$] directions are commonly considered to be the two senses of one direction, and not two separate directions. However, if desired, we could list the negative indices of each of the above directions. ◀

Example 3–6.3   (a) What is the angle between the [111] and [001] directions in a *cubic* crystal? (b) [111] $\measuredangle$ [$\bar{1}\bar{1}$1]?

Solution: From the *cubic* analog of Fig. 3–6.1(b),

a) $$\cos [111] \measuredangle [001] = a/a \sqrt{3},$$

$$[111] < [001] = 54.75°.$$

b)   Observe that $[001]$ bisects $[111] \times [\bar{1}\bar{1}1]$; therefore,

$$[111] \times [\bar{1}\bar{1}1] = 2(54.75°) = 109.5°;$$

or, by the dot product (*since the crystal is cubic*):

$$\cos [111] \times [\bar{1}\bar{1}1] = -\tfrac{1}{3},$$

$$[111] \times [\bar{1}\bar{1}1] = 109.5°.$$

Comment. Compare the results here with Fig. 2–3.3(a) where all four of the bond angles are 109.5°. ◀

● **Example 3–6.4** The lattice constant a is 0.357 nm for diamond, which is cubic. (a) What is the linear density of equivalent sites in the $[11\bar{1}]$ direction? (b) Of atoms?

Solution: Refer to Fig. 2–2.4. Use the *upper*, rear, left corner as the origin.
a)   In the $[11\bar{1}]$ direction, the repeating distance 0,0,0 to 1,1,−1 is $a\sqrt{3}$, or 0.618 nm; therefore,

$$\text{linear density} = 1/0.618 \times 10^{-6} \text{ mm} = 1.6 \times 10^6/\text{mm}.$$

b)   There are atoms at 0,0,0 and $\tfrac{1}{4},\tfrac{1}{4},-\tfrac{1}{4}$; therefore,

$$2 \text{ atoms}/a\sqrt{3} = 3.2 \times 10^6/\text{mm}.$$

Comment. The location of the $\tfrac{1}{4},\tfrac{1}{4},-\tfrac{1}{4}$ atom can be checked by relating it to its four neighbors at

$$0,0,0 \quad \tfrac{1}{2},\tfrac{1}{2},0 \quad \tfrac{1}{2},0,-\tfrac{1}{2} \quad 0,\tfrac{1}{2},-\tfrac{1}{2}. \quad ◀$$

3–7 CRYSTAL PLANES

A crystal contains planes of atoms; these influence the properties and behavior of a material. Thus it will be advantageous to identify various planes within crystals.

The lattice planes most readily visualized are those that outline the unit cell, but there are many other planes. The more important planes for our purposes are those sketched in Figs. 3–7.1, 3–7.2, and 3–7.3. These are labeled (010), (110), and ($\bar{1}$11) respectively, where the numbers within the parentheses (*hkl*) are called the *Miller indices*.

Miller indices   We may use the plane with the darker tint in Fig. 3–7.4 to explain how (*hkl*) numbers are obtained. The plane intercepts the *x*-, *y*-, and *z*-axes at $1a$, $1b$, and $0.5c$. The Miller indices are simply the reciprocals of these intercepts: (112). The plane of lighter tint in Fig. 3–7.4 is the (111) plane, since it intercepts the axes at $1a, 1b$, and $1c$. Returning to the earlier figures, we have

Figure	Plane	Intercepts	Miller indices
3–7.1(a)	Middle	∞a, $1b$, ∞c	(010)
3–7.2(a)	Left	$1a$, $1b$, ∞c	(110)
3–7.3(a)	Middle	$-1a$, $1b$, $1c$	($\bar{1}$11)

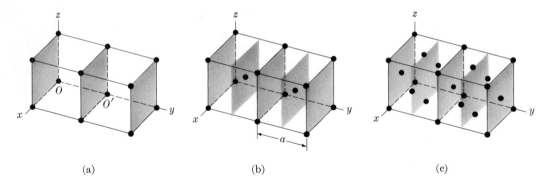

Fig. 3–7.1   (010) planes in cubic structures. (a) Simple cubic. (b) Bcc. (c) Fcc. (Note that the (020) planes included for bcc and fcc are comparable to (010) planes.)

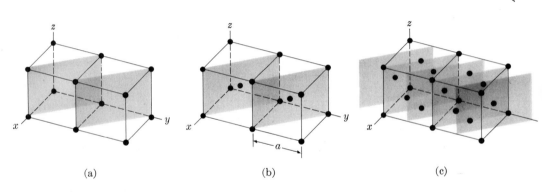

Fig. 3–7.2   (110) planes in cubic structures. (a) Simple cubic. (b) Bcc. (c) Fcc. (The (220) planes included for fcc are comparable to (110) planes.)

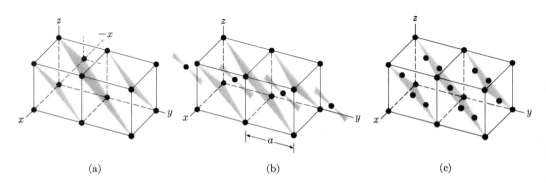

Fig. 3–7.3   ($\bar{1}$11) planes in cubic structures. (a) Simple cubic. (b) Bcc. (c) Fcc. Negative intercepts are indicated by bars above the index. The ($\bar{2}$22) planes included for bcc are comparable ($\bar{1}$11) planes.)

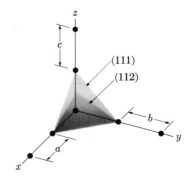

Fig. 3–7.4 Miller indices. The (112) plane cuts the three axes at 1, 1, and 1/2 unit distances.

Note that a *minus intercept* is handled readily with an overbar. Furthermore, observe that we use parentheses (*hkl*) to denote planes (and no commas), in order to avoid confusion with individual directions that were denoted in Section 3–6 with square brackets: [*uvw*].

All parallel planes are identified with the same indices. Figure 3–7.1(a) will explain this:

Plane	Origin	Intercepts	Miller indices
Right	0	∞a, $2b$, ∞c	$(0\frac{1}{2}0)$
Right	0′	∞a, $1b$, ∞c	(010)
Left	0′	∞a, $-1b$, ∞c	$(0\bar{1}0)$

Since the choice of the origin, O or O', is arbitrary, we could use either $(0\frac{1}{2}0)$ or (010) to index the right plane. The (010) indices are simpler to say and write; therefore we use them. This is entirely permissible since these three shaded planes are geometrically identical and behave the same during plastic deformation (Chapter 6) and under other circumstances. We can carry this one step further in Fig. 3–7.1(c):

Plane	Intercepts	Miller indices
3rd from left	∞a, $1b$, ∞c	(010)
2nd from left	∞a, $0.5b$, ∞c	$(020) = (010)$

Again, these planes are parallel and *identical* (but offset $\frac{1}{2}$ in the *x*-direction and $\frac{1}{2}$ in the *z*-direction). Ordinarily, we do not need to distinguish between the (010), $(0\frac{1}{2}0)$, and (020) planes. Furthermore, in Fig. 3–7.2(c), the (220) plane is fully comparable to the (110) plane, and the $(\bar{2}22)$ plane is the same as the $(\bar{1}11)$ plane (Fig. 3–7.3b). As a result, we use the indices with the lowest set of integers. A more rigorous definition of *Miller indices* is now possible. They are *the reciprocals of the three axial intercepts for a plane, cleared of fractions and common multipliers.* *

---

* This definition will not apply to *interplanar spacings* (d_{020}, etc.) in Section 3–8, since these spacings will *not* be planes, but distances between pairs of planes.

Families of planes   Depending on the crystal system, two or more planes may belong to the same family of planes. In the cubic system, an example of multiple planes includes the following, which constitute a *family of planes*, or *form*:

$$\frac{(100)\quad(010)\quad(001)}{(\bar{1}00)\quad(0\bar{1}0)\quad(00\bar{1})} = \{100\}.$$

The collective notation for a family of planes is $\{hkl\}$. Figure 3–1.1 indicates for us the form of the $\{100\}$ family that has the six planes listed above. Each face is identical except for the consequences of our arbitrary choice of axis labels and directions. The reader is asked to verify that the $\{111\}$ family includes eight planes, and the $\{110\}$ family includes 12 planes, when all of the permutations and combinations of individual planes are included.

Indices for planes in hexagonal crystals (*hkil*)   The three Miller indices (hkl) can describe all possible planes through any crystal. In hexagonal systems, however, it is frequently useful to establish four axes, three of them coplanar (Fig. 3–3.1a). This leads to four intercepts and ($hkil$) indices.* The fourth index i is an additional index that mathematically is related to the sum of the first two:

$$h + k = -i. \tag{3–7.1}$$

These optional ($hkil$) indices are generally favored because they reveal hexagonal symmetry more clearly. Although partially redundant, they are used almost exclusively, in preference to the equivalent (hkl) indices, in scientific papers that consider hexagonal planes.

Planar densities of atoms   When we consider plastic deformation later, we will need to know the density of atoms on a crystal plane. Example 3–7.3 shows how we may calculate these by means of the relationship:

$$\text{Planar density} = \frac{\text{Atoms}}{\text{Unit area}}. \tag{3–7.2}$$

As we did in dealing with linear density, we may also calculate the planar density of equivalent points.

Study aids (crystal directions and planes)   If your available time for study was too short to understand the origin and significance of crystal indices, you are strongly urged to use the paperback *Study Aids for Introductory Materials Courses*.[†] Directions and Miller indices are presented in greater detail by a series of annotated sketches. A separate section presents the four indices used for hexagonal crystals. Optional sketches address the topics of families of direction and planes and the use of dot and cross products.

---

* Called *Miller–Bravais* indices.
[†] *Study Aids for Introductory Materials Courses*, Reading, Mass.: Addison-Wesley, 1977.

Example 3–7.1 Sketch a (111) plane through a unit cell of a simple tetragonal crystal having a c/a ratio of 0.62.

Solution: Figure 3–7.5 shows this plane (shaded). The (111) plane cuts the three axes at unit distances. However, the unit distance along the z-axis is shorter than the unit distances on the x- and y-axes. ◀

Example 3–7.2 Sketch two planes with (122) indices on the axes of Fig. 3–7.5.

Solution: The intercepts will be the reciprocals of (122)—$1a$, $0.5b$, and $0.5c$. This is shown by the inner set of dashed lines. A plane with intercepts—$2a$, $1b$, and $1c$—is parallel and therefore has the same indices (outer set of dashed lines). ◀

Fig. 3–7.5 Noncubic intercepts (tetragonal structure). The shaded (111) plane cuts the three axes of any crystal at equal unit distances. However, since c may not equal a, the actual intercepting distances are different. (The dashed lines refer to Example 3–7.2.)

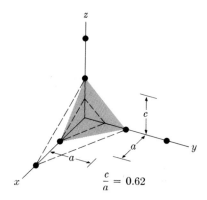

$$\frac{c}{a} = 0.62$$

Example 3–7.3 How many atoms per mm² are there on the (100) and (111) planes of lead (fcc)?

Solution: Pb radius = 0.1750 nm (from Appendix B),

$$a_{Pb} = \frac{4R}{\sqrt{2}} = \frac{4(0.1750 \text{ nm})}{1.414} = 0.495 \text{ nm}.$$

Figure 3–7.6 shows that the (100) plane contains two atoms per unit-cell face.

(100): $$\frac{\text{atoms}}{\text{mm}^2} = \frac{2 \text{ atoms}}{(0.495 \times 10^{-6} \text{ mm})^2} = 8.2 \times 10^{12} \text{ atoms/mm}^2.$$

Fig. 3–7.6 (100) atom concentration (fcc). A (100) plane in an fcc structure has two atoms per a^2.

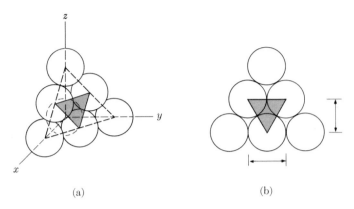

(a)   (b)

Fig. 3–7.7   (111) atom concentration (fcc). A (111) plane has one-half atom per $R^2\sqrt{3}$.

Figure 3–7.7(b) shows that the (111) plane contains three one-sixth atoms in the triangular shaded area.

(111):
$$\frac{\text{atoms}}{\text{nm}^2} = \frac{\frac{3}{6}}{\frac{1}{2}bh} = \frac{\frac{3}{6}}{\frac{1}{2}(2)(0.1750 \text{ nm})(\sqrt{3})(0.1750 \text{ nm})}$$
$$= 9.4 \text{ atoms/nm}^2 = 9.4 \times 10^{12} \text{ atoms/mm}^2. \quad \blacktriangleleft$$

Example 3–7.4   A plane includes locations 0,0,0 and $\frac{1}{2},\frac{1}{4},0$ and $\frac{1}{2},0,\frac{1}{2}$. What are its Miller indices?

Solution: Make a sketch (Fig. 3–7.8).

Since the plane passes through the origin, shift the origin, for example, 1 unit in the x-direction. The intercepts are now $-1a$, **0.5a**, and $+1a$; therefore, ($\bar{1}21$).

Comment. Had we shifted the origin one unit upward, the intercepts would have been $+1, -\frac{1}{2}, -1$ to give us ($1\bar{2}\bar{1}$). This is parallel to ($\bar{1}21$), and therefore equivalent. $\blacktriangleleft$

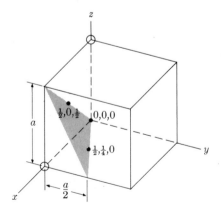

Fig. 3–7.8   The ($1\bar{2}\bar{1}$) plane (see Example 3–7.4).

Example 3–7.5 What direction is the line of intersection of the (111) and (112) planes?

Solution: Refer to Fig. 3–7.4: $[1\bar{1}0]$, (or $[\bar{1}10]$).

Comment. For our purposes, the line of intersection is most readily obtained by inspection. However, it may also be obtained as a cross product of the indices of the two planes. ◀

•**Example 3–7.6** Refer to Fig. 3–3.1(a). Provide an index for the plane facing the reader.

Solution

	a_1	a_2	a_3	c
Axis:				
Intercept:	1	∞	-1	∞

Therefore, the $(hkil)$ indices are $(10\bar{1}0)$, since we use reciprocals.

Comment. If we had used the alternate presentation of Fig. 3–3.1(b), the (hkl) indices would be (100). The additional index, i, is related to h and k by Eq. (3–7.1). ◀

Example 3–7.7 List the individual planes that belong to the {111} family in cubic crystals.

Solution

$$(111) \quad (\bar{1}11) \quad (1\bar{1}1) \quad (11\bar{1});$$

$$(\bar{1}\bar{1}\bar{1}) \quad (1\bar{1}\bar{1}) \quad (\bar{1}1\bar{1}) \quad (\bar{1}\bar{1}1).$$

Comment. In reality, there are only four sets of planes (rather than 8) because those of the second set are parallel to those of the first set. However, unlike the negative directions (Example 3–6.2), we commonly list the redundant planes because the eight planes (in this case) enclose a volume and produce a crystal *form*. The form for {111} is a bipyramid. ◀

•**3–8   X-RAY DIFFRACTION**   OMIT

Excellent experimental verification for the crystal structures that we have been discussing is available through x-ray diffraction. When these high-frequency electromagnetic waves are selected to have a wavelength slightly greater than the *interplanar* spacings of crystals, they are diffracted according to very exacting physical laws. The angles of diffraction let us decipher crystal structures with a high degree of accuracy. In turn, one can readily determine the interplanar spacings (and therefore atomic radii) in metals to four significant figures, and even more precisely, if necessary. Let us first examine the spacings between planes. Then we will turn to diffraction.

Interplanar spacings   Recall, from Section 3–7, that all parallel planes bear the same (hkl) notation. Thus the several (110) planes of Fig. 3–7.2 have still another (110) plane that passes directly through the origin. As a result, if we measure a *perpendicular* distance from the origin to the next adjacent (110) plane, we have measured the interplanar distance, d. We will observe that in the simple cubic structures of Figs. 3–7.1(a), 3–7.2(a), and 3–7.3(a), the interplanar distances are a, $(a\sqrt{2})/2$, and $(a\sqrt{3})/3$ for d_{010}, d_{110}, and $d_{\bar{1}11}$, respectively. That is, there is one spacing per cell edge a for d_{010}; two spacings per face diagonal, $a\sqrt{2}$, for d_{110}; and three spacings per body diagonal, $a\sqrt{3}$, for $d_{\bar{1}11}$. We may formulate a general rule for d-spacings in *cubic* crystals:

$$d_{hkl} = \frac{a}{\sqrt{h^2 + k^2 + l^2}}, \tag{3–8.1}$$

where a is the lattice constant and h, k, and l are the indices of the planes.* The inter-planar spacings for noncubic crystals may be expressed with equations that are related to Eq. (3–8.1), but take into account the variables of Table 3–1.1.

Bragg's law   When x-rays encounter a crystalline material, they are diffracted by the planes of the atoms (or ions) within the crystal. The *diffraction angle* θ depends upon the wavelength λ of the x-rays and the distance d between the planes:

$$n\lambda = 2d \sin \theta. \tag{3–8.2}$$

Consider the parallel planes of atoms in Fig. 3–8.1, from which the wave is diffracted. The waves may be "reflected" from the atom at H or H' and remain in phase at K. However, x-rays are reflected not only from the surface plane but also from the adjacent subsurface planes. If these reflections are to remain in phase and be *coherent*, the distance $MH''P$ must equal one or more integer wavelengths of the rays. The value n of Eq. (3–8.2) is the integer number of waves that occur in the distance $MH''P$.

Diffraction analyses   The most common procedure for making x-ray diffraction analyses utilizes very fine powder of the material in question. It is mixed with a plastic cement and formed into a very thin filament that is placed at the center of a circular camera (Fig. 3–8.2). A collimated beam of x-rays is directed at the powder. Since there is a very large number of powder particles with essentially all possible orientations, the diffracted beam emerges as a cone of radiation at an angle 2θ from the initial beam. (Observe in Fig. 3–8.1 that the diffracted beam is 2θ away from the initial beam.)

The diffraction cone exposes the film strip in the camera at two places; each is 2θ from the straight-through exit port. There is a separate cone (or pair of *diffraction lines*) for each d_{hkl} value of interplanar spacings. Thus, the diffraction lines may be measured and the d-spacings calculated from Eq. (3–8.2). All fcc metals will have a similar set of diffraction lines, but with differing 2θ values, since they have different lattice constants; for example, $a_{Cu} = 0.3615$ nm; $a_{Al} = 0.4049$ nm; $a_{Pb} = 0.495$ nm, etc. Thus, we can differentiate between various fcc metals.

In Fig. 3–8.3, we see x-ray diffraction films for copper (fcc), tungsten (bcc) and zinc (hcp). It is immediately apparent that the sequences of diffraction lines are different for the three types of crystals. Since we lack space here to explain these differences, we will have to simply observe that, with different "fingerprints," it is

---

* The reciprocal nature of Miller indices permits this type of simplified calculation. Likewise, the cross product of the Miller indices for two intersecting planes gives the direction indices for the line of intersection. Finally, in order for a direction to lie in a plane, the dot product of the indices for that direction and for the plane must equal zero. In brief, there is purpose in using the "down-side up" Miller indices.

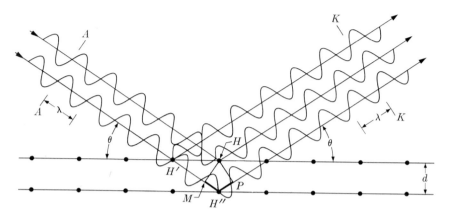

Fig. 3–8.1 X-ray diffraction.

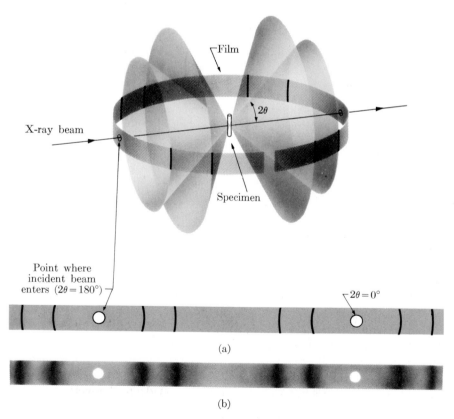

Fig. 3–8.2 The exposure of x-ray diffraction patterns. Angle 2θ is precisely fixed by the lattice spacing d and the wavelength λ as shown in Eq. (3–8.2). Every cone of reflection is recorded in two places on the strip of film. (B. D. Cullity, *Elements of X-ray Diffraction*, 1st ed.)

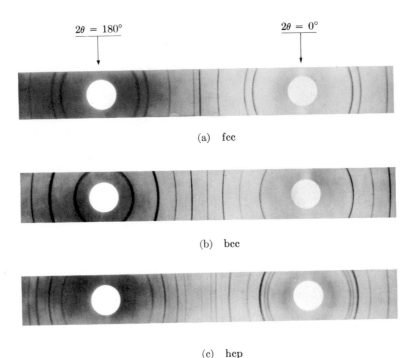

$2\theta = 180°$ $2\theta = 0°$

(a) fcc

(b) bcc

(c) hcp

Fig. 3–8.3   X-ray diffraction patterns for (a) copper, fcc, (b) tungsten, bcc, and (c) zinc, hcp. The crystal structure and the lattice constants may be calculated from patterns such as these. (B. D. Cullity, *Elements of X-ray Diffraction*, 2d ed.)

possible not only to determine the size of the lattice constants, with utmost precision, but also to identify the crystal lattice. X-ray diffraction is an extremely powerful tool in the study of the internal structure of materials.

Example 3–8.1   X-rays of an unknown wavelength are diffracted 43.4° by copper (fcc) whose lattice constant a is 0.3615 nm. Separate determinations indicate that this diffraction line for copper is the first-order ($n = 1$) line for d_{111}.

a)   What is the wavelength of the x-rays?
b)   The same x-rays are used to analyze tungsten (bcc). What is the angle, 2θ, for the second-order ($n = 2$) diffraction lines of the d_{010} spacings?

Solution

a)   Since $2\theta = 43.4°$, and by using Eqs. (3–8.1) and (3–8.2), $n\,\lambda = 2\,d\,\sin\theta$

$$(1)\lambda = 2\left[\frac{0.3615 \text{ nm}}{\sqrt{1^2 + 1^2 + 1^2}}\right]\sin 21.7°$$

$$= 0.1543 \text{ nm}.$$

b)   From Appendix B,

$$R_w = 0.1367 \text{ nm}.$$

From Eq. (3–2.1),

$$a_w = 4(0.1367 \text{ nm})/\sqrt{3} = 0.3157 \text{ nm}.$$

From Eqs. (3–8.1) and (3–8.2),

$$\sin \theta = 2(0.154 \text{ nm})(\sqrt{0 + 1 + 0})/(2)(0.3157 \text{ nm}),$$

$$2\theta = 58.4°.$$

Comments. The second-order diffraction of d_{010} is equivalent to the first-order diffraction of a d_{020} spacing; i.e., the perpendicular distance from the origin to a plane that cuts the x-, y-, and z-axes at ∞, $\frac{1}{2}$, and ∞, respectively. This may be checked with Eqs. (3–8.1) and (3–8.2).

The d_{hkl}-spacings for all planes of the same form are equal. For example, $d_{100} = d_{010} = d_{001}$ in the cubic system. ◄

REVIEW AND STUDY

In Chapter 2 we considered the bonding of atoms with their immediate neighbors and the resulting coordination. This chapter carries our understanding of internal structure one step further by looking at longer-range geometric order. Since all atoms of one kind have similar coordination requirements, it is not surprising that we find the same patterns repeated time and time again throughout the material.

SUMMARY

We may divide crystalline solids into unit cells with each having all the characteristics of atomic coordination found throughout the crystal. Comparable positions in each unit cell have the *same neighbors* in the *same directions* and at the *same distances.* When we consider properties, this simplifies our description of the internal structure, because we only have to describe the unit cell.

There are seven crystal systems based on the geometry of the unit cell (unit-cell dimensions and axial angles). These seven systems possess a number of lattices, based on the internal arrangement of the equivalent sites within the unit cell. Our attention was directed primarily toward the body-centered cubic (bcc), face-centered cubic (fcc) and a common structure of the hexagonal lattice called the hexagonal close-packed (hcp). On occasion, we will also consider body-centered tetragonal (bct) and simple cubic (sc) lattices. Every point within a body-centered lattice is separated from an equivalent site by a $\pm\frac{1}{2},\pm\frac{1}{2},\pm\frac{1}{2}$ translation. Each point within a face-centered lattice is separated from three additional equivalent sites by translations of $\pm\frac{1}{2},\pm\frac{1}{2},0$; of $\pm\frac{1}{2},0,\pm\frac{1}{2}$; and of $0,\pm\frac{1}{2},\pm\frac{1}{2}$. Points within a simple lattice are not duplicated within that unit cell.

Crystal directions are indexed on the basis of the unit cell dimensions. Thus, a [111] direction is a ray from the origin, 0,0,0, through the center of the unit cell and out to the far corner, regardless of the crystal system. Families of directions $\langle hkl \rangle$ include all directions that are identical except for our arbitrary choice of coordinate references.

Crystal planes are indexed by the reciprocal of the axial intercepts (followed by the elimination of fractions). We will give the greatest, but not exclusive, attention to the (100), (110), and (111) planes; therefore, the reader should be familiar with the following facts:

1. Any plane of the {100} family parallels two of the coordinate axes and cuts the third;

2. Any plane of the {110} family parallels one axis and cuts the other two with equal intercept coefficients; and

3. The planes of the {111} family cut all three axes with equal intercept coefficients.

Since this is not a course in crystallography, we did not discuss in detail how crystal structures are established and verified in the laboratory. However, we did note that use is made of x-ray diffraction to study crystal lattices. The diffraction angles are interpreted by Bragg's law. There are few if any structural characteristics that can be measured with a greater degree of certainty than these crystal parameters.

KEY TERMS AND CONCEPTS

Atomic packing factor, APF

• Bragg's law

Body-centered cubic, bcc

Crystal

Crystal direction $[uvw]$

Crystal plane (hkl)

Crystal system

• Diffraction lines

Equivalent sites

Face-centered cubic, fcc

Family of directions $\langle uvw \rangle$

Family of planes $\{hkl\}$

Hexagonal close-packed, hcp

• Interplanar spacing, d_{hkl}

Lattice constant

• Linear density

Long-range order

Miller indices

NaCl structure

• Planar density

Polymorphism

Simple cubic, sc

• Translation

Unit cell

• X-ray diffraction

FOR CLASS DISCUSSION

A_3 What is meant by long-range order?

B_3 Examine a tile-covered floor and pick a point at random. Explain what is meant by "repeat distance."

C_3 List the several crystal systems and indicate their distinguishing characteristics.

D_3 The corner site of a cubic unit cell is associated with how many unit cells? The corner site of a rectangular area of a plane belongs to how many "unit areas"?

E_3 From Fig. 3–2.4, show that CN = 12 for fcc.

F_3 Examine wallpaper that has an intricate design and pick a point at random. Find three nearby equivalent sites. What are the translation distances between the equivalent sites?

G_3 For 2-D patterns, there are five possible unit areas: simple square, simple rectangle, area-centered rectangle, parallelogram, and hexagon. How many equivalent sites does each have? Why is there no area-centered square listed? The area-centered rectangle is sometimes called a simple rhombus. Explain.

H_3 An fcc lattice may be presented alternatively as a rhombohedral lattice. (a) How do the two lattices compare? (b) What is the axial angle? (c) What is the ratio of unit cell volumes? •(d) Why do we prefer the fcc over the rhombohedral?

I_3 Among 3-D lattices, we find bcc and bct (i.e., body-centered tetragonal). Also there is an fcc, but no fct. Explain why.

J_3 What is the ratio of volumes for the two presentations of the hexagonal unit cell in Fig. 3–3.1? The ratio of equivalent sites?

K_3 Compare the size and "shape" of the holes that are centered at $\frac{1}{2},\frac{1}{2},0$ and at $\frac{1}{2},\frac{1}{4},0$ of a bcc metal.

L_3 Compare the size and "shape" of the holes that are centered at $\frac{1}{2},0,0$ and $\frac{1}{4},\frac{1}{4},\frac{1}{4}$ of an fcc metal.

M_3 Diamond requires a higher pressure, to be stable, than does graphite. What does this indicate about the densities of diamond and graphite?

N_3 Cobalt changes during heating from hcp to fcc at 1120°C. How will its volume be altered (a) compared with iron which contracts on heating from the low to the high temperature forms (Example 3–4.1), and (b) with titanium that contracts on cooling (Study Problem 3–4.3), when it changes structure from the high to the low temperature form?

O_3 From Fig. 3–2.4, show that fcc iron could be categorized as a body-centered tetragonal (bct) with a c/a ratio of 1.414. Why do we use fcc rather than bct?

P_3 Gray tin has the structure of diamond (Fig. 2–2.4). White tin has a body-centered tetragonal structure with atoms at

$$0,0,0 \quad \frac{1}{2},\frac{1}{2},\frac{1}{2} \quad \frac{1}{2},0,\frac{1}{4} \quad 0,\frac{1}{2},\frac{3}{4}.$$

Sketch the unit cell. Show that the second and third sites listed above are not equivalent, but the third and fourth sites are equivalent.

Q_3 How many directions are in the $\langle 012 \rangle$ family of a cubic crystal? (We usually consider $[uvw]$ and $[\bar{u}\bar{v}\bar{w}]$ to be the same, since a reversal of all indices simply follows the same ray but in the opposite sense; that is, $[\bar{u}\bar{v}\bar{w}] = -[uvw]$.)

- R$_3$ Repeat Q$_3$ but for a tetragonal crystal. (Recall ($c \neq a$) in tetragonal; therefore we cannot permutate the 2 with the 0 and 1, as we can in the cubic system.)

  S$_3$ The hcp unit cell of Fig. 3–3.2(b) can be redrawn as a rhombic lattice. (Cf. Fig. 3–3.1 b.) How many equivalent sites are there per unit cell? (Do the sites you picked have identical neighbors in identical directions and at identical distances? If not, they are not equivalent sites. Work Study Problem 3–5.10.)

  T$_3$ Which of the following directions lie in the (110) plane?

$$[112] \quad [1\bar{1}0] \quad [001] \quad [1\bar{1}2] \quad [8\bar{8}9]$$

- U$_3$ Show that the dot product of $[uvw]$ and (hkl) equals zero when the direction lies within the plane.

- V$_3$ The $[111]$ direction is normal to the (111) plane in a *cubic* crystal, but not in a tetragonal crystal. Why?

- W$_3$ With Fig. 3–7.2(c), prove to yourself that second-order ($n = 2$) diffraction for d_{110} equals first-order ($n = 1$) diffraction for d_{220}.

  X$_3$ The atomic packing is independent of atom size in metals. Under what conditions would the packing factor for a sand within a container be independent of sand size?

  Y$_3$ As stated in Section 3–1, the majority of metals are cubic; however, few molecular solids are cubic. Suggest a reason why.

- Z$_3$ We consider (020) to be comparable to (010); but we differentiate between d_{020} and d_{010}. Why?

STUDY PROBLEMS

3–1.1 The unit cell of aluminum is cubic with $a = 0.4049$ nm. From its density, calculate how many atoms there are per unit cell.

Answer: 4

3–1.2 There are two atoms per unit cell in zinc. What is the volume of the unit cell?

3–1.3 Tin is tetragonal with $c/a = 0.546$. There are 4 atoms/unit cell. (a) What is the unit cell volume? (b) What are the lattice dimensions?

Answer: $c = 0.318$ nm, $a = 0.582$ nm

3–1.4 Gold foil is 0.08 mm thick and 670 mm^2 in area. (a) It is cubic with $a = 0.4076$ nm. How many unit cells are there in the foil? (b) What is the mass of each unit cell if the density is 19.32 Mg/m^3 ($= 19.32$ g/cm^3)?

3–2.1 Calculate the atomic packing factor of a bcc metal.

Answer: 0.68

3–2.2 The volume of the unit cell of chromium in Example 3–1.1 is 24×10^{-30} m^3, or 0.024 nm^3. Based on the data used in that example problem, plus PF$_{bcc\ metal} = 0.68$, calculate the radius of the chromium atom to verify that of Appendix B.

3–2.3 Silver is fcc and its atomic radius is 0.1444 nm. How large is the side of its unit cell?

Answer: 0.4084 nm

3–2.4 Nickel is fcc with a density of 8.9 Mg/m³ (= 8.9 g/cm³). (a) What is the volume per unit cell based on the density? (b) From your answer in part (a), calculate the radius of the nickel atom. Check this with the radius listed in Appendix B.

3–2.5 Titanium is bcc at high temperatures and its atomic radius is 0.145 nm. (a) How large is the edge of the unit cell? (b) Calculate the density.

Answer: (a) 0.335 nm   (b) 4.24 Mg/m³ (= 4.24 g/cm³)

3–2.6 Lead is fcc and its atomic radius is 1.750×10^{-10} m. What is the volume of its unit cell?

3–2.7 Verify the density of tungsten from the other data in Appendix B.

Answer: 19.41 Mg/m³ (= 19.41 g/cm³)

3–2.8 Barium is bcc with a density of 3.6 Mg/m³ (= 3.6 g/cm³). (a) Calculate the center-to-center distance between closest atoms. (b) What is the edge dimension of the unit cell? (Its atomic number is 56 and its atomic weight is 137.3 amu.)

3–2.9 Magnesium oxide (MgO) has the same structure as NaCl (Fig. 3–1.1). (a) What is its lattice constant a? (b) Its density? (*Note:* MgO is an ionic compound; therefore, we must use ionic radii rather than atomic radii.)

Answer: 3.8 Mg/m³ (= 3.8 g/cm³)

3–2.10 Calculate the density of NaCl.

3–2.11 Refer to Fig. 3–1.1. (a) What is the distance between the centers of the closest Cl^- ions? (b) What is the distance from the center of the Na^+ ion to the center of the *second-closest* Cl^- ion? (c) What is the distance between the surfaces of the closest Cl^- ions? The closest Na^+ ions?

Answer: (a) 0.39 nm   (b) 0.48 nm   (c) 0.03 nm, 0.2 nm

3–2.12 Show in tabular form the relationship between atom (or ionic) radii and unit-cell dimensions for face-centered and body-centered.

	bcc_{metal}	fcc_{metal}	fcc_{NaCl}
Side of unit cell	$a = 4R/\sqrt{3}$		
Face diagonal			
Body diagonal			

3–2.13 (a) How many atoms are there per mm³ in solid strontium? (b) What is the atomic packing factor? (c) It is cubic. What is its space lattice? (Atomic number = 38; atomic mass = 87.62 amu; atomic radius = 0.215 nm; ionic radius = 0.127 nm; density = 2.6 Mg/m³.)

Answer: (a) 1.78×10^{19}/mm³   (b) 0.74

3–2.14 (a) How many atoms are there per mm³ in solid tantalum? (b) What is the atomic packing factor? (c) It is cubic. What is its space lattice? (Atomic number = 73; atomic mass = 180.95 amu; atomic radius = 0.1429 nm; ionic radius = 0.068 nm; density = 16.6 Mg/m³.)

3–3.1 From the data for cobalt (hcp) in Appendix B, calculate the volume of the unit cell (Fig. 3–3.2).

Answer: 0.066 nm^3

3–3.2 Osmium is hcp with an average radius of 0.137 nm. Calculate its unit cell volume.

3–3.3 The atomic mass of zirconium is 91.22 amu, and the average atomic radius of this hcp metal is 0.16 nm. (a) What is the volume associated with each atom (sphere + interstitial space)? (b) From the answer in part (a), calculate the density.

Answer: (a) 23.2 × 10^{-30} m^3 (or 0.0232 nm^3) (b) 6.5 Mg/m^3 (=6.5 g/cm^3)

3–3.4 Zinc has an hcp structure. The height of the unit cell is 0.494 nm. The centers of the atoms in the base of the unit cell are 0.2665 nm apart. (a) How many atoms are there per hexagonal unit cell? (Show reasoning.) (b) What is the volume of the hexagonal unit cell? (c) Is the calculated density greater or less than the actual density of 7.135 Mg/m^3?

3–3.5 Refer to the data of Study Problem 3–3.4. (a) What is the c/a ratio for zinc? ●(b) What are the *two* interatomic distances between adjacent atoms?

Answer: (a) 1.85 (b) 0.2665 nm, 0.291 nm

3–3.6 Magnesium is hcp with nearly spherical atoms having a radius of ~0.161 nm. Its c/a ratio is 1.62. (a) Calculate the volume of the unit cell from these data. (b) Check your answer by using it to calculate the density of magnesium.

3–4.1 The lattice constant a for diamond (Fig. 2–2.4b) is 0.357 nm. What percent volume change occurs when it transforms to graphite (ρ = 2.25 Mg/m^3, or 2.25 g/cm^3)?

Answer: 56 v/o expansion

3–4.2 The volume of a unit cell of bcc iron is 0.02464 nm^3 at 912°C. The volume of a unit cell of fcc iron is 0.0486 nm^3 at the same temperature. What is the percent change in density as the iron transforms from bcc to fcc?

3–4.3 Titanium is bcc in its high-temperature form. The radius increases 2% when the bcc changes to hcp during cooling. What is the percentage volume change? (*Recall* that there will be a change in atomic packing factor.)

Answer: −2.5 v/o (=−0.8 l/o)

3–4.4 Metallic tin has a tetragonal structure with a = 0.5820 nm and c = 0.3175 nm and with four atoms per unit cell. Another form of tin (gray) has the cubic structure of diamond (Fig. 2–2.4b) with a = 0.649 nm. What is the volume change as tin transforms from gray to metallic?

3–4.5 The carbon layers in graphite (Fig. 2–2.6) are 0.3348 nm apart. Within the layers, the closest center-to-center distance of carbon atoms is 0.142 nm. (a) Will the density of graphite be less than, or greater than, the density of diamond (Fig. 2–2.4b)? (b) What percent? ($a_{diamond}$ = 0.357 nm.) (*Hint:* Sketch the plan view of one sheet of graphite and lay out a unit area.)

Answer: (a) ρ_{gr} = 2.25 Mg/m^3 (or 2.25 g/cm^3) (b) 64% of diamond

3–5.1 Refer to Fig. 3–2.1. We may identify the structure of a bcc metal with only two locations. What are they?

Answer: 0,0,0 and $\frac{1}{2},\frac{1}{2},\frac{1}{2}$. (The remaining atoms are at redundant translations.)

3–5.2 Refer to Fig. 3–2.3. We may identify the structure of an fcc metal with four locations. What are they?

3–5.3 Copper is fcc and has a lattice constant of 0.3615 nm. What is the distance between the $1,0,0$ and $0,\frac{1}{2},\frac{1}{2}$ locations?

Answer: 0.4427 nm

3–5.4 MgO has the same structure as NaCl (Fig. 3–1.1). Its lattice constant is ~ 0.42 nm. What is the distance from the Mg^{2+} ion at $0,0,\frac{1}{2}$ and (a) its nearest Mg^{2+} neighbor? (b) its second-nearest Mg^{2+} neighbor?

3–5.5 When a copper atom is located at the origin of an fcc unit cell, a small interstitial hole is centered at $\frac{1}{4},\frac{1}{4},\frac{1}{4}$. Where are there other equivalent holes?

Answer: $\frac{3}{4},\frac{3}{4},\frac{1}{4}$ $\frac{3}{4},\frac{1}{4},\frac{3}{4}$ $\frac{1}{4},\frac{3}{4},\frac{3}{4}$ for a total of four per unit cell if we count $\frac{1}{4},\frac{1}{4},\frac{1}{4}$.

3–5.6 Repeat Study Problem 3–5.5, but start with $\frac{3}{4},\frac{3}{4},\frac{3}{4}$ rather than $\frac{1}{4},\frac{1}{4},\frac{1}{4}$.

3–5.7 Each atom in bcc iron has eight nearest neighbors. (a) How many *second* nearest neighbors are there? (b) If a_{bcc} for iron is 0.286 nm, what are the distances to their second nearest neighbors?

Answer: (a) 6   (b) 0.286 nm

3–5.8 Refer to Study Problem 3–5.5, and Fig. 3–2.3. Observe that locations $\frac{1}{4},\frac{1}{4},\frac{1}{4}$ and $\frac{3}{4},\frac{3}{4},\frac{3}{4}$ both lie among four immediate neighbors. Are they equivalent locations? In other words, why was $\frac{3}{4},\frac{3}{4},\frac{3}{4}$ not among the answers for Study Problem 3–5.5?

3–5.9 The structure of β' brass is simple cubic as shown in Fig. 3–5.4. Estimate its density if the radii of the Cu and Zn atoms are 0.13 nm and 0.14 nm, respectively.

Answer: 7.1 Mg/m³ ($= 7.1$ g/cm³)

3–5.10 Sketch a 2-D plan view (looking down from above) of the atoms in the lower layer of Fig. 3–3.2. Place a dot at locations on your plan view for each atom in the middle layer. Place a × at locations on your plan view for each atom in the upper layer. Do the dots and the ×'s occupy equivalent sites in the crystal?

3–5.11 Refer to Study Problem 3–5.10. What are the a_1, a_2, and c unit-cell locations for the front atom in the middle layer?

Answer: $a_1 = \frac{2}{3}$, $a_2 = \frac{1}{3}$, $c = \frac{1}{2}$

3–5.12 Sketch a 2-D plan view (looking down the [111] direction) of the (111) plane of an fcc crystal. Place a dot at locations on your plan view for each atom in the overlying (111) plane. Place a × at locations in your plan view for each atom in the second overlying layer. Do the dots and the ×'s occupy equivalent sites in the crystal? Compare your answer with Study Problem 3–5.10.

3–6.1 (a) A ray in the [122] direction passes through the origin. Where does it leave the reference unit cell? (b) Another parallel [122] direction leaves the reference unit cell at $1,1,1$. Where did it enter the reference unit cell?

Answer: (a) $\frac{1}{2},1,1$ (b) $\frac{1}{2},0,0$

3–6.2 A ray in the [111] direction passes through location $\frac{1}{2},\frac{1}{2},0$. What are two other locations along its path?

3–6.3 Draw a line from $\frac{1}{2},\frac{1}{2},0$ to the center of the *next* unit cell, which could be indexed $\frac{1}{2},\frac{3}{2},\frac{1}{2}$. What is the direction?

Answer: [021]

3–6.4 A line is drawn through a tetragonal unit cell from location $\frac{1}{2},0,0$ to $\frac{1}{8},\frac{3}{4},\frac{3}{4}$. What is the index of that direction?

3–6.5 (a) In a cubic crystal, what is the tangent of the angle between the [110] direction and the [111] direction? (b) The [001] direction and the [112] direction?

Answer: (a) $a/(a\sqrt{2}) = 0.71$ (b) $(a\sqrt{2})/(2a) = 0.71$

3–6.6 (a) In a cubic crystal, what is cos [113] ⋇ [110]? (b) What is sin [001] ⋇ [112]?

3–6.7 What is the angle between [111] and [1$\bar{1}$1] in a cubic crystal? (*Hint:*[101] bisects this angle.)

Answer: 70.5°

3–6.8 (a) What is the angle between [101] and [$\bar{1}$01] in a cubic crystal? (b) In a tetragonal crystal where $c/a = 0.55$?

● **3–6.9** What is the linear density of atoms in (a) the ⟨100⟩ directions of copper (fcc, $a = 0.361$ nm)? (b) The ⟨100⟩ directions of iron (bcc, $a = 0.286$ nm)?

Answer: (a) 2.77×10^6 Cu/mm   (b) 3.50×10^6 Fe/mm

● **3–6.10** What is the linear density of atoms in (a) the [221] direction of copper? (b) The [221] direction of iron? (See Study Problem 3–6.9 for lattice constants.)

3–6.11 Draw a line from location $\frac{1}{4},\frac{1}{4},\frac{3}{4}$ to the center of the next unit cell in the rear ($-x$ direction). (a) What is the [*uvw*] direction between these two points? (b) The unit cell is tetragonal with $a = 0.31$ nm and $c = 0.33$ nm. What is the distance between the two points in part (a)?

Answer: (a) [$\bar{3}$1$\bar{1}$]   (b) 0.259 nm

3–6.12 Draw a line from the $\frac{1}{2},0,\frac{1}{2}$ location of aluminum through the center of the base of next unit cell (*x*-direction). (a) What is the direction? ●(b) What is the linear density of atoms in that direction? (c) On a sketch, show a parallel direction that passes through the 0,0,1 location. Indicate two other unit cell locations through which this parallel line passes.

● **3–6.13** Metallic tin is bct with $a = 0.5820$ nm and $c = 0.3175$ nm. (a) What is the linear density of equivalent sites in the ⟨101⟩ directions? (b) The ⟨110⟩ directions? (c) The ⟨201⟩ directions?

Answer: (a) 1.50×10^6/mm   (b) 1.21×10^6/mm   (c) 0.83×10^6/mm

● **3–6.14** Refer to the data of Study Problem 3–6.13. There are four atoms per unit cell of metallic tin. These may be described as being at 0,0,0 and $\frac{1}{2},0,\frac{1}{4}$, and their $\pm\frac{a}{2},\pm\frac{a}{2},\pm\frac{c}{2}$ translations. (a) Make a sketch of the metallic tin unit cell showing the location of all atoms. (b) How far apart are the centers of the closest tin atoms?

● **3–6.15** Refer to Study Problems 3–6.13 and 3–6.14. What is the linear density of tin atoms in (a) the ⟨110⟩ directions? (b) the ⟨201⟩ directions?

Answer: (a) 1.21×10^6/mm   (b) 1.66×10^6/mm

3–6.16 How many directions are there in (a) the ⟨100⟩ family of a cubic crystal? (b) The ⟨110⟩ family? (c) The ⟨111⟩ family?

3–6.17 (a) How many directions are there in the ⟨210⟩ family of a cubic crystal? (b) Of a tetragonal crystal?

Answer: (a) 12   (b) 4

3–7.1 A plane intercepts the crystal axes at $a = 0.5$ and $b = 1$. It is parallel to the z-axis. What are the Miller indices?

Answer: (210)

3–7.2 A plane intercepts axes at $a = 2$, $b = 1$, and $c = 1$. What are the Miller indices?

3–7.3 Give the indices for a plane with intercepts at $a = -2$, $b = \frac{2}{3}$, and $c = \frac{3}{2}$.

Answer: ($\bar{3}$94)

3–7.4 What are the indices for a plane with intercepts at $a = \frac{1}{2}$, $b = -\frac{3}{2}$, and $c = \frac{1}{3}$.

3–7.5 What are the axial intercepts for a (211) plane that passes through the location, 0,2,0?

Answer: $a = 1$, $b = 2$, $c = 2$

3–7.6 A (120) plane contains the location 1,1,1. Where does it intercept the three axes?

3–7.7 What are the axial intercepts for a (111) plane that passes through the center of the unit cell, that is, $\frac{1}{2},\frac{1}{2},\frac{1}{2}$?

Answer: $a = 1.5$, $b = 1.5$, $c = 1.5$

3–7.8 A ($33\bar{1}$) plane contains locations 0,0,0 and $-1,1,0$. What is another crystal location (of many) that lies in this plane?

3–7.9 A plane includes unit-cell locations 0,0,0 and 0,1,0 and $\frac{1}{2},1,\frac{1}{2}$. What are its Miller indices?

Answer: ($\bar{1}$01), or (10$\bar{1}$)

3–7.10 A plane includes unit-cell locations $\frac{1}{2},\frac{1}{2},0$; 1,0,0; and $0,\frac{1}{2},\frac{1}{2}$. What are the Miller indices?

3–7.11 (a) How many atoms are there per mm^2 in the (100) plane of silver? (b) The (110) plane? (c) The (111) plane?

Answer: (a) 12×10^{12}/mm^2 (b) 8.5×10^{12}/mm^2 (c) 14×10^{12}/mm^2

3–7.12 (a) How many atoms are there per mm^2 in the (100) plane of nickel? (b) The (110) plane? (c) The (111) plane?

3–7.13 (a) How many atoms are there per mm^2 in the (111) plane of iron? (b) The (210) plane of silver?

Answer: (a) 7×10^{12}/mm^2 (b) 5.4×10^{12}/mm^2

3–7.14 (a) Sketch a unit cell of silver (fcc) and shade the (012) plane. (b) What is the planar density of atoms?

3–7.15 (a) Sketch a unit cell of chromium (bcc) and shade the (102) plane. (b) What is the planar density of atoms?

Answer: (b) 5.4×10^{12}/mm^2

• **3–7.16** (a) How many atoms are there per mm^2 in the (110) plane of diamond? (b) The (111) plane? (See Fig. 2–2.4(b); $a = 0.357$ nm.)

• *3–7.17* (a) What is the line of intersection between the (110) plane and ($1\bar{1}1$) plane of a cubic crystal? (b) Of a tetragonal crystal?

Answer: (a) [$\bar{1}$12] (b) [$\bar{1}$12]

• **3–7.18** (a) What is the line of intersection between the (112) plane and the (010) plane? (b) The (112) plane and the (110) plane?

• 3–7.19 What are the (hkil) indices that intercept axes at $a_1 = 1$; $a_2 = 1$, and $c = 0.5$?

Answer: $(11\bar{2}2)$

• 3–7.20 Which of the atoms sketched in Fig. 3–3.2(a) lie in the $(10\bar{1}1)$ plane that intercepts the vertical axis at $c = 1$? (Be very careful. It is easy to mislead yourself in answering this question.)

• 3–7.21 (a) List the planes that belong to the {100} family in tetragonal crystals. (b) The {001} family.

Answer: (a) $(100) (010) (\bar{1}00) (0\bar{1}0)$
(b) $(001) (00\bar{1})$

3–7.22 List the twelve planes of the {110} family in cubic crystals.

3–7.23 (a) What ⟨110⟩ directions lie in the (111) plane of copper? (b) In the $(1\bar{1}1)$ plane?

Answer: (a) $[1\bar{1}0], [\bar{1}01], [01\bar{1}]$, and their inverses
(b) $[110], [\bar{1}01], [011]$, and their inverses

3–7.24 (a) What ⟨111⟩ directions lie in the (110) plane of iron? (b) In the $(1\bar{1}0)$ plane?

• 3–8.1 The lattice constant for a unit cell of aluminum is 0.4049 nm. (a) What is d_{220}? (b) d_{111}? (c) d_{200}?

Answer: (a) 0.1431 nm   (b) 0.2337 nm   (c) 0.2024 nm

• 3–8.2 Nickel is face-centered cubic with an atom radius of 0.1246 nm. (a) What is the d_{200} spacing? (b) The d_{220} spacing? (c) The d_{111} spacing?

• 3–8.3 The distance between (110) planes in a body-centered cubic metal is 0.203 nm. (a) What is the size of the unit cell? (b) What is the radius of the atoms? (c) What might the metal be?

Answer: (a) 0.287 nm   (b) 0.124 nm   (c) bcc iron or Cr (not Ni)

• 3–8.4 The d_{111} interplanar spacing in an fcc metal is 0.235 nm. (a) What is the size of the unit cell? (b) What is the radius of the atoms? (c) What might the metal be?

• 3–8.5 X-rays with a wavelength of 0.058 nm are used for calculating d_{200} in nickel. The diffraction angle 2θ is 19°. What is the size of the unit cell? ($n = 1$.)

Answer: 0.35 nm

• 3–8.6 A sodium chloride crystal is used to measure the wavelength of some x-rays. The diffraction angle 2θ is 10.2° for the d_{111} spacing of the chloride ions. What is the wavelength? (The lattice constant is 0.563 nm.)

• 3–8.7 The first line (lowest 2θ) of Fig. 3–8.3(b) is for the d_{110} spacing in tungsten. (a) Determine the diffraction angle, 2θ, graphically. (b) The radius of the tungsten atom is 0.1367 nm. What wavelength was used? (c) The second diffraction line is for what d_{hkl} spacing? ($n = 1$.)

Answer: (a) 41°   (b) 0.156 nm   (c) $h^2 + k^2 + l^2 = 4$, $\therefore d_{200}$

• 3–8.8 The first line (lowest 2θ) of Fig. 3–8.3(a) is for the d_{111} spacing of copper. (a) What is its 2θ value? (b) The radius of the copper atom is 0.1278 nm. Calculate the wavelength of the x-rays. (c) The second diffraction line is for what d_{hkl} spacing?

CHAPTER 4

Atomic Disorder in Solids

PREVIEW

Nature is not perfect: inherently there is some disorder present. Thus the structures just described in Chapter 3 are subject to exceptions. Often these exceptions are minor: maybe one atom out of 10^{10} is out of place. Even so, they can become important. Imperfections account for the behavior of semiconductors, for the ductility of metals, and for the strengthening of alloys. Imperfections account for the color of sapphire. They also permit the movement of atoms during heat treating, so that new structures and enhanced properties may be realized.

This chapter considers *impurities* first, then *crystalline imperfections*. Some solids are so disordered that we do not detect crystallinity. Finally we will look at *atomic vibrations and movements* within solids.

CONTENTS

4–1 **Impurities in Solids.**

4–2 **Solid Solutions in Metals:**
substitutional, interstitial.

•4–3 **Solid Solutions in Compounds:**
nonstoichiometric compounds.

4–4 **Imperfections in Crystals:**
point defects, line defects (dislocations), surfaces, grain boundaries,
grain-boundary area and grain size.

4–5 **Noncrystalline Materials:**
liquids, glasses, phases.

4–6 **Atomic Vibrations:**
thermal expansion, thermal energy distribution.

4–7 **Atomic Diffusion:**
diffusivity, diffusivities versus temperature.

•4–8 **Diffusion Processes.**

STUDY OBJECTIVES

1 To modify the knowledge you obtained in Chemistry on liquid solutions to
new concepts on solid solutions. (These include substitutional and interstitial
solid solutions and nonstoichiometry.)

2 To identify imperfections as 0-dimensional (vacancies and interstitials), as
1-dimensional (dislocations), or as 2-dimensional (boundaries and surfaces),
and to explain the accompanying energy requirements in terms of interatomic
distances.

3 To be able to index quantitatively the grain boundary area within a solid
(and to be familiar with the empirical ASTM "grain-size index").

4 To differentiate between the changes that accompany (a) the
solidification of a crystalline material, and (b) the solidification of an
amorphous (noncrystalline) material.

5 To know the common factors that affect the diffusivity of atom
movements. These are significant in manufacturing processes and heat
treatments.

4–1 IMPURITIES IN SOLIDS

We admire "the real thing"; thus we commonly prefer pure wool, refined sugar, and like to think of 24-carat gold. Although these ideals may be noble, there are instances where, because of cost, availability, or properties, it is desirable to have impurities present. An example is *sterling silver*, which contains 7.5% copper and only 92.5% silver (Example 2–1.1). This material, which we rate highly, could be refined to well over 99% purity.* It would cost more; however, it would be of inferior quality. Without altering its appearance, the 7.5% Cu makes the silver stronger, harder, and therefore more durable—at a lower cost!

Of course, we must consider the properties pertinent to our design. Zinc added to copper produces *brass*, again at a lower cost than the pure copper. Brass is harder, stronger, and more ductile than copper. On the other hand, brass has lower electrical conductivity than copper, and so we use the more expensive pure copper for electrical wiring and similar applications where conductivity is important.

Alloys are combinations of two or more metals into one material. These combinations may be *mixtures* of two kinds of crystalline structures (e.g., bcc iron and Fe_3C in a steel bridge beam). Alternatively, alloys may involve *solid solutions*, which will be exemplified in the next section by brass. Although the term of alloy is generally not used specifically, various combinations of two or more oxide components may be incorporated advantageously into ceramic products (e.g., in the sparkplug insulator of Fig. 2–7.3). Likewise, the plastic telephone casing of Fig. 2–7.2 contains a combination of several types of molecules.

4–2 SOLID SOLUTIONS IN METALS

Solid solutions form most readily when the solvent and solute atoms have similar sizes and comparable electron structures. For example, the individual metals of brass—copper and zinc—have atomic radii of 0.1278 nm and 0.139 nm, respectively. They both have 28 subvalent electrons and they each form crystal structures of their own with a coordination number of 12. Thus, when zinc is added to copper, it substitutes readily for the copper within the fcc lattice, until a maximum of nearly 40 percent of the copper atoms has been replaced. In this solid solution of copper and zinc, the distribution of zinc is entirely random (see Fig. 4–2.1).

Substitutional solid solutions The solid solution described above is called a *substitutional* solid solution because the zinc atoms substitute for copper atoms in the crystal structure. This type of solid solution is quite common among various metal systems. The solution of copper and nickel to form cupronickel is another example. Any fraction of the atoms in the original copper structure may be replaced by nickel. Copper–nickel solid solutions may range from no nickel and 100 percent copper, to 100 percent nickel and no copper. All copper–nickel alloys are face-centered cubic.

---

* As stated in the footnote with Example 2–1.1, weight percent (w/o) is implied in liquids and solids *unless specifically stated otherwise.*

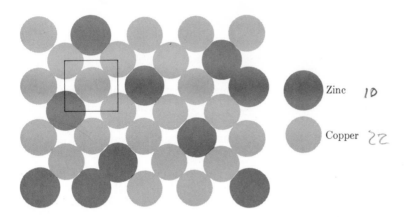

Zinc *10*

Copper *22*

Fig. 4–2.1 Random substitutional solid solution (zinc in copper, i.e., brass). The crystal pattern is not altered.

On the other hand, there is a very definite limit to the amount of tin that may replace copper to form *bronze*, and retain the face-centered cubic structure of the copper. Tin in excess of the maximum amount of *solid solubility* must form another phase. This *solubility limit* will be considered in more detail in Chapter 9.

If there is to be extensive replacement in a substitutional type of solid solution, the atoms must be nearly the same size. Nickel and copper have a complete range of solutions because both of their individual structures are fcc and their radii are 0.1246 nm and 0.1278 nm, respectively. As the difference in size increases, less substitution can occur. Only 20 percent of copper atoms can be replaced by aluminum, because the latter has a radius of 0.1431 nm as compared with only 0.1278 nm for copper. Extensive solid solubility rarely occurs if there is more than about 15 percent difference in radius between the two kinds of atoms. There is further restriction in solubility when the two components have different crystal structures or valences.

The limiting factor is the *number* of substituted atoms rather than the *weight* of the atoms that are substituted. However, engineers ordinarily express composition as weight percent. It is therefore necessary to know how to convert weight percent to atomic percent, and vice versa (see Example 2–1.1).

• **Ordered solid solutions**   Figure 4–2.1 shows a *random substitution* of one atom for another in a crystal structure. In such a solution, the chance of one element occupying any particular atomic site in the crystal is equal to the atomic percent of that element in the alloy. In that case, there is no *order* in the substitution of the two elements.

However, it is not unusual to find an *ordering* of the two types of atoms into a specific arrangement. Figure 4–2.2 shows an ordered structure in which most dark "atoms" are surrounded by light "atoms." Such ordering is less common at higher temperatures, since greater thermal agitation tends to destroy the orderly arrangement.

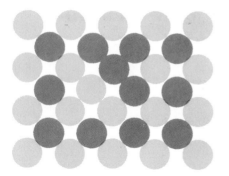

Fig. 4–2.2 Ordered substitutional solid solution. The majority (but not all) of the atoms are coordinated among atoms unlike themselves. If ordering is complete, a compound is formed (Section 4–3).

Interstitial solid solutions   In another type of solid solution, illustrated in Fig. 4–2.3, a small atom may be located in the interstices between larger atoms. Carbon in iron is an example. At temperatures below 912°C (1673°F), pure iron occurs as a body-centered cubic structure. Above 912°C (1673°F), there is a temperature range in which iron has a face-centered cubic structure. In the face-centered cubic lattice, a relatively large *interstice*, or "hole," exists in the center of the unit cell. Carbon, being an extremely small atom, can move into this hole to produce a solid solution of iron and carbon. At lower temperatures, where the iron has a body-centered cubic structure, the interstices between the iron atoms are much smaller. Consequently the solubility of carbon in body-centered cubic iron is very limited (Section 9–6).

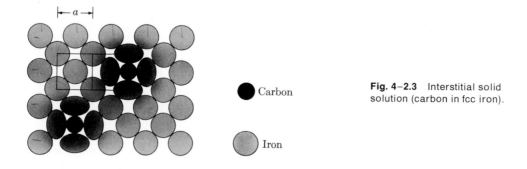

Carbon

Iron

Fig. 4–2.3 Interstitial solid solution (carbon in fcc iron).

Example 4–2.1 Bronze is a solid-solution alloy of copper and tin in which 3%, more or less, of the copper atoms are replaced by tin atoms. The fcc unit cell of copper is retained, but expanded a bit because the tin atoms have a radius of approximately 0.151 nm. (a) What is the weight percent in a 3 a/o tin bronze? (b) Assuming the lattice constant increases linearly with the atomic fraction of tin, what is the density of this bronze?

Solution: Basis: 100 atoms = 25 fcc unit cells.

a)   Mass of copper: 97(63.54 amu) = 6163 amu = 0.945 (or 94.5 w/o);

     mass of tin: 3(118.69 amu) = $\underline{\ 356}$ amu = 0.054 (or 5.4 w/o).

                                                   6519

b)   Average radius = 0.97(0.1278 nm) + 0.03(0.151 nm) = 0.1285 nm;

     $\bar{r}_{cc}$ $a = (4)(0.1285)/\sqrt{2} = 0.3634$ nm.

$$\rho = \frac{6519 \text{ amu}/(0.602 \times 10^{24} \text{ amu/g})}{25(0.3634 \times 10^{-9} \text{ m})^3} = 9.0 \text{ Mg/m}^3 \ (=9.0 \text{ g/cm}^3). \ \blacktriangleleft$$

Example 4–2.2 The maximum solubility limit of tin in bronze is 15.8 w/o at 586°C. What is the atom percent tin?

Solution: Basis: 100,000 amu = 15,800 amu Sn + 84,200 amu Cu.

Number of Cu atoms: 84,200/63.54 = 1325, which is 90.9 a/o Cu;

Number of Sn atoms: 15,800/118.69 = $\underline{\ 133}$, which is 9.1 a/o Sn. $\blacktriangleleft$

                                                  1458

Example 4–2.3 At 1000°C, there can be 1.7 w/o carbon in solid solution with fcc iron (Fig. 4–2.3). How many carbon atoms will there be for every 100 unit cells?

Solution: Basis: 100 unit cells = 400 Fe atoms;

                     (400 Fe)(55.85 amu/Fe) = 22,340 amu.

For 1.7% C,   22,340 (1.7/98.3) = 386.3 amu C;
                                           w/o / w/o
                   386.3 amu/(12.01 amu/C atom) = 32 carbon atoms.

Comment. The carbon atom sits at $\frac{1}{2},\frac{1}{2},\frac{1}{2}$ locations in about one third of the unit cells. Since the carbon atom is slightly larger than the hole, it is not possible to have carbon atoms at all equivalent locations. $\blacktriangleleft$

• 4–3 SOLID SOLUTIONS IN COMPOUNDS

Substitutional solid solutions can occur in ionic phases as well as in metals. In ionic phases, just as in the case of solid metals, atom or ion size is important. A simple example of an ionic solid solution is shown in Fig. 4–3.1. The structure is that of MgO (Fig. 3–1.1) in which the Mg^{2+} ions have been replaced by Fe^{2+} ions. Inasmuch as the radii of the two ions are 0.066 nm and 0.074 nm, complete substitution is possible. On the other hand, Ca^{2+} ions cannot be similarly substituted for Mg^{2+} because their radius of 0.099 nm is comparatively large.*

An additional requirement, which is more stringent for solid solutions of ceramic compounds than for solid solutions of metals, is that the valence charges on the

---

* See Appendix B for ionic radii.

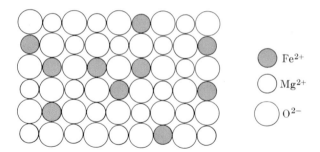

Fig. 4–3.1 Substitutional solid solution in a compound. Fe^{2+} is substituted for Mg^{2+} in the MgO structure.

replaced ion and the new ion must be identical. For example, it would be difficult to replace the Mg^{2+} in MgO with an Li^+, although the two have identical radii, because there would be a net deficiency of charges. Such a substitution could be accomplished only if there were other compensating changes in charge. (See Section 8–4.)

Nonstoichiometric compounds Many compounds have exact ratios of elements (e.g., H_2O, CH_4, MgO, Al_2O_3, Fe_3C, to name but a few). They have a fixed ratio of atoms; thus they are *stoichiometric*. Bonds form between unlike atoms. As a result, the structure becomes even more perfectly ordered than we saw in Fig. 4–2.2.

Other compounds deviate from specific integer ratios for the two (or more) elements which are present. Thus we find that "Cu_2Al" ranges from 31 a/o to 37 a/o Al (16 to 20 w/o Al) rather than being exactly $33\frac{1}{3}$ a/o Al. Likewise, at 1000°C "FeO" ranges from 51 to 53 a/o oxygen, rather than being exactly 50 a/o. We call these compounds *nonstoichiometric* because they do not have a fixed ratio of atoms.

Nonstoichiometric compounds always involve some solid solution. In the Cu_2Al example cited above, the atoms are nearly enough the same size and sufficiently comparable electronically so that with excess aluminum, some of the copper atoms are replaced by aluminum atoms (up to the maximum of 37 a/o Al). Conversely, in the presence of excess copper, the Cu/Al atom ratio reaches 69/31, by substituting a few copper atoms into the aluminum sites of Cu_2Al.

The nonstoichiometry of $Fe_{1-x}O$ arises from a different origin. The iron ions and the oxygen ions are too different to permit any measurable substitution. However, iron compounds always include some ferric (Fe^{3+}) ions along with the ferrous (Fe^{2+}) ions. Thus in order to balance charges there always must be more than 50 a/o oxygen. In fact, each two Fe^{3+} ions require an extra O^{2-} ion; or conversely, each pair of Fe^{3+} ions must be accompanied by a cation vacancy (Fig. 4–3.2). This is called a *defect structure* since there are irregularities in the atom packing. At 1000°C and with iron saturation, the composition is $Fe_{0.96}O$; with oxygen saturation, the composition is $Fe_{0.88}O$.

These defect structures will be important to us in the next chapter when we discuss electrical conduction. Also, we will observe later in this chapter that the atoms or ions can diffuse more readily through the crystalline solid when vacancies are present.

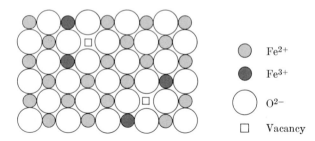

⬤ (light)	Fe^{2+}
⬤ (dark)	Fe^{3+}
◯	O^{2-}
☐	Vacancy

Fig. 4–3.2 Defect structure ($Fe_{1-x}O$). This structure is the same as NaCl (Fig. 3–1.1) except for some iron ion vacancies. Since a fraction of the iron ions are Fe^{3+} rather than Fe^{2+}, the vancancies are necessary to balance the charge. The value of x ranges from 0.04 to 0.16, depending on temperature and the amount of available oxygen.

Example 4–3.1 An iron oxide (Fig. 4–3.2) contains 52 a/o oxygen and has a lattice constant of 0.429 nm. (a) What is the Fe^{2+}/Fe^{3+} ion ratio? (b) What is the density? (This structure is like the NaCl's except for the vacancies, ☐.)

Solution: Basis: 100 ions = 52 O^{2-} + 48 Fe + 4 ☐.

a) Charge balance: $52(2-) + (y)(3+) + (48 - y)(2+) = 0$,

$$y = 8\ Fe^{3+}; \qquad 48 - y = 40\ Fe^{2+};$$

$$Fe^{2+}/Fe^{3+} = 5.$$

b) With 4 anions per unit cell (see Fig. 3–1.1): 13 unit cells.

$$\rho = \frac{[48(55.8) + 52(16)]\ \text{amu}/13\ \text{u.c.}}{(0.602 \times 10^{24}\ \text{amu/g})(0.429 \times 10^{-9}\ \text{m})^3/\text{u.c.}} = 5.7\ \text{Mg/m}^3\ (=5.7\ \text{g/cm}^3).\ \blacktriangleleft$$

• **Example 4–3.2** A β'-brass is nominally an intermetallic compound, CuZn, with the simple cubic structure shown in Fig. 3–5.4. It may also be called a *partially ordered solid solution*, particularly since it is nonstoichiometric with a range of 46 to 50 a/o zinc at 450°C. Assume 90% of the $\frac{1}{2},\frac{1}{2},\frac{1}{2}$ sites of Fig. 3–5.4 are occupied by copper atoms in a 46 a/o Zn–54 a/o Cu alloy. What per cent of the 0,0,0 sites are occupied by copper atoms?

Solution: Basis: 50 unit cells = 50 0,0,0 sites (and 50 $\frac{1}{2},\frac{1}{2},\frac{1}{2}$ sites)

$$= 100\ \text{atoms} = 54\ \text{Cu} + 46\ \text{Zn}.$$

$\frac{1}{2},\frac{1}{2},\frac{1}{2}$ sites: 45 Cu + 5 Zn;   90% Cu (50 sites) = 45 Cu

0,0,0 sites: 9 Cu + 41 Zn.

Therefore, 9 of the 50, or 18% of the 0,0,0 sites, are occupied by copper.

Comments. At low temperatures almost all of the neighbors of zinc atoms are copper atoms, and *vice versa*. However, as the temperature is increased, the atoms start to disorder. In this problem, at 450°C, the 0,0,0 sites are 82 Zn–18 Cu, while the $\frac{1}{2},\frac{1}{2},\frac{1}{2}$ sites are 10 Zn–90 Cu (all atom percents). Thus the two sites are not equivalent, and the structure is simple cubic.

Fig
pg. 84

Above 470°C, this alloy becomes fully random, with no preference for copper atoms to be surrounded by zinc atoms. Under those conditions, the substitutional solid solution (called β-brass rather than β'-brass) is bcc because the unit-cell centers and corners have equal probability for the same average composition. ◀

4–4 IMPERFECTIONS IN CRYSTALS

We have just seen a type of imperfection in crystals where a vacancy is required to accommodate a charge imbalance (Fig. 4–3.2). When imperfections such as vacancies involve one or a few atoms, we call them *point defects*. Other imperfections may be lineal through the crystal; hence, the term *line defects*. They become particularly significant when crystals are plastically deformed by shear stresses. In fact, a small number of these cause metal crystals to be 1000 times more ductile than would be possible in their absence. When present in large numbers, these lineal imperfections increase the strength of the metal. Finally, other imperfections may be two-dimensional in concept and involve external *surfaces* or internal *boundaries*.

Point defects   The simplest point defect is a *vacancy*, which involves a missing atom (Fig. 4–4.1a) within a crystal. Such defects can be a result of imperfect packing during the original crystallization, or they may arise from thermal vibrations of the atoms at elevated temperatures (Section 4–6), because as thermal energy is increased there is an increased probability that individual atoms will jump out of their position

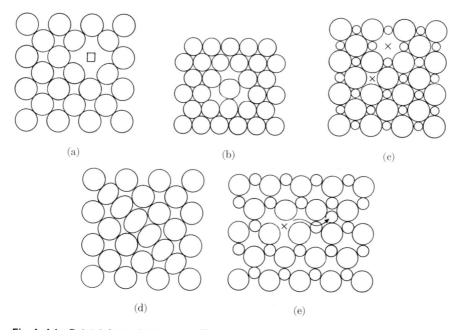

(a)   (b)   (c)

(d)   (e)

Fig. 4–4.1   Point defects. (a) Vacancy, □. (b) Di-vacancy (two missing atoms). (c) Ion-pair vacancy (Schottky defect). (d) Interstitialcy. (e) Displaced ion (Frenkel defect).

of lowest energy. Vacancies may be single, as shown in Fig. 4–4.1(a), or two or more of them may condense into a di-vacancy (Fig. 4–4.1b) or a tri-vacancy.

Ion-pair vacancies (called Schottky imperfections) are found in compounds that must maintain a charge balance (Fig. 4–4.1c). They involve vacancies of pairs of ions of opposite charges. Ion-pair vacancies, like single vacancies, facilitate atomic diffusion (Section 4–7).

An extra atom may be lodged within a crystal structure, particularly if the atomic packing factor is low. Such an imperfection, called an *interstitialcy*, produces atomic distortion (Fig. 4–4.1d).

A *displaced ion* from the lattice into an interstitial site (Fig. 4–4.1e) is called a Frenkel defect. Close-packed structures have fewer interstitialcies and displaced ions than vacancies, because additional energy is required to force the atoms into the interstitial positions.

Line defects (dislocations)   The most common type of line defect within a crystal is a dislocation. An *edge dislocation* is shown in Fig. 4–4.2. It may be described as an edge of an extra plane of atoms within a crystal structure. Zones of compression and of tension accompany an edge dislocation (Fig. 4–4.3) so that there is a net increase in energy along the dislocation. The displacement distance for atoms around the dislocation is called the *slip vector*, **b**.* This vector is at right angles to the edge dislocation line.

A *screw dislocation* is like a spiral ramp with an imperfection line down its axis (Fig. 4–4.4). Its slip vector is parallel to the defect line. Shear stresses are associated with the atoms adjacent to the screw dislocation; therefore, extra energy is involved here as with the previously cited edge dislocations.

Dislocations of both types may originate during crystallization. Edge dis-locations, for example, arise when there is a slight mismatch in the orientation of adjacent parts of the growing crystal so that an extra row of atoms is introduced or

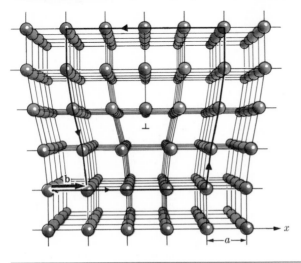

Fig. 4–4.2   Edge dislocation, $\perp$. A linear defect occurs at the edge of an extra plane of atoms. The slip vector **b** is the resulting displacement. (Guy and Hren, *Elements of Physical Metallurgy.*)

---

* Also called a *Burgers vector*.

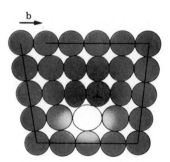

Fig. 4–4.3 Dislocation energy. Atoms are under compression (darker) and tension (light) adjacent to the dislocation.

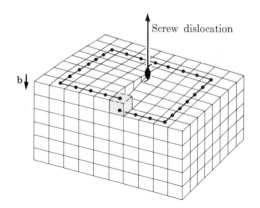

Fig. 4–4.4 Screw dislocation (unit cells shown). The slip vector **b** is parallel to the linear defect.

Fig. 4–4.5 Dislocation formation by shear. (a) The *dislocation line*, D, expands through the crystal until displacement is complete. (b) This defect forms a screw dislocation where the line is parallel to the shear direction. (c) The linear defect is an edge dislocation where the line is perpendicular to the shear direction.

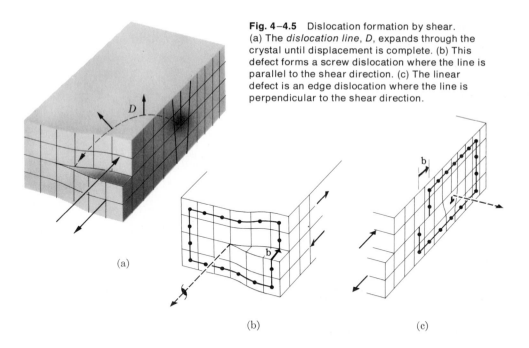

eliminated. As shown in Fig. 4–4.4, a screw dislocation provides for easy crystal growth because additional atoms and unit cells can be added to the "step" of the screw. Thus the term screw is very apt, because the step swings around the axis as growth proceeds.

Dislocations more commonly originate during deformation. We see this in Fig. 4–4.5, where shear is seen to introduce both edge dislocations and screw dislocations.

Both lead to the same final displacement and are in fact related through the dislocation line which forms.

Surfaces   Crystalline imperfections may extend in two dimensions as a boundary. The most obvious boundary is the external *surface*. Although we may visualize a surface as simply a terminus of the crystal structure, we should quickly appreciate the fact that atomic coordination at the surface is not fully comparable to the atoms within a crystal. The surface atoms have neighbors on only one side (Fig. 4–4.6); therefore they have higher energy and are less firmly bonded than the internal atoms. This energy may be rationalized with Fig. 2–5.2 by noting that if additional atoms were to be deposited onto the surface atoms, energy would be released just as it was for the combination of two individual atoms. We find our best visible evidence of this surface energy in the case of liquid drops that have spherical shape to minimize the surface area (and therefore the surface energy) per unit volume. Surface adsorption provides additional evidence of the energy differential at the surface.

Grain boundaries   Although a material such as copper in an electric wire may contain only one phase, i.e., only one structure (fcc), it contains many crystals of various orientations. These individual crystals are called *grains*. The shape of a grain in a solid is usually controlled by the presence of surrounding grains. Within any particular grain, all of the unit cells are arranged with one orientation and one pattern. However, at the *grain boundary* between two adjacent grains there is a transition zone that is not aligned with either grain (Fig. 4–4.7).

Surface

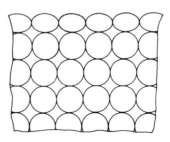

Fig. 4–4.6   Surface atoms (schematic). Since these atoms are not entirely surrounded by others, they possess more energy than internal atoms.

Although we cannot see the individual atoms illustrated in Fig. 4–4.7, we can quite readily locate grain boundaries in a metal under a microscope, if the metal has been treated by *etching*. First the metal is smoothly polished so that a plane, mirrorlike surface is obtained, and then it is chemically attacked for a short period of time. The atoms in the area of transition between one grain and the next will dissolve more readily than other atoms and will leave a line that can be seen with the microscope (Fig. 4–4.8); the etched grain boundary does not act as a perfect mirror as does the remainder of the grain (Fig. 4–4.9).

We may consider that the grain boundary is two-dimensional, although it may be curved, and actually has a finite thickness of 2 or 3 atomic distances. The mismatch of the orientation of adjacent grains produces a less efficient packing of the atoms along the boundary. Thus the atoms along the boundary have a higher energy

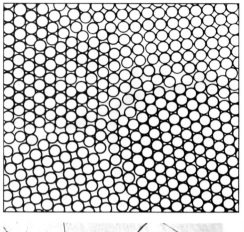

Fig. 4–4.7 Grain boundaries. Note the area of disorder at the boundary. (Reproduced by permission from Clyde W. Mason, *Introduction to Physical Metallurgy*, American Society for Metals, Chapter 3.)

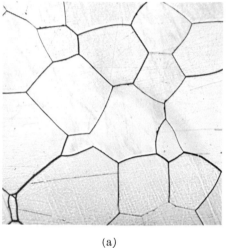

(a)

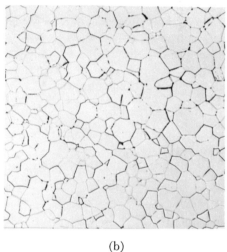

(b)

Fig. 4–4.8 Grain boundaries.
(a) Molybdenum (× 250) (O. K. Riegger).
(b) High-density periclase, MgO (× 250) R. E. Gardner and G. W. Robinson, Jr., *J. Amer. Ceram. Soc.*

To eyepiece

Plane glass

Rays from illuminator

Lens

Reflected portion of rays

Object

Fig. 4–4.9 Grain boundary observation. The metal has been polished and etched. The corroded boundary does not reflect light through the microscope. (Reproduced by permission from B. Rogers, *The Nature of Metals*, 2d ed., American Society for Metals, and Iowa State University Press, Chapter 2.)

than those within the grains. This accounts for the more rapid etching along the boundaries described above. The higher energy of the boundary atoms is also important for the nucleation of polymorphic phase changes (Section 3–4). The lower atomic packing along the boundary favors atomic diffusion (Section 4–7), and the mismatch between adjacent grains modifies the progression of dislocation movements (Fig. 4–4.5). Thus the grain boundary modifies plastic strain of a material.*

Grain boundary area and grain size   It is obvious that the two *microstructures* of Fig. 4–4.8 are different. The grains in Fig. 4–4.8(a) are larger than are the grains in Fig. 4–4.8(b). Conversely, this MgO has more grain-boundary area than does the molybdenum. (Both are at ×250, i.e., their lineal dimensions have been magnified 250 times.)

Since grain boundaries affect a material in a number of ways, it is useful to know the amount of grain-boundary surface per unit volume, S_V. This area can be readily estimated if we place a line randomly across the microstructure. Of course, this line will intercept more grain boundaries in a *fine-grained* material than in a *coarse-grained* material. The relationship is

$$S_V = 2\,P_L, \tag{4–4.1}$$

where P_L is the number of points of intersections per unit length between the line and the boundaries.

We shall not prove the above relationship, but we can demonstrate it in Fig. 4–4.10 where a 50-mm circle has been placed randomly on the previous photo-

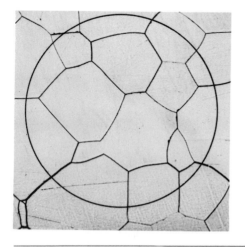

Fig. 4–4.10 Grain-boundary area calculation. Since the magnification is × 250, the length of the circle is π(50mm)/250, or 0.63 mm (0.025 in.). It intersects 11 boundaries in that distance. Therefore, there are 2(11/0.63 mm), or 35 mm², of boundary area per mm³ (see Eq. (4–4.1). (O. K. Riegger.)

---

* At normal temperatures the grain boundaries interfere with slip. Therefore a *fine-grained* material is stronger than a *coarse-grained* material. At elevated temperatures, the boundaries can accommodate the dislocations. As a result the situation is reversed, and creep results (Section 6–7).

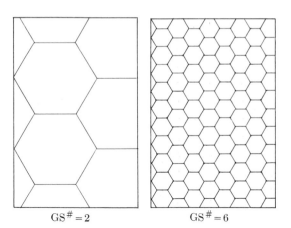

Fig. 4–4.11 Grain-size numbers (ASTM). A large grain-size number, GS#, (Eq. 4–4.2) indicates more grains and more grain-boundary area per unit volume (× 100).

$$GS^{\#} = 2 \qquad\qquad GS^{\#} = 6$$

micrograph of molybdenum. There are 11 intersections. Since the magnification is × 250, the circumference on the metal is actually $50\pi/250 = 0.63$ mm, and the value of P_L is 11/0.63 mm = 17.5/mm. Based on Eq. (4–4.1), the surface area per unit volume S_V is 35 mm^2/mm^3. This may be compared with the boundary area of the MgO in Fig. 4–4.8(b) in Example 4–4.1.

Although the boundaries are the microstructural features that relate to property behavior, it is common to refer to *grain size*. A method to determine a grain-size number has been standardized by the American Society for Testing and Materials (ASTM). Although empirical, it is a quantitative and reproducible index. This index uses **2** as a base:

$$N_{(0.0645\ mm^2)} = 2^{n-1}. \tag{4–4.2}$$

The term N is the number of grains observed in an area of 0.0645 mm^2.* The value n is the *grain-size number* (G.S.#). Example 4–4.2 calculates the G.S.# for the molybdenum shown in Fig. 4–4.8(a). Figure 4–4.11 shows two of a series of grain-size nets that can be used for quick visual assignments of a grain-size number to × 100 photomicrographs. These grain-size numbers are pertinent to the heat treating of steels (Section 11–6) and for the transition temperatures of steels (Section 6–8).

Example 4–4.1 Estimate the grain-boundary area per unit volume in the MgO of Fig. 4–4.8(b).

Solution: Lay a 50-mm straight-edge at random across the figure and count the grain boundaries intersected. If, by repeating this for a total of 5 times, you get counts of 13, 17, 12, 14, and 12, you would have 68 counts per 250 mm: However, the magnification is × 250. Therefore, you have 68/mm.

$$S_V = 2(68/mm) = \sim 140/mm \quad (or\ 140\ mm^2/mm^3).$$

---

* This area equals 1 in^2 with × 100 magnification. The procedure was originally standardized to use a microscope with × 100 lenses, and to count the grains within a 1 in. × 1 in. area (0.0001 in.^{2})

Comments. There is about four times as much grain-boundary area per unit volume in the MgO of Fig. 4–4.8 as in the molybdenum. Both of these are *estimates*. However, with reasonable care, one can be accurate within $\pm 10\%$, which is sufficient for most purposes. ◀

Example 4–4.2 Assign an ASTM G.S.# to the molybdenum of Fig. 4–4.8(a).

Solution: The area of this enlarged figure is $(59 \text{ mm})^2$:

$$\text{Actual area} = (59 \text{ mm}/250)^2 = 0.056 \text{ mm}^2.$$

There are ~ 17 grains in this area. (See comments.)

$$\frac{\sim 17}{0.056 \text{ mm}^2} = \frac{N}{0.0645 \text{ mm}^2};$$

$$N = \sim 20 = 2^{n-1};$$

$$n = 5^+.$$

Comments. The photomicrograph in Fig. 4–4.8 is, itself, a sample and therefore subject to statistical variations. As a result, we should not expect utmost precision. Furthermore, it is seldom necessary to be more specific in our estimates of the grain-size number than ± 0.5.

The number of grains in the area which was sampled is most readily obtained by (1) counting the grains that lie entirely within the area, (2) adding to the count half of the grains at the edges (since these are shared by adjacent areas), and (3) then adding a fourth of each of the four-corner grains. If you took another sample of the metal in Fig. 4–4.8(a), and your counts were as low as 15 or as high as 20, your grain-size estimate would still be $n = 5^+$. The count must double to shift from one G.S.# to the next. ◀

Example 4–4.3 A very accurate measurement was made to four significant figures of the density of aluminum. When cooled rapidly from 650°C, $\rho = 2.698 \text{ Mg/m}^3$. Compare that value with the theoretical density obtained from diffraction analyses where a was determined to be 0.4049 nm.

Solution: Since the atomic weight is 26.98 amu and aluminum is fcc,

$$\rho_{\text{theor.}} = \frac{4(26.98 \text{ amu})/(0.6022 \times 10^{24} \text{ amu/g})}{(0.4049 \times 10^{-9} \text{ m})^3} = 2.700 \text{ Mg/m}^3;$$

$$\frac{\rho}{\rho_{\text{theor.}}} = 2.698/2.700 = 0.999; \quad \text{or } \sim 1 \text{ vacancy per 1000 atoms.}$$

Comments. The match is close. Aluminum, like most metals just below their melting temperatures, has about 1 vacancy in 10^3 atoms. The equilibrium fraction drops to approximately 1 vacancy in 10^7 atoms at half of the absolute melting temperature. ◀

4–5 NONCRYSTALLINE MATERIALS (AMORPHOUS)

Long-range order is absent in some materials of major engineering and scientific importance. Included are all liquids, glass, the majority of plastics, and a few metals if the latter are cooled extremely rapidly from their liquid. In principle, we can view this lack of repetitive structure as a volume, or 3-dimensional, disorder, and as a continuation of our sequence: point defects, lineal defects, and 2-dimensional boundaries. We call these materials *amorphous* (literally "without form") in contrast to crystalline materials.

Liquids   For the most part, liquids are *fluids* (i.e., they *flow* under their own mass). However, just as "molasses in January" gets semirigid, various liquids of technical importance can become very viscous and even solid, without crystallizing.

First let us look at the disorder that occurs in a single-component metal as it approaches melting, then transforms into a liquid. We can use aluminum for our example. As was implied in the last section, the greater thermal energy at higher temperatures introduces not only greater thermal vibrations, but also some vacancies. Just short of the melting point, crystalline aluminum may contain up to 0.1% vacancies in its lattice. When the vacancies approach one percent in a close-packed structure, "turmoil reigns." The regular 12-fold coordination is destroyed and the long-range order of the crystal structure disappears (Fig. 4–5.1).

Fig. 4–5.1   Melting (metal). (a) Crystalline metal with CN = 12. (The 6 in the plane plus 3 above and 3 below.) (b) Liquid metal. Long-range order is lost; CN < 12, and the average interatomic distance increases slightly.

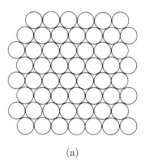
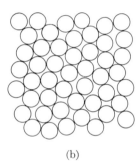

(a)   (b)

This disorder of melting increases the volume of most materials* (Fig. 4–5.2). With this disorder, the number of nearest neighbors drops from 12 to only 11 or 10; however, these are not in a regular pattern, so the space per atom and the average interatomic distance is increased a few percent. Each material requires a characteristic amount of energy to be melted. We call this energy the *heat of fusion*. Since the heat of fusion, ΔH_f, is the energy required to disorganize a mole of atoms, and the melting temperature, T_m, is a measure of the atomic bond strength, we find a general correlation between the two (Table 4–5.1).[†]

Glasses   As indicated earlier, glasses are sometimes considered to be very viscous liquids, inasmuch as they are noncrystalline. However, only a few liquids can actually form glasses. Therefore, in order to make a distinction, we must look at the structure of glass more critically.

---

* It increases the volume of all those materials that are close-packed and that do not have directional bonds. A *few* materials with low packing factors and stereospecific bonds (Chapter 2) collapse into denser structures when they are thermally excited. Water is the prime example of this exception. (And are we lucky!!)

[†] For the correlation to be good, we must have comparable materials, e.g., among metals.

Table 4–5.1

Heats of fusion of metals

Metal	Melting temperature, °C (°F)	Heat of fusion, joule/mole*
Tungsten, W	3387 (6129)	32,000
Molybdenum, Mo	2623 (4753)	28,000
Chromium, Cr	1863 (3385)	21,000
Titanium, Ti	1672 (3042)	21,000
Iron, Fe	1538 (2800)	15,300
Nickel, Ni	1455 (2651)	17,900
Copper, Cu	1084 (1984)	13,500
Aluminum, Al	660 (1221)	10,500
Magnesium, Mg	649 (1200)	9,000
Zinc, Zn	419 (787)	6,600
Lead, Pb	327 (621)	5,400
Mercury, Hg	−38.9 (−38)	2,340

* joule/0.6×10^{24} atoms; 4.18 J = 1 cal.

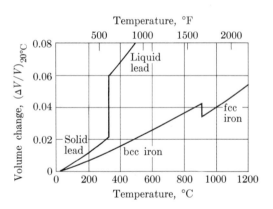

Fig. 4–5.2 Volume changes (based on the volume at 20°C). (a) Lead with ∼4 v/o expansion between fcc and liquid. (b) At 912°C, iron has a volume change between bcc and fcc (Example 3–4.1).

At high temperatures glasses form true liquids. The atoms have freedom to move around and respond to shear stresses. When a commercial glass at its liquid temperature is supercooled, there is thermal contraction caused by atomic rearrangements that produce more efficient packing of the atoms. This contraction (Fig. 4–5.3) is typical of all liquid phases; however, with more extensive cooling, there is an abrupt change in the expansion coefficient of glasses. Below a certain temperature, called the glass transition temperature, or more simply the *glass temperature*, T_g, there are no further rearrangements of the atoms and the only contraction is a result of smaller thermal vibrations. This lower coefficient is comparable to the thermal coefficient in crystals where thermal vibrations are the only factor causing contraction, and no rearrangement occurs.

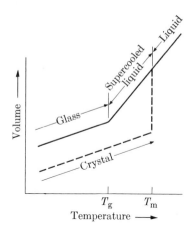

Fig. 4–5.3 Volume changes in supercooled liquids and glasses. When a liquid is cooled, it contracts rapidly and continuously because with decreased thermal agitation, the atoms develop more efficient packing arrangements. In the absence of crystallization the contraction continues below T_m to the glass transition temperature, T_g, where the material becomes a rigid glass. Below T_g, no further rearrangements occur, and the only further contraction is caused by reduced thermal vibrations of the atoms in their established locations.

The term *glass* applies to those materials that have the expansion characteristics of Fig. 4–5.3. Glasses may be either inorganic or organic and are characterized by a short-range order (and an absence of long-range order). Figure 4–5.4 presents one of the simplest glasses (B_2O_3), in which each small boron atom fits among three larger oxygen atoms. Since boron has a valence of three and oxygen a valence of two, electrical balance is maintained if each oxygen atom is located between two boron atoms. As a result, a continuous structure of strongly bonded atoms is developed. Below the glass transition temperature, where the atoms are not readily rearranged, the fluid characteristics are lost and a noncrystalline solid exists. Such a solid has a significant resistance to shear stresses and therefore cannot be considered a true liquid.

Fig. 4–5.4 Structure of B_2O_3 glass. Although there is no long-range crystalline order, there is a short-range coordinational order. Each boron atom is among three oxygen atoms. Each oxygen atom is coordinated with two boron atoms.

The temperature–volume characteristics of Fig. 4–5.3 were first observed in silicate glasses (Section 8–5). However, it soon became apparent that these characteristics have major significance in polymeric materials. Thus in Chapter 7, we will look at the glass transition temperature, T_g, closely. Below the T_g, polymers are hard and brittle and have low dielectric constants. Above the T_g, a plastic becomes flexible and even rubbery, with concurrent changes in its dielectric constant.

Phases Reference has already been made to phases and will be made again numerous times throughout this text. We are now in a position to define a phase as that part of a material that is *distinct from others in structure and/or composition.* Consider "ice–water." While of the same composition, the ice is a crystalline solid with a hexagonal lattice; the water is a liquid. The *phase boundary* between the two locates a discontinuity in structure: they are *separate phases*. Or consider silver-plated copper. Both silver and copper are fcc; however, the silver atoms are sufficiently larger than the copper atoms* so that there is nearly complete composition discontinuity at room temperature. Thus, they form two separate phases.

Commonly, two phases of a material have distinct differences in both composition and structure, e.g., a plastic with a fiberglass reinforcement. In contrast, some phases lose their distinctiveness and dissolve, e.g., after dissolving in coffee, sugar is no longer a separate phase. The same is true of zinc, which, by itself hcp, dissolves in copper (fcc) to produce brass, a single-phase solid solution.

In terms of the above discussion, a *solution* (liquid or solid) is a phase with more than one *component*. A *mixture* is a material with more than one *phase*.

There are many crystalline phases because there are innumerable permutations and combinations of atoms, or groups of atoms. There are relatively few amorphous phases because, lacking long-range order, their atomic arrangements are less definite and permit a greater range of solution than do crystals.[†] There is only one gaseous phase. The atoms or molecules are far apart and randomly distributed; as a result additional vapor components may be introduced into one "structure." No discontinuities are observed in a gas other than at the atomic or molecular level.

Example 4–5.1 From Fig. 4–5.2, calculate the packing factor (a) of solid lead at 326°C; (b) of liquid lead at 328°C. (Assume the hard-ball radius of 0.1750 nm is retained.)

Solution: Lead is fcc with a calculated PF of 0.74 (see Example 3–2.1a); thus an increase in volume of 2.1% at 326°C and 6% at 328°C decreases the PF accordingly.

a) $$PF_{326°C} = 0.74/1.021 = 0.725;$$

b) $$PF_{328°C} = 0.74/1.06 = 0.698.$$

Comment. One could suggest that the radius increases by 0.7 l/o from 0.1750 nm at 20°C to 0.1762 nm at 326°C to give the 2.1 v/o expansion. If so, the packing factor at 326°C would remain at 74%, and the packing factor of the liquid becomes 0.713. One cannot argue against that suggestion, since one definition of radius is one half of the closest interatomic distance. In either event, there is an abrupt discontinuity in packing efficiency at the melting temperature. ◀

4–6 ATOMIC VIBRATIONS

The atoms of a material become static only at 0 K ($-273°C$ or $-460°F$). Under that condition the atoms settle down to their lowest energy positions among their neighbors (Fig. 2–5.2b). As the temperature is increased, the increased energy permits the atoms to vibrate into greater and shorter interatomic distances. However, as

---

* $R_{Cu} = 0.1278$ nm; $R_{Ag} = 0.1444$ nm.
† The fiberglass and plastic just cited are both amorphous. While their structures are not sufficiently similar to produce a single phase, each can be a solvent for large quantities of solutes.

can be noted by the energy well of that earlier figure, the displacements in the two directions are not symmetric; i.e., for a given energy level (temperature), the atoms can move farther apart more readily than they can be pushed together. This produces a *thermal expansion* because the mean interatomic distance is increased.

Even at a given temperature, not all atoms (or molecules) have identical energies at any specific *instant* of time. Rather there is a spectrum of energies among the atoms that extends from near zero to very high values. Of course the majority of the atoms have energies somewhere near the mean value. Conversely, over a period of time, a specific *atom* will experience a range of energies that extends from near zero to very high values. The majority of the time, however, its energy is somewhere near the mean value. We will need to look at this *energy spectrum* so we can predict the probability of an atom having enough energy to break its bond and jump to new locations, a process we call *diffusion*.

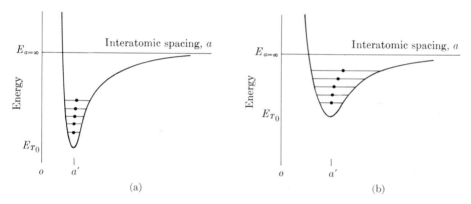

Fig. 4–6.1   Energy and expansion. (a) Strongly bonded solid. (b) Weakly bonded solid. With equal additions of thermal energy, above absolute zero, T_0, the mean interatomic spacing changes less in a material with a deeper energy trough (cf. Fig. 2–5.2b). The expansion becomes more pronounced at higher temperatures (higher energy).

Thermal expansion   In the vicinity of our ambient temperature, the heat capacity of many solids is essentially constant. Therefore, we can assign a succession of equal temperature intervals to the energy trough (Fig. 4–6.1a). With successive energy increases, the median interatomic spacing increases; hence, the thermal expansion we spoke of two paragraphs back. Note two features, however. There is less dimensional change per ΔT in a strongly bonded (high-melting) material characterized by a deep energy trough than in a weakly bonded material (Fig. 4–6.1b). This is illustrated in Fig. 4–6.2, a graphical presentation of the data given in generalization (5) in the Review section of Chapter 2. Figure 4–6.2 emphasizes that our generalization is limited to comparable materials.

The second feature of Fig. 4–6.1 is that, as the temperature increases, the change in dimension becomes relatively more pronounced. The median curves of Fig. 4–6.1

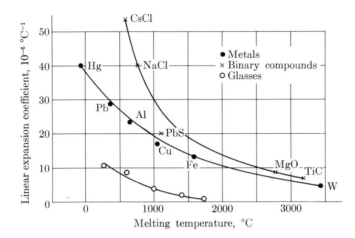

Fig. 4–6.2   Melting temperatures and expansion coefficients (20°C). The more strongly bonded, higher melting materials have lower expansion coefficients (cf. Fig. 4–6.1). Comparisons must be made between similar materials.

bend to the right. This explains the data of Fig. 1–3.1 where the expansion coefficients of copper and iron increased with temperature.

Thermal expansion is *isotropic* only in cubic and amorphous materials. In other crystals, the amount of expansion varies with orientation. Consider graphite (Fig. 2–2.6); the structure is markedly *anisotropic*. As a result, the thermal agitation expands the lattice more in the vertical direction than in the two horizontal directions.* This is not surprising because the bonds between layers are weaker than those within the layers.

Thermal energy distribution   The total kinetic energy, K.E., of a mole of gas increases in proportion to the temperature T, so that the equation

$$\text{K.E.} = \tfrac{3}{2}RT \tag{4–6.1}$$

is appropriate. The R of this equation is the same gas constant encountered in introductory chemistry courses, where its units are commonly reported as 1.987 cal/mole·K. If we switch to joules of the SI units and pay attention to the individuals instead of moles (or 0.602×10^{24}), the value becomes 13.8×10^{-24} J/K. This is called *Boltzmann's constant* and is identified by k. Thus, the average K.E. of an individual molecule of gas is

$$\text{K.E.} = \tfrac{3}{2}kT. \tag{4–6.2}$$

---

* At 700°C, for example, the thermal-expansion coefficient of graphite is only 0.8×10^{-6}/°C within the plane of the layers, but 29×10^{-6}/°C perpendicular to the layers. Both of these values decrease at lower temperatures, and in fact, the first coefficient passes through zero at 400°C and becomes slightly negative at room temperature. The second value ($\perp$ to layers) is 26×10^{-6}/°C at 20°C, so that the overall volume coefficient $\alpha_V \simeq \alpha_x + \alpha_y + \alpha_z$ is positive. (See Example 4–6.1.)

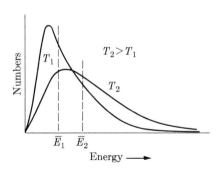

Fig. 4–6.3 Energy distribution. Both the average energy $\bar{E}$ and the fraction with energies in excess of a specified level are increased as the temperature T is increased.

However, as discussed in the second paragraph of this section, this does not imply that all molecules of air in this room have this same energy. Rather, there will be a statistical distribution of energies as indicated in Fig. 4–6.3. At any particular instant of time, a very few molecules will have nearly zero energy; many molecules will have energies near to the average energy, and some molecules will have extremely high energies. As the temperature increases, there is (1) an increase in the average energy of the molecules, and (2) an increase in the number of molecules with energies in excess of any specified value.

The above applies to the kinetic-energy distribution of molecules in a gas. However, the same principle applies to the distribution of vibrational energy of atoms in a liquid or solid. Specifically, at any particular instant of time, a negligible number of atoms will have zero energy; many atoms will have energies near the average energy; and some atoms will have extremely high energies.

Our interest will be directed toward those atoms that have high energies. Very often we should like to know the probability of atoms possessing more than a specified amount of energy, e.g., what fraction of the atoms has energy greater than E of Fig. 4–6.4. The statistical solution to this problem has been worked out by Boltzmann as follows:

$$\frac{n}{N_{\text{tot}}} \propto e^{-(E-\bar{E})/kT}, \tag{4–6.3a}$$

where k is the previously described Boltzmann's constant. The number n of atoms with an energy greater than E, out of the total number N_{tot} present, is a function of

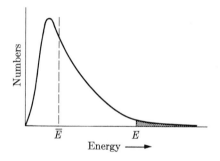

Fig. 4–6.4 Energies. The ratio of the number of high-energy atoms (shaded) to total number of atoms is an exponential function ($-E/kT$) when $E \gg \bar{E}$ (Eq. 4–6.3b).

the temperature T. When E is considerably in excess of the average energy $\bar{E}$, the equation reduces to

$$\frac{n}{N_{tot}} = Me^{-E/kT}, \qquad (4\text{-}6.3b)$$

where M is the proportionality constant. The value of E is normally expressed as joules/atom; thus, k is 13.8×10^{-24} J/atom · K. However, conversions may be made from other units by means of Appendix A.

Example 4–6.1 At 700°C, the linear-expansion coefficients, α, for graphite are as follows:

$$\alpha_\perp = 29 \times 10^{-6}/°C,$$

$$\alpha_{||} = \alpha_{||} = 0.8 \times 10^{-6}/°C$$

where $\alpha_\perp$ is the expansion coefficient perpendicular to the graphite layers of Fig. 2–2.6, and $\alpha_{||}$ is the coefficient in the two directions parallel to the layers. What is the volume increase in graphite between 600°C and 800°C?

Solution: Since $V = L^3$, and $V + \Delta V = (L + \Delta L_\perp)(L + \Delta L_{||})^2$,

$$\Delta V/V = \Delta L_\perp/L + 2\Delta L_{||}/L + \cdots,$$

$$\alpha_V \, \Delta T \simeq (\alpha_\perp + 2\alpha_{||}) \, \Delta T = (29 + 1.6)(10^{-6}/°C)(200°C)$$

$$= 0.006 \qquad \text{(or 0.6 v/o)}.$$

Comments. In general,

$$\alpha_V \simeq \alpha_x + \alpha_y + \alpha_z, \qquad (4\text{-}6.4)$$

where the three subscripts at the right refer to the linear-expansion coefficient of the three co-ordinate directions (Eq. 1–3.2a). ◀

Example 4–6.2 At 500°C (773 K), one out of 10^{10} atoms has the energy required to jump out of its lattice site into an interstitial position. At 600°C (873 K), this fraction is increased to 10^{-9}. (a) What is the energy required for this jump? (b) What fraction has enough energy at 700°C (973 K)?

Solution: From Eq. (4–6.3b),

$$\ln\left(\frac{n}{N_{tot}}\right) = \ln M - \frac{E}{kT} \qquad (4\text{-}6.5)$$

a) $\qquad \ln 10^{-10} = -23 = \ln M - E/(13.8 \times 10^{-24} \text{ J/atom·K})(773 \text{ K}),$

and

$\qquad \ln 10^{-9} = -20.7 = \ln M - E/(13.8 \times 10^{-24} \text{ J/atom·K})(873 \text{ K}).$

Solving simultaneously,

$$\ln M = -2.92,$$

and

$$E = 0.214 \times 10^{-18} \text{ J/atom};$$

or, in terms of a mole,

$$E = 129{,}000 \text{ J/mole} \quad (=30{,}900 \text{ cal/mole}).$$

b) $\ln\left(\dfrac{n}{N_{\text{tot}}}\right) = -2.92 - (0.214 \times 10^{-18}\ \text{J/atom})/(13.8 \times 10^{-24}\ \text{J/atom}\cdot\text{K})(973\ \text{K}),$

$$\dfrac{n}{N_{\text{tot}}} = \sim 6 \times 10^{-9}.$$

Comments. The relationship shown in Eq. (4–6.5) has a logarithmic value which is linear with reciprocal temperature, $1/T$,

$$y = C - Bx,$$

or

$$\ln\dfrac{n}{N} = C - \dfrac{B}{T}, \qquad\qquad (4\text{–}6.6)$$

where $y = \ln n/N$, and $x = (1/T)$. The slope B is E/k. Equation (4–6.6) is called an *Arrhenius* equation. It is widely encountered in any situation where temperature is the driving force. Examples include the intrinsic conductivity of semiconductors, the catalytic reaction of emission control, the diffusion processes in metals, the creep in plastics, and the viscosity of fluids. We will encounter this relationship again. ◀

4–7   ATOMIC DIFFUSION

As the temperature is increased and the atoms vibrate more energetically, a small fraction of the atoms will relocate themselves in the lattice. Example 4–6.2 related this fraction to temperature. Of course, the fraction depends not only on temperature, but also on how tightly the atoms are bonded in position. The energy requirement for an atom to change position is called the *activation energy*. This energy may be expressed as calories/mole, Q; as J/atom, E (Table 4–7.2); or as eV/atom.

Let us use Fig. 4–7.1 to illustrate activation energy schematically. A carbon atom is small ($r \simeq 0.07$ nm) and can sit interstitially among a number of fcc iron atoms. If it has enough energy,* it can squeeze between the iron atoms to the next interstice when it vibrates in that direction. At 20°C there is only a small probability that it will have that much energy. At higher temperatures, the probability increases (cf. Example 4–6.2).

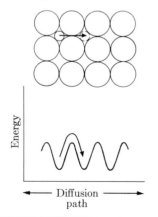

Fig. 4–7.1 Atom movements. Interstitial mechanism. Additional energy is required because the normal interatomic distances between the large atoms are altered when the interstitial atom is moving to the next site.

Energy

Diffusion
path

---

* Approximately 34,000 cal/mole, 0.24×10^{-18} J/atom, or 1.5 eV/atom.

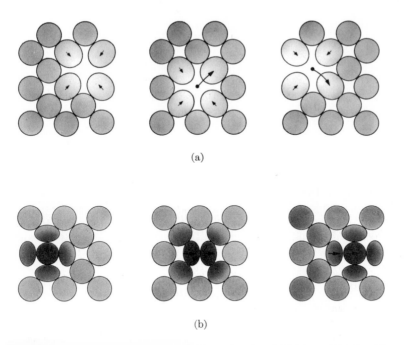

(a)

(b)

Fig. 4–7.2 Diffusion mechanisms. (a) By vacancies. (b) By interstitialcies. The vacancies move in the opposite direction from the diffusing atoms. (In the next chapter we shall observe an analogy for the movements of electron holes.) The movements follow "random-walk" statistics.

Other diffusion mechanisms are sketched in Fig. 4–7.2. When all the atoms are the same size, or nearly so, the vacancy mechanism becomes predominant. The vacancies may be present, either as part of a defect structure (Fig. 4–3.2) or because of extensive thermal agitation (e.g., Example 4–4.3, where aluminum was just 10° below its melting point).

Diffusivity   When an atom moves into a vacancy, a new hole is opened. In turn, this may receive an atom from *any* of the neighboring sites. As a result, a vacancy makes a "random walk" through a crystal. The same random-walk mechanism may be described for a carbon atom moving among iron atoms from interstice to interstice. However, we frequently encounter *concentration gradients*. As an example, assume there is one carbon atom per 20 unit cells of fcc iron at point (1), and only one carbon atom per 30 unit cells at point (2), which is one millimeter away. Now, since there are random movements of carbon atoms at *each* point, we will find a net flux of carbon atoms from point (1) to point (2), simply because there are half again as many atoms jumping in the vicinity of point (1). (See Example 4–8.1.) This net flow of atoms (or molecules) is called *diffusion*.

$$\frac{dc}{dx} = m^A$$

Table 4–7.1

Atomic diffusivity*

Solute	Solvent (host structure)	Diffusivity, m²/sec	
		500°C (930°F)	1000°C (1830°F)
1. Carbon	fcc iron	$(5 \times 10^{-15})^{\dagger}$	3×10^{-11}
2. Carbon	bcc iron	10^{-12}	(2×10^{-9})
3. Iron	fcc iron	(2×10^{-23})	2×10^{-16}
4. Iron	bcc iron	10^{-20}	(3×10^{-14})
5. Nickel	fcc iron	10^{-23}	2×10^{-16}
6. Manganese	fcc iron	(3×10^{-24})	10^{-16}
7. Zinc	Copper	4×10^{-18}	5×10^{-13}
8. Copper	Aluminum	4×10^{-14}	$10^{-10}\,M^{\ddagger}$
9. Copper	Copper	10^{-18}	2×10^{-13}
10. Silver	Silver (crystal)	10^{-17}	$10^{-12}\,M$
11. Silver	Silver (grain boundary)	10^{-11}	—
12. Carbon	hcp titanium	3×10^{-16}	(2×10^{-11})

* Calculated from data in Table 4–7.2.
† Parentheses indicate that the phase is metastable.
‡ M—Calculated, although temperature is above melting point.

The *flux J* of atoms (expressed in atoms/m²·sec) is proportional to the concentration gradient, $(C_2 - C_1)/(x_2 - x_1)$. In mathematical terms,

$$J = -D \frac{dC}{dx}. \tag{4-7.1a}*$$

The proportionality constant D is called *diffusivity*, or diffusion coefficient. The negative sign indicates that the flux is in the down-hill gradient direction. The units are

$$\frac{atoms}{(m^2)(sec)} = \left[\frac{m^2}{sec} \right] \left[\frac{atoms/m^3}{m} \right]. \tag{4-7.1b}$$

Diffusivity varies with the nature of the solute atoms, with the nature of the solid structure, and with changes in temperature. Several examples are given in Table 4–7.1. Some reasons for the various values of Table 4–7.1 are as follows:

1. Higher temperatures provide higher diffusivities, because the atoms have higher thermal energies and therefore greater probabilities of being activated over the energy barrier between atoms (Fig. 4–7.1).

---

* This is called *Fick's first law*. There is also a Fick's second law,

$$\frac{\partial C}{\partial t} = D \left(\frac{\partial^2 C}{\partial x^2} \right), \tag{4-7.2}$$

which shows the rate at which the concentration will change with time. In fact, the values of $\partial C/\partial t$ and $\partial^2 C/\partial x^2$ were determined experimentally in the laboratory to calculate the values of D found in Table 4–7.1. [See Van Vlack, *Materials Science for Engineers*, Reading, Mass.: Addison-Wesley (1970), p. 171f.]

2.   Carbon has a higher diffusivity in iron than does nickel in iron because the carbon atom is a small one (Appendix B).

3.   Copper diffuses more readily in aluminum than in copper because the Cu–Cu bonds are stronger than the Al–Al bonds (as evidenced by their melting temperatures).

4.   Atoms have higher diffusivity in bcc iron than in fcc iron because the former has a lower atomic packing factor (0.68 versus 0.74). (We shall observe later that the fcc structure has larger interstitial holes; however, the passageways between the holes are smaller than in the bcc structure.)

5.   The diffusion proceeds more rapidly along the grain boundaries because this is a zone of crystal imperfections (Fig. 4–4.7).

Diffusivities versus temperature   The discussion of Section 4–6 related the distributions of thermal energy to temperature. Boltzmann was able to quantify this with Eq. (4–6.3), which showed that the number of atoms that have more than a specified amount of energy increases in proportion to an exponential function that includes that energy and the reciprocal of temperature. With diffusion, the *activation energy* for atom movements corresponds to the energy E of Boltzmann's equation. Thus,

$$D = D_0 e^{-E/kT}, \tag{4-7.3}$$

where D_0 is the proportionality constant independent of temperature that includes M of Eq. (4–6.3b). The logarithm of the diffusivity is related to the reciprocal of the temperature, $1/T$:

$$\ln D = \ln D_0 - \frac{E}{kT}. \tag{4-7.4a}$$

The other term k is the same Boltzmann constant as in Eq. (4–6.3); that is, 13.8×10^{-24} J/atom $\cdot$ K.

Table 4–7.2 lists the values of D_0 and the activation energy for a number of diffusion reactions. Since the chemist prefers molar and calorie units, we can rewrite the above equation:

$$\ln D = \ln D_0 - \frac{Q}{RT}. \tag{4-7.4b}$$

Energy is expressed as Q (cal/mole). The gas constant R is 1.987 cal/mole $\cdot$ K.

The twelve sets of data in Tables 4–7.1 and 4–7.2 are plotted in Fig. 4–7.3. These are Arrhenius-type plots.*

Study aids (atom movements)   A series of 16 sketches are included in the paperback *Study Aids for Introductory Materials Courses*. The factors affecting the diffusivity are emphasized.

---

* See the comments after Example 4–6.2.

Table 4–7.2

Constants for diffusivity calculations* $(\ln D = \ln D_0 - Q/RT = \ln D_0 - E/kT)^{\dagger}$

Solute	Solvent (host structure)	D_0, m²/sec	Q, cal/mole	E, J/atom
1. Carbon	fcc iron	0.2×10^{-4}	34,000	0.236×10^{-18}
2. Carbon	bcc iron	2.2×10^{-4}	29,300‡	0.204×10^{-18}
3. Iron	fcc iron	0.22×10^{-4}	64,000	0.445×10^{-18}
4. Iron	bcc iron	2.0×10^{-4}	57,500	0.400×10^{-18}
5. Nickel	fcc iron	0.77×10^{-4}	67,000	0.465×10^{-18}
6. Manganese	fcc iron	0.35×10^{-4}	67,500	0.469×10^{-18}
7. Zinc	Copper	0.34×10^{-4}	45,600	0.317×10^{-18}
8. Copper	Aluminum	0.15×10^{-4}	30,200	0.210×10^{-18}
9. Copper	Copper	0.2×10^{-4}	47,100	0.327×10^{-18}
10. Silver	Silver (crystal)	0.4×10^{-4}	44,100	0.306×10^{-18}
11. Silver	Silver (grain boundary)	0.14×10^{-4}	21,500	0.149×10^{-18}
12. Carbon	hcp titanium	5.1×10^{-4}	43,500	0.302×10^{-18}

* See J. Askill, *Tracer Diffusion Data for Metals, Alloys, and Simple Oxides*, New York: Plenum (1970),
for a more complete listing of diffusion data. $T = {}^{\circ}K$
† $R = 1.987$ cal/mole·K; $k = 13.8 \times 10^{-24}$ J/atom·K.
‡ Lower below 400°C.

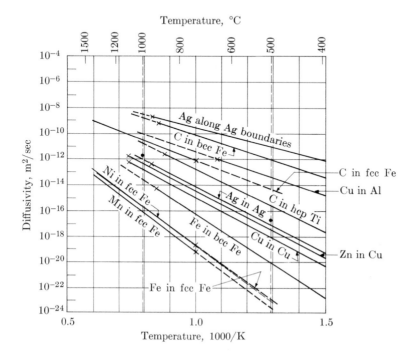

Fig. 4–7.3 Diffusivity versus temperature. (See Tables 4–7.1 and 4–7.2.)

Example 4–7.1 There are 0.19 a/o copper at the surface of some aluminum and 0.18 a/o copper, 1.2 mm underneath the surface. What will the flux of copper atoms be from the surface inward at 500°C? (The aluminum is fcc, and $a = 0.4049$ nm.)

Solution

$$\text{Atoms/mm}^3 = 4/(0.4049 \times 10^{-6} \text{ mm})^3$$
$$= 6 \times 10^{19}/\text{mm}^3.$$

$$(dC/dx)_{Cu} = [(0.0018 - 0.0019)(6 \times 10^{19}/\text{mm}^3)]/1.2 \text{ mm} \cong$$
$$= -5 \times 10^{15} \text{ Cu/mm}^4.$$

From Table 4–7.2 and Eq. (4–7.3), $D = D_o\, e^{-E/kT}$

Diffusivity $= (0.15 \times 10^{-4} \text{ m}^2/\text{sec}) \exp(-0.210 \times 10^{-18} \text{ J}/(13.8 \times 10^{-24} \text{ J/K})(773 \text{ K}))$
$$= {\sim}4 \times 10^{-14} \text{ m}^2/\text{sec}.$$

Flux $= -(4 \times 10^{-14} \text{ m}^2/\text{sec})(-5 \times 10^{15} \text{ Cu/mm}^4)(10^6 \text{ mm}^2/\text{m}^2)$ $J = -D \dfrac{dc}{dx}$
$$= 2 \times 10^8 \text{ Cu/mm}^2 \cdot \text{sec}.$$

Comments. The value of copper diffusivity may also be obtained graphically from Fig. 4–7.3 (and for 500°C, from Table 4–7.1). Since the activation energy enters the exponent of the equation, its three significant figures do not carry over into the value of diffusivity. Usually diffusivity values are significant to only the first or second figure. ◀

Example 4–7.2 The diffusivity of aluminum in copper is 2.6×10^{-17} m^2/sec at 500°C and 1.6×10^{-12} m^2/sec at 1000°C. (a) Determine the values of D_0, Q, and E for this diffusion couple. (b) What is the diffusivity at 750°C?

Solution: With Eq. (4–7.4a),

a) $$\ln (2.6 \times 10^{-17}) = \ln D_0 - \frac{E}{13.8 \times 10^{-24} \,(773)};$$

$$\ln 1.6 \times 10^{-12} = \ln D_0 - \frac{E}{13.8 \times 10^{-24} \,(1273)}.$$

Solving simultaneously, we get

$$D_0 = 4 \times 10^{-5} \text{ m}^2/\text{sec},$$

and

$$E = 0.3 \times 10^{-18} \text{ J/atom.}$$

Alternatively, with Eq. (4–7.4b),

$$Q = 43{,}000 \text{ cal/mole.}$$

b) From Eq. (4–7.4a),

$$\ln D = \ln 4 \times 10^{-5} - (0.3 \times 10^{-18} \text{ J})/(13.8 \times 10^{-24} \text{ J/K})(1023 \text{ K});$$

and therefore,

$$D = {\sim}2.5 \times 10^{-14} \text{ m}^2/\text{sec} \qquad (\text{or } 10^{-13.6} \text{ m}^2/\text{sec}).$$

This value is also obtainable graphically with Fig. 4–7.3 by inserting the known values (dots) on the Arrhenius plot and interpolating. (However, care must be taken to read the nonlinear scales correctly.)

Comments. Observe that the diffusivity of copper through aluminum is higher than for aluminum through copper. This is to be expected from our knowledge of the bond strength of the host metals:

$$(T_m \text{ of Cu}) > (T_m \text{ of Al}); \text{ therefore, } D_{Al/Cu} < D_{Cu/Al}. \ \blacktriangleleft$$

Example 4–7.3 A steel contains 8.5 w/o Ni at the center x of a grain of fcc iron, and 8.8 w/o Ni at the edge e of the grain. The two points are separated by 40 μm. What is the flux of atoms between x and e at 1200°C? ($a = 0.365$ nm.)

Solution: Basis: 100 amu. At x,

$$(100 \text{ amu})(0.085)/(58.71 \text{ amu/Ni}) = 0.1448 = \ 8.1 \text{ a/o Ni};$$
$$(100 \text{ amu})(0.915)/(55.85 \text{ amu/Fe}) = \underline{1.638} \ = 91.9 \text{ a/o Fe.}$$
$$\phantom{(100 \text{ amu})(0.915)/(55.85 \text{ amu/Fe}) = } 1.783$$

At e,

$$\sim 8.4 \text{ a/o Ni.}$$

Per unit cell,

$$C_x = (4 \text{ atoms})(0.081)/(0.365 \times 10^{-9} \text{ m})^3 = 6.66 \times 10^{27}/\text{m}^3;$$
$$C_e = (4 \text{ atoms})(0.084) /(0.365 \times 10^{-9} \text{ m})^3 = 6.91 \times 10^{27}/\text{m}^3.$$

$$\ln D = \ln D_o - E/KT$$

$$\ln D = \ln (0.77 \times 10^{-4} \text{ m}^2/\text{s}) - (0.465 \times 10^{-18} \text{ J})/(13.8 \times 10^{-24} \text{ J/K})(1473 \text{ K});$$
$$D = 9 \times 10^{-15} \text{ m}^2/\text{sec.}$$
$$\text{Flux} = -(9 \times 10^{-15} \text{ m}^2/\text{s})(6.66 - 6.91)(10^{27}/\text{m}^3)/(40 \times 10^{-6} \text{ m}) = -D(\Delta \text{Atoms}/\text{uc})/d$$
$$= 5.6 \times 10^{16} \text{ atoms/m}^2 \cdot \text{s.}$$

Comment. We must work with a/o rather than w/o because the flux involves atoms rather than mass. $\blacktriangleleft$

• 4–8 DIFFUSION PROCESSES

There are numerous applications of diffusion to technical processes. *Carburization* of steel may be familiar to the reader. In this process, a low-carbon steel (which is relatively tough, but soft) is heated in a carbon-containing atmosphere so that carbon diffuses into the steel, producing a hard, carbon-enriched *case* (Fig. 4–8.1). Another example among commercial diffusion processes is encountered in making semiconductors. Boron can be diffused into silicon to provide a *p*-type region of a junction device (Chapter 5).

We will not detail these processes here but will note that simultaneous consideration may be given to diffusivity and processing time. If we multiply each side of Eq. (4–7.1) by time, we get an integrated flux:

$$\text{atoms/cm}^2 = Dt(-dC/dx). \tag{4–8.1}$$

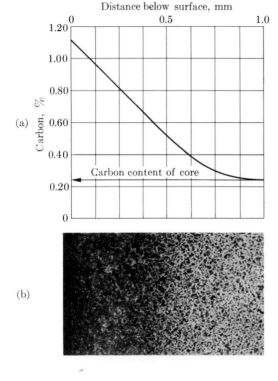

(a)

(b)

Fig. 4–8.1 Carburized steel (surface of the bar is at the left). This 25-mm (1-in.) steel bar initially had 0.24% carbon throughout. The steel was heated to 870°C (1600°F) in the presence of excess carbon to allow carbon to diffuse inward from the surface. (The carbon-enriched areas were darkened by etching.) (F. Harris, Buick Motor Div., General Motors Corporation.)

Thus, the total number of atoms entering or leaving a material through a unit area and down a fixed concentration gradient is proportional to Dt, the product of diffusivity and time. This is a convenient relationship because it lets us anticipate processing time as a function of temperature. For example, assume a 1.0-mm carburized case is formed on a steel (fcc iron) in 100 minutes at 800°C, where the diffusivity is $\sim 2.4 \times 10^{-12}$ m²/sec. With the same concentration gradient, it would take only 8 minutes at 1000°C where the diffusivity is 12.5 times as great ($\sim 3 \times 10^{-11}$ m²/sec from Table 4–7.1).

The thickness of a case increases with $\sqrt{Dt}$.*

$$x \propto \sqrt{Dt}. \tag{4–8.2}$$

From this and from the examples of the previous paragraph, a case that is ~ 1.9 mm thick would develop in a 30-min period at 1000°C.

$$\frac{[(2.4 \times 10^{-12} \text{ m}^2/\text{s})(6{,}000 \text{ s})]^{1/2}}{1 \text{ mm}} = \frac{\sqrt{Dt}}{x}.$$

$$\frac{\sqrt{Dt}}{x} = \frac{[(3 \times 10^{-11} \text{ m}^2/\text{s})(1{,}800 \text{ s})]^{1/2}}{\sim 1.9 \text{ mm}}.$$

---

* This is derivable from Fick's second law (Eq. 4–7.2) in the footnote in the previous section.

Example 4–8.1 At the surface of a steel bar there is one carbon atom per 20 unit cells of iron. One millimeter behind the surface, there is one carbon atom per 30 unit cells. The diffusivity at 1000°C is 3×10^{-11} m²/sec. The structure is fcc at 1000°C ($a = 0.365$ nm). How many carbon atoms diffuse through each unit cell per minute?

Solution: Carbon concentrations are

$$C_2 = 1/[30(0.365 \times 10^{-9} \text{ m})^3] = \frac{1 \text{ CARBON}}{30 \text{ ATOMS}} (a)^3$$
$$= 0.68 \times 10^{27}/\text{m}^3;$$

$$C_1 = 1/[20(0.365 \times 10^{-9} \text{ m})^3]$$
$$= 1.03 \times 10^{27}/\text{m}^3.$$

From Eq. (4–7.1a),

$$J = -(3 \times 10^{-11} \text{ m}^2/\text{sec})(0.68 - 1.03)(10^{27}/\text{m}^3)/(10^{-3} \text{ m})$$
$$= 1.05 \times 10^{19}/\text{m}^2 \cdot \text{sec} \qquad (\text{or} \sim 10^{13}/\text{mm}^2 \cdot \text{sec}).$$

Each unit cell has an area of $(0.365 \times 10^{-9}$ m$)^2$. Therefore,

$$J_{\text{u.c.}} = (1.05 \times 10^{19}/\text{m}^2 \cdot \text{sec})(0.365 \times 10^{-9} \text{ m})^2 (60 \text{ sec/min})$$
$$= 84 \text{ atoms/min.}$$

Comments. It is apparent that a piece of steel is not a dormant material; rather, numerous changes occur within it.

We use the above process to *carburize* steel. In Chapter 11, we will see how this can be used to advantage to modify the surface hardness of a steel. ◀

Example 4–8.2 Radioactive nickel has been diffused into the surface of iron for 20 minutes at 1200°C. The limit of detection of the radioactivity is at a position that is 1.5 mm behind the surface. A companion sample was heated for 1 hour at 1100°C. Will the limit of detection be greater or less than the 1.5 mm observed in the first sample?

Solution: The iron is fcc at these temperatures. From Table 4–7.2, Eq. (4–7.3 or 4–7.4), and Eq. (4–8.2),

$$D_{1200} = (0.77 \times 10^{-4} \text{ m}^2/\text{s}) \exp{(-67{,}000 \text{ cal/mole})/(1.987 \text{ cal/mole} \cdot \text{K})(1473 \text{ K})}$$
$$= 9 \times 10^{-15} \text{ m}^2/\text{sec.}$$

$$D_{1100} = 1.7 \times 10^{-15} \text{ m}^2/\text{sec.}$$

Based on $x \propto \sqrt{Dt}$, or $\sqrt{Dt}/x = $ constant:

$$\left(\frac{\sqrt{(9 \times 10^{-15} \text{ m}^2/\text{s})(1200 \text{ s})}}{1.5 \text{ mm}}\right)_{1200°C} = \left(\frac{\sqrt{(1.7 \times 10^{-15} \text{ m}^2/\text{s})(3600 \text{ s})}}{x}\right)_{1100°C};$$

$$x = 1.1 \text{ mm} \qquad (<1.5 \text{ mm}). \blacktriangleleft$$

REVIEW AND STUDY

In Chapter 3 attention was given to the preciseness and regularity of the atomic structure of crystals. Perfect crystals serve as a basis for considering many properties and characteristics of materials, such as density, anisotropy, slip planes, phase stability, piezoelectricity, and semiconducting compounds. At the same time, crystals are not always perfect, and many important properties and behaviors of materials arise from the irregularities. We cited some of these in the preview of the chapter. We must consider the *disorder* of materials as well as their order.

SUMMARY

Impurities are found in all materials unless special means are used to reduce them to a low level. An alloy is a metal with impurities that are intentionally present. Solid solutions contain atoms (or molecules) of a second component as a solute. These may be present either interstitially among the solvent atoms or as a substitute replacement within the crystal lattice. In order for solid solution to be extensive in crystals, the solute and solvent must be comparable in size and electronic behavior.

Imperfections are found in all crystals unless special means are used to reduce them to a low level. It is convenient to categorize them by geometry:

1. Point defects, that include vacancies and/or interstitials;

2. Lineal defects, commonly called dislocations; and

3. Boundaries, which may be external surfaces or internal separations between grains or phases. The latter may be treated as two-dimensional defects.

Liquids generally lack the long-range order that characterizes crystals. For some materials, it is possible to avoid crystallization and retain the amorphous character of a liquid into a solid (1) if the anticipated crystal structure is complex, or (2) by rapid cooling. We label these amorphous solids by the term *glass*. Although lacking a freezing temperature, amorphous materials possess a glass temperature T_g, which will be very important to us when we study the behavior of plastics.

A *phase* was defined as that part of a material that is distinct from others in structure and/or composition.

At any instant of time, most atoms or molecules possess near-average energy. However, some will possess very little energy, and some will possess abnormally high energies. We are interested in the high end of this statistical distribution, because those atoms may be activated to break bonds, and to diffuse to new locations within the material. It is only through this mechanism that the internal structure (and hence the properties) of the material may be modified. Diffusion occurs more readily (1) at high temperatures, (2) when the diffusing atom is small (e.g., carbon in iron), (3) when the packing factor of the host structure is low (e.g., bcc versus fcc), (4) when the bonds of the host structure are weak (e.g., low-melting materials), and (5) when there are defects in the material (e.g., vacancies or grain boundaries).

KEY TERMS AND CONCEPTS

- Activation energy, E or Q
  Alloy
  Amorphous
  Anisotropic
  Arrhenius equation
  ASTM G.S.#, n
  Boltzmann's constant, k
  Brass
  Bronze
  Component
  Concentration gradient
  Defect structures
  Diffusion
  Diffusivity, D
  Dislocation
  Energy distribution
  Fick's first law
  Flux (diffusion), J
  Glass
  Grain

  Grain boundary
  Grain boundary area, S_v
  Heat of fusion
  Imperfection
  Interstice
  Mixture
  Nonstoichiometric compounds
  Phase
  Phase boundary
  Point defects
  Slip vector, **b**
  Solid solutions,
      interstitial
      ordered
      substitutional
  Solubility limit
  Stoichiometric
  Thermal expansion
  Vacancy

FOR CLASS DISCUSSION

A_4 What is the difference between brass and bronze?

B_4 Distinguish between an alloy that is a mixture of phases and one that is a solid solution.

C_4 Distinguish between solvent and solute, as applied to solid solutions.

D_4 Distinguish between interstitial and substitutional solid solutions.

E_4 From Appendix B, select elements that have a favorable size for substitutional solid solution with metallic iron. Select those ions that might substitute for Fe^{2+} ions.

F_4 There is a good argument for considering that an ordered solid solution is a compound. Discuss.

G_4 Nonstoichiometric AlMg does not possess a significant number of vacancies; nonstoichiometric FeO does. Explain the difference.

H_4 Why do Schottky defects occur in pairs?

I_4 Explain the origin of surface energy; of grain-boundary energy.

J_4 Cite various ways grain boundaries affect the behavior of materials.

K_4 Magnification is always expressed on the basis of the lineal multiplication factor. With $\times 100$, what is the area multiplication factor?

L_4 G.S. #5 has ＿＿ times as many grains per square inch in a two-dimensional section as G.S. #3. The linear dimensions of the former are ＿＿ times as large. The grain boundary area is ＿＿ times as much per grain, but ＿＿ times as much per unit volume.

M_4 Compare and contrast the change that occurs at 380°C in $AuCu_3$ (Study Problem 4–2.6) with melting.

N_4 Why is energy given off during the freezing of water (80 cal/g, or 6 kJ/mole)?

O_4 The volume thermal-expansion coefficient of a liquid is almost always higher than that of the corresponding solid. Why?

P_4 Explain the glass temperature, T_g, to a friend who is not in this course.

• Q_4 Which will be higher, the *mean energy* or the *median energy* of the gas molecules in this room?

R_4 What phases are present after mixing 10 g of oil with 100 g of water? 10 g of salt with 100 g of water? Give an example of a materials "system" with three phases present.

S_4 Refer to Fig. 4–4.1(d). This defect is more common than the presence of a single atom in a specific hole of the fcc lattice, e.g., at $\frac{1}{2}$,0,0. Suggest a reason.

T_4 What is meant by activation energy? By phase?

• U_4 The term D_0 in Eq. (4–7.3) includes M of Eq. (4–6.3b) but does not equal M. Suggest why.

V_4 Why must the T of Arrhenius-type equations be in absolute temperature, K? We can use °C for thermal expansion and thermal conductivity calculations.

W_4 Refer to Table 4–7.1. Why are the values higher for couple #2 than for couple #1? For couple #2 than for couple #4? For couple #11 than for couple #10? For couple #8 than for couple #9?

• X_4 Fluidity (the reciprocal of viscosity) varies with temperature according to Arrhenius-type equations (Eqs. 4–6.6 and 4–7.4). Compare that property with the mechanism for diffusion.

Y_4 Self-diffusion is the movement of atoms within their own structure. Explain how this can occur. Suggest a way to measure it in the absence of a concentration gradient, through the use of radioactive isotopes.

• Z_4 Thermal diffusivity h is sometimes used in calculations instead of thermal conductivity k, where $h = k/c\rho$. (The terms in the denominator are heat capacity, c, (J/g·°C) and density, ρ.) What are the units of h? Recast Eqs. (4–8.1) and (4–8.2) for use in heat transfer calculations using h. Give an example of a problem using one of these equations.

STUDY PROBLEMS

4–2.1 An alloy contains 85 w/o copper and 15 w/o tin. Calculate the atomic percent (a/o) of each element.

Answer: 8.6 a/o Sn; 91.4 a/o Cu

4–2.2 There is 5 a/o magnesium in an Al–Mg alloy. Calculate the w/o magnesium.

4–2.3 Consider Fig. 4–2.3 to be an interstitial solution of carbon in fcc iron. What is the w/o carbon present?

Answer: 1.3 w/o carbon

4–2.4 Consider Fig. 4–2.1 to be a substitutional solid solution of copper and gold. What is the w/o Cu present if (a) Cu is the more prevalent atom? (b) Au is the more prevalent atom?

4–2.5 The interstitial solid solution of carbon in bcc iron is 1C/500 unit cells. What is the w/o carbon?

Answer: 0.02 w/o

4–2.6 An alloy with 25 a/o gold and 75 a/o copper forms an fcc solid solution which is random above 380°C. Below that temperature, it becomes ordered with gold atoms at the corners of the unit cell and copper atoms at the center of each face. (a) What is the weight percent gold? (b) Which of the cubic lattices does it possess below 380°C? Simple cubic - moving ½, ½, 0 from origin does not lead to equiv. site

4–2.7 Refer to Study Problem 4–2.6. (a) Estimate the lattice constant of that alloy. (b) Calculate its density.

Answer: (a) 0.373 nm   (b) 12.4 g/cm^3 ($=$12.4 Mg/m^3)

4–2.8 Silicon that is used as a semiconductor contains 10^{21} aluminum atoms per m^3. (a) What w/o Al is present? (b) How many unit cells of silicon per aluminum atom? (Si is cubic with 8 atoms/unit cell; $a = 0.543$ nm.)

4–2.9 An alloy contains 80 w/o Ni and 20 w/o Cu in substitutional fcc solid solution. Calculate the density of this alloy.

Answer: 8.92 g/cm^3 ($=$8.92 Mg/m^3)

4–3.1 (a) What is the w/o FeO in the solid solution of Fig. 4–3.1? (b) The w/o Fe^{2+}? (c) Of O^{2-}?

Answer: (a) 51 w/o   (b) 39.8 w/o   (c) 30.8 w/o

4–3.2 If all the iron ions of Fig. 4–3.1 were changed to Ni ions, what would be the w/o MgO?

4–3.3 What is the density of $Fe_{<1}O$, if the Fe^{3+}/Fe^{2+} ratio is 0.14? ($Fe_{<1}O$ has the structure of NaCl; and ($r_{Fe} + R_O$) averages 0.215 nm.)

Answer: 5.73 Mg/m^3 ($=$5.73 g/cm^3)

4–3.4 An intermetallic compound of aluminum and magnesium varies from 52Mg–48Al to 56 Mg–44Al (weight bases). What are the atom ratios of these compositions?

4–4.1 Calculate the radius of the largest atom that can exist interstitially in fcc iron without crowding. (*Hint:* Sketch the (100) face of several adjacent unit cells.)

Answer: 0.053 nm

4–4.2 Calculate the radius of the largest atom that can fit interstitially into fcc silver without crowding.

4–4.3 Determine the radius of the largest atom that can be located in the interstices of bcc iron without crowding. (*Hint:* The center of the largest hole is located at $\frac{1}{2},\frac{1}{4},0$.)

Answer: 0.036 nm

4–4.4 (a) What is the coordination number of the interstitial site in Study Problem 4–4.1? (b) What structure would result if *every* such site were occupied by a smaller atom or ion?

4–4.5 (a) What is the coordination number for the interstitial site in Study Problem 4–4.3? (b) How many of these sites are there per unit cell?

Answer: (a) 4   (b) 12

4–4.6 In copper at 1000°C, one out of every 473 lattice sites is vacant. If these vacancies remain in the copper when it is cooled to 20°C, what will be the density of the copper?

4–4.7 The number of vacancies increases at higher temperature. Between 20°C and 1020°C, the lattice constant of a bcc metal increased 0.5 1/o from thermal expansion. In the same temperature range, the density decreased 2.0%. Assuming there was one vacancy per 1000 unit cells in this metal at 20°C, estimate how many vacancies there are per 1000 unit cells at 1020°C.

Answer: 11/1000 unit cells

4–4.8 Estimate the grain-boundary area per unit volume for the iron in Fig. 9–8.2(a). The magnification is × 500.

4–4.9 (a) Assume that the G.S. #6 of Fig. 4–4.11 represents a two-dimensional cut through a polycrystalline solid. Estimate the corresponding grain-boundary area. (b) Repeat for G.S. #2.

Answer: (a) 60 mm²/mm³ (or 1500 in.²/in.³)   (b) 15 mm²/mm³ (or 380 in.²/in.³)

4–4.10 How many grains are observed in a microscope per square inch (at × 100) for (a) G.S. #8? (b) G.S. #5? (c) Repeat for × 200.

4–4.11 Assume Fig. 4–4.10 is at × 100 rather than × 250. (a) Estimate the boundary area per mm³. (b) Per in.³. (c) Estimate the grain-size number.

Answer: (a) 14 mm²/mm³   (b) 350 in.²/in.³   (c) Slightly less than #3

4–4.12 Joe Moe concluded that the ASTM grain size of a metal was #2. However, he had neglected to observe that the magnification was × 300 and not × 100. (a) What was the correct grain-size number? (b) Refer to Fig. 9–8.2(a). What is the ASTM G.S.#?

4–5.1 The density of liquid aluminum at its melting point is 2.37 Mg/m³. Assume a constant atom size; calculate its atomic packing factor.

Answer: 0.65

4–5.2 (a) Estimate the heat of fusion of sodium from Table 4–5.1. (b) Of silver.

4–5.3 Based on the slopes of the curves in Fig. 4–5.2, estimate the volume expansion coefficient, α_V, of (a) solid lead and (b) liquid lead at the melting point of lead.

Answer: (a) 100×10^{-6}/°C   (b) 120×10^{-6}/°C

4–6.1 An aluminum wire is stretched between two rigid supports at 35°C. It cools to 15°C. What stress is developed?

Answer: 31 MPa (or 4500 psi).

4–6.2 Estimate the linear expansion coefficient of bcc iron at 900°C from the data in Fig. 4–5.2. Compare your answer with the data in Fig. 1–3.1.

4–6.3 Quartz (SiO_2) is hexagonal. Mean values of its two thermal expansion coefficients are $\alpha_a = 15 \times 10^{-6}/°C$ and $\alpha_c = 10 \times 10^{-6}/°C$. What is the percent volume expansion between 20°C and 570°C of a quartz crystal?

Answer: ~ 2.2 v/o

• 4–6.4 Plot the answers to Study Problem 2–5.5. Plot the middle of the energy trough as a function of energy. Relate qualitatively your results to thermal expansion.

4–6.5 Refer to Example 4–6.2. At what temperature does 1 out of 10^8 atoms have enough energy to jump out of its atom site?

Answer: 1000 K (727°C)

4–6.6 Refer to Example 4–6.2. What fraction of the atoms have sufficient energy to jump out of their sites at 1000°C?

4–6.7 At 800°C, 1 out of 10^{10} atoms, and at 900°C, 1 out of 10^9 atoms, have appropriate energy for movements within a solid. (a) What is the activation energy in J/atom? (b) In cal/mole?

Answer: 0.4×10^{-18} J/atom (or 57,600 cal/mol)

4–6.8 Refer to Study Problem 4–6.7. At what temperature will 1 out of 10^8 atoms have the required amount of energy?

4–6.9 At 500°C, 1 of every 10^{12} solute atoms has enough energy to relocate. At 600°C, the fraction increases to 1 of every 10^{10} solute atoms. (a) What is the activation energy (in J/atom, in cal/mole)? (b) What fraction has enough energy to relocate at 700°C?

Answer: (a) 0.43×10^{-18} J/atom, 61,800 cal/mole (b) 4×10^{-9}

4–6.10 An activation energy of 2.0 eV (or 0.32×10^{-18} J) is required to form a vacancy in a metal. At 800°C there is one vacancy for every 10^4 atoms. At what temperature will there be one vacancy for every 1000 atoms?

4–7.1 A solid solution of copper in aluminum has 10^{26} atoms of copper per m^3 at point X, and 10^{24} copper atoms per m^3 at point Y. Points X and Y are 10 micrometers apart. What will be the diffusion flux of copper atoms from X to Y at 500°C?

Answer: 4×10^{17} atoms/m^2·sec

4–7.2 (a) What is the ratio of diffusivities for carbon in bcc iron to carbon in fcc iron at 500°C? (b) Of carbon in fcc iron to nickel in fcc iron at 1000°C? (c) Of carbon in fcc iron at 1000°C to carbon in fcc iron at 500°C? (d) Why are the ratios greater than 1?

4–7.3 The inward flux of carbon atoms in fcc iron is $10^{19}/m^2$·sec at 1000°C. What is the concentration gradient?

Answer: $-3.3 \times 10^{29}/m^4$

4–7.4 There are 4 a/o carbon at the surface of the iron in Study Problem 4–7.3. What is the a/o carbon at 1 mm behind the surface? ($a = \sim 0.365$ nm at 1000°C.)

4–7.5 (a) Using the data of Table 4–7.2, calculate the diffusivity of copper in aluminum at 400°C. (b) Check your answer with Fig. 4–7.3.

Answer: 2×10^{-15} m²/sec

4–7.6 Zinc is moving into copper. At point X there are 2.5×10^{17} Zn/mm³. What concentration is required at point Y (2 mm from X) to diffuse 60 Zn atoms/mm²·min at 300°C.

4–7.7 What is the weight percent of zinc in the copper at point X of Study Problem 4–7.6?

Answer: 0.3 w/o

4–7.8 A zinc gradient in a copper alloy is 10 times greater than the aluminum gradient in a copper alloy. Compare the flux of solute atoms/m²·sec in the two alloys at 500°C. (The data for $D_{Al \text{ in } Cu}$ are in Example 4–7.2.)

4–7.9 Refer to Study Problem 4–7.1. (a) What is the diffusion coefficient of copper in aluminum at 100°C? (b) What will be the diffusion flux of copper atoms from X to Y at 100°C?

Answer: (a) 3×10^{-23} m²/sec (b) 300 atoms/mm²·sec

4–7.10 Aluminum is to be diffused into a silicon single crystal. At what temperature will the diffusion coefficient be 10^{-14} m²/sec? ($Q = 73,000$ cal/mole and $D_0 = 1.55 \times 10^{-4}$ m²/sec.)

4–7.11 The Ni-in-fcc Fe and the Fe-in-fcc Fe curves of Fig. 4–7.3 cross. (a) By calculation, determine the temperature where they cross. (b) Suggest why the two curves are nearly coincident.

Answer: (a) ~ 1200 K (~ 900°C); a more precise answer is not warranted.

4–7.12 The diffusion of carbon in tungsten has an activation energy of 0.78×10^{-18} J/atom (112,000 cal/mole) and D_0 of 0.275 m²/sec. Where does the diffusivity curve lie on Fig. 4–7.3? Suggest why it lies where it does in comparison to the other curves for carbon diffusion.

4–7.13 How much should the concentration gradient be for nickel in iron if a flux of 100 nickel atoms/mm²·sec is to be realized (a) at 1000°C? (b) at 1400°C?

Answer: (a) $-(5 \times 10^{23}/\text{m}^3)/\text{m}$ (b) $-(7 \times 10^{20}/\text{m}^3)/\text{m}$

4–7.14 Compare the diffusivities of iron (fcc), copper, and silver at 60% of their melting temperatures, $0.6\ T_m$.

● **4–8.1** A suitable carburized case on steel (fcc iron) was obtained in 1 hour at 820°C (1510°F). How long would it take to double the thickness of the case at the same temperature?

Answer: 4 hours

● **4–8.2** A suitable carburized case on steel (fcc iron) was obtained in 1 hour at 820°C (1510°F). How long would it take to obtain a case of the same thickness at 850°C (1560°F)?

● **4–8.3** Refer to Study Problem 4–8.1. (a) What temperature is required to double the thickness in one hour? (b) In two hours?

Answer: (a) 1200 K (927°C)   (b) 1144 K (871°C)

● **4–8.4** Refer to Example 4–8.1. How many atoms diffuse per minute through each cell at 950°C?

CHAPTER **5**

Electron
Transport
in Solids*

PREVIEW

Atoms and their arrangements have received the bulk of
our attention to date. In this chapter we will focus
attention on electrons and their freedom to move among
atoms.

*Metals, with their weak hold on valence electrons,
are good conductors of both electricity and heat.* This
conductivity occurs because very little energy is required
to activate delocalized electrons into conduction levels. In
contrast, electrons must be raised across a *large energy
gap in an insulator. Semiconductors have small energy
gaps* so that a useful number of electrons are available
for conduction.

Simplified concepts of many devices are possible
with an introductory understanding of energy gaps and
junctions.

---

* If desired, the instructor may choose to defer this chapter until later in the course with a minimum of
adjustment in the intervening chapters. However, in view of the major technological role that electronic
materials play in our society, the instructor is urged not to let a class skip this chapter.

CONTENTS

STUDY OBJECTIVES

1 To handle simple calculations for conductivity based on the concept of the electron as a negative charge carrier. Conversely, a missing electron, called an electron hole, is a positive charge carrier.

2 To relate resistivity (and conductivity) changes to the effect of impurities and temperature changes by the concept of the mean-free-path. (This will permit us to automatically evaluate the effect of cold work on resistivity in Chapter 6.)

3 To associate energy bands and energy gaps qualitatively with the conductivity of metals, semiconductors, and insulators.

4 To solve introductory problems (a) on the conductivity of intrinsic semiconductors based on their charge mobility, and the size of their energy gap, (b) on the conductivity of extrinsic semiconductors based on the doping concentration, and (c) on the progress of recombination based on relaxation times.

5 To understand the bases for simple semiconductor devices, such as the thermistor, photoconductor, L.E.D., and junction rectifier.

6 To develop sufficient familiarity with terms used with electronic materials to communicate as required with design engineers.

If your assignments skipped Sections 1–3 and 1–4 earlier, return to them now. You will use the electronic charge frequently, so it is suggested that you learn its value ($q = 0.16 \times 10^{-18}$ A·s). The Arrhenius equation of Chapter 4 appears again in this chapter.

5–1   CHARGE CARRIERS

Various materials that are available to the engineer or scientist exhibit a wide range of conductivities (or resistivities, since $\sigma = 1/\rho$). As shown in Fig. 5–1.1, we commonly divide materials into three categories, *conductors*, *semiconductors*, and *insulators*. Metals fall in the first category, since they have delocalized electrons that are free to move throughout the structure (Sections 2–2 and 2–7). Insulators include those ceramics and polymeric materials with strongly held electrons and nondiffusing ions. Their function is to isolate neighboring conductors. It was not very long ago that only the two ends of the spectrum of Fig. 5–1.1 were considered to be useful. Today, however, the middle, semiconducting category has become exceedingly important and will, in fact, be the chief subject of this chapter.

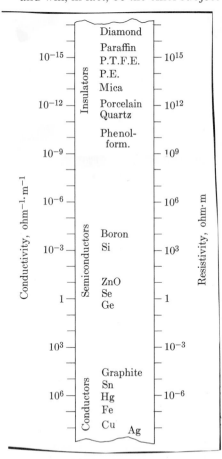

Fig. 5–1.1   Spectrum of conductivity (and resistivity). Commercial semiconductors lie between 10^{+4} and 10^{-4} ohm$^{-1} \cdot$ m^{-1}.

In those types of materials that conduct electricity, the charge is carried in modules of 0.16×10^{-18} coul, this being the charge on an individual electron. In metals, it is the individual electron that moves. In ionic materials, charge can be carried by diffusing ions. However, their charge is simply an integer number of electron charges

(− or +, for anions and cations, respectively). Thus, an SO_4^{2-} ion carries 0.32×10^{-18} coul of charge within a storage battery, and all Pb^{2+} ions have an absence of two electron charges as they move in the opposite direction.

Electrons and anions are *negative charge carriers*. In contrast, a cation such as Pb^{2+} is a *positive charge carrier* because, as we have just seen, it has an *absence* of electrons. There is another positive charge carrier that becomes important in semiconductors, viz., an *electron hole*. It is an absence of an electron within the energy band for the delocalized electrons discussed in Section 2–2. We will come back to these in Section 5–4.

Conductivity σ and *resistivity* ρ values for a material depend upon the number n of charge carriers, the charge q on each, and their *mobility* μ according to Eq. (1–4.1a):

$$\sigma = \frac{1}{\rho} = nq\mu. \qquad (5\text{–}1.1)$$

As in Chapter 1, n, q, and μ have units of m^{-3}, coul, and $m^2/V \cdot s$, respectively.*

In this chapter, we shall look at atomic and structural factors that affect n and μ. This will let us anticipate factors that influence our choices of materials for design and service behavior.

We can calculate the *drift velocity*, $\bar{v}$, of a charge carrier from Eq. (1–4.2),

$$\bar{v} = \mu\mathscr{E}. \qquad (5\text{–}1.2)$$

Note that the drift velocity is proportional to the *electric field* $\mathscr{E}$, or volts/m. Thus, in the absence of a voltage gradient, there is no net, or drift, velocity. This does not indicate that the charged carriers do not move; rather the movements are random, and so charge carried in one direction is balanced by charge being carried in the opposite direction.

Example 5–1.1 A semiconductor with 10^{21} charge carriers/m^3 has a resistivity of 0.1 ohm·m at 20°C. What is the drift velocity of the electrons if one ampere of current is carried across a gradient of 0.15 volts/mm?

Solution: From Eqs. (5–1.1) and (5–1.2),

$$\bar{v} = \mathscr{E}/\rho n q$$

$$= \frac{(150 \text{ V/m})}{(0.1 \text{ ohm·m})(10^{21}/m^3)(0.16 \times 10^{-18} \text{ amp·sec})} = 9.4 \text{ m/sec.} \qquad (5\text{–}1.3)$$

Comment. From Eqs. (1–4.2) and (5–1.2), the mobility of the charge carriers in this material is 0.0625 (m/sec)/(volt/m). ◄

5–2 METALLIC CONDUCTIVITY

The metallic bond was described in Section 2–2 in terms of *delocalized electrons*. Specifically, the valence electrons are able to move throughout the metal as standing waves. Thus, there is no net charge transport in the absence of the electronic field.

---

* If Section 1–4 was skipped earlier, the reader is advised to refer to it now as a review of introductory electrical properties.

If the metal is placed in an electrical circuit, the electrons moving toward the positive electrode *acquire more energy* and gain velocity in that direction. Conversely, those electrons moving toward the negative electrode *reduce their energy* and velocity. As a result, the *drift velocity* of Eq. (5–1.2) is developed.

Mean free path   Waves move through periodic structures without interruption. A well-ordered crystal (Chapter 3) provides one of the most regular of the periodic structures available. Thus a metallic crystal lattice provides an excellent medium for electron movements. However, any irregularity in the repetitive structures through which a wave travels may deflect the wave. Thus, if an electron had been traveling toward the positive electrode and was then deflected, it would no longer continue to gain velocity in that direction. The net effect is to *reduce the drift velocity* just cited, even though we have not altered the electric field. In brief, irregularities in the lattice *decrease* the mobility of Eq. (5–1.2); therefore, they *decrease* the conductivity and *increase* the resistivity (Eq. 5–1.1).

The average distance that an electron can travel in its wavelike pattern without deflection is called the *mean free path*. We will want to identify irregularities that deflect electron movements, because that will help us understand why resistivities of metals are not all the same. We can identify two effects on the basis of Chapter 4.

Resistivity versus temperature   The resistivity of a metal increases with temperature (Fig. 5–2.1). To a first approximation it is linear (except near absolute zero). We have no basis on which to conclude that the n of Eq. (5–1.1) decreases significantly with increased temperature in a metal;* rather we must look at the mobility μ. Thermal agitation (Section 4–6) increases in intensity in proportion to increased temperature (except at very low temperatures). This increased agitation decreases the mean free path of the electrons by decreasing the regularity of the crystal, and therefore decreases

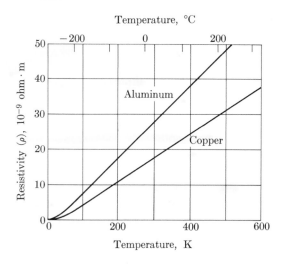

Fig. 5–2.1 Resistivity versus temperature (metals). The resistivity of metals is linear with temperature under normally encountered conditions.

---

* The number of charge carriers increases with increased temperature in a semiconductor (Section 5–4).

Table 5–2.1
Temperature resistivity coefficients

Metal		Resistivity at 0°C*, ohm·nm	Temperature resistivity coefficient, y_T, °C^{-1}
Aluminum		27	0.0039
Copper		16	0.0039
Gold		23	0.0034
Iron		90	0.0045
Lead		190	0.0039
Magnesium		42	0.004
Nickel		69	0.006
Silver		15	0.0038
Tungsten		50	0.0045
Zinc		53	0.0037
Brass	(Cu-Zn)	~60	0.002
Bronze	(Cu–Sn)	~100	0.001
Constantan	(Cu–Ni)	~500	0.00001
Monel	(Ni–Cu)	~450	0.002
Nichrome	(Ni–Cr)	~1000	0.0004

* These values will not agree with those in Appendix C, since they are based on different reference temperatures.

the mobility of electrons in a metal. The consequent change in resistivity is important to the engineer who is designing electrical equipment. In some cases, compensation must be introduced into a circuit to avoid an unwanted temperature sensitivity. In other cases, this temperature sensitivity provides a useful "brake." Example 5–2.3 will point this out in a familiar application (the toaster).

We may determine the ρ-versus-T relationship with a *temperature resistivity coefficient* y_T as follows:

$$\rho_T = \rho_{0°C}(1 + y_T \, \Delta T), \qquad\qquad (5-2.1)*$$

where $\rho_{0°C}$ is the resistivity at 0°C, and ΔT is $(T - 0°C)$. The value of this coefficient is approximately 0.004/°C for pure metals (Table 5–2.1). This suggests that the mean free path of electrons is reduced by a factor of two between 0°C and 250°C.

Resistivity in solid solutions   Another factor that can reduce the mean free path of electrons in a metal is the presence of solute atoms. A solid-solution alloy always has a higher resistivity than do its pure component metals.[†]

The reason for this generalization is that an electron encounters an irregularity in the potential field of the crystal lattice when it approaches an impurity atom. In the first place, the lattice is slightly distorted in an alloy such as brass, because the atomic radii differ a few percent; in addition, a zinc atom has 30 protons rather than

* The decreases in resistivity (increase in conductivity) we are describing are not related to *superconductivity* that appears near absolute zero. That involves another phenomenon (quantum mechanical), which is beyond the scope of this book.
† Examine the data for metals and alloys in either Table 5–2.1 or Appendix C.

the 29 in copper. This also alters the local field. Although these differences seem small, they deflect additional electrons and reduce the mean free path. Since brass (70 Cu–30 Zn) has a resistivity 3 or 4 times as great as that of pure copper, we can assume that the mean free path for electrons is only 25–30% as long in brass as in pure copper. If high conductivity is paramount in a design, the engineer will turn to pure metals (Section 6–1).

Energy bands   Recall from Fig. 2–2.7, that electrons of isolated atoms occupy only specific orbitals or energy levels, and that forbidden-energy gaps exist between these levels. In effect, the electrons establish standing waves around an individual atom. This pattern is also found in the inner or subvalence electrons of metals; however, the outer or valence electrons are delocalized when we have a large number of co-ordinated atoms. As a result, the valence orbitals form a band (Fig. 2–2.7b), and the standing wave is influenced by every atom that is involved. A consequence of this fact is that a band possesses as many standing wave forms and discrete energy *levels* as there are atoms in the system. Since the number is exceedingly great, and the energy bands are usually only a few electron volts wide, it follows that the energy levels within a band are so infinitesimally separated that we may pretend the band forms a continuum.

A physical principle* states that only two electrons may occupy the same level (and these two must be of opposite magnetic spin). Thus with its multitude of levels, *a band may contain twice as many electrons as there are atoms.* As a result, a monovalent metal such as sodium has its first valence band only half-filled (Fig. 5–2.2a). Since aluminum has three valence electrons per atom, its first valence band is filled and its second band is half full (Fig. 5–2.2c). Naturally the lower energy levels of the band fill first.

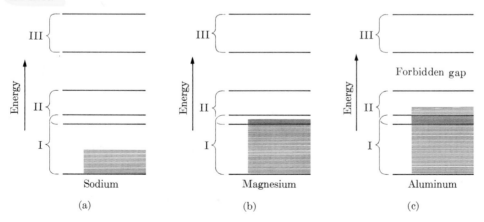

Fig. 5–2.2   Energy bands. (a) Sodium. Since it has only one valence electron per atom, its first valence band (I) is only half filled. (b) Magnesium. Its first band would be full, except that its second band (II) overlaps, to contain a few electrons. (c) Aluminum. With three valence electrons, its first band is filled, and its second band is half full. All of these metals have empty levels in the upper part of their valence bands.

---

* Pauli exclusion principle.

Definition of a metallic conductor   We characterized metals in Section 2–7 by their "ability to give up valence electrons," and pointed out that they thus were conductors. We now have a better definition of metallic conductors, viz., they have *unfilled valence bands*. Figures 5–2.2(a) and (c) show this schematically for sodium and aluminum.

The empty energy levels within a band are important for conduction because they permit an electron to rise to a higher energy level when it moves toward the positive electrode. This would not be possible if the energy band were completely filled and an overlying forbidden-energy gap were present.

Magnesium, with its two valence electrons per atom, is expected to fill the first valence band. It so happens, however, that the first and second bands overlap (Fig. 5–2.2b). Thus some of the $2N$ electrons (where N is the number of atoms) spill over into the second band, where there are plenty of vacant levels to receive the accelerating electrons. As a result, magnesium is metallic.

Silicon, however, presents another story because its four valence electrons per atom completely fill the first two valence bands (Fig. 5–2.3). Furthermore, there is a forbidden-energy gap above the second band. Thus, electrons cannot be energized within these valence bands, and there is a large gap below the energy levels of the third band (III). Silicon is not a metallic conductor (with pure materials, at 20°C, $\rho_{Si} = 2 \times 10^3$ ohm·m; $\rho_{Cu} = 17 \times 10^{-9}$ ohm·m—a ratio of about 10^{11}). We shall see, in the next two sections, that the difference between insulators and semiconductors is related to the size of the *energy gap* E_g that overlies the filled *valence band* (Fig. 5–2.4). Silicon is a semiconductor (Fig. 5–1.1) because its energy gap is of such size

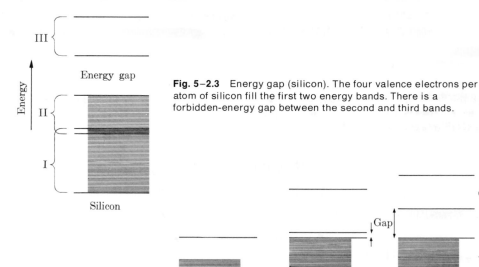

Fig. 5–2.3   Energy gap (silicon). The four valence electrons per atom of silicon fill the first two energy bands. There is a forbidden-energy gap between the second and third bands.

Fig. 5–2.4   Metals, semiconductors, and insulators. Metals have unfilled energy bands. Semiconductors have a narrow forbidden gap above the top filled valence band (VB). A few electrons can "jump the gap" to the conduction band (CB). insulators have a wide energy gap, which is a barrier to the electrons.

that some highly energized electrons are able to "jump the gap" into the *conduction band* (Section 5–4).

Study aids (metallic conductivity)   Energy bands, mean free path, and factors that affect metallic resistivity are presented by a series of nine sketches in the paperback *Study Aids for Introductory Materials Courses*. These are recommended for the student who finds difficulty with the visual concepts. In addition, Fermi energy and energy distributions are introduced for a more detailed understanding.

Example 5–2.1   Calculate the resistivity of silver at $-40°C$.   EQ 5-2.1   Pg 154

Solution: From Table 5–2.1, $\rho = 15$ ohm·nm at $0°C$, and $y_T = 0.0038/°C$.

$$\rho_{-40°C} = 15 \text{ ohm·nm} [1 + (0.0038/°C)(-40°C)]$$
$$= 13 \text{ ohm·nm}.$$

Example 5–2.2   Each atom percent of tin increases the resistivity of copper 30 ohm·nm regardless of temperature. What is the resistivity at $100°C$ of an alloy containing 99Cu–1Sn (weight basis)?

Solution: Basis: 100 amu

$$\text{Sn: } 1 \text{ amu}/(118.7 \text{ amu/atom}) = 0.0084 = 0.54 \text{ a/o.}$$
$$\text{Cu: } 99 \text{ amu}/(63.54) \qquad\quad = \underline{1.5581}$$
$$1.5665$$

The sum of the alloy and temperature resistivities are

$$\rho = (0.54 \text{ a/o})(30 \text{ ohm·nm per a/o}) + 16 \text{ ohm·nm} [1 + (0.0039/°C)(100°C)]$$
$$= 38 \text{ ohm·nm} \qquad (\text{or } 38 \times 10^{-9} \text{ ohm·m}).$$

Comment.   The increase in resistivity is nearly linear with composition at low solute concentrations. ◀

Example 5–2.3   A toaster uses 300 watts when it is in operation and the nichrome element is at $870°C$. It operates off a 110-V line. (a) How many amperes does it draw when it is hot? (b) When the switch is first snapped on?

Solution

a) $I = 300 \text{ W}/110 \text{ V} = 2.7 \text{ amp.}$

b) $R_{870°C} = 110 \text{ V}/2.7 \text{ A} = 40 \text{ ohm.}$

Since dimension changes are minor (and partially compensating),

$$R_{20°C}/R_{870°C} = \rho_{20°C}/\rho_{870°C}.$$

From Eq. (5–2.1) and Table 5–2.1,

$$R_{20} = R_{870}\left[\frac{\rho_0(1 + y_T 20°C)}{\rho_0(1 + y_T 870°C)}\right]$$

$$= 40 \text{ ohm} [1 + 0.0004(20)]/[1 + 0.0004(870)]$$
$$= 40 \text{ ohm} [1.008/1.35] = 30 \text{ ohms;}$$
$$I_{20} = 110 \text{ V}/30 \, \Omega = 3.7 \text{ amp.}$$

Comments. Had the element continued to draw 3.7 amperes, the temperature would continue to rise beyond 870°C, subjecting it to faster oxidation and related service deterioration.

Note that the temperature coefficients of resistivity of alloys are less than for pure metals. This is due in part to the fact that the mean free path for the electron is already short and the resistivity is initially higher. ◀

5–3 INSULATORS

In terms of energy bands, an insulator is a material with a large energy gap between the highest filled valence band and the next empty band (Fig. 5–2.4). The gap is so large that for all intents and purposes we can state that electrons are trapped in the lower band. Their number n in the conduction band for Eq. (5–1.1) is insignificantly low.

We commonly describe the valence electrons as being bound within the negative ions (or in a covalent bond). Approximately 7 eV ($=1.1 \times 10^{-18}$ J) of energy would be required to break an electron loose from the Cl^- ions in NaCl, and about 6 eV of energy to separate an electron from the covalent bond of diamond. These activation energies of 7 eV and 6 eV are also the dimensions of the energy gaps and may be compared to those we will encounter for silicon and germanium in the next section of 1.1 eV and 0.7 eV, respectively. The physicist considers an energy gap of about 4 eV ($=0.64 \times 10^{-18}$ J) as an arbitrary distinction between semiconductors and insulators. This is consistent with the fact that NaCl and diamond are electronic insulators, and silicon and germanium are semiconductors* (Fig. 5–1.1).

Example 5–3.1 An 0.1-mm film of polyethylene (PE) is used as a dielectric to separate two electrodes at 110 V. Based on Fig. 5–1.1, what is the electron flux through the film?

Solution: Basis = 1 m² and 1 sec.

$$R = \frac{(10^{14} \text{ ohm·m})(10^{-4} \text{ m})}{(1 \text{ m}^2)} = 10^{10} \text{ ohm};$$

$$I = \frac{110 \text{ V}}{10^{10} \text{ ohm}} = 11 \times 10^{-9} \text{ amp}.$$

$$\frac{(11 \times 10^{-9} \text{ A/m}^2)}{(0.16 \times 10^{-18} \text{ A·s/el})} = 7 \times 10^{10} \text{ electrons/sec·m}^2.$$

Comments. This nanoampere current is small, but measurable. Very commonly, extraneous factors such as impurities, pinhole porosity, surface leakage, must be considered when making measurements. ◀

5–4 INTRINSIC SEMICONDUCTORS

Semiconductors and insulators are differentiated on the basis of the size of their forbidden-energy gaps (see Fig. 5–2.4). In a semiconductor, the energy gap is such

---

* Diamond can be an electronic semiconductor if impurities are present; and NaCl can be an ionic semiconductor if conditions are favorable for sodium *ion* diffusion.

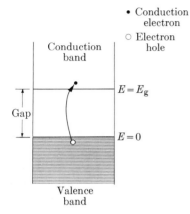

Fig. 5-4.1 Semiconduction. In semiconductors, a useful fraction of the valence electrons can jump the forbidden-energy gap. The electron is a negative carrier in the conduction band. The electron hole is a positive carrier in the valence band.

Table 5-4.1
Energy gaps in semiconducting elements

At 20°C (68°F)

Element	Energy gap E_g,		Fraction of valence electrons with energy $> E_g$	Conductivity σ, ohm$^{-1} \cdot$m^{-1}
	10^{-18} J	eV		
C(diamond)	0.96	~6	~$1/30 \times 10^{21}$	$< 10^{-16}$
Si	0.176	1.1	~$1/10^{13}$	5×10^{-4}
Ge	0.112	0.7	~$1/10^{10}$	2
Sn(gray)	0.016	0.1	~$1/5000$	10^6

that usable numbers of electrons are able to jump the gap from the filled valence band to the empty conduction band (Fig. 5-4.1). Those energized electrons can now carry a charge toward the positive electrode; furthermore, the resulting electron holes in the valence band become available for conduction because electrons deeper in the band can move up into those vacated levels.

Figure 5-4.2 shows the energy gap schematically for C(diamond), Si, Ge, and Sn(gray). The gap is too large in diamond to provide a usable number of charge carriers, so diamond is categorized as an insulator (Table 5-4.1). The number of carriers increases as we move down through Group IV of the periodic table to silicon, germanium, and tin; as a result, the conductivity increases, as shown in the accompanying table. This conductivity is an inherent property of these materials and does not arise from impurities. Therefore, it is called *intrinsic semiconductivity*.

The crystal structure of diamond is repeated (from Chapter 2) in Fig. 5-4.3(a). Each carbon atom has a coordination number of 4, and each neighboring pair of

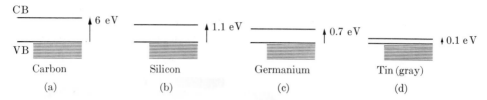

Fig. 5–4.2 Energy gaps in Group IV elements (schematic). All these elements can have the same structure, and they all have filled bands. Because tin has the smallest energy gap, it has the most electrons in the conduction band (CB) at normal temperatures, and therefore the highest conductivity. (Cf. Table 5–4.1.)

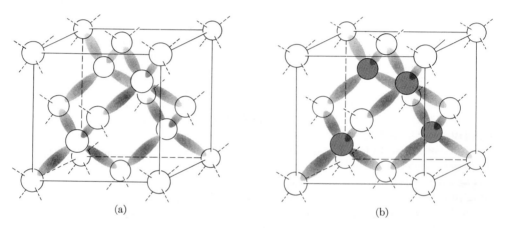

Fig. 5–4.3 Crystal structures of familiar semiconductors. (a) Diamond, silicon, germanium, gray tin. (b) ZnS, GaP, GaAs, InP, etc. (Cf. Fig. 8–2.3.) The two structures are similar, except that two types of atoms are in alternate positions in the semiconducting compounds. All atoms have CN = 4; each material has an average of four valence electrons per atom, and two electrons per bond.

atoms shares a pair of electrons (Section 2–2). Silicon, germanium, and gray tin have the same structure.* Figure 5–4.4 uses germanium to represent schematically the mechanism of semiconductivity in these elements.

The above four Group IV elements are the only elements that obtain semiconductivity from the structure of Fig. 5–4.3(a). However, a number of III–V *compounds* are based on the same structure (Fig. 5–4.3b). Atoms of elements from Group III of the periodic table (B, Al, Ga, In) alternate with atoms of elements from Group V of the periodic table (N, P, As, Sb). Most of the 16 III–V compounds that can form from these elements are semiconductors because every atom has four neighbors, and the average number of shared valence electrons is *four*. This matches exactly the situation for silicon and germanium, our predominant semiconductors.

---

* White tin is the more familiar polymorph. It is stable above 13°C (but may be supercooled to lower temperatures). White tin (bct) is denser than gray tin ($\rho_w = 7.3$ Mg/m^3 = 7.3 g/cm^3, while $\rho_g = 5.7$ Mg/m^3); therefore, the energy bands of white tin overlap, and this phase is a metallic conductor (cf. Fig. 5–2.2b).

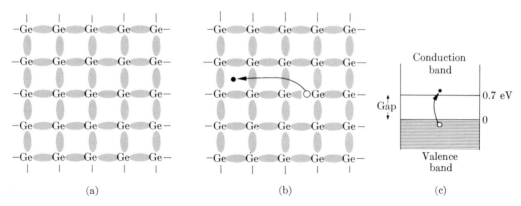

Fig. 5-4.4 Intrinsic semiconductor (germanium). (a) Schematic presentation showing electrons in their covalent bonds (and their valence bands). (b) Electron-hole pair. (Positive electrode at the left.) (c) Energy gap, across which an electron must be raised to provide conduction. For each conduction electron, there is a hole produced among the valence electrons.

Charge mobility   Our introductory equation on conductivity (Eq. 1-4.1) must now be modified to match Fig. 5-4.1, since an intrinsic semiconductor has both negative and positive carriers.

The *electrons* that jump into the conduction band are the *negative*-type carriers. The conductivity they produce depends on their mobility μ_n through the conduction band of the semiconductor. The *electron holes* that are formed in the valence band are the *positive*-type carriers.* The conductivity they produce depends on their mobility μ_p through the valence band of the semiconductor. The total conductivity arises from both contributors.

$$\sigma = n_n q \mu_n + n_p q \mu_p. \tag{5-4.1}$$

Of course, both the hole and the electron carry the same basic charge unit of 0.16×10^{-18} coul. In an intrinsic semiconductor, where there is a one-for-one formation of conduction electrons and electron holes, $n_n = n_p$; thus we could simplify Eq. (5-4.1). However, let's leave it in its present form because n_n does not equal n_p for *extrinsic* semiconductors (Section 5-5).

Table 5-4.2 summarizes the properties of a number of semiconductors. Note that we can make two generalizations.

1.   The size of the energy gap commonly decreases as we move down in the periodic table (C $\rightarrow$ Si $\rightarrow$ Ge $\rightarrow$ Sn), or (GaP $\rightarrow$ GaAs $\rightarrow$ GaSb), or (AlSb $\rightarrow$ GaSb $\rightarrow$ InSb).

2.   The mobility of electrons within a given semiconductor is greater than the mobility of electron holes in the same semiconductor.[†] The latter difference will be important when considering the use of *n*-type semiconductors in contrast to *p*-type semiconductors.

---

* Comparably, anions, with extra electrons, are negative-type, and cations, which are deficient in electrons, are positive-type.
† This relationship exists for all of the semiconductors of Table 5-4.2, with the possible exception of AlSb, where the mobility data have not been accurately determined.

Table 5–4.2
Properties of common semiconductors (20°C)*

Material	Energy gap E_g,		Mobilities, $m^2/volt \cdot sec$		Intrinsic conductivity, $ohm^{-1} \cdot m^{-1}$	Lattice constant, a, nm
	10^{-18} J	eV	Electron, μ_n	Hole, μ_p		
Elements						
C(diamond)	0.96	~6	0.17	0.12	$<10^{-16}$	0.357
Silicon	0.176	1.1	0.19	0.0425	5×10^{-4}	0.543
Germanium	0.112	0.7	0.36	0.23	2	0.566
Tin (gray)	0.016	0.1	0.20	0.10	10^6	0.649
Compounds						
AlSb	0.26	1.6	0.02	—	—	0.613
GaP	0.37	2.3	0.019	0.012	—	0.545
GaAs	0.22	1.4	0.88	0.04	10^{-6}	0.565
GaSb	0.11	0.7	0.60	0.08	—	0.612
InP	0.21	1.3	0.47	0.015	500	0.587
InAs	0.058	0.36	2.26	0.026	10^4	0.604
InSb	0.029	0.18	8.2	0.17	—	0.648
ZnS	0.59	3.7	0.014	0.0005	—	—
SiC (hex)	0.48	3	0.01	0.002	—	—

*   Revised data collected by B. Mattes.

Semiconductivity (intrinsic) versus temperature   Unlike metals, which have increased resistivity and decreased conductivity at higher temperatures (Fig. 5–2.1), the conductivity of intrinsic semiconductors *increases* at higher temperatures. The explanation is straightforward when one considers that the number of charge carriers, n, increases directly with the number of electrons that jump the gap (Fig. 5–4.1). At 0 K, *no* electron would have the necessary energy to do this; however, as the temperature rises, the electrons receive energy, just as the atoms do. At 20°C, a useful fraction of the valence electrons in silicon, germanium, and tin have energy in excess of E_g, the energy gap (Table 5–4.1). The same is true for compound semiconductors.

By analogy with Eq. (4–6.3) the distribution of these thermally energetic electrons is

$$n_i \propto e^{-(E - \bar{E})/kT}, \qquad (5\text{–}4.2a)*$$

where n_i is the number of electrons per m^3 in the conduction band (and the number of holes per m^3 in the valence band). Within the forbidden-energy gap of an intrinsic

---

*   Equations (5–4.2a) and (4–6.3a) are analogous at the upper end of the energy range only. In that range, the Pauli exclusion principle of two electrons per quantum state is not restrictive, since the probabilities of occupancy are very low. Of course, our interest is in the upper end of the range and in the initial electrons to jump the gap.

semiconductor, the average energy $\bar{E}$ is at mid-gap, $E_g/2$. Therefore,

$$n_i \propto e^{-E_g/2kT}. \tag{5-4.2b}$$

As in Chapter 4, T is the absolute temperature (K), and k is Boltzmann's constant, often expressed as 86.1×10^{-6} eV/K rather than 13.8×10^{-24} J/K.

Conductivity σ is directly proportional to the number of carriers n; therefore,

$$\sigma = \sigma_0 e^{-E_g/2kT}, \tag{5-4.3a}$$

where σ_0 is the proportionality constant that includes, among other factors, both q and μ of Eq. (5-1.1). Admittedly, the mobility μ varies some with temperature; however, its variation within the normal working range of most semiconductors is small compared with the exponential variation of the number of carriers n. Thus, we can rewrite this last equation in an Arrhenius form:

$$\ln \sigma = \ln \sigma_0 - E_g/2kT. \tag{5-4.3b}$$

If we measure the conductivity (or resistivity) of a semiconductor in the laboratory and plot $\ln \sigma$ versus $1/T$, we can calculate E_g from the slope of the curve, i.e., slope $= -E_g/2k$. Conversely, from E_g and one known value of σ, we may calculate σ at a second temperature (Example 5-4.4).

Photoconduction There is only a small probability that an electron in the valence band of silicon may be raised across the energy gap by thermal activation into the conduction band (~ 1 out of 10^{13}, according to Table 5-4.1). In contrast, a photon of red light (wavelength = 660 nm) has 1.9 eV of energy, which is more than enough to cause an electron to jump the 1.1 eV energy gap in silicon (Fig. 5-4.5). Thus the conductivity of silicon increases markedly by photoactivation when it is exposed to light.

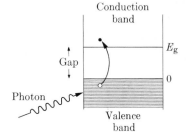

Fig. 5-4.5 Photoconduction. A photon (i.e., light energy) raises the electron across the energy gap, producing a "conduction electron + valence hole" pair, forming charge carriers. Recombination (Eq. 5-4.4b) occurs when the electron drops back to the valence band.

Recombination The reaction that produces an *electron–hole pair*, as shown in Fig. 5-4.5, may be written as

$$E \to n + p \tag{5-4.4a}$$

where E is energy, n is the conduction electron, and p is the hole in the valence band. In this case the energy came from light, but it could have come from other energy sources such as heat or fast-moving electrons.

Since all materials are more stable when they reduce their energies, electron–hole pairs recombine sooner or later:

$$n + p \rightarrow E. \qquad (5\text{–}4.4b)$$

In effect, the electron drops from the conduction band back to the valence band, just the reverse of Fig. 5–4.4(c). Were it not for the fact that light or some other energy source continually produces additional electron–hole pairs, the conduction band would soon become depleted.

The time required for recombination varies from material to material. However, it follows a regular pattern because, within a specific material, every conduction electron has the same probability of recombining within the next second (or minute). This leads to the relationship

$$N = N_0 e^{-t/\tau} \qquad (5\text{–}4.5a)$$

which we usually rearrange to

$$\ln (N_0/N) = t/\tau. \qquad (5\text{–}4.5b)^*$$

In these equations, N_0 is the number of electrons in the conduction band at a particular moment of time (say, when the source of light is turned off). After an additional time t, the number of remaining conduction electrons is N. The term τ is called the *relaxation* or *recombination time*, and is characteristic of the material.

• **Luminescence**   The energy released in Eq. (5–4.4b) may appear as heat. It may also appear as light. When it does, we speak of *luminescence* (Fig. 5–4.6). Sometimes we subdivide luminescence into several categories. *Photoluminescence* is the light emitted after electrons have been activated to the conduction band by light photons. *Chemoluminescence* is the word used when the initial activation is due to chemical reactions. Probably *electroluminescence* is best known, because this is what occurs in a TV tube, where a stream of electrons (cathode rays) scans the screen, activating the electrons in the phosphor to their conduction band. Almost immediately, however, the electrons and holes recombine, emitting energy as visible light.

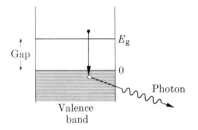

Fig. 5–4.6   Luminescence. Each millisecond, a fraction of the electrons energized to the conduction band return to the valence band. As the electron drops across the gap, the energy may be released as a photon of light.

---

* Equation (5–4.5) can be derived through calculus (by those who wish to do so) from the information stated above:

$$dN/dt = -N/\tau. \qquad (5\text{–}4.6)$$

Rearranging,

$$dN/N = -dt/\tau;$$

then integrating we get,

$$\ln N/N_0 = -t/\tau. \qquad (5\text{–}4.5c)$$

Since the recombination rate is proportional to the number of activated electrons, the intensity I of luminescence also follows Eq. (5–4.5b):

$$\ln (I_0/I) = t/\tau. \qquad (5\text{–}4.7)$$

For a TV tube, the engineer chooses a phosphor with a relaxation time such that light continues to be emitted as the next scan comes across. Thus our eyes do not see a light–dark flickering. However, the light intensity from the previous trace should be weak enough so that it does not compete with the new scan that follows one-thirtieth of a second later. (See Example 5–4.5.)

Study aids (intrinsic semiconductors)   The paperback *Study Aids for Introductory Materials Courses* has a section on "intrinsic semiconduction" that reviews energy bands; contrasts conductors, semiconductors, and insulators; then relates conductivity to gap size and temperature. Optional sketches describe several intrinsic semiconductor devices.

Example 5–4.1 (a) What fraction of the charge in intrinsic silicon is carried by electrons? (b) By electron holes?

Solution: From Eq. (5–4.1) and Table 5–4.2, and with $n_n = n_p$ in an intrinsic semiconductor,

$$\sigma_n/\sigma = \mu_n/(\mu_n + \mu_p)$$
$$= (0.19 \ m^2/V \cdot s)/(0.2325 \ m^2/V \cdot s) = 0.82.$$

$$\sigma_p/\sigma = 0.18.$$

Comment. Typically, conductivity is higher in the conduction band than from the electron holes in the valence band. ◀

Example 5–4.2 From Table 5–4.2, the compound gallium arsenide has an intrinsic conductivity of $10^{-6} \ ohm^{-1} \cdot m^{-1}$ at 20°C. How many electrons have jumped the energy gap?

Solution: From Eq. (5–4.1),

$$n = (10^{-6} \ ohm^{-1} \cdot m^{-1})/(0.16 \times 10^{-18} \ amp \cdot sec)(0.88 + 0.04 \ m^2/V \cdot sec)$$
$$= 6.8 \times 10^{12}/m^3.$$

Comment. There will be 1.36×10^{13} carriers/m^3 since an electron hole remains for each electron activated across the energy gap. ◀

Example 5–4.3 Each gray tin atom has four valence electrons. The unit cell size (Fig. 5–4.3a) is 0.649 nm. Separate calculations indicate that there are 2×10^{25} conduction electrons per m^3. What fraction of the electrons have been activated to the conduction band?

Solution: Basing our calculations on Fig. 5–4.3, we find that there are 8 tin atoms per unit cell.

$$\text{Valence electrons/cm}^3 = \frac{(8 \ \text{atoms/uc})(4 \ \text{el/atom})}{(0.649 \times 10^{-9} \ m)^3/uc} = 1.17 \times 10^{29}/m^3.$$

$$\text{Fraction activated} = \frac{2 \times 10^{25}}{1.17 \times 10^{29}} \simeq 0.0002. \quad ◀$$

Example 5–4.4 The resistivity of germanium at 20°C (68°F) is 0.5 ohm·m. What is its resistivity at 40°C (104°F)?

Solution: Based on Eq. (5–4.3) and an energy gap of 0.7 eV (Table 5–4.2):

$$\frac{\sigma_2}{\sigma_1} = \frac{\rho_1}{\rho_2} = \frac{\sigma_0 e^{-E_g/2kT_2}}{\sigma_0 e^{-E_g/2kT_1}},$$

$$\ln \rho_1/\rho_2 = \left(\frac{E_g}{2k}\right)\left[\frac{1}{T_1} - \frac{1}{T_2}\right], \tag{5–4.8}$$

$$= \frac{0.7 \text{ eV}}{2(86.1 \times 10^{-6} \text{ eV/K})}\left[\frac{1}{293 \text{ K}} - \frac{1}{313 \text{ K}}\right] = 0.9.$$

$$\rho_1/\rho_2 = \sim 2.5.$$

Thus, if $\rho_{20°} = 0.5$ ohm·m, $\rho_{40°C} = 0.2$ ohm·m.

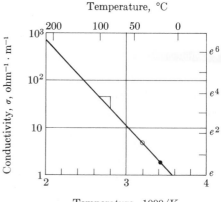

Fig. 5–4.7 Semiconduction versus temperature (intrinsic germanium). The slope is $-E_g/2k$ of Eq. (5–4.3) when the ordinate is ln σ and the abscissa is reciprocal temperature, K^{-1}. See Example 5–4.4.

Alternate solutions from the slope (Fig. 5–4.7):

$$\text{Slope} = -0.7 \text{ eV}/2(86.1 \times 10^{-6} \text{ eV/K}) = -4060 \text{ K};$$

or

$$\text{slope} = -(0.112 \times 10^{-18} \text{ J})/2(13.8 \times 10^{-24} \text{ J/K}) = -4060 \text{ K}.$$

At 20°C (●),

$$\ln \sigma = \ln (1/0.5 \text{ ohm·m}) = 0.693 \qquad (\text{or } \log_{10} = 0.301);$$

$$1/T = 1/293 \text{ K} = 0.00341/\text{K} \qquad (\text{or } 1000/T = 3.41).$$

At 40°C (○)

$$1/T = 1/313 \text{ K} = 0.00319/\text{K} \qquad (\text{or } 1000/T = 3.19).$$

$$-4060 \text{ K} = \frac{0.693 - \ln \sigma_{40°C}}{(0.00341 - 0.00319)/\text{K}};$$

$$\ln \sigma_{40°C} = 1.6 \qquad (\text{or } \log \sigma_{40°C} = 0.7);$$

$$\sigma_{40°C} = 5 \text{ ohm}^{-1} \text{ m}^{-1} \qquad (\text{or } \rho = 0.2 \text{ ohm·m}).$$

Comments. It is possible to measure resistance changes (and therefore resistivity changes) of <0.1%. Therefore one can measure temperature changes of a small fraction of a degree. (See Study Problem 5–6.1.) ◀

• **Example 5–4.5** The scanning beam of a television tube covers the screen with 30 frames per second. What must the relaxation time for the activated electrons of the phosphor be if only 20% of the intensity is to remain when the following frame is scanned?

Solution: Refer to Eq. (5–4.7):

$$\ln (1.00/0.20) = (0.033 \text{ sec})/\tau,$$

$$\tau = 0.02 \text{ sec}.$$

Comments. We use the term *fluorescence* when the relaxation time is short compared to the time of our visual perception. If the luminescence has a noticeable afterglow, we use the term *phosphorescence.* ◀

5–5 EXTRINSIC SEMICONDUCTORS

N-type semiconductors Impurities alter the semiconducting characteristics of materials by introducing excess electrons or excess electron holes. Consider, for example, some silicon containing an atom of phosphorus. Phosphorus has five valence electrons rather than the four that are found with silicon. In Fig. 5–5.1(a), the extra electron is present independently of the electron pairs that serve as bonds between neighboring atoms. This electron can carry a charge toward the positive electrode (Fig. 5–5.1b). Alternatively, in Fig. 5–5.1(c), the extra electron—which cannot reside in the valence band because that is already full—is located near the top of the energy gap. From

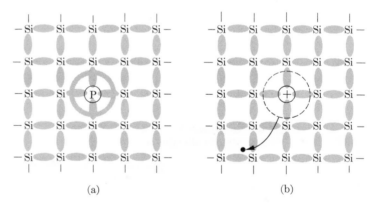

(a)   (b)

Fig. 5–5.1 Extrinsic semiconductors (*n*-type). A Group V atom has an extra valence electron beyond the average of four sketched in Fig. 5–4.3. This fifth electron can be pulled away from its parent atom with very little added energy, and "donated" to the conduction band, to become a charge carrier. We observe the donor energy level, E_d, as being just below the top of the energy gap. (a) An *n*-type impurity, such as phosphorus. (b) Ionized phosphorus atom. (Positive electrode at left.) (c) Band model.

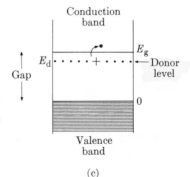

(c)

this position—called a *donor* level E_d—the extra electron can easily be activated into the conduction band. Regardless of which model is used, Fig. 5–5.1(b) or 5–5.1(c), we can see that atoms from Group V (N, P, As, and Sb) of the periodic table (Fig. 2–1.1) can supply negative, or *n*-type, charge carriers to semiconductors.

P-type semiconductors   Group III elements (B, Al, Ga, and In) have only three valence electrons. Therefore, when such elements are added to silicon as impurities, electron holes come into being. As shown in Fig. 5–5.2(a) and (b), each aluminum atom can accept one electron. In the process a positive charge moves toward the negative electrode. Using the band model (Fig. 5–5.2c), we note that the energy difference for electrons to move from the valence band to the *acceptor level*, E_a, is much less than the full energy gap. Therefore, electrons are more readily activated into the acceptor sites than into the conduction band. The electron holes remaining in the valence band are available as positive carriers for *p*-type semiconduction.

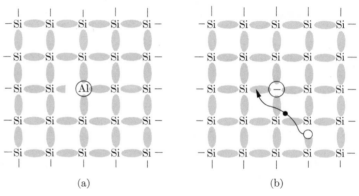

(a)   (b)

Fig. 5–5.2   Extrinsic semiconductors (*p*-type). A Group III atom has one less valence electron than the average of four sketched in Fig. 5–4.3. This atom can accept an electron from the valence band, thus leaving an electron hole as a charge carrier. The acceptor energy level, E_a, is just above the bottom of the energy gap. (a) A *p*-type impurity such as aluminum. (b) Ionized aluminum atom. (Negative electrode at right.) (c) Band model.

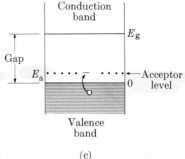

(c)

Donor exhaustion (and acceptor saturation)   Since donated electrons have only a small jump to the conduction band, they initiate *extrinsic* conductivity at relatively low temperatures. As the temperature is increased, the slope of the Arrhenius curve is $-(E_g - E_d)/k$ as shown at the right in Fig. 5–5.3.

   If the donor impurities are limited in number (e.g., 10^{21} P/m^3 within silicon), essentially all of the donated electrons have moved into the conduction band at

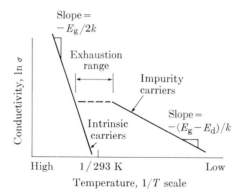

Fig. 5-5.3   Donor exhaustion. Intrinsic (left-hand curve) and extrinsic (right-hand curve) conductivity require energies of E_g and $(E_g - E_d)$, respectively, to raise electrons into the conduction band. At lower temperatures, donor electrons provide most of the conductivity. Exhaustion occurs when all of the donor electrons have entered the conduction band, and before the temperature is raised high enough for valence electrons to jump the energy gap. The conductivity is nearly constant in this temperature range.

temperatures below that of normal usage. This supply has been *exhausted*. In the above example of $10^{21}/m^3$, the extrinsic conductivity is

$$\sigma_{ex} = (10^{21}/m^3)(0.16 \times 10^{-18} \text{ A·sec})(0.19 \text{ m}^2/\text{V·sec})$$
$$= 30 \text{ ohm}^{-1} \cdot m^{-1}.$$

The extrinsic conductivity will not continue to rise with further temperature increases and there is a conductivity plateau.*

In the meantime, the *intrinsic* conductivity is very low in a semiconductor such as silicon (5×10^{-4} ohm^{-1}·m^{-1} at 20°C according to Table 5-4.1). Its Arrhenius curve is at the left in Fig. 5-5.3 with the intrinsic slope of $-E_g/2k$, that is, -1.1 eV/$2k$ = 6400 K. Only at elevated temperatures does the total conductivity increase above the exhaustion plateau. (Cf. Example 5-5.3.)

Donor exhaustion of n-type semiconductors has its parallel in *acceptor saturation* of p-type semiconductors. (The reader is asked to paraphrase the previous paragraphs for the saturation analog.) Donor exhaustion and acceptor saturation are important to materials and electrical engineers, since these situations provide a region of essentially constant conductivity. This means that it is less necessary to compensate for temperature changes in electrical circuits than it would be if the log σ-versus-$1/T$ characteristics followed an ever-ascending line.

• **Defect semiconductors**   The iron oxide of Fig. 4-3.2 possessed Fe^{3+} ions in addition to the regular Fe^{2+} ions. A similar situation occurs in Fig. 5-5.4(a) when NiO is oxidized to give some Ni^{3+} ions and, in fact, is relatively common among transition-metal oxides that have multiple valences. In nickel oxide, three Ni^{2+} are replaced

---

*   With constant n, experiments can detect a slight decrease in μ that results from the shorter mean free path that accompanies increased temperatures (Section 5-2).

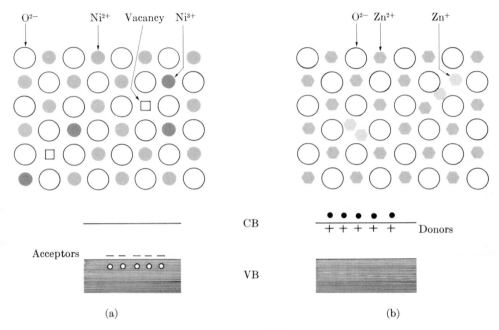

Fig. 5-5.4 Defect semiconductors. (a) $Ni_{1-x}O$. The Ni^{3+} ions serve as electron acceptors, so that holes, $\circ$, form in the valence band. (b) $Zn_{1+y}O$. The Zn^{+} ions are donors of electrons, $\bullet$, to the conduction band for n-type semiconduction.

by $2\,Ni^{3+}$ and a vacancy, $\square$. This maintains the charge balance; it also permits easier diffusion and therefore some ionic conductivity. More important, however, is the fact that electrons can hop from an Ni^{2+} ion into acceptor sites in Ni^{3+} ions. Conversely, an electron hole moves from one nickel ion to another as it migrates toward the negative electrode. Nickel oxide and other oxides with $M_{1-x}O$ defect structures are p-type semiconductors.

There are also n-type oxides. Zinc oxide, when exposed to a reducing atmosphere, produces $Zn_{1+y}O$ by removing some oxygen. However, in this case, an oxygen vacancy does not develop. Rather, a zinc ion moves into an interstitial position (Fig. 5-5.4b). The Zn^{+} ions that arise to balance the charge have one electron more than the bulk of the Zn^{2+} ions. These can donate electrons to the conduction band for n-type semiconductivity.

Study aids (extrinsic semiconductors)   A series of eleven sketches is available as an alternate approach to extrinsic semiconduction in *Study Aids for Introductory Materials Courses*. By using this alternate approach, it is possible to be more specific about junction devices, such as the Zener diode and transistors.

Example 5-5.1 Silicon, according to Table 5-4.2, has a conductivity of 5×10^{-4} ohm$^{-1}\cdot$m^{-1} when pure. An engineer wants it to have a conductivity of 200 ohm$^{-1}\cdot$m^{-1} when it contains aluminum as an impurity. How many aluminum atoms are required per m^3?

Solution: Since the intrinsic conductivity is negligible compared with 200 ohm$^{-1}\cdot$m^{-1}, essentially all of the conductivity must come from holes originating by the presence of acceptor atoms:

$$n_p = (200 \text{ ohm}^{-1}\cdot\text{m}^{-1})/(0.16 \times 10^{-18} \text{ amp}\cdot\text{sec})(0.0425 \text{ m}^2/\text{volt}\cdot\text{sec})$$
$$= 3 \times 10^{22}/\text{m}^3.$$

Comments. Each aluminum atom contributes one acceptor site and hence one electron hole. Therefore 3×10^{22} aluminum atoms are required per m^3. This is, of course, a large number; however, it is still small (0.6 ppm) when compared with the number of silicon atoms per m^3. (See Study Problem 5-5.1a.) ◀

Example 5-5.2 Early transistors used germanium with an extrinsic resistivity of 0.02 ohm$\cdot$m and a conduction electron concentration of $0.87 \times 10^{21}/\text{m}^3$. (a) What is the mobility of the electrons in the germanium? (b) What impurity elements could be added to the germanium to donate the conduction electrons?

Solution: Since we are considering extrinsic conductivity from electrons, i.e., *n*-type,

a)
$$\mu_n = 1/(0.02 \text{ ohm}\cdot\text{m})(0.87 \times 10^{21}/\text{m}^3)(0.16 \times 10^{-18} \text{ amp}\cdot\text{sec})$$
$$= 0.36 \text{ m}^2/\text{volt}\cdot\text{sec};$$

b)   Group V elements: N, P, As, Sb.

Comments. Note that the electron mobility does not depend on which of these Group V elements is added, since the electron, once in the conduction band, moves through the germanium lattice independent of its donor.

Group VI elements could also be added. Since they have a second additional electron (beyond the four necessary for bonding), it would take only $0.4 \times 10^{21}/\text{m}^3$ of these atoms to supply $0.8 \times 10^{21}/\text{m}^3$ conduction electrons. ◀

Example 5-5.3 The residual phosphorus content of purified silicon is 0.1 part per billion (by weight). Will the resulting conductivity exceed the intrinsic conductivity of silicon?

Solution: Based on density (Appendix B): 1 m^3 silicon $= 2.33 \times 10^6$ g Si $= 2.33 \times 10^{-4}$ g P.

$$(2.33 \times 10^{-4} \text{ g/m}^3)/(30.97 \text{ g}/0.6 \times 10^{24}) = 4.5 \times 10^{18}/\text{m}^3.$$

$$\sigma_{ex} = (4.5 \times 10^{18}/\text{m}^3)(0.16 \times 10^{-18} \text{ amp}\cdot\text{sec})(0.19 \text{ m}^2/\text{V}\cdot\text{s})$$

$\sigma = nq\mu$

$$= 0.14 \text{ ohm}^{-1}\cdot\text{m}^{-1};$$

versus

$$\sigma_{in} = 5 \times 10^{-4} \text{ ohm}^{-1}\cdot\text{m}^{-1}$$

(from Table 5-4.2).

Comment. It was necessary to develop entirely new processing procedures in order to achieve the low impurity levels required for semiconduction production. ◀

Example 5-5.4 There are 10^{22} Al/m^3 in silicon to produce a *p*-type semiconductor. At what temperature will the intrinsic conductivity of silicon equal the maximum extrinsic conductivity?

Solution: At saturation,

$$\sigma_{ex} = (10^{22}/\text{m}^3)(0.16 \times 10^{-18} \text{ A}\cdot\text{sec})(0.0425 \text{ m}^2/\text{V}\cdot\text{sec})$$
$$= 68 \text{ ohm}^{-1}\cdot\text{m}^{-1}.$$

From Table 5–4.2, $\sigma_{in} = 5 \times 10^{-4}$ ohm^{-1}·m^{-1} at 20°C.

Eq. (5–4.3):
$$\frac{5 \times 10^{-4} \text{ ohm}^{-1} \text{ m}^{-1}}{68 \text{ ohm}^{-1} \text{ m}^{-1}} = \frac{\sigma_0 e^{-1.1/2k(293\,K)}}{\sigma_0 e^{-1.1/2kT}};$$

$$\ln(5 \times 10^{-4}/68) = -11.8 = \frac{-1.1 \text{ eV}}{2(86.1 \times 10^{-6} \text{ eV/K})}\left[\frac{1}{293 \text{ K}} - \frac{1}{T}\right];$$

$$T = 640 \text{ K} \qquad (\text{or } 367°\text{C}).$$

Comments. The most general equation for conductivity in semiconductors is

$$\sigma = \sigma_{in} + (\sigma_n)_{ex} + (\sigma_p)_{ex}$$
$$= (n_{in}q)(\mu_n + \mu_p) + (n_n q \mu_n)_{ex} + (n_p q \mu_p)_{ex}. \qquad (5–5.1)$$

Typically, only one of the three terms is significant at a time, and the others can be ignored. In this example, only $(\sigma_p)_{ex}$ is significant at 20°C. ◀

• 5–6   SEMICONDUCTING DEVICES

There are many electronic devices that use semiconductors. We shall consider but a few.

Conduction and resistance devices   We have already seen that the conductivity of a *photoconductor* will vary directly with the amount of incident light. This capability leads to *light sensing* devices. The radiation does not have to be visible—it may also be ultraviolet or infrared, providing the photons have energy comparable to or greater than the energy gap.

A second device is a *thermistor.* It is simply a semiconductor that has had its resistance calibrated against temperature. If the energy gap is large, so that the ln σ versus $1/T$ curve is steep, it is possible to design a thermistor that will detect temperature changes of 10^{-4}°C.*

Because many semiconducting devices have low packing factors, they have a high compressibility. Experiments show that as the volume is compressed, the size of the energy gap is measurably reduced; this, of course, increases the number of electrons that can jump the energy gap. Thus pressure can be calibrated against resistance for *pressure gauges*.

A *photomultiplier* device makes use of electron activation, first by photons *and* then by the electrons themselves. Assume, for example, that a very weak light source, even just one photon, were to hit a valence electron. Our eye would not have been able to detect it. However, if that electron is raised to the conduction band, and simultaneously the semiconductor is within a very strong electric field, that electron

---

* In technical practice, thermistors are better than other types of thermometers for measuring small temperature *changes*. However, thermocouples, etc., are more convenient for measuring the temperature itself. The measurement of temperature changes is important in microcalorimetric studies involving chemical or biological reactions.

would be accelerated to high velocities and high energies. In turn, it could activate one or more additional electrons that would also respond to the very strong field. The multiplying effect may be used to advantage. A very weak light signal may be amplified. With appropriate focusing, the image in nearly complete darkness may be brought into visible display.

Junction devices (diodes)   A number of devices utilize junctions between n-type and p-type semiconductors. The most familiar of these is the *light-emitting diode* (LED). We see it used in the digital displays (red) that are placed in many hand calculators. An LED operates on the principle shown schematically in Fig. 5–6.1. The charge carriers on the n-side and p-side of the junction are electrons and holes, respectively. If a current is passed through the device in the direction shown, the holes of the valence band move through the junction, into the n-type material; conversely, the electrons of the conduction band cross into the p-type material. Adjacent to the junction, there are excess carriers that recombine and produce luminescence:

$$n + p \rightarrow \text{photon.} \tag{5–6.1}$$

When GaAs is used, the photons emitted in the recombination zone are red; GaP gives green photons.

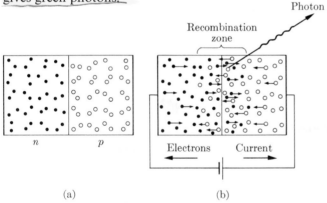

(a)   (b)

Fig. 5–6.1   Light-emitting diode (schematic). (a) An LED is a junction device between n-type and p-type semiconductors. (b) When a forward bias is placed across the junction, carriers of both types cross the junction where they recombine, emitting a photon (Eq. (5–4.4b) and Fig. 5–4.6).

     The junction of Fig. 5–6.1 can also serve as a *rectifier*; i.e., it is an electrical "check-valve" that lets current pass one way and not the other. Figure 5–6.2 extends the concept of Fig. 5–6.1 by introducing a reverse voltage, or bias. In Fig. 5–6.2(b) the carriers are pulled away from each side to leave a carrier-free, "insulating zone" at the junction. The conductivity of the zone is low. If a greater voltage is applied, it simply widens the insulating zone. Usable current passes only with a forward bias (Fig. 5–6.2a).

     The junction of Fig. 5–6.2(b) passes only very minor amounts of current with increased voltage—up to a point. With a sufficiently high reverse bias, the top of the

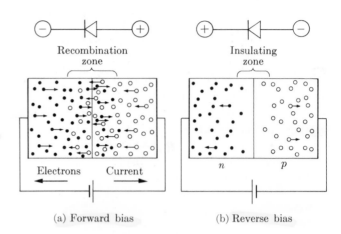

Fig. 5–6.2 Rectifier (schematic). (a) Current flows with a forward bias because charge carriers pass the junction. (b) With a reverse bias, charge carriers are depleted from the junction region. Extrinsic conductivity disappears from the junction region, and only a small amount of intrinsic conductivity remains.

valence band in the *p*-type material may become higher than the bottom of the conduction band in the *n*-type material, with the result that valence-band electrons may pass from the *p* to the *n* (Fig. 5–6.3). With such a voltage, the "reverse" current can be very high. In effect, we have a valve that opens at a definite voltage. Silicon carbide (SiC) is an early material used in this manner for *lightning arrestors*. The size of its energy gap makes it a very poor conductor, except when hit with a stroke of lightning. At that voltage, current passes to the ground. Once grounded, it is an insulator again. It is an electrical "safety valve."

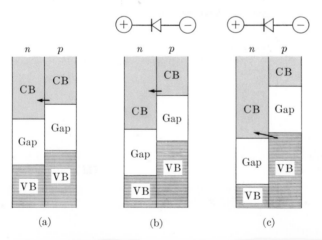

Fig. 5–6.3 Zener diode (schematic). A reverse bias lowers the gap on the *n*-side of the junction and raises the gap on the *p*-side of the junction. Little current flows in parts (a) and (b) because there are few electrons in the conduction band (CB) of the *p*-side to serve as carriers. (c) When the bias becomes great enough, an avalanche of electrons can move directly from the valence band (VB) of the *p*-side of the junction to the conduction band of the *n*-side. If the reverse bias is decreased, and the situation returns to (b) and (a), negligible current flows. This breakdown voltage can be tailor-made by controlling the doping levels.

Diodes, based on the above principle, may be designed for a large variety of breakdown voltages ranging from thousands down to a few volts. Called *Zener diodes*, these low-voltage devices can serve advantageously as gates and filters in electrical circuits.

Transistors   As the holes move across the junction with a forward bias (Fig. 5–6.1b), they recombine with the electrons in the *n*-type material according to Eq. (5–4.4b). Likewise, the electrons combine with the holes as the electrons move into *p*-type material. These reactions do not occur immediately, however. In fact, an excess number of positive and negative carriers may move measurable distances beyond the junction. The numbers of excess, unrecombined carriers are an exponential function of the applied voltage *V*, and become important to transistors.

Transistors revolutionized engineering design. We can explore their operation by means of a simplified model. A transistor has two junctions in series. They may be *p-n-p* or *n-p-n*. The former has been somewhat more common in the past; however, we shall consider the *n-p-n* transistor, since it's a little easier for us to envision the movements of electrons than the movements of holes. The principles behind each type are the same, though.

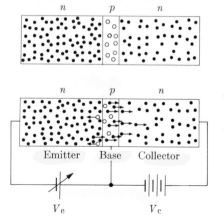

Fig. 5–6.4   Transistor (*n–p–n*). The number of electrons crossing from the emitter-base junction is highly sensitive to the emitter voltage. If the base is narrow, these carriers move to the base–collector junction, and beyond, before recombination. The total current flux, emitter to base, is highly magnified, or amplified, by fluctuations in the voltage of the emitter.

A transistor consists of an *emitter*, a *base*, and a *collector* (Fig. 5–6.4). For the moment, consider only the *emitter junction*, which is biased so that electrons move into the base (and toward the collector). As discussed a moment ago, the number of electrons that cross this junction and move into the *p*-type material is an exponential function of the emitter voltage, V_e. Of course, these electrons at once start to combine with the holes in the base; however, if the base is narrow, or if the recombination time is long (τ of Eq. 5–4.5), the electrons keep on moving through the thickness of the base. Once they are at the second or *collector junction*, the electrons have free sailing, because the collector is an *n*-type semiconductor. The total current that moves through the collector is controlled by the emitter voltage, V_e. As the emitter voltage fluctuates, the collector current, I_c, changes exponentially. Written logarithmically,

$$\ln I_c \simeq \ln I_0 + V_e/B, \tag{5–6.1a}$$

or

$$I_c = I_0 e^{V_e/B},\qquad\qquad (5\text{–}6.1\text{b})$$

where I_0 and B are constants for any given temperature. Thus, if the voltage in the emitter is increased even slightly, the amount of current is increased markedly. It is because of these relationships that a transistor serves as an amplifier.

Example 5–6.1 Zinc sulfide is used as a thermistor. To what fraction sensitivity, δ, must the resistance be measured to detect $0.001°C$ change at $20°C$?

Solution: Since the geometry is constant,

$$\delta = \frac{R_1 - R_2}{R_1} = \frac{\rho_1 - \rho_2}{\rho_1} = 1 - \frac{\rho_2}{\rho_1} = 1 - \frac{\sigma_1}{\sigma_2};$$

$$1 - \delta = \frac{\sigma_1}{\sigma_2} = \left[\frac{\sigma_0 e^{-3.7\text{eV}/2kT}}{\sigma_0 e^{-3.7\text{eV}/2k(T+0.001)}} \right];$$

$$\ln(1 - \delta) = \frac{-3.7\text{ eV}}{2(86.1 \times 10^{-6}\text{ eV/K})} \left[\frac{1}{T} - \frac{1}{T + 10^{-3}} \right].$$

Since

$$\frac{1}{T} - \frac{1}{T + 10^{-3}} = \frac{(T + 10^{-3}) - T}{T(T + 10^{-3})} = \sim \frac{10^{-3}}{T^2},$$

$$\ln(1 - \delta) = (-21{,}500\text{ K})(10^{-3}\text{ K}/[293\text{ K}]^2)$$
$$= -2.5 \times 10^{-4}.$$

$$\delta = 0.00025 \qquad (\text{or } 0.025\%).$$

Comment. A bridge-type instrument would be required. ◀

Example 5–6.2 A transistor has a collector current of 4.7 milliamperes when the emitter voltage is 17 millivolts. At 28 millivolts, the current is 27.5 milliamperes. Given that the emitter voltage is 39 millivolts, estimate the current.

Solution: Based on Eq. (5–6.1),

$$\ln 4.7 \simeq \ln I_0 + 17/B = 1.55,$$

$$\ln 27.5 \simeq \ln I_0 + 28/B = 3.31.$$

Solving simultaneously, using *milli*units, we have

$$\ln I_0 \simeq -1.17 \quad \text{and} \quad B \simeq 6.25$$

At 39 millivolts,

$$\ln I_c \simeq -1.17 + 39/6.25 \simeq 5.07,$$

$$I_c \simeq 160 \text{ milliamp.}$$

Comments. The electrical engineer modifies Eq. (5–6.1) to take care of added current effects. These, however, do not change the basic relationship: The variation of the collector current is much greater than the variation of the signal voltage. ◀

REVIEW AND STUDY

SUMMARY

In contrast to atomic order found in crystals (Chapter 3) and the atomic disorder of crystalline imperfections (Chapter 4), this chapter focused on electrons and their freedom to move among the atoms. In metals the charge is carried by electrons ($q = 0.16 \times 10^{-18}$ coul); in ionic materials it is carried by either anions, which are atoms with extra electrons, or cations, which have an electron deficiency; in semiconductors the charge carriers are both electrons and electron holes. Comparable to cations, electron holes carry a positive charge.

Metals have unfilled valence energy bands. In both insulators and semiconductors, the valence bands are filled. In semiconductors, however, the overlying gap is small enough so that electrons can be energized into the conduction band.

Since many electrons are available for conductivity in metals, their conductivity is established by the mobility of the electrons. Any factors such as thermal vibrations or impurities that introduce crystal imperfections will reduce the mean free path of the electrons and thereby decrease the conductivity (increase the resistivity).

Semiconductors include two main categories, intrinsic and extrinsic. The former gain their conductivity by electron activation across the energy gap, independent of impurities. Thermal activation is most common. Since the numbers of charge carriers increase with temperature in the exponent, the changing values of n in Eq. (5–4.1) overshadow the effects of temperature on carrier mobility.

It is possible to activate the electrons into the conduction band, not only by heat but also by light and other types of radiation. Each activated electron in an intrinsic semiconductor leaves an electron hole, so that each member of the electron–hole pair contributes to the conductivity. In general the electrons in the conduction band have greater mobility than the holes in the valence band. Devices such as thermistors and photoconductors make use of intrinsic conductivity that arises from electron activation across the energy gap.

Extrinsic semiconductors gain their conductivity from (1) impurities that donate electrons to the conduction band (n-type), or (2) impurities that accept electrons from the valence band (p-type). For silicon and similar Group IV elements, n-type conductivity requires Group V (or VI) elements; p-type conductivity requires Group III (or II) elements. Extrinsic semiconductors have permitted the materials scientist and electrical engineer to design a variety of junction devices. Rectifiers and transistors are two of many such devices.

Defect semiconduction represents a special class of extrinsic semiconductors that originate from nonstoichiometry in multivalent ions. Thus, numerous metallic oxides can possess either n-type or p-type conductivity.

KEY TERMS AND CONCEPTS

Acceptor level	Charge carrier, n
Acceptor saturation	Conductivity, σ

Conduction band

Conductors

Donor

Donor exhaustion

Drift velocity, $\bar{v}$

Electric field, $\mathscr{E}$

Electron charge, q

Electron hole

Electron–hole pair

Energy band

Energy gap, E_g

• Fluorescence

Insulator

• Junction

• Light-emitting diode, LED

• Luminescence

Mean free path

Metallic conductor

Mobility, μ

• Phosphorescence

Photoconductors

Recombination

• Rectifier

Recombination time, τ

Resistivity, ρ

  vs. solid solution

  vs. temperature

Semiconductors

  compound

  • defect

  extrinsic

  intrinsic

  n-type

  p-type

Temperature resistivity coefficient, y_T

• Thermistor

• Transistor

Valence band

• Zener diode

FOR CLASS DISCUSSION

A_5 Cite and compare the various types of charge carriers.

B_5 Discuss various factors that affect the drift velocity of electrons in solids.

C_5 The conductivity mechanism is sometimes associated with diffusivity because we are concerned with transport processes. Based on their units, cite the similarity and a difference between mobility and diffusivity.

D_5 Why does brass have a higher resistivity than copper?

E_5 Why does the conductivity of metals decrease at elevated temperatures?

F_5 Why does the conductivity of semiconductors increase at elevated temperatures?

G_5 Explain to a classmate why there are energies that are forbidden to electrons that are associated with individual atoms.

H_5 Differentiate between metallic conductors, insulators, and semiconductors.

I_5 Assume that the first and second bands of magnesium do not overlap. How would the properties of magnesium differ (a) if the gap were $<1\,\text{eV}$? (b) If $>4\,\text{eV}$?

J_5 A bubble rises through a vertical water pipe. Make an analogy with an electron hole with respect to the force field and the direction of transport.

K_5 Differentiate between intrinsic and extrinsic semiconductors.

L_5 Predict whether boron nitride, with the structure of Fig. 5–4.3(b), will be an insulator or semiconductor.

M_5 Explain how an electric eye works.

N_5 What fraction of electrons remain in the conduction band after $t = \tau$? After $t = 2\tau$?

• O_5 Distinguish between fluorescence and phosphorescence.

P_5 ZnS may have the structure shown in Fig. 5–4.3(b). Assume some phosphorous atoms replace sulfur atoms. Will the compound be n-type or p-type?

Q_5 Differentiate between n-type and p-type semiconductivity.

R_5 Differentiate between acceptor and donor impurities.

S_5 Why is the slope for intrinsic carriers in Fig. 5–5.3 steeper than for impurity carriers?

T_5 Detail the mechanism of acceptor saturation in terms of Fig. 5–5.3.

U_5 There are 10^{11} magnesium atoms, which replace the same number of silicon atoms, in a mm^3 of silicon. How will this affect the conductivity compared with an impurity of 10^{11} aluminum atoms?

V_5 The mean free paths of electrons in semiconductors are shortened at higher temperatures. However, this effect of temperature on conductivity can only be detected in the exhaustion (or saturation) range. Why?

• W_5 Cu_2O contains predominantly Cu^+ ions, but some Cu^{2+} ions. Will it be n-type or p-type?

• X_5 Describe the principle of the light-emitting diode.

• Y_5 Explain how a silicon rectifier operates.

• Z_5 Describe the principle of the photomultiplier tube.

STUDY PROBLEMS

5–1.1 Seventy millivolts are placed across the 0.5-mm dimension of a semiconductor with a carrier mobility of $0.23 \ m^2/V \cdot sec$. What drift velocity develops?

Answer: $32 \ m/sec$

5–1.2 A semiconductor, which has a resistivity of $0.0313 \ ohm \cdot m$, develops a drift velocity of $6.7 \ m/sec$ when 1.2 volts are applied across a 9-mm piece. How many carriers are there per m^3?

5–1.3 Laboratory measurements indicate that the drift velocity of electrons in a semiconductor is $149 \ m/sec$ when the voltage gradient is $15 \ V/mm$. The resistivity is $0.07 \ ohm \cdot m$. What is the carrier concentration?

Answer: $9 \times 10^{21}/m^3$

5–1.4 A flashlight bulb has a resistance of 5 ohms when it is used in a 2-cell flashlight battery. Assume 3 volts. How many electrons move through the filament per minute?

5–2.1 Determine the temperature at which the resistivity of silver is 10 ohm·nm.

Answer: $-88°C$

5–2.2 At what temperature does silver have the same resistivity as gold does at 50°C?

5–2.3 The resistivity of copper doubles between 20°C and 300°C. At what temperature does the resistivity of aluminum equal the higher value for copper?

Answer: 70°C

5–2.4 Based on the data of Table 5–2.1 and Appendix C, estimate the effect of copper on the resistivity of silver (ohm·nm per a/o).

5–2.5 Based on the data of Example 5–2.2, estimate the resistivity of a 95 Cu–5 Sn (weight basis) bronze at 0°C.

Answer: 97×10^{-9} ohm·m ($=97$ ohm·nm)

5–2.6 A 6% variation (maximum) is permitted in resistance between 0° and 25°C. Which metals of Table 5–2.1 meet the specification?

5–2.7 An iron wire (1.07 m long and 0.53 mm in diameter) is connected to 25 volts at 20°C. (a) What is the initial amperage? The amperage at 910°C, just below the bcc $\leftrightarrow$ fcc transition temperature? (b) What is the approximate change in length between 20° and 910°C? Between 20° and 915°C? (Review Example 3–4.1. The thermal expansion for bcc iron will be directly proportional to the change in radius.)

Answer: (a) 53A, 11A (b) 16 mm, 11 mm (Use this Study Problem for Discussion Topic V_9.)

5–4.1 A silicon chip is 1 mm × 1 mm × 0.1 mm. How fast do electrons drift through its short dimension if 37 millivolts are applied?

Answer: 70 m/sec

5–4.2 Pure germanium has a conductivity of 2 ohm^{-1}·m^{-1} with equal numbers of negative carriers, n_n, and positive carriers, n_p. What fraction of the conductivity is due to electrons and what fraction is due to electron holes?

5–4.3 What fraction of the charge is carried by electron holes in intrinsic gallium arsenide (GaAs)?

Answer: 0.043 (or 4.3%)

5–4.4 How many electron carriers (and electron holes) does intrinsic silicon require to provide a conductivity of 1.1 ohm^{-1}·m^{-1}?

5–4.5 The resistivity of a semiconductor that possesses 10^{21} negative carriers/m^3 (and few positive carriers) is 0.016 ohm·m. (a) What is the conductivity? (b) What is the electron mobility? (c) What is the drift velocity when the potential gradient is 5 mV/mm? 0.5 V/m?

Answer: (a) 62.5 ohm^{-1}·m^{-1} (b) 0.39 m^2/V·sec (c) 1.95 m/sec, 0.195 m/sec

5–4.6 Based on the data in Table 5–4.2, which is larger, (1) the conductivity from electrons in intrinsic InP, or (2) the conductivity from holes in intrinsic InAs?

5–4.7 Pure silicon has 32 valence electrons per unit cell (8 atoms with 4 electrons each). Its resistivity is 2×10^3 ohm·m. What fraction of the valence electrons are conductors?

Answer: 1 out of 1.5×10^{13}

5–4.8 At an elevated temperature, 1 of every 10^9 valence electrons in germanium is in the conduction band. What is the conductivity? (Germanium has the same structure as silicon, but with $a = 0.566$ nm).

5–4.9 The mobility of electrons in silicon is 0.19 m^2/volt·sec. (a) What voltage is required across a 2-mm chip of Si to produce a drift velocity of the electrons of 0.7 m/sec? (b) What electron concentration must be in the conduction band to produce a conductivity from negative carriers of 20 ohm^{-1}·m^{-1}? (c) What would the total conductivity for this silicon be if no impurities are present?

Answer: (a) 7.4 mV   (b) 6.6×10^{20}/m^3 (c) 24.5 ohm^{-1}·m^{-1}

5–4.10 The conductivity of silicon is 5×10^{-4} ohm^{-1}·m^{-1} at 20°C (68°F). Estimate the conductivity at 30°C.

5–4.11 To what temperature must germanium be cooled in order for its conductivity to be reduced by a factor of two below its 20°C (68°F) value?

Answer: 6°C

5–4.12 A semiconductor has a conductivity of 7 ohm^{-1}·m^{-1} at 20°C and 20 ohm^{-1}·m^{-1} at 60°C. What is the size of the energy gap?

5–4.13 An intrinsic semiconductor has a conductivity of 111 ohm^{-1}·m^{-1} at 10°C and 172 ohm^{-1}·m^{-1} at 17°C. (a) What is the energy gap? (b) What is the conductivity at 13.5°C?

Answer: (a) 0.88 eV   (b) 138 ohm^{-1}·m^{-1}

5–4.14 An intrinsic semiconductor has a conductivity of 390 ohm^{-1}·m^{-1} at 5°C and 1010 ohm^{-1}·m^{-1} at 25°C. (a) What is the size of the energy gap? (b) What is the conductivity at 15°C?

• *5–4.15* Refer to Example 5–4.5. Assume a phosphor is used with a recombination time of 0.04 sec. What fraction of the light intensity will remain when the next scan is made?

Answer: 0.44

• *5–4.16* A phosphorescent material is exposed to ultraviolet light. The intensity of emitted light decreased 20% in the first 37 min after the ultraviolet light was removed. (a) How long will it be after the uv light has been removed before the emitted light has only 20% of the original intensity (a decrease of 80%)? (b) Only 1%?

5–4.17 A phosphorescent material must have an intensity of 50 (arbitrary units) after 24 hrs and 20 after 48 hrs. Based on these figures, what initial intensity is required? (Solve *without* a calculator.)

Answer: 125

5–4.18 With a calculator, determine the recombination time for the material in the previous problem.

5–5.1 Silicon has a density of 2.33 Mg/m^3 ($=2.33$ g/cm^3). (a) What is the concentration of silicon atoms per m^3? (b) Phosphorus is added to silicon to make it an *n*-type semiconductor with a conductivity of 100 mho/m and an electron mobility of 0.19 m^2/volt·sec. What is the concentration of donor electrons per m^3?

Answer: (a) 5×10^{28} Si/m^3 (b) 3.3×10^{21}/m^3

5–5.2 How many silicon atoms are there for each aluminum atom in Example 5–5.1?

5–5.3 Extrinsic germanium is formed by melting 3.22×10^{-6} g of antimony (Sb) with 100 g of germanium. (a) Will the semiconductor be *n*-type or *p*-type? (b) Calculate the concentration of antimony (in atoms/cm^3) in germanium. (The density of germanium is 5.35 Mg/m^3, or 5.35 g/cm^3.)

Answer: (a) *n*-type   (b) $8.5 \times 10^{20}/m^3$

5–5.4 Gallium arsenide is made extrinsic by adding 0.000001 a/o phosphorus (and keeping a stoichiometric Ga/As ratio). Calculate the extrinsic conductivity at exhaustion.

5–5.5 Aluminum is a critical impurity in making silicon for semiconductors. Assume only 10 ppb (0.000001 a/o) remain. Will the resulting extrinsic conductivity be greater or less than the intrinsic conductivity of silicon at 20°C?

Answer: Greater (3.4 versus 5×10^{-4} ohm$^{-1} \cdot$m^{-1})

5–5.6 Refer to Study Problem 5–5.5. At what temperature will 1% of the conductivity be intrinsic?

5–5.7 Three grams of *n*-type silicon that had been doped with phosphorus to produce a conductivity of 600 ohm$^{-1} \cdot$m^{-1} are melted with three grams of *p*-type silicon that had been doped with aluminum to produce a conductivity of 600 ohm$^{-1} \cdot$m^{-1}. (a) What is the resulting conductivity? (b) Will it be *p*-type or *n*-type?

Answer: (a) 230 ohm$^{-1} \cdot$m^{-1} (b) *p*-type

5–5.8 Silicon ($10^{21}/m^3$) is in solid solution in GaAs. Assume it replaces equal numbers of Ga and As atoms. (a) What is the anticipated conductivity? (b) Assume the 10^{21} Si/m^3 replaced only arsenic atoms. What is the anticipated conductivity?

5–5.9 Refer to Example 4–3.1. (a) Is the oxide *n*-type or *p*-type? (b) How many charge carriers are there per mm^3?

Answer: (a) *p*-type   (b) $7.8 \times 10^{18}/mm^3$

5–5.10 Experiments indicate that 99% of the charge is carried by electrons in Fe$_{<1}$O. (The rest is carried by ions.) If the oxide in Study Problem 5–5.9 has a conductivity of 93 ohm$^{-1} \cdot$m^{-1}, what is the mobility of the electron holes?

● **5–6.1** The resistance of a certain silicon wafer is 1031 ohms at 25.1°C. With no change in measurement procedure, the resistance decreases to 1029 ohms. What is the temperature change? The energy gap of silicon is 1.1 eV.

Answer: +0.03°C.

● *5–6.2* To what temperature must you raise InP in order to make its resistivity half what it is at 0°C? Its energy gap is 1.3 eV.

● **5–6.3** A transistor operates between 10 millivolts and 100 millivolts across the emitter. At the lower voltage, the collector current is 6 milliamps; at the higher voltage, 600 milliamps. Estimate the current when the emitter voltage is 50 mV.

Answer: 46 mA

● *5–6.4* Refer to Example 5–6.2. If the emitter voltage is doubled from 17 to 34 millivolts, by what factor is the collector current increased?

CHAPTER 6

Single-Phase Metals

PREVIEW

Metals are somewhat simpler than the other two principal categories of materials—polymers and ceramics. This is true because a large number of metals contain only one kind of atom (or are a solid solution such as brass in which zinc proxies for copper without a change in structure). Therefore, we will examine how properties relate to the structure of metals first. We will then turn to polymers and ceramics in Chapters 7 and 8, respectively.

We must consider both crystalline structure and microstructure as we look at *elastic* and *plastic deformation* and at eventual *fracture*. A new feature will be presented; that is, the opportunity to *anneal* the metal and remove previously introduced structural imperfections. This can affect properties significantly.

CONTENTS

STUDY OBJECTIVES

1 To familiarize yourself with the magnitudes of commonly encountered
properties of annealed metals, such as copper and nickel, and of annealed
alloys, such as brasses and Cu–Ni.

2 To describe various single-phase microstructures and to understand the
basis for grain growth.

3 To extend your knowledge of elastic behavior beyond that of Section 1–2
to include (a) the three principal types of moduli (Young's, shear, and bulk),
and (b) the variations with temperature, bond strength, and crystal orientation.

4 To know the origin of solution hardening, of strain hardening, and of
annealing in terms of dislocations, and of recrystallization.

5 To apply your knowledge of solution hardening, strain hardening and
recrystallization to the selection of materials and processes to meet
specifications.

6 To anticipate the behavior of metals in service where high temperatures,
impact loading, cyclic loading, and radiation environments are encountered.

6–1 SINGLE-PHASE ALLOYS

Many widely used metals have only a single phase. These may be commercially pure metals with only one component. Examples of such metals include copper for electric wiring, zinc for the coating on galvanized steel, and the aluminum used for housewares. However, in numerous other cases a second component is intentionally added to the metal in order to improve the properties. Any such combination of metals is called an *alloy*.

Alloys are single-phase metals if the solid solubility limit is not exceeded. Brass (a single-phase alloy of copper and zinc), bronze (a similar alloy of copper and tin), and copper–nickel alloys are typical of the single-phase alloys we shall study in this chapter. Multiphase, or polyphase, alloys contain additional phases because the solid solubility limit is exceeded. The majority of our steels, as well as many other metals, are multiphase alloys. They will be discussed in later chapters.

Properties of single-phase alloys   The properties of alloys are different from those of pure metals, as shown by Figs. 6–1.2 and 6–1.3 (pp. 186–187) for brass and Cu–Ni solid solutions. The increases in strength and hardness are due to the presence of solute atoms that interfere with the movements of dislocations in the crystals during plastic deformation. We shall observe later (Section 6–4) that this interference arises because the dislocations cannot readily move past the alloying atoms.

Very small amounts of impurities reduce the electrical conductivity of a metal, because foreign atoms introduce nonuniformities in the electrical field within the crystal lattice. Therefore, the electrons experience more deflections and reflections, with a consequent reduction in the length of the mean free path (Section 5–2).

In a metal, electrons carry the majority of the energy for thermal conduction. Thus there is a correspondence between the thermal and electrical conductivity. (Compare (e) and (f) of Figs. 6–1.2 and 6–1.3.) In fact, it was pointed out some time ago that k/σ in most pure metals is about 7×10^{-6} watt·ohm/°C at normal temperatures (~ 20°C) when thermal conductivity k and electrical conductivity σ are

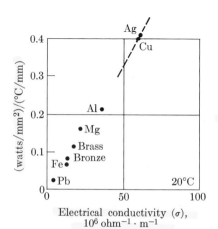

Fig. 6–1.1 Metal conductivity (thermal versus electrical at 20°C). Since electrons transport thermal energy, a good electrical conductor is a good thermal conductor. (The W–F ratio (k/σ) is proportional to absolute temperature. This relationship applies best to pure metals. It does not include nonmetals.)

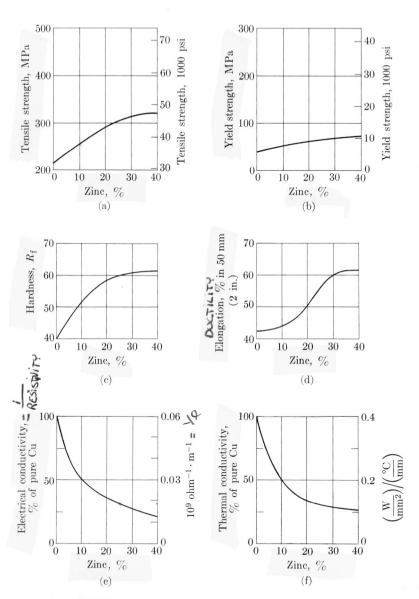

Fig. 6-1.2   Mechanical and physical properties of annealed brasses. The solubility limit of zinc in fcc copper is near 40%. (Adapted from ASM data.)

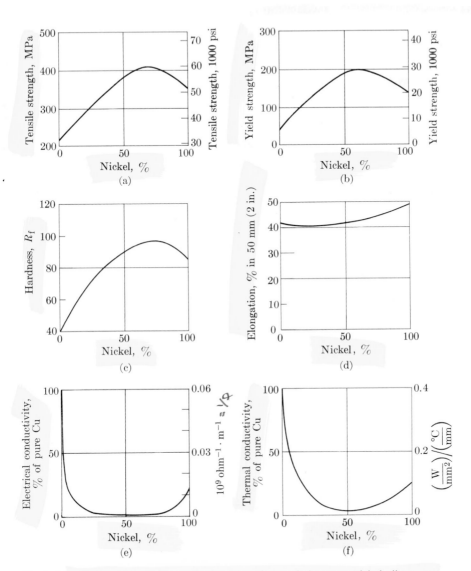

Fig. 6–1.3 Mechanical and physical properties of annealed copper–nickel alloys. Copper and nickel form a complete series of solid solutions. (Adapted from ASM data.)

expressed in (watts/m²)/(°C/m) and ohm⁻¹·m⁻¹, respectively (Fig. 6–1.1). This is called the *Wiedemann–Franz* (W–F) ratio. Since data are readily available on the electrical resistivity values for metals at various temperatures, the W–F ratio provides a convenient rule of thumb for less readily available thermal conductivity values.

Microstructures of single-phase alloys   Grains were defined in Section 4–4 as individual crystals. Materials with many grains are described as polygranular, or more commonly, *polycrystalline*. Adjacent crystals have dissimilar orientations so that a grain boundary is present (Fig. 4–4.7). The microstructures of single-phase metals can be varied by changes in *size, shape,* and *orientation* of the grains (Fig. 6–1.4). These aspects are not wholly independent, because the shape and size of grains are both consequences of grain growth. Likewise, grain shape is commonly dependent on the crystalline orientation of grains during growth.

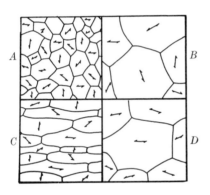

Fig. 6–1.4   Microstructural variables of single-phase metals. (*A* versus *B*) grain size. (*A* versus *C*) grain shape. (*B* versus *D*) preferred orientation.

Although it is common to speak of grain size in terms of diameter, few if any grains of a single-phase metal are spherical. Rather, the grains must completely fill space and also maintain a minimum of total boundary area. This was shown in Figs. 4–4.8 and 6–1.4(A), where the grains are described as being *equiaxed* because they have approximately equal dimensions in the three coordinate directions. In addition to equiaxed grains, other commonly encountered *grain shapes* may be plate-like (Fig. 6–1.4C), columnar, or dendritic (i.e., tree-like). No attempt will be made to systematize them in this book.

The *orientation* of grains within a metal is typically quite random (Fig. 6–1.4A). However, there are exceptions, that can be important from the standpoint of engineering properties. For example, the [100] directions of iron have a higher magnetic permeability than other directions. Therefore, if the grains within a polycrystalline transformer sheet are not random, but are processed to have a *preferred orientation* so that the [100] direction is preferentially aligned with the magnetic field, a significantly more efficient performance may be obtained from the transformer. The metallurgist has learned how to do this, with the result of billions of dollars worth of savings in electrical power distribution.

Example 6–1.1 A copper-nickel alloy must have a tensile strength greater than 290 MPa (42,000 psi), and a thermal conductivity greater than 0.056 (W/mm²)/(°C/mm). Select an alloy from Fig. 6–1.3.

Solution   Tensile strength, S_t 19% $< x <$ 100% Ni

                          Thermal conductivity, k 0% $< x <$ 25% Ni

                                          and

                                          85% $< x <$ 100% Ni

        Specification ranges   19%–25% Ni

                                          and

                                          85%–100% Ni.

Specify 80 Cu–20 Ni, since nickel is more expensive than copper.

Comments. If the coins taken from one's pocket are examined, the reader will realize that nickel is more expensive than copper.
     Specifications seldom are set at the extreme edge of the range. ◀

Example 6–1.2 Iron is to be used as a thermal conductor at 300°C. Its conductivity, in Appendix C, is for 20°C. Estimate $k_{300°C}$ if k/σ is proportional to K.

Solution: From Table 5–2.1 and Eq. (5–2.1),

$$\sigma_{300°C} = 1/\rho_{300°C} = 1/([90 \times 10^{-9} \text{ ohm} \cdot \text{m}][1 + (0.0045/°C)(300°C)])$$
$$= 4.7 \times 10^6 \text{ ohm}^{-1} \cdot \text{m}^{-1}.$$

Since the W–F ratio at 20°C is $\sim 7 \times 10^{-6}$ W·ohm/°C,

$$\text{W–F}_{300°C} = (\sim 7 \times 10^{-6} \text{ W} \cdot \text{ohm}/°C)(573 \text{ K}/293 \text{ K})$$
$$= \sim 13.7 \times 10^{-6} \text{ W} \cdot \text{ohm}/°C.$$

$$k_{300°C} \approx \sigma(\text{W–F}_{300°C}) = (4.7 \times 10^6 \text{ ohm}^{-1} \cdot \text{m}^{-1})(\sim 13.7 \times 10^{-6} \text{ W} \cdot \text{ohm}/°C)/(10^3 \text{ mm/m})$$
$$= \sim 0.064 \text{ (W/mm}^2)/(°C/\text{mm}).$$

Comments. The experimental value reported in metals handbooks is 0.06 (W/mm²)/(°C/mm) at 300°C. The W–F ratio does not apply to nonmetallic materials where electrons are not available to transfer thermal energy. ◀

Example 6–1.3 When a rod is loaded with 250 kg (550 lb), it must have a resistance of less than 0.1 ohm/m and an elastic strain of less than 0.001 m/m; it must not deform plastically. What is the minimum required diameter for a rod made from monel metal (70 Ni–30 Cu)? Made from brass (70 Cu–30 Zn)?

Solution: Data from Appendix C and Figs. 6–1.2 and 6–1.3:

	70 Ni–30 Cu	70 Cu–30 Zn
S_y, MPa	195	70
ρ, ohm·m	480×10^{-9}	62×10^{-9}
E, MPa	180,000	110,000

	70 Ni–30 Cu	70 Cu–30 Zn

For 0.1 ohm/m and rearranging Eq. (1–4.3),

$$d = \sqrt{(4\rho L)/(\pi\ 0.1)}$$

	2.5 mm	0.89 mm

For 0.001 m/m and with $F = (250 \times 9.8\ N)$,

$$d = \sqrt{(4F)/\pi E(0.001)}$$

	4.2 mm	5.3 mm

For no plastic deformation and with $F = (250 \times 9.8\ N)$,

$$d = \sqrt{4F/\pi S_y}$$

	4.0 mm	**6.7 mm**

Comment. The minimum diameter is shown in heavy type. Anything smaller would not meet the three requirements. ◄

• 6–2 PROCESSING OF SINGLE-PHASE ALLOYS

Most metals are processed initially by melting. While molten, they will be chemically refined to remove unwanted impurities. If alloys such as brass or bronze are being produced, zinc or tin, respectively, are added to molten copper. Being molten, the zinc (or tin) readily dissolves and becomes uniformly distributed within the liquid metal.

Next, the typical metal is cast (poured) into a mold to solidify. The mold may be the shape of the final product, as in the case of a large bronze bell; or the solidified *casting* may require some machining for final shaping (Fig. 6–2.1). In other cases, the mold is an *ingot*, which is simply a solidified block of metal (or other material)

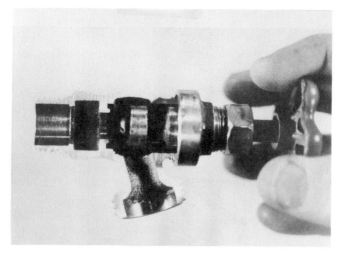

Fig. 6–2.1 Casting (cut-away section of a brass faucet). By solidifying the molten metal in a sand mold, the shape was obtained that required only a small amount of machining.

that is subsequently deformed by *mechanical working* or *forming* to produce a rod, wire, tube, plate, forging, etc. During the mechanical processing, the shape is changed. As a result, stresses must be applied that are above the yield strength described in Chapter 1. Consequently, the mechanical processing on the initial large ingot is commonly performed at high temperatures because the material is typically softer and more ductile. At those temperatures less energy is required for deformation and there is less chance for fracturing during processing. *Rolling, forging,* and *extrusion* (Fig. 6–2.2) are among the common hot deformation processes.

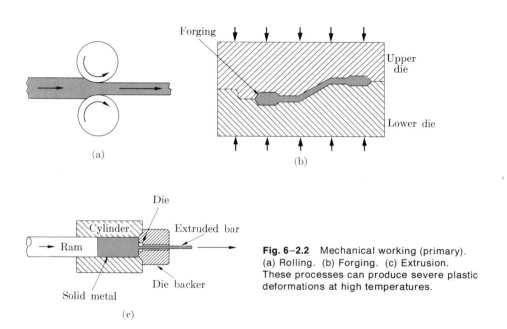

Fig. 6–2.2 Mechanical working (primary). (a) Rolling. (b) Forging. (c) Extrusion. These processes can produce severe plastic deformations at high temperatures.

Following the primary deformation steps, it is often desirable to continue the processing at ambient temperatures. This is possible because the dimensions are now smaller and the required forces and energy reduced accordingly. Also furnaces and fuel costs are avoided. Further, most metals oxidize rapidly at high temperatures. This is generally avoided at low temperatures. Finally, we shall see in this chapter that strengths can commonly be increased if the deformation is performed while the metal is at normal temperatures. Wire *drawing, spinning* (Fig. 6–2.3), and *stamping* are among the common secondary deformation processes performed at ambient temperatures.

In order to understand the behavior of metals during either the processing steps or subsequent service, it will be necessary to be more complete in our knowledge of elastic deformation than we were in Chapter 1, and to visualize the mechanisms of

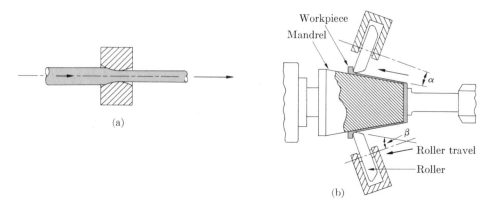

Fig. 6–2.3 Mechanical working (secondary). (a) Wire drawing. (b) Spinning. Most metals are strengthened as they are deformed at ambient temperatures. Therefore, these processes are generally limited to products with smaller cross sections than the primary processes of Fig. 6–2.2.

plastic deformation. Finally, it will be necessary to show the consequences of elevated temperatures on materials because a metal may change its structure (and therefore properties) during high-temperature processes and/or service. These topics will follow in turn.

6–3 ELASTIC DEFORMATION

Elastic deformation occurs when a stress is placed on a piece of metal or, for that matter, on any solid material (Section 1–2). When the load is applied in tension, the piece becomes slightly longer; removal of the load permits the specimen to return to its original dimension. Conversely, when a load is applied in compression, the piece becomes slightly shorter. Elastic strain is a result of a slight elongation of the unit cell in the direction of the tensile load, or a slight contraction of the unit cell in the direction of the compressive load (Fig. 6–3.1).

When only elastic deformation occurs, the strain is essentially proportional to the stress. This ratio between stress and strain is the *modulus of elasticity* (Young's modulus) and is a characteristic of the type of metal. The greater the forces of attraction between atoms in a metal, the higher the modulus of elasticity.

Any lengthening or compression of the crystal structure in one direction, due to a uniaxial force, produces an adjustment in the dimensions at right angles to the force. In Fig. 6–3.1(a), for example, a small contraction is indicated at right angles to the tensile force. The negative ratio between the lateral strain e_y and the direct tensile strain e_z is called *Poisson's ratio* v:

$$v = -\frac{e_y}{e_z}. \quad = -\frac{LATERAL\ \epsilon}{AXIAL\ \epsilon} \quad\quad (6\text{–}3.1)$$

$$e_z = \frac{\sigma}{E} = \epsilon_{AXIAL}$$

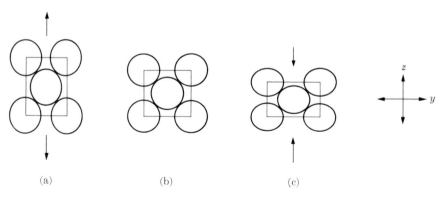

Fig. 6-3.1 Elastic normal strain (greatly exaggerated). Atoms are not permanently displaced from their original neighbors. (a) Tension. (b) No strain. (c) Compression.

Engineering materials may be loaded in shear as well as in tension (and compression). In shear loading, the two forces are parallel but are not aligned (Fig. 6-3.2b). As a result, the *shear stress*, τ, is that force, F_s divided by the sheared area, A_s:

$$\tau = F_s/A_s. \tag{6-3.2}$$

A shear stress produces an angular displacement, α. We define *shear strain*, γ, as the tangent of that angle, i.e., as x/y in Fig. 6-3.2(b). The recoverable or elastic shear strain is proportional to the shear stress:

$$G = \tau/\gamma, = \frac{E}{2(1+\gamma)} \tag{6-3.3}$$

where G is the *shear modulus*. Also called the modulus of rigidity, the shear modulus is different from the modulus of elasticity E; however, the two are related at small strains by a relationship that may be expressed by

$$E = 2G(1 + v). = \frac{\sigma}{\epsilon} \tag{6-3.4}$$

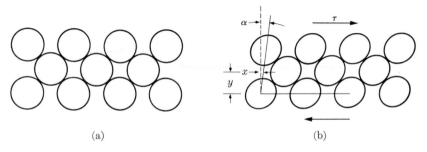

(a)   (b)

Fig. 6-3.2 Elastic shear strain. Shear couples produce a relative displacement of one plane of atoms past the next. This strain is elastic so long as atoms keep their original neighbors. (a) No strain. (b) Shear strain.

Since Poisson's ratio v is normally between 0.25 and 0.5, the value of G is approximately 35% of E.

A third elastic modulus is the *bulk modulus, K*. It is the reciprocal of the compressibility β of the material and is equal to the hydrostatic pressure P_h per unit of volume compression, $\Delta V/V$:

$$K = \frac{P_h V}{\Delta V} = \frac{1}{\beta}. \qquad (6-3.5)$$

The bulk modulus is related to the modulus of elasticity as follows:

$$K = \frac{E}{3(1 - 2v)}. \qquad (6-3.6)$$

The reader may derive this equation in Study Problem 6–3.6.

Elastic moduli versus temperature   Elastic moduli decrease as temperature increases, as shown in Fig. 6–3.3 for four common metals. In terms of Fig. 2–5.2(a), a thermal expansion reduces the value of dF/da, and therefore the modulus of elasticity. The discontinuity in the curve for iron in Fig. 6–3.3 is due to the change from bcc to fcc at 912°C (1673°F). Not surprisingly, the more densely packed fcc polymorph requires greater stresses for a given strain, i.e., the elastic modulus increases for fcc. Also note from Fig. 6–3.3 that higher-melting metals have greater elastic moduli.

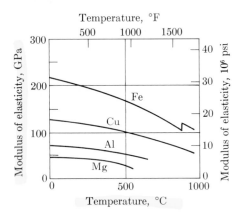

Fig. 6–3.3 Modulus of elasticity versus temperature. (Adapted from A. G. Guy, *Elements of Physical Metallurgy*.)

Elastic moduli versus crystal direction   Elastic moduli are *anisotropic* within materials; that is, they vary with crystallographic direction. As an example, iron has an average modulus of elasticity of about 205 GPa (30,000,000 psi); however, the actual modulus of a crystal of iron varies from 280 GPa (41,000,000 psi) in the [111] direction to only 125 GPa (18,000,000 psi) in the [100] direction (Table 6–3.1). The consequence of any such anisotropy becomes significant in polycrystalline materials. Assume, for example, that Fig. 6–3.4(a) represents the cross section of a steel wire in which the average stress is 205 MPa (30,000 psi). If the grains are randomly oriented, the elastic strain is 0.001, because the average modulus of elasticity is 205

Table 6-3.1
Moduli of elasticity (Young's Modulus)* E

Metal	Maximum GPa	Maximum 10^6 psi	Minimum GPa	Minimum 10^6 psi	Random GPa	Random 10^6 psi
Aluminum	75	11	60	9	70	10
Gold	110	16	40	6	80	12
Copper	195	28	70	10	110	16
Iron (bcc)	280	41	125	18	205	30
Tungsten	345	50	345	50	345	50

* Adapted from E. Schmid and W. Boas, *Plasticity in Crystals*. English translation, London: Hughes and Co.

GPa (30,000,000 psi). However, in reality, the stress varies from 125 MPa (18,000 psi) to 280 MPa (41,000 psi) as shown in Fig. 6–3.4(b), because grains have different orientations, but each is strained equally (0.001). Of course, this means that some grains will exceed their yield strength before other grains reach their yield strength.

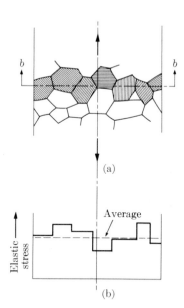

(a)

(b)

Average

Elastic stress

b   b

Fig. 6–3.4 Stress heterogeneities (schematic). Elastic stresses vary with grain orientation, because the moduli of elasticity are not isotropic.

Example 6–3.1 A plate of steel has a 100.0-cm × 100.0-cm square scribed on its surface. It is loaded in one direction (perpendicular to opposite edges of the square) with a 200-MPa (29,000-psi) stress. (a) What are the dimensions of the scribed area? (Poisson's ratio of steel = 0.29.)

Without removing the initial stress, a second tension stress of 410 MPa (60,000 psi) is applied at right angles to the first, i.e., perpendicular to the other edges of the square. (b) What are the new dimensions of the scribed area?

Solution: Since no preferred orientation is indicated, we will assume the random grain value of Young's modulus (Table 6–3.1). $e_z = \sigma / E$

a) With Eq. (1–2.1): $e_z = $ 200 MPa/205,000 MPa = 0.000975;
    with Eq. (6–3.1): $e_y = -0.29(0.000975) = -0.00028$;
    1000 mm $(1 + 0.000975) \times$ 1000 mm $(1 - 0.00028) = 1001.0$ mm $\times$ 999.7 mm.

b) $e_y = -0.00028 + $ 410 MPa/205,000 MPa = 0.00172;
    $e_z = 0.000975 - 0.29(410/205,000) = 0.00040$;
    1000 mm $(1 + 0.0004) \times$ 1000 mm $(1 + 0.00172) = 1000.4$ mm $\times$ 1001.7 mm

Comment. We can write a general equation for elastic deformation in three dimensions from Eqs. (1–2.1) and (6–3.1):

$$e_x = \frac{s_x}{E} - \frac{v s_y}{E} - \frac{v s_z}{E}. \qquad (6-3.7) \blacktriangleleft$$

Example 6–3.2 What is the percentage volume change in iron if it is hydrostatically compressed with 1400 MPa (200,000 psi)? (Poisson's ratio = 0.29.)

Solution: From Eq. (6–3.6) and with $E_{Fe} = 205,000$ MPa,

$$K = (205,000 \text{ MPa})/(3(1 - 0.58))$$
$$= 162,700 \text{ MPa}.$$

$$\Delta V/V = -1400 \text{ MPa}/162,700 \text{ MPa} = -0.86 \text{ v/o}.$$

Alternative solution: Since $s_x = s_y = s_z = P_h$, Eq. (6–3.7) becomes

$$e_x = (1 - 2v)(-1400 \text{ MPa})/(205,000 \text{ MPa})$$
$$= -0.00287 = e_y = e_z;$$
$$1 + \Delta V/V = (1 + e)^3,$$
$$\Delta V/V = -0.86 \text{ v/o}.$$

Comment. Equation (6–3.6) is most readily derived by using the special case of Eq. (6–3.7), where $s_x = s_y = s_z = P_h$. $\blacktriangleleft$

6–4 PLASTIC DEFORMATION WITHIN SINGLE CRYSTALS

Cubic metals and their nonordered alloys deform predominantly by *plastic shear*, or *slip*, where one plane of atoms slides over the next adjacent plane. This is also one of the methods of deformation in hexagonal metals. Shear deformation occurs even when compression or tension stresses are applied, because these stresses may be resolved into shear stresses.

Critical shear stress, and slip systems   Slip occurs more readily along certain crystal directions and planes than along others. This is illustrated in Fig. 6–4.1, where a single crystal of an hcp metal was deformed plastically. The shear stress required to produce slip on a crystal plane is called the *critical shear stress* τ_c.

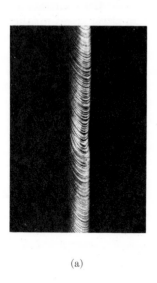

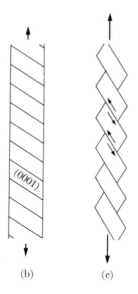

Fig. 6–4.1 Slip in a single crystal (hcp). Slip paralleled the (0001) plane,* which contains the shortest slip vector. (Constance Elam, *Distortion of Metal Crystals,* Oxford: Clarendon Press).

(a)   (b)   (c)

Table 6–4.1
Predominant slip systems in metals

Structure	Examples	Slip planes	Slip directions	Number of independent slip systems
bcc	α-Fe, Mo, Na, W	$\{101\}$	$\langle\bar{1}11\rangle$	12
bcc	α-Fe, Mo, Na, W	$\{211\}$	$\langle\bar{1}11\rangle$	12
fcc	Ag, Al, Cu, γ-Fe, Ni, Pb	$\{111\}$	$\langle\bar{1}10\rangle$	12
hcp	Cd, Mg, α-Ti, Zn	$\{0001\}$*	$\langle11\bar{2}0\rangle$	3
hcp	α-Ti	$\{10\bar{1}0\}$*	$\langle11\bar{2}0\rangle$	3

We can summarize the predominant sets of *slip systems* in several familiar metals in Table 6–4.1. A slip system includes the *slip plane* (*hkl*),* and a *slip direction* [*uvw*]. There are a number of slip systems, because of the multiple planes in a family of planes and multiple directions in a family of directions (Sections 3–6 and 3–7). Two facts stand out in Table 6–4.1.

1. The slip direction in each metal crystal is the direction with the highest linear density of equivalent points, or the shortest distance between equivalent points.

2. The slip planes are planes that have wide interplanar spacings (Section 3–8) and therefore, high planar densities (Section 3–7).

---

*   See Section 3–7 for (*hkil*) indices of hexagonal crystals.

Resolved shear stresses   The axial stress (tension or compression) that is required to produce plastic deformation depends not only on the critical shear stress for the slip system of each material, but also on the orientation of the applied stress with respect to the slip system (Fig. 6–4.2). For one thing, the shear area, A_s, will not equal the cross-sectional area, A, perpendicular to the tension or compression forces. Secondly, the forces must be resolved from the force F in the axial direction to the force F_s in the shear direction. The area increases as the plane is rotated away from the axis of applied stress; thus, $A_s = A/\cos \phi$, where ϕ is the angle between the axial direction and the normal to the slip plane (Fig. 6–4.2). The resolved force is $F_s = F \cos \lambda$, where λ is the angle between the two force directions. Since the shear stress, τ, is the shear force, F_s, per unit shear area, A_s,

$$\tau = F_s/A_s = (F \cos \lambda)/(A/\cos \phi) = s \cos \lambda \cos \phi. \qquad (6\text{--}4.1)$$

This is *Schmid's law*, which relates the resolved shear stress, τ, to the axial stress, s, $(= F/A)$.

Slip occurs with the minimum axial force when both λ and ϕ are 45°. Under these conditions τ is equal to one half the axial stress F/A. The resolved shear stress is less in relation to the axial stress for any other crystal orientation, dropping to zero as either λ or ϕ approaches 90°.

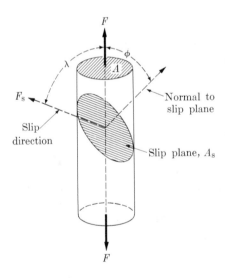

Fig. 6–4.2   Resolved stresses. The axial stress σ equals F/A. The resolved shear stress τ equals F_s/A_s, where $F_s = F \cos \lambda$, and $A_s = A/\cos \phi$. This leads to Eq. (6–4.1).

Mechanism of slip   Figure 6–4.3 shows a simplified mechanism of slip. If we attempted to calculate the strength of metals on this basis, the result would indicate that the strength of metals should be approximately $G/6$, where G is the shear modulus. Since metals are not that strong, a different slip mechanism must be operative. All experimental evidence supports a mechanism involving dislocation movements. If

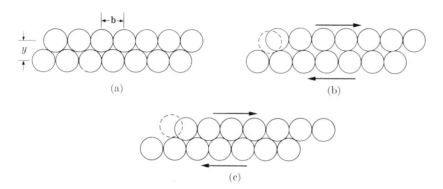

(a)   (b)

(c)

Fig. 6–4.3   An assumed mechanism of slip (simplified). Metals actually deform with less shear stress than this mechanism would require.

we use Fig. 6–4.4 as a model of a dislocation and place a shear stress along the horizontal direction, the dislocation can be moved (Fig. 6–4.5) with a shearing displacement within the crystal. (See also Fig. 4–4.5.) The shear stress required for this type of deformation is a small fraction of the previously cited value of $G/6$, and matches observed shear strengths.

• The mechanism of slip requires the growth and movement of a dislocation line; therefore, energy is required. The energy, E, of a dislocation line is proportional to the length of the dislocation line, l, the product of the shear modulus, G, and the square of the unit slip vector $\mathbf{b}$ (Section 4–4):

$$E \propto lG\mathbf{b}^2. \qquad (6\text{–}4.2)$$

This automatically means that the easiest dislocations to generate and expand for plastic deformation are those with the shortest unit slip vector, $\mathbf{b}$ (Fig. 4–4.3), particularly since the $\mathbf{b}$-term is squared. The directions in a metal with the shortest slip vector will be the directions with the greatest lineal density of atoms. The lowest value for the shear modulus, G, accompanies the planes that are farthest apart and hence have the greatest planar density of atoms. Thus, we may develop the rule-of-

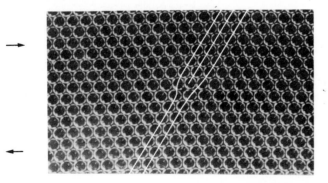

Fig. 6–4.4   Edge dislocation. "Bubble-raft" model of an imperfection in a crystal structure. Note the extra row of atoms. (Bragg and Nye, *Proc. Roy. Soc. (London)* A190, 1947, p. 474.)

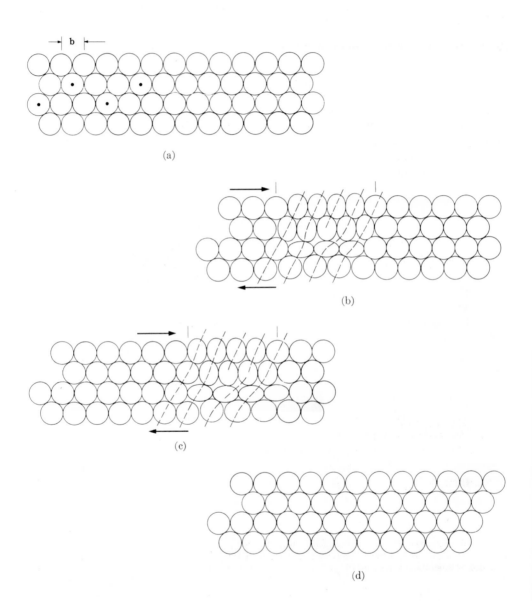

Fig. 6–4.5 Slip by dislocations. In this model only a few atoms at a time are moved from their low-energy positions. Less stress is therefore required to produce slip than if all of the atoms moved at once as proposed in Fig. 6–4.3.

thumb that predicts the lowest critical shear stresses *on the most densely packed planes and in the most densely packed directions.**

Dislocation movements in solid solutions   The energy associated with an edge dislocation (Fig. 4–4.3) is the same, whether the dislocation is located at (b) or (c) of Fig. 6–4.5. Therefore, no net energy is required for the movement between the two.[†] Such is not the case when solute atoms are present. As shown in Fig. 6–4.6, when an impurity atom is present, the energy associated with a dislocation is less than it is in a pure metal. Thus, when a dislocation encounters foreign atoms, its movement is restrained because energy must be supplied to release it for further slip. As a result, solid-solution metals always have higher strengths than do pure metals (Figs. 6–1.2 and 6–1.3). We call this *solution hardening.*

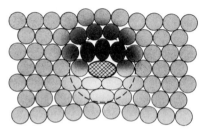

(a) Larger impurity atom

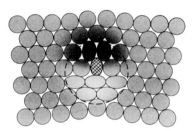

(b) Smaller impurity atom

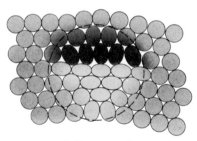

(c) Pure metal

Fig. 6–4.6 Solid solution and dislocations. An odd-sized atom decreases the stress around a dislocation. As a result, energy must be supplied and additional stress applied to detach the dislocation from the solute atom. This accounts for solution hardening.

---

*   •Apparent exceptions to the rule generally arise because the planes and/or directions of easy slip are oriented to receive a low resolved shear stress. As an example, slip cannot occur on $(0\bar{1}1)[\bar{1}11]$ of a bcc metal from a force applied in the [011] direction because $\cos\phi = 0$. However, this leads to the possibility of slip in the $(211)[\bar{1}11]$ system, where $\cos\phi$ and $\cos\lambda$ are $1/\sqrt{3}$ and $\sqrt{\frac{2}{3}}$, respectively; and therefore $\tau = 0.47(F/A)$.

[†]   This statement does not apply if (1) the movement includes an increase in the length of the dislocation, or (2) there is a pile-up of dislocations (Fig. 6–7.6).

OMIT

● **Intermetallic compounds** Phases such as Al_5Mg_3, Cu_2Al, $CuAl_2$, and β' brass (CuZn) receive mention elsewhere in this text. As is true with ceramics (Chapter 8) or any 3-D compound that contains neighbors of unlike atoms, these phases resist shear, because slip would destroy the preferred order among the unlike atoms (Fig. 4–2.2). They are, therefore, hard and also nonductile. As a result, they are brittle.

These hard phases can not be used readily by themselves; however, they strengthen metals when they are present as very fine particles in alloys that are otherwise soft (Chapter 11).

Example 6–4.1 What force does it take in the $[1\bar{1}0]$ direction to have a resolved force of 130 N (29 lbs$_f$) in the $[100]$ direction of a cubic crystal?

Solution

$$F_{[1\bar{1}0]}/F_{[100]} = \cos [1\bar{1}0] \measuredangle [100].$$

$$F_{[100]} = 130 \text{ N}/\cos 45° = 184 \text{ N}$$

$$= 184 \text{ N} \qquad (\text{or } 41 \text{ lbs}_f). \quad \blacktriangleleft$$

Example 6–4.2 The critical shear stress τ_c for the $\langle\bar{1}10\rangle\{111\}$ slip systems of *pure* copper was found to be 1 MPa (145 psi). (a) What stress s must be applied in the $[001]$ direction to produce slip in the $[101]$ direction on the $(\bar{1}11)$ plane? (b) In the $[110]$ direction on the $(\bar{1}11)$ plane?

Solution: From Fig. 3–7.3(c),

a) $\dfrac{[001]}{[111]}$ $\cos\phi = \dfrac{\text{Edge of unit cell}}{\text{Body diagonal of unit cell}} = \dfrac{a}{a\sqrt{3}} = 0.577;$

$\dfrac{[001]}{[101]}$ $\cos\lambda = \dfrac{\text{Edge of unit cell}}{\text{Face diagonal of unit cell}} = \dfrac{a}{a\sqrt{2}} = 0.707;$

$$s = \frac{F}{A} = \frac{1 \text{ MPa}}{(0.577)(0.707)} = 2.45 \text{ MPa} (= 355 \text{ psi}).$$

b) Since $\lambda = 90°$ by inspection, therefore $\cos\lambda = 0$, and $s = \infty$. (Slip cannot occur in this direction when the stress is applied in the $[001]$ direction.)

Comment. The dot product is useful here but can only be used for *cubic* crystals. $\quad\blacktriangleleft$

● **Example 6–4.3** With a sketch show the 12 slip systems that are included in the $\langle\bar{1}10\rangle\{111\}$ slip systems of fcc metals.

Solution: See Fig. 6–4.7. Each of the four planes of the $\{111\}$ family has three slip directions. Thus we have

$[\bar{1}10](111)$	$[1\bar{1}0](11\bar{1})$	$[110](1\bar{1}1)$	$[110](\bar{1}11)$
$[10\bar{1}](111)$	$[101](11\bar{1})$	$[10\bar{1}](1\bar{1}1)$	$[101](\bar{1}11)$
$[0\bar{1}1](111)$	$[011](11\bar{1})$	$[011](1\bar{1}1)$	$[01\bar{1}](\bar{1}11).$

Comment. The rear planes are parallel to the front planes and therefore involve the same slip systems, for example, $[110](\bar{1}11)$ is the same slip system as $[110](1\bar{1}\bar{1})$. $\quad\blacktriangleleft$

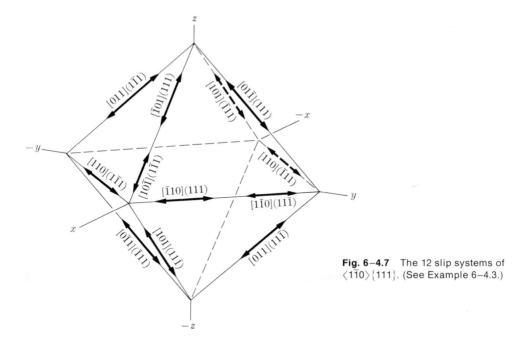

Fig. 6–4.7   The 12 slip systems of $\langle 1\bar{1}0 \rangle \{111\}$. (See Example 6–4.3.)

● **Example 6–4.4** Both $[01\bar{1}]$ and $[11\bar{2}]$ lie in the (111) plane of fcc aluminum. Therefore, both $[01\bar{1}](111)$ and $[11\bar{2}](111)$ slip are conceivable. (a) Make a sketch of the (111) plane and show the $[01\bar{1}]$ and $[11\bar{2}]$ unit slip vectors. (b) Compare the energies of the dislocation lines that have these two displacement vectors.

Solution: (a) See sketch (Fig. 6–4.8).

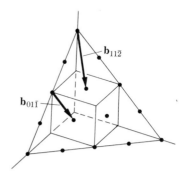

Fig. 6–4.8   See Example 6–4.4.

(b) Refer to Eq. (6–4.2):

$$\frac{E_{01\bar{1}}}{E_{11\bar{2}}} = \frac{lGb_{01\bar{1}}^2}{lGb_{11\bar{2}}^2}.$$

Both involve the same slip plane (111); therefore, the same shear modulus applies. Based on unit length l,

$$E_{01\bar{1}}/E_{11\bar{2}} = (\mathbf{b}_{01\bar{1}}/\mathbf{b}_{11\bar{2}})^2$$
$$= [(a/\sqrt{2})/(a\sqrt{6}/2)]^2 = \tfrac{1}{3};$$

$$E_{01\bar{1}} = \tfrac{1}{3}E_{11\bar{2}}.$$

Comment. With this $\frac{1}{3}$ ratio, slip occurs appreciably more readily with the first of the two potential slip systems for fcc metals. ◀

6–5 PROPERTIES OF PLASTICALLY DEFORMED METALS

Plastic deformation changes the internal structure of a metal; therefore, it is to be expected that deformation also changes the *properties* of a metal. Evidence of such property changes may be obtained through resistivity measurements. The distorted structure reduces the mean free path of electron movements (Section 5–2) and therefore increases the resistivity (Fig. 6–5.1).

A second property change that has much greater engineering significance is strength. Plastically deformed metals become stronger, as we saw in Chapter 1. There, the s/e curve of Fig. 1–2.1(c) continued to rise after the yield strength was exceeded.

In Fig. 6–5.1, as well as in other cases, it is convenient to refer to the percent of *cold work* as an index of plastic deformation. Cold work is the amount of plastic strain introduced during processing, expressed by the percent decrease in cross-sectional area from deformation; that is,

$$\% \text{ CW} = \left[\frac{A_O - A_f}{A_O} \right] 100, \qquad (6\text{–}5.1)$$

where A_O and A_f are the original and final areas respectively.

Fig. 6–5.1 Electric resistivity versus cold work (wrought aluminum alloys); 1100 = 99.9% Al; 3003 = 1.2% Mn, balance Al. (See Eq. (6–5.1) for the definition of cold work.)

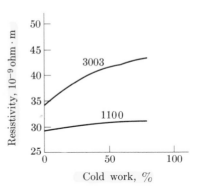

Strain hardening The copper of Fig. 6–5.2 has been plastically deformed. The deformation shows up as traces of slip planes on a previously polished surface. The effect of plastic deformation may also be revealed in an electron microscope. Fig. 6–5.3 shows dislocations in a stainless steel that had been severely cold-worked. The entanglement of lines are dislocations, the numbers and lengths of which increase greatly with additional cold work. As seen in Fig. 6–5.4, the total length of dislocation lines markedly affects the shear stress. We may conclude that *although dislocations account for plastic deformation* (Fig. 6–4.5), *they interfere with the movements of other dislocations*. The dislocation entanglements, or "traffic jams," increase the critical shear stress, τ_c, and therefore the strength of the material.

Data of the type presented in Fig. 6–5.4 are shown in a more practical engineering format in Figs. 6–5.5, 6–5.6 and 6–5.7.

Fig. 6–5.2 Plastically deformed, polycrystalline copper (X25). The traces of the slip planes are revealed at the polished surface of the metal. (National Bureau of Standards. Reproduced by permission from B. Rogers, *The Nature of Metals*, 2nd ed., American Society for Metals, and Iowa State University Press, Chapter 13.)

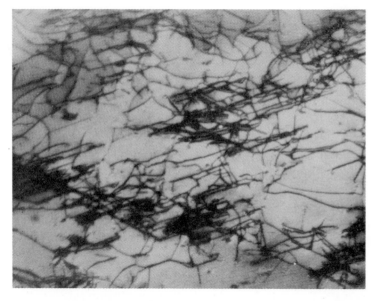

Fig. 6–5.3 Dislocations in plastically deformed metal (stainless steel, × 30,000). Additional deformation introduces more dislocations. Also, additional dislocations interfere with further deformation. *Strain hardening* is the result. (M.J. Whelan, *Proc. Roy. Soc.*)

The increase in hardness that results from plastic deformation is called *strain hardening*. Laboratory tests show that an increase in both tensile strength and yield strength accompanies this increase in hardness. On the other hand, strain hardening reduces ductility, because part of the deformation occurred during cold-working, before the gage marks (Fig. 1–2.2) are placed on the test bar. Thus, less ductility is observed during testing. The process of strain hardening increases yield strength more than tensile strength (Fig. 6–5.7), and the two approach the true breaking strength as the amount of cold work is increased.

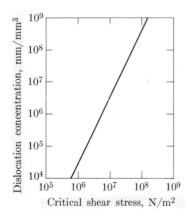

Fig. 6–5.4 Dislocation density versus critical shear stress. An increased number of dislocations provides interference to dislocation movements. (Adapted from Wiedersich, *Jour. of Metals.*)

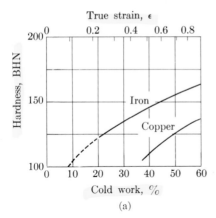

(a)

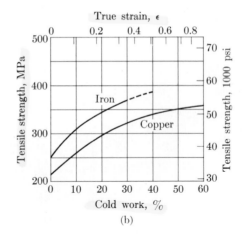

(b)

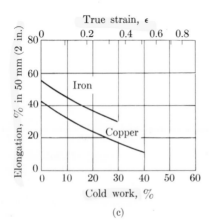

(c)

Fig. 6–5.5 Cold work versus mechanical properties (iron and copper). Cold work is the amount of plastic strain, expressed as the decrease in cross-sectional area (Eq. 6–5.1).

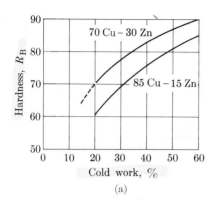

(a)

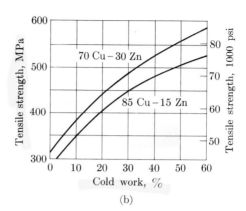

(b)

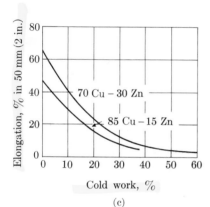

(c)

Fig. 6–5.6 Cold work versus mechanical properties (brasses).

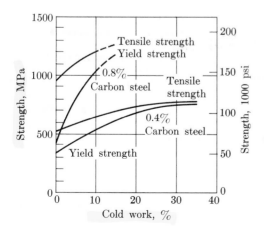

Fig. 6–5.7 Cold work versus strength of plain-carbon steels.

Example 6-5.1 An annealed iron wire was originally 1.23 mm (0.048 in.) in diameter. It is cold-worked by drawing it through a die with a hole that is 1.04 mm (0.041 in.) in diameter. (a) What was the ductility of the wire before drawing? (b) After drawing?

Solution: (a) From Fig. 6-5.5(c),

$$55\% \text{ Elongation (50 mm).}$$

b) $$\text{Cold work} = \frac{(\pi/4)(1.23)^2 - (\pi/4)(1.04)^2}{(\pi/4)(1.23)^2} = 0.285 \quad \text{(or 28.5\%).}$$

From Fig. 6-5.5(c),

$$30\% \text{ Elongation (50 mm).}$$

Comment. A limited amount of extrapolation is generally permissible. Also one may interpolate between curves to obtain estimates of data for other alloys, for example, 80 Cu-20 Zn in Fig. 6-5.6, or 90 Cu-10 Zn, if Figs. 6-5.6 and 6-5.5 are used together. ◀

Example 6-5.2 A cold-worked copper or a brass may be used in an application with the specifications of $S_t \geq 345$ MPa ($\geq 50{,}000$ psi), and a ductility of greater than 20% elongation in 50 mm (2 in.). Choose a metal.

Solution: The cold-work data provide the following specification ranges:

	Copper	Brass (85-15)	Brass (70-30)
Figure	6-5.5	6-5.6	6-5.6
Tensile strength	$\geq 40\%$ CW	$\geq 9\%$ CW	$\geq 4\%$ CW
Ductility (% Elong.)	$\leq 24\%$ CW	$\leq 16\%$ CW	$\leq 23\%$ CW
Range	—	9%-16% CW	4%-23% CW

Comments. The designer has a choice of brasses. Other factors equal, he or she would choose the higher-zinc brass (70 Cu-30 Zn) because zinc is cheaper than copper. Furthermore, the specification range of 4%-23% CW gives more flexibility in processing the metal. ◀

6-6 RECRYSTALLIZATION

This is a process of growing new crystals from previously deformed crystals.

Crystals that have been plastically deformed, like those shown in Figs. 6-5.2 and 6-5.3, have more energy than unstrained crystals because they are loaded with dislocations and point imperfections. Given a chance, the atoms will move to form a more perfect, unstrained array. Such an opportunity arises when the crystals are subjected to elevated temperatures, through the process called *annealing.* The greater thermal vibrations of the lattice at high temperatures permit a reordering of the atoms into less distorted grains. Figure 6-6.1 shows the progress of this *recrystallization.*

Brass was cold-worked 33% and therefore strain-hardened (Fig. 6-6.1a). Several samples were heated to 580°C (1076°F) for a few seconds. New grains, i.e., new fcc crystals of brass, are detected in the sample that was heated for only 3 sec (Fig. 6-6.1b). The initial recrystallization started along traces of the earlier slip planes.

In 4 sec, the brass has been nearly half recrystallized (Fig. 6–6.1c) and completely recrystallized in 8 sec (Fig. 6–6.1d). Were we to examine the new crystals of that last figure by an electron microscope, we would observe a greatly reduced number of dislocations. Also the hardness has dropped significantly from 165 BHN initially to less than 100 BHN.

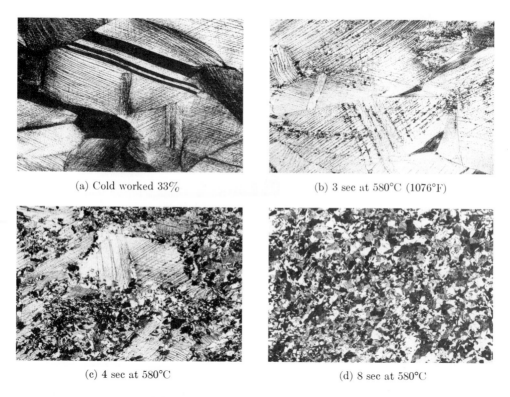

(a) Cold worked 33%   (b) 3 sec at 580°C (1076°F)

(c) 4 sec at 580°C   (d) 8 sec at 580°C

Fig. 6–6.1 Recrystallization of strain-hardened brass (X40). (Courtesy of J. E. Burke, General Electric Co.)

Recrystallization temperatures The recrystallization process requires atom movements and rearrangements. These rearrangements for recrystallization occur more readily at high temperatures. In fact, we will observe in Fig. 6–6.2 that a marked decrease in strength has occurred in a sample held for one hour at 300°C (570°F). Initially cold-worked 75%, this sample was almost completely recrystallized in that period of time. In contrast, samples held for an hour at temperatures below 200°C (∼400 °F) retained almost all of their higher strength, which was obtained during

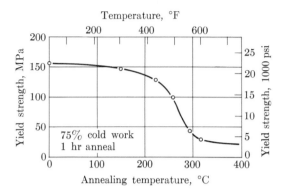

Fig. 6-6.2 Yield strength versus recrystallization (aluminum). Initially cold-worked 75%, the metal was reheated to the indicated temperatures for 1 hour. This was enough time to recrystallize the metal at 300°C and above. The yield strength decreases and the strain hardening disappears with the development of the new grains. Cf. Fig. 6-6.1. (Adapted from *Aluminum*, Vol. 1, American Society for Metals.)

the 75% cold work. We thus speak of a *recrystallization temperature*, T_R, in this case approximately 270°C where the microstructure and strength change drastically.*

   The temperature for recrystallization is dictated by several factors. A "ball-park" figure is that T_R lies between three tenths and six tenths of the absolute, K, melting temperature, that is, $0.3T_m$ to $0.6T_m$. The rationale for this generality is that the

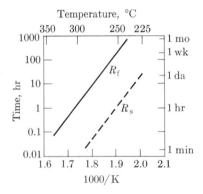

Fig. 6-6.3 Recrystallization time versus temperature (aluminum, 75% cold-worked). Dashed line: start of recrystallization, R_s. Solid line: recrystallization finished. An Arrhenius relationship (ln t versus $1/T$) is followed because recrystallization requires atom movements. (Adapted from *Aluminum*, Vol. 1, American Society for Metals.)

---

* There is a slight loss of strength and a major recovery of electrical conductivity at temperatures slightly below the recrystallization temperature. Called *recovery*, these lower-temperature changes come about because the point imperfections (vacancies, interstitials, etc., that were introduced by cold work) move to dislocation edges within the strained crystals. The point imperfections do not affect deformation significantly; therefore, only slight softening occurs during this recovery stage. The mean free paths of the electrons are significantly lengthened by the disappearance of the point defects; therefore, the resistivity drops with their removal.

diffusivity D for self-diffusion is directly related to the melting temperature for the metal.* However, *time* is a second factor that affects recrystallization temperatures. For example, as shown in Fig. 6–6.3, the recrystallization of commercially pure aluminum alloy that had been cold-worked 75% was completed in 1 minute at 350°C (623 K and 662°F), but required 60 minutes at 300°C and 40 days at 230°C. We expect this, because the flux of atoms is proportional to the diffusivity (Eq. 4–7.1a), which in turn is a function of temperature (Eq. 4–7.3). This permits us to relate the recrystallization temperature to time on an <u>Arrhenius plot</u> (ln t versus $1/T_R$) as shown in Fig. 6–6.3.

A third factor that affects the recrystallization temperature is the amount of *strain hardening*. As shown by the hardness data in Fig. 6–6.4, the recrystallization temperature drops from above 320°C (>600°F) for a 65–35 brass with 20% cold work to approximately 280°C (~535°F) for the same brass with 60% cold work (each with one hour of annealing time). An explanation takes cognizance of the fact that a highly strain-hardened metal has more stored energy in the form of vacancies and dislocations (Fig. 4–4.3) than does one with little cold work. With this energy available, it takes less thermal energy for atoms to rearrange themselves into an annealed grain. That is, recrystallization can occur at lower temperatures.

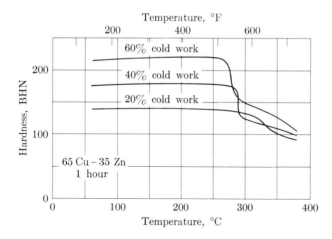

Fig. 6–6.4 Softening during annealing (65Cu–35Zn brass). The hardnesses were determined at 20°C after heating to the indicated temperatures for 1 hour. The more highly strain-hardened brass softens at a lower temperature and with less thermal energy. (ASM data.)

Finally, *pure* materials recrystallize at lower temperatures than solid solutions. Thus, highly pure electrical copper wire anneals much more readily than a comparably deformed brass wire, even though the melting temperature would predict a small opposite relationship. The rationale, which involves dislocation movements, will not be detailed here.

Hot-working versus cold-working of metals   In the production operations described in Figs. 6–2.2 and 6–2.3, the distinction between *hot-working* and *cold-working* does

---

* A solution of Study Problem 4–7.14 gives diffusivity values at $0.6T_m$ for iron, copper, and silver of $10^{-17.5}$, $10^{-17.3}$, and $10^{-17.4}$ m²/sec, respectively. (Each of these calculations were for fcc structures and, therefore, can be compared directly.)

not rest on temperature alone, but on the relationship of the processing temperature to the recrystallization temperature. Hot-working refers to shaping processes that are performed above the recrystallization temperature; cold-working refers to shaping processes that are performed below it. Thus the temperature for cold-working copper may be higher than that for hot-working lead.

The choice of the recrystallization temperature as the point for distinguishing between hot- and cold-working is quite logical from the production point of view. Below the recrystallization temperature, the metal becomes harder and less ductile with additional deformation during processing. More power is required for deformation and there is a greater chance for cracking during the process. Above the recrystallization temperature, the metal will anneal itself during, or immediately after, the mechanical working process. Thus, it remains soft and relatively ductile.

Engineering significance of cold-working and annealing Strain hardening by cold work is of prime importance to the design engineer. It permits the use of smaller parts with greater strength. Of course, the product must not be used at temperatures that will anneal the metals.

Cold work reduces the amount of plastic deformation that a metal can subsequently undergo during a shaping operation. The hardened, less ductile, cold-worked metal requires more power for further working and is subject to cracking. Therefore, *cold-work–anneal cycles* are used to assist production. Example 6–6.1 describes such a process.

The loss of ductility during cold-working has a useful side-effect in machining. With less ductility, the chips break more readily (Fig. 6–6.5) and facilitate the cutting operation. As a result, the engineer may commonly specify cold-working for "screw-stock" metal, i.e., metals that require cutting in automatic screw machines.

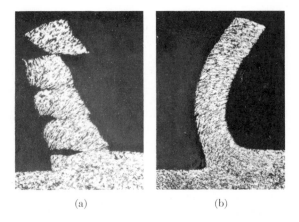

(a)   (b)

Fig. 6–6.5 The cutting of metal turnings by a machine tool. The strain-hardened metal in (a) formed the more desirable chips, while the annealed metal in (b) formed continuous turnings. (Hans Ernst, Cincinnati Milacron, Inc.)

Example 6–6.1 A 70–30 brass rod is required to have a diameter of 5.0 mm (0.197 in.), a tensile strength of more than 420 MPa (61,000 psi), and a 50-mm elongation of more than 18%. The rod is to be drawn from a larger 9.0-mm (0.355-in.) rod. Specify the final processing steps for making the 5.0-mm rod.

Solution: From Fig. 6–5.6,

$$CW > 16\% \text{ for } S_t; \qquad CW < 24\% \text{ for elongation.}$$

Use 20% cold work as *last* drawing step. By Eq. (6–5.1),

$$0.20 = \frac{d^2\pi/4 - (5.0)^2\pi/4}{d^2\pi/4},$$

$$d = 5.6 \text{ mm } (=0.22 \text{ in.}).$$

Comments. Hot work from 9.0 mm to 5.6 mm (or cold work and anneal in one or more cycles). The rod should be annealed at 5.6-mm diameter. Cold draw 20% to 5.0-mm diameter. ◀

Example 6–6.2 The time-temperature relationship for the completed recrystallization of the aluminum in Fig. 6–6.3 follows an Arrhenius-type relationship. Establish an appropriate empirical equation.

Solution: From the curve of Fig. 6–6.3, at 250°C (523 K),

$$T^{-1} = 1.91 \times 10^{-3} \quad \text{and} \quad t = 200 \text{ hr,}$$

$$\ln 200 = C + B/523.$$

At 327°C (600 K),

$$T^{-1} = 1.67 \times 10^{-3} \quad \text{and} \quad t = 0.14 \text{ hr,}$$

$$\ln 0.14 = C + B/600.$$

Solving simultaneously,

$$C = -52; \qquad B = 30,000 \text{ K};$$

$$\ln t = -52 + 30,000 \text{ K}/T.$$

Comment. Just as the various diffusion couples of Table 4–7.2 had their own values of D_0 and E, these values of C and B apply only to this commercially pure aluminum that had been cold-worked 75%. (See Eqs. (10–1.1) and (10–1.2) for greater detail.) ◀

6–7   BEHAVIOR OF POLYCRYSTALLINE METALS

The solidified metal in either a casting (Fig. 6–7.1) or an ingot contains many grains, each an individual crystal. The same is true for rods, bars, sheets, tubes, etc. The size of the grains has an effect on the deformation of material. In brief, those metals with small grains are stronger at low temperatures and weaker at elevated temperatures than are metals with large grains. In each case, the effect is primarily a consequence of the amount of grain boundary area. Recall from Section 4–4 that there is an inverse relationship between the average grain dimension and the amount of boundary area.

Fig. 6–7.1   Coarse-grained metal in cast turbine blades. These blades are used at high temperatures; therefore the grain-boundary area is minimized by the development of coarse grains in processing. In contrast, grain boundaries increase the strength at low temperatures. (Courtesy of Pratt and Whitney Aircraft Group, United Technologies.)

Since the engineer has some control over the grain size, we will look first at grain growth, then at the effects of grain sizes on deformation both at low temperatures and at high temperatures.

Grain growth in metals   The average grain size of a single-phase metal increases with time if the temperature produces significant atom movements (Section 4–7). The driving force for *grain growth* is the energy released as an atom moves across the boundary from the grain with the convex surface to the grain with the concave surface. There, the atom is, on the average, coordinated with a larger number of neighbors at equilibrium interatomic spacings (Fig. 6–7.2). As a result, the boundary moves toward the center of curvature. Since small grains tend to have surfaces of sharper convexity than large grains, they disappear because they feed the larger grains (Fig. 6–7.3). The net effect is grain growth (Fig. 6–7.4).

All crystalline materials, metals and nonmetals, exhibit this characteristic of grain growth. An interesting example of grain growth can be seen in the ice of a snow bank. Snowflakes start out as numerous small ice crystals, lose their identity with time, and are replaced by larger granular ice crystals. A few of the crystals grow at the expense of the many smaller crystals.

The growth rate depends heavily upon temperature. An increase in temperature increases the thermal vibrational energy, which in turn *accelerates* the net diffusion of atoms across the boundary from small to large grains. A subsequent decrease in temperature slows down the boundary movement, but *does not reverse it*. The only way to reduce (refine) the grain size in alloys that always have a single phase is to plastically deform the grains and start new grains by recrystallization (Section 6–6).

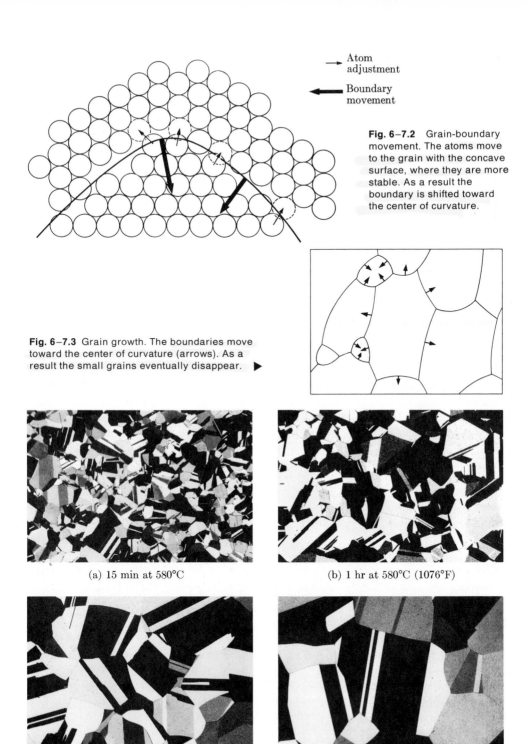

→ Atom adjustment

⟵ Boundary movement

Fig. 6–7.2 Grain-boundary movement. The atoms move to the grain with the concave surface, where they are more stable. As a result the boundary is shifted toward the center of curvature.

Fig. 6–7.3 Grain growth. The boundaries move toward the center of curvature (arrows). As a result the small grains eventually disappear. ▶

(a) 15 min at 580°C

(b) 1 hr at 580°C (1076°F)

(c) 10 min at 700°C

(d) 1 hr at 700°C (1292°F)

Fig. 6–7.4 Grain growth (brass at X40). (Courtesy of J. E. Burke, General Electric Co.)

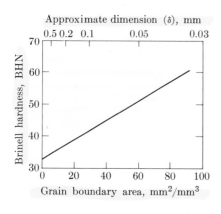

Fig. 6–7.5 Hardness versus grain growth (annealed 70–30 brass). Larger grains with less grain boundary area are softer. (After F. Rhines, *Metal Progress*, 1977.)

Deformation at low temperatures   Figure 6–7.5 plots the hardness of an annealed 70–30 brass as a function of grain-boundary area. The boundaries interfere with slip because the dislocation movements must be redirected as they try to enter the new grain. The adjacent grain is invariably tilted or rotated with respect to the grain undergoing slip. (If the two were not misaligned, there would be no boundary.) Thus, it takes a greater force to continue the slip across a grain boundary. Commonly, there is a pile-up of dislocations that have the same effect as a traffic jam on an arterial highway (Fig. 6–7.6).

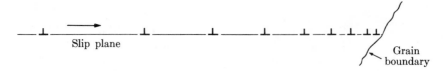

Fig. 6–7.6   Dislocation pile-up. A boundary or surface hinders continued dislocation movements at low temperatures. ⊥ = edge dislocation.

• Polycrystalline metals deform differently from single crystals of metals in another respect. The single crystal of Fig. 6–4.1 was not constrained by adjacent crystals. In contrast, observe Fig. 6–5.2 where it is evident that the large grain of copper in the center did not yield independently of its neighboring grains. The metallurgist can show that at least five slip systems (Table 6–4.1) must operate simultaneously within a grain if that grain is to deform in concert with its neighbors and not introduce cracks or gaps. Since not all slip planes are favorably oriented (Example 6–4.2b), we can appreciate why bcc and fcc metals, with a larger number of slip systems, are ductile metals; and hcp metals are less ductile (Table 6–4.1).

• **Deformation at high temperatures**   We just observed in Fig. 6–7.5 that fine-grained metals are harder and therefore stronger than coarse-grained metals—*at low temperatures*. At elevated temperatures, the situation is reversed. Above temperatures at which atoms start to move significantly, the grain boundary is a source of weakness to a material. We may understand this better with Fig. 6–7.7, which shows schematically several grains loaded vertically in tension. With tension in one direction, there is a perpendicular contraction. (Cf. Poisson's ratio, Section 6–3.) Thus, atoms along vertically oriented boundaries are crowded; atoms along horizontally oriented boundaries have an increased amount of space. This induces diffusion from the vertical boundaries to the horizontal boundaries, with the net effect that there is a gradual change in the shape of the metal.*

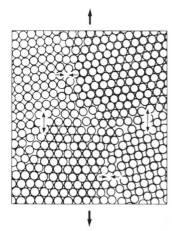

Fig. 6–7.7   Grain boundaries and deformation (schematic). Under tension, vertical boundaries are crowded and lose atoms by diffusion; the horizontal boundaries receive relocated atoms because they develop extra space. The resulting creep becomes a significant factor in the use of materials at high temperatures, where the atoms diffuse readily. (Base sketch from Fig. 4–4.7.)

The strain just described is one mechanism of *creep*. With smaller grains and, therefore, more grain boundary area, creep is more rapid. There are more "sinks" for atoms along horizontal boundaries, and more "sources" of atoms from vertical boundaries (Fig. 6–7.7). Even more important is the fact that the diffusion distances are shorter in fine-grained materials.

Of course, this mechanism of creep does not occur at low temperatures where there are negligible atom movements, but increases exponentially with the increases in the values of diffusivity, D, for self-diffusion (Section 4–7). As with recrystallization, the temperature for the reversal of these grain-size effects is a function of time, bond strength, impurities, etc.

• **High-temperature service**   Implied by the name, *creep* is a slow process of strain. The rates range from a few percent per hour at high stress levels or at high temperatures, down to less than $10^{-4}\%/hr^\dagger$ (Fig. 6–7.8). These are small; however, consider their importance when designing a steam power plant or nuclear reactor, which must be in high-temperature service for many years. Likewise, creep becomes important in gas turbines and in equipment that must be operated without change in dimensions at high stresses and elevated temperature to maximize energy-conversion efficiencies.

---

* Compressive forces induce similar, but opposite, shape changes.
† ∼1%/year.

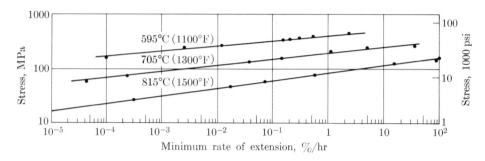

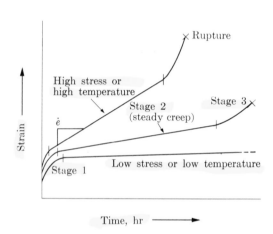

Fig. 6-7.8 Creep data (type 316 stainless steel). Higher stresses and higher temperatures increase the creep rate. (From Shank, in McClintock and Argon, *Mechanical Behavior of Metals*, Addison-Wesley.)

Fig. 6-7.9 Creep. The steady rate of creep in the second stage determines the useful life of the material.
◄

We may plot strain as a function of time for various stresses and temperatures. Figure 6–7.9 is schematic. When a metal is stressed, it undergoes immediate elastic deformation, which is greater when either the stress or the temperature is high. In the first short period of time (Stage 1), it makes additional, relatively rapid, plastic adjustments at points of stress concentrating along grain boundaries and at internal flaws. These initial plastic adjustments give way to a slow, nearly steady rate of strain that we define as the *creep rate* $\dot{e}$. This second stage of *steady creep* continues over an extended period of time, until sufficient strain has developed so that a necking-down and area-reduction occurs. With this area change at a constant load, the rate of strain accelerates until rupture occurs (Stage 3). If the load could be adjusted to match the reduction in area and thus maintain a constant stress, the creep rate of Stage 2 would continue until rupture.

In Fig. 6–7.9 the following relationships are shown schematically: (1) The steady creep rate increases with both increased temperature and stress. (2) The total strain at rupture also increases with these variables. (3) The time before eventual failure by *stress-rupture* is decreased as the temperatures and applied stresses are increased (Fig. 6–7.10).

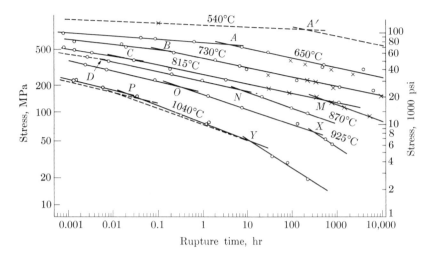

Fig. 6–7.10   Stress-rupture data. The time to rupture is less with higher stresses and higher temperatures. (After N. J. Grant, American Society for Metals.)

• 6–8   FRACTURE

The end of the stress-strain test (Fig. 1–2.2) is punctuated by fracture. This may or may not have been preceded by some plastic deformation. When there was plastic deformation, we speak of *ductile fracture*; otherwise *brittle fracture*. The relative toughness can be indicated by (1) ductility measurements (either percent elongation or percent reduction of area), or (2) the energy that is absorbed in an impact test* (Fig. 1–2.4).

Transition temperatures   Many materials exhibit a brittle–ductile transition and have a characteristic temperature of change (Figs. 6–8.1 and 6–8.2). At low temperatures, a crack can propagate faster than the plastic deformation can occur, so little energy is absorbed. At higher temperatures cracking is preceded by energy-consuming deformation. This discontinuity in energy absorption is particularly characteristic of bcc metals; as a result the designer should be alert to the behavior of steels at low temperatures.[†]

The transition temperature varies with the rate of loading. Thus a slowly deformed steel may fail ductilely, while it would fail in a brittle manner under impact, because there is no opportunity for plastic deformation to occur. The data of Fig. 6–8.2 are for impact loading. These data are for samples of standardized geometry, since we also find that the energy requirements depend on the three-dimensional distribution of stresses in the neighborhood of the crack. Although the exact transition

---

* The results of the latter are sometimes erroneously called "impact strength," rather than impact energy.
† Metals such as copper and aluminum (fcc) do not change abruptly in toughness as temperature is reduced.

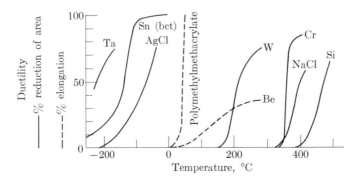

Fig. 6–8.1 Ductility versus temperature (tensile tests). Except for fcc metals, most materials abruptly lose ductility at decreasing temperatures. For a given material the transition temperature is higher for higher strain rates, e.g., impact loading. (After data by A. H. Cottrell, *The Mechanical Properties of Matter*, Wiley.)

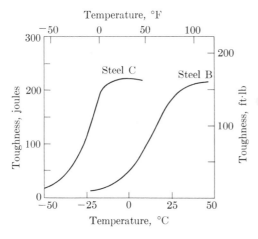

Fig. 6–8.2 Toughness transitions. For each steel, there is a marked decrease in toughness at lower temperatures. The transition temperature is significantly lower for Steel C (fine-grained) than for Steel B (rimmed). (Adapted from Leslie, Rickett, and Lafferty, *Trans. AIME*.)

temperature varies with geometry, naval architects will prefer Steel C in Fig. 6–8.2 over Steel B when designing a welded ship for use in North Atlantic winter waters. A crack, once started in a steel with a high transition temperature, could continue to propagate with a low-energy fracture until the ship has broken apart. There were a number of unfortunate naval catastrophes that occurred before the designers learned how to make appropriate corrections, and metallurgists found that fine-grained steels had lower transition temperatures than coarse-grained steels.

Fatigue There are many documented examples of eventual failure of rotating shafts on power turbines and on other mechanical equipment, which had initially performed satisfactorily for long periods of time. Figure 6–8.3 shows one such fracture.*

---

* The early explanation was that the metal became "tired" and failed from fatigue. We now know that fatigue fracturing is a result of very localized microstructural movements that lead to crack propagation.

Fig. 6–8.3 Fatigue fracture (14-cm (5½-in.) steel shaft). Fracture progressed slowly from the set-screw hole at the top through nearly 90% of the cross section before the final rapid fracture (bottom). (Courtesy of H. Mindlin, Battelle Memorial Institute.)

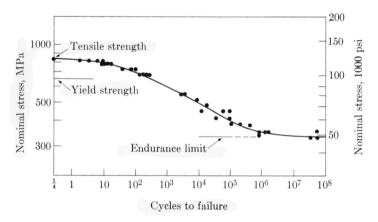

Fig. 6–8.4 *S–N* curve (SAE 4140 normalized steel). *S–N* = cyclic *stress* versus *number* of cycles to failure. At the endurance limit, the number of cycles becomes indeterminately large. (Adapted from R. E. Peterson, *ASTM Materials Research and Standards*.)

The stresses that a material can tolerate under cyclic loadings are much less than under static loading. The tensile strength can be used as a guide in design only for structures that are in service under static loading. The number of cycles N that a metal will endure decreases with increased stresses S. Figure 6–8.4 is a typical *S–N* curve for fatigue fracture of steels. In designing for unlimited cyclic loading, it is necessary to restrict the stresses to values below the *endurance limit* of this curve.

Figure 6–8.5 shows three examples of cyclic loading. The axle of a train has many sinusoidal stress cycles. Examples of low-cycle stresses are found in the rotor

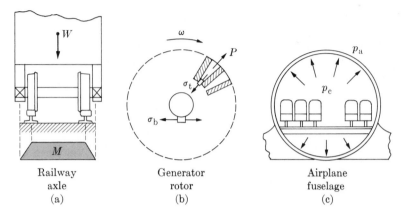

Fig. 6–8.5 Examples of cyclic loading. (a) Axle of rail car. (b) Rotor of generator during starting and stopping. (c) Pressurization and depressurization of plane. The latter may encounter only a few thousand cycles; however, as indicated by Fig. 6–8.4, the yield strength cannot be used by designers. (Courtesy of *ASTM*, R. E. Peterson, "Fatigue in Metals," *Materials Research and Standards*.)

of a generator that is used as a "topping" unit to meet peak demand for electricity. It starts and stops up to 1000 times per year, each time introducing a stress cycle at the base of the winding slot and at the arbor. Likewise an airplane fuselage has stresses imposed during pressurization following each take-off; these are removed as it returns to the ground.

Fatigue cracks usually start at the surface where bending or torsion cause the highest stresses to occur and where surface irregularities introduce stress concentrations. As a result, the endurance limit is very sensitive to surface finish (Table 6–8.1).

Table 6–8.1
Surface finish versus endurance limit
(SAE 4063 steel, quenched and tempered to 44R$_c$)*

Type of finish	Surface roughness, μm	μin.	Endurance limit, MPa	psi
Circumferential grind	0.4–0.6	16–25	630	91,300
Machine lapped	0.3–0.5	12–20	720	104,700
Longitudinal grind	0.2–0.3	8–12	770	112,000
Superfinished (polished)	0.08–0.15	3–6	785	114,000
Superfinished (polished)	0.01–0.05	0.5–2	805	116,750

* Adapted from M. F. Garwood, H. H. Zurburg, and M. A. Erickson, "Correlation of Laboratory Tests and Service Performance," *Interpretation of Tests and Correlation with Service*, Amer. Soc. Metals.

A close examination of the early stages of crack development shows that microscopic and irreversible slip occurs within individual grains.* There is a gradual reduction in ductility in the regions of these slip planes, causing microscopic cracks to form. This may be after only 10% or 20% of the eventual fatigue life. The cracks progress slowly during the remaining cycles. Eventually they reduce the cross-sectional area sufficiently so that a final catastrophic failure occurs (Fig. 6–8.3).

Any design factor that concentrates stresses can lead to premature failure. We have already seen, in Table 6–8.1, that surface finish is important. Keyways and other notches (Fig. 6–8.3) are also critical. Finally the generous use of fillets is recommended in engineering design as shown in Fig. 6–8.6.

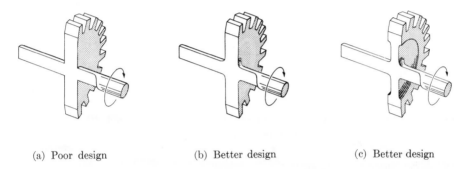

(a) Poor design   (b) Better design   (c) Better design

Fig. 6–8.6 Design of fillet. The use of generous fillets is recommended in mechanical engineering design. It should be observed that (c) is a better design than (a), even with some additional material removed. Of course, if too much metal is removed, failure may occur by mechanisms other than fatigue.

• 6–9   RADIATION DAMAGE omit

Materials selection is the key to successful nuclear-reactor design. First, they must be "fail-safe" in case there are any service problems. Secondly, the radiation environment will alter the internal structure of many materials. Since these alterations generally introduce undesired property changes, we speak of *radiation damage*. We will limit our discussion in this chapter to neutron radiation and its effect on metals; however, other types of radiation such as α-particles (He^{2+} ions), β-rays (energetic electrons), and γ-rays are also present. We will see, in Chapter 7, that polymers are affected by radiation too.

---

* This will cause extrusions and intrusions on the external surface of the grains even when highly polished. Then these further concentrate stresses. See Fig. 21–20, Van Vlack, *Materials Science for Engineers*, Addison-Wesley.

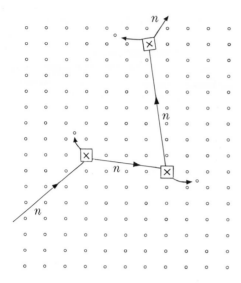

Fig. 6–9.1   Atom displacements by neutrons. When a neutron collides with an atom, part of the energy may be used to remove the atom from its lattice site into an interstitial site.

○   Atom

☒   Vacancy from
    displaced atom

Collisions of neutrons with atoms   Since neutrons are not charged, if they are given sufficient energy they can move directly through a material without being preferentially attracted to ions within the material, as are β-rays, protons, or α-particles. Neutrons interact with the atoms in a material only when they happen to "collide" with a nucleus, and such collisions occur only after many atoms have been passed. When a collision does occur in a crystal, a neutron is deflected and the atom (or ion) may be displaced from its position in the crystal to produce a vacancy and interstitialcy (Fig. 6–9.1). Because each collision slows down the neutron, collisions become more frequent until the neutron is finally captured* by the atom.

The displacement of an atom produces defects and distortion in the structure of the solid. The resulting changes in properties are somewhat similar to those arising from cold work and strain-hardening.

Figure 6–9.2 shows the effect of neutron exposure on the mechanical properties of a carbon–silicon steel. Each of the altered properties is a consequence of a decrease in the ease of slip in the distorted lattice structure.

Irradiation as a method of hardening or strengthening metals looks attractive at first sight. However, there are several inherent disadvantages. First, the effect of irradiation is logarithmic with exposure, as indicated in Fig. 6–9.3 for type 347 stainless steel. The exposure required for each succeeding increment of hardness is progressively greater. Secondly, the neutron capture and gamma-ray activation which accompanies irradiation may produce a radiologically "hot" material.

---

* Capture entails an isotopic change in the capturing atom. For example, when a manganese nucleus with 30 neutrons and 25 protons captures a neutron, it will contain 31 neutrons and 25 protons. This particular isotope of manganese happens to be unstable and will sooner or later (half-life = 2.59 hr) lose an electron (β-ray) from a neutron in the nucleus to form iron, which contains 30 neutrons and 26 protons: $n \rightarrow p^+ + e^-$.

Fig. 6–9.2 Radiation damage to steel
(ASTM A–212–B carbon–silicon steel).
(Adapted from C. O. Smith, ORSORT, Oak
Ridge, Tenn.)

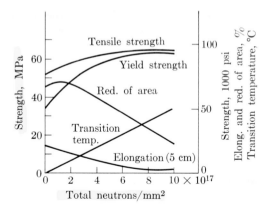

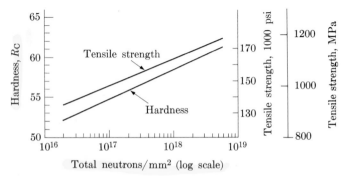

Fig. 6–9.3   Radiation hardening (Type 347 stainless steel). Neutrons dislodge atoms and
therefore restrict slip in metals. Since the neutron flux is plotted on a logarithmic scale, each
cycle requires appreciably longer exposures. (Adapted from C. O. Smith, ORSORT, Oak Ridge,
Tenn.)

Thermal and electrical resistivities increase with neutron irradiation. Figure
6–9.4 illustrates the effect on thermal resistivity. The increase of both resistivities could
be predicted from knowledge of the altered electron mobility in disordered solids
(Section 5–2).

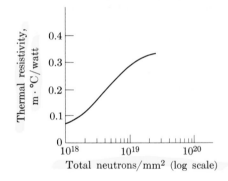

Fig. 6–9.4   Resistivity versus neutron radiation
(graphite parallel to the direction of preferred
crystal orientation). (Adapted from J. F. Fletcher
and W. A. Snyder, "Use of Graphite in the Atomic
Energy Program," *Bulletin Amer. Cer. Soc.*)

Recovery from radiation damage   Radiation damage may be erased by appropriate annealing at elevated temperatures. The mechanism of damage removal is somewhat analogous to the mechanism of recrystallization (Section 6–6). However, the required temperature is usually lower, apparently because the atoms were displaced individually into positions of higher energy than is the average position along a dislocation.

REVIEW AND STUDY

Metals have high electrical conductivity and thermal conductivity because of their delocalized electrons. They also have very useful strengths, commonly with a high degree of ductility. This permits them to be shaped into components of industrial and consumer products. Thus in this chapter we have focused on the mechanical behavior of metals as a function of alloy content, deformation, and elevated temperatures. Our attention was limited to metals with only one phase. Multiphase materials will be considered in later chapters.

SUMMARY

Single-phase alloys are stronger than either metal component alone. We call this solution hardening. It is one of two prime means we have for strengthening copper and related alloys, and is also used to increase the strength of high-temperature alloys. Solution hardening occurs because impurity atoms interfere with dislocation movements along slip planes.

The variables of single-phase microstructures include (1) grain size, (2) grain shape, and (3) grain orientation. Grain growth occurs at all temperatures, but is insignificantly slow until the temperature exceeds one half of the melting temperature. At low temperatures, grain boundaries interfere with plastic deformation; therefore, fine-grained steels are somewhat stronger than coarse-grained steels. The situation is reversed at high temperatures, when atoms can move more readily within the solid metal.

The shear modulus is the ratio of shear stress to elastic shear strain. The bulk modulus relates hydrostatic pressure to elastic volume change. Both of these moduli can be calculated from Young's modulus (Section 1–2) through Poisson's ratio. Elastic moduli are commonly anisotropic.

Metals deform by shear. Thus it is the resolved shear component of axial stresses that is important for plastic deformation. A critical shear stress is associated with each slip system.

Strain hardening is a consequence of prior plastic deformation. Any cold-forming process therefore alters the mechanical properties. Along with solution hardening this is the second major way we have to modify the properties of metals. Unfortunately there is generally a concurrent decrease in ductility. At higher temperatures, atoms can rearrange themselves and eliminate the effects of cold work, thereby softening the metal and making it more ductile. We call this annealing and recrystallization because new crystals form. Deformation processes above the recrystallization temperature are called hot-working processes. At these elevated temperatures, the metal is annealed as fast as it is deformed; thus it does not show strain hardening.

Service conditions may produce creep at high temperatures, or fatigue with cyclic loading. Fracturing is brittle in the absence of plastic deformation. It is ductile when it is accompanied by energy-consuming plastic deformation. Neutron radiation introduces point defects into materials that alter the properties, generally becoming harder, stronger, and less ductile. The various factors cited in this paragraph must be taken into account in specifying metals for products and structures.

KEY TERMS AND CONCEPTS

Alloy, single-phase
Anisotropic
Annealing
• Casting
Cold work
• Creep
• Creep rate
Deformation
  elastic
  plastic
• Endurance limit
• Fatigue
• Forming processes
  drawing
  extrusion
  forging
  rolling
  spinning
  stamping
• Fracture
  brittle
  ductile
Grain growth
Hot work

•Ingot
•Intermetallic phase
Microstructures
Modulus
  bulk, K
  shear, G
  Young's, E
Poisson's ratio, v
•Radiation damage
•Recovery
Recrystallization
Recrystallization temperature
Schmid's law
Shear strain, γ
Shear stress, τ
  critical, τ_c
  resolved
Slip direction
Slip plane
Slip system
Solution hardening
Strain hardening
•Transition temperature
Wiedemann–Franz ratio

FOR CLASS DISCUSSION

A_6 Why does the Wiedemann–Franz ratio not apply to nonmetallic materials?

B_6 Distinguish between grain shape and grain orientation in single-phase materials.

C_6 Why do alloys of copper and nickel have lower conductivity than either metal alone?

• D_6 Compare the mechanical forming processes of Figs. 6–2.2 and 6–2.3. Which will more uniformly deform the metal? Least?

E_6 Distinguish between the three moduli of elasticity.

F_6 The engineer commonly defines the stress for yield strength after an 0.2% offset strain has occurred (Fig. 1–2.1c). Based on Fig. 6–3.4, why is this necessary?

G_6 Assume a material deforms elastically in tension with no change in volume, that is, $(1 + e_x)(1 + e_y)(1 + e_z) = 1$. Calculate Poisson's ratio.

H_6 Alloying elements affect elastic moduli only slightly if at all. (Cf. copper and brass in Appendix C.) However, the strengths are significantly affected (Fig. 6–1.2). Discuss reasons for this difference.

I_6 At what combination of angles λ and ϕ (Eq. 6–4.1) is the resolved shear stress the greatest?

J_6 Differentiate between resolved shear stress and critical shear stress.

K_6 Alloys of zinc and of magnesium are more widely used as casting alloys than as wrought (plastically deformed) alloys. Suggest a valid reason.

L_6 Why are alloys of copper and nickel stronger than either metal alone?

M_6 An impurity atom stops a dislocation movement. Explain.

N_6 A rod is cold-drawn through a die to twice its original length. Why is its ductility decreased? How much is it cold worked? Will its percent reduction in area be increased or decreased? (This question was asked in order to review the difference between *lengthening* as a process and *elongation* as a property, and between decrease in area arising from *cold work* and *reduction-in-area* as a property.)

O_6 Use the data of Figs. 6–1.2(b), 6–5.6, and 6–5.7. Sketch a plausible curve of yield strength versus cold work for a 70–30 brass.

P_6 How will longer times (>1 hour) affect the curves of Fig. 6–6.4? How will the curves differ for an 85–15 brass?

Q_6 As a rule of thumb, one may consider that the recrystallization temperature is one half of the melting temperature. Why must this rule be taken with caution?

R_6 Distinguish between hot-working and cold-working of tin; of tungsten.

• S_6 Explain the basis for the various stages of the creep shown in Fig. 6–7.9.

T_6 How does the fact that an atom has lower energy when it coordinates with the grain that has the concave surface lead to boundary movement?

U_6 Explain why a single-crystal metal is softer and more plastic than a polycrystalline metal.

V_6 Grains increase in their average size at higher temperatures. Why do they not decrease at lower temperatures?

W_6 Why are coarse-grained materials stronger than fine-grained materials at high temperatures?

• X_6 Other things being equal, which will have the lowest creep rate: (a) steel in service with a high tensile stress and low temperature, (b) steel in service with a low tensile stress and high temperature, (c) steel in service with a high tensile stress and high temperature, (d) steel in service with a low tensile stress and low temperature? Why?

• Y_6 Give examples where ductile fracture provides a "fail-safe" design feature.

• Z_6 Transition temperatures were not given the same attention in the design of riveted ships as is now given in the design of welded ships. Why?

STUDY PROBLEMS

6–1.1   Based on Fig. 6–1.2, what is the electrical resistivity of annealed 70–30 brass?

Answer: 62×10^{-9} ohm·m (or 62 ohm·nm)

6–1.2   Refer to Fig. 6–1.2. Can the same W–F ratio be used for brasses as for pure metals?

6–1.3   Estimate the thermal conductivity of copper at 90°C if, as in Example 6–1.2, the W–F ratio is proportional to temperature.

Answer: 400 $(W/m^2)/(°C/m)$ [or 0.40 $(W/mm^2)/(°C/mm)$]

6–1.4   A copper wire has a resistance of 0.5 ohm per 100 m. Consideration is being given to the use of a 75–25 brass wire instead of copper. What would be the resistance, if the brass wire were the same size?

6–1.5   A brass alloy is to be used in an application that will have tensile strength of more than 275 MPa (40,000 psi) and an electrical resistivity of less than 50×10^{-9} ohm·m (resistivity of Cu = 17×10^{-9} ohm·m). What percent zinc should the brass have?

Answer: 15 to 23% zinc

6–1.6   A copper alloy (brass, or Cu–Ni) is required that has a tensile strength of at least 245 MPa (35,000 psi) and a resistivity of less than 40×10^{-9} ohm·m. (a) Select a suitable alloy bearing in mind that $\$_{Ni} > \$_{Cu} > \$_{Zn}$. (b) What is the ductility of the alloy you chose?

6–1.7   A certain application requires a piece of metal having a yield strength greater than 100 MPa and a thermal conductivity greater than 0.04 $(W/mm^2)/(°C/mm)$. Specify either an annealed brass or an annealed Cu–Ni alloy that could be used.

Answer: Use 80 Cu–20Ni

6–1.8 A motorboat requires a seat brace. Iron is excluded because it rusts. Select the most appropriate alloy from Figs. 6–1.2 and 6–1.3. The requirements include a tensile strength of at least 310 MPa (45,000 psi); a ductility of at least 45% elongation (in 50 mm); and low cost. (*Note:* Zinc is cheaper than copper.)

6–1.9 A brass wire must carry a load of 45 N (10 lb) without yielding and have a resistance of less than 0.033 ohm/m (0.01 ohm per foot). (a) What is the smallest wire that can be used if it is made of 60–40 brass? (b) 80–20 brass? (c) 100% Cu?

Answer: (a) 1.8-mm (0.07-in.) dia.   (b) 1.4-mm (0.055-in.) dia.   (c) 1.2-mm (0.05-in.) dia.

6–2.1 Two 20-m (65-ft) aluminum rods are each 14.0 mm (0.551 in.) in diameter. One bar is drawn through a 12.7-mm (0.50-in.) die. (a) What are the new dimensions of that bar? (b) Assume identical test samples are made from each bar (deformed and undeformed) and marked with 50-mm gage lengths. Which one, if either, of the bars will have the greater ductility? The greater yield strength?

Answer: (a) 24.3 m   (b) Review Section 1–2.

6–2.2 The carbon content of 225,000 kg of liquid steel must be lowered from 3 w/o to 0.25 w/o by oxidizing it to CO. Each mole of gas occupies 0.0224 m^3 under standard conditions. How much gas is evolved from the refining process?

6–3.1 What is the volume change of a rod of brass when it is loaded axially by a force of 233 MPa (33,800 psi)? Its yield strength is 270 MPa (39,200 psi), and its Poisson's ratio is 0.3.

Answer: 0.085 v/o

6–3.2 What is the bulk modulus of the brass in the previous problem?

6–3.3 A test bar 12.83 mm (0.5051 in.) in dia. with a 50-mm gage length is loaded elastically with 156 kN (35,000 lb) and is stretched 0.356 mm (0.014 in.). Its diameter is 12.80 mm (0.5040 in.) under load. (a) What is the bulk modulus of the bar? (b) The shear modulus?

Answer: (a) ~1.5×10^5 MPa (~20×10^6 psi)   (b) 65×10^3 MPa (10^7 psi)

6–3.4 Copper that has a modulus of elasticity of 110,000 MPa (16,000,000 psi) and a Poisson's ratio of 0.3 is under a hydrostatic pressure of 83 MPa (12,000 psi). What are the dimensions of the unit cell?

6–3.5 If copper has an axial stress of 97 MPa (14,000 psi), what will be the highest local stress within a polycrystalline copper bar?

Answer: 170 MPa (24,500 psi)

6–3.6 A precisely ground cube (100.00 mm)^3 of an aluminum alloy is compressed in the x-direction 70 MPa. (a) What are the new dimensions if Poisson's ratio is 0.3? (b) A second compression of 70 MPa is concurrently applied in the y-direction. Now, what are the dimensions? (c) Derive Eq. (6–3.6). (*Hint:* Consider the cube under a stress $s_x = s_y = s_z = P_h$ and recall that $(1 + \Delta V/V) = (1 + \Delta l/l)^3 = 1 + 3e + \cdots$.)

6–3.7 When iron is compressed hydrostatically with 205 MPa (30,000 psi), its volume is changed by 0.13%. How much will its volume change when it is stressed axially with 415 MPa (60,000 psi)?

Answer: 0.09 v/o

6–4.1 A force of 660 N (150 lb$_f$) is applied in the [111] direction of a cubic crystal. What is the resolved force in the [110] direction?

Answer: 540 N (120 lb$_f$)

6–4.2 What force must be applied in the [112] direction of copper to produce a resolved force of 410 N (92 lb$_f$) in the [110] direction?

6–4.3 A force of 3800 N (850 lb$_f$) occurs in the [111] direction of tin. What is the resolved force in the [110] direction? (Tin is bcc with $a = 0.5820$ nm and $c = 0.3175$ nm.)

Answer: 3550 N (795 lb$_f$)

6–4.4 An axial stress of 123 MPa (17,800 psi) exists in the [110] direction of bcc iron. What is the resolved shear stress in the [101] direction on the (010) plane?

Answer: 43.5 MPa (6300 psi)

6–4.5 An aluminum crystal slips on the (111) plane and in the [1$\bar{1}$0] direction with a 3.5 MPa (500 psi) stress applied in the [1$\bar{1}$1] direction. What is the critical resolved shear stress?

Answer: 0.95 MPa (138 psi)

6–4.6 A single crystal lies completely across a copper wire (1.0-mm dia) that is being pulled in tension. The loading direction is lengthwise of the wire and very close to the [$\bar{1}$01] direction. The load is 1 kg. (a) What is the axial stress on a plane that cuts directly across the wire? (b) What is the shear stress in the [$\bar{1}$10] direction on the ($\bar{1}\bar{1}$1) plane?

6–4.7 Refer to Example 3–7.3. (a) If there are 8.2×10^{12} atoms/mm^2 on the {100} planes of lead, how far apart are the adjacent planes? (b) If there are 9.4×10^{12} atoms/mm^2 on the {111} planes of lead, how far apart are the adjacent planes? Solve via atoms/mm^3 and without regard to unit-cell geometry. (c) On which set will slip be expected to occur more readily?

Answer: (a) 0.248 nm (b) 0.285 nm

6–4.8 What are the twelve $\langle\bar{1}11\rangle\{101\}$ slip systems in bcc metals? The twelve $\langle\bar{1}11\rangle\{211\}$ slip systems?

• **6–4.9** Refer to Fig. 6–4.8. The shortest slip vector on the {111} plane is $b_{01\bar{1}}$ (or its permutations such as $b_{10\bar{1}}$). The second shortest is $b_{11\bar{2}}$ (or its permutations). (a) Excluding the directions of these slip vectors, which direction contains the next shortest slip vector? (b) How would its relative energy requirements compare with $b_{01\bar{1}}$?

Answer: (a) $b_{\bar{3}21}$ (b) $\times 7$

6–4.10 A flat brass plate must conduct more than 2.7 cal/mm^2·sec (or 11.3 W/mm^2) when its two faces have a temperature difference of 125°C. It must also have a tensile strength-thickness product of more than 0.55 MPa (m) that is, $(S_T)(t) > 0.55$ MPa (m), (> 3140 lb$_f$/in). Consider the 90–10, 80–20, and 70–30 brasses of Fig. 6–1.2, and select the alloy that will meet the requirements least expensively.

6–5.1 How much cold work was performed on the aluminum rod in Study Problem 6–2.1?

Answer: 18%

6–5.2 Change the metal in Study Problem 6–2.1 from aluminum to copper. What are the tensile strength and the ductility for the deformed metal?

6–5.3 A copper wire 2.5 mm (0.10 in.) in diameter was annealed before cold-drawing it through a die 2.0 mm (0.08 in.) in diameter. What tensile strength does the wire have after cold-drawing?

Answer: 330 MPa (48,000 psi)

6–5.4 A pure iron sheet 2.5-mm (0.10-in.) thick is annealed before cold-rolling. It is rolled to 2.0 mm (0.08 in.), with negligible change in width. What will be the ductility of the iron after cold-rolling?

6–5.5 Change the metal in Study Problem 6–2.1 to 80 Cu–20 Zn. What are the ductilities for the two rods?

Answer: Annealed—50% elongation; 18% CW—20% elongation

6–5.6 Copper is to be used as a wire with at least 310 MPa (45,000 psi) tensile strength and at least 18% elongation (50 mm, or 2 in.). How much cold work should the copper receive?

6–5.7 Iron is to have a BHN of at least 125 and an elongation of at least 32% (50 mm, or 2 in.). How much cold work should the iron receive?

Answer: Use 25% cold work

6–5.8 (a) How much zinc should be in a cold-worked brass to give a ductility of at least 20% elongation, *and* a hardness of at least $70R_B$? (b) How much should it be cold-worked? (Your answer should not exceed 36% Zn because of processing complications.)

6–6.1 A copper wire must have a diameter of 0.7 mm and a tensile strength of 345 MPa (50,000 psi). It is to be processed from a 10-mm copper rod. What should be the diameter for annealing prior to the final cold-draw?

Answer: 0.95 mm

6–6.2 A round bar of 85 Cu–15 Zn alloy, 5 mm in diameter, is to be cold-reduced to a rod 1.25 mm in dia. Suggest a procedure to be followed if a final tensile strength of 415 MPa (or greater) is to be achieved along with a final ductility of at least 10% elongation (in 50-mm gage length).

6–6.3 A 70–30 brass (Fig. 6–5.6) wire with a tensile strength of more than 415 MPa (60,000 psi), a hardness of less than $75R_B$, and an elongation (50 mm, or 2 in.) of more than 25% is to be made by cold-drawing. The diameter, as received, is 2.5 mm (0.1 in.). The diameter of the final product is to be 1.0 mm (0.04 in.). Prescribe a procedure for obtaining these specifications.

Answer: 15% to 19% cold work; therefore it should be annealed when the diameter is 1.1 mm (0.043 in.) before the final 17% cold work.

6–6.4 A round bar of brass (85% Cu, 15% Zn), 5 mm (0.20 in.) in diameter, is to be cold-drawn to wire that is 2.5 mm (0.10 in.) in diameter. Specify a procedure for the drawing process such that the wire will have a hardness less than $72R_B$, a tensile strength greater than 415 MPa (60,000 psi), and a ductility of greater than 10% elongation in the standard gage length.

6–6.5 A rolled 66 Cu–34 Zn brass plate 12.7-mm (0.500-in.) thick had a ductility of 2% elongation (in 50-mm (2-in.) gage length) when it was received from the supplier. This plate is to be rolled to a sheet with a final thickness of 3.2 mm (0.125 in.). In this final form, it is to have a tensile strength of at least 483 MPa (70,000 psi) and a ductility of at least 7% elongation (in 50-mm gage length). Assuming that the rolling process that reduces the plate to a sheet does not change the width, specify *all steps* (including temperature, times, thickness, etc.) that are required.

6–6.6 Aluminum sheet has been shaped into a cake pan by spinning (Fig. 6–2.3). Assume the data from Figure 6–6.3 apply to this cold-worked aluminum. Will it start to recrystallize while in a 180°C (350°F) oven to bake a cake? Solve, both graphically and mathematically.

6–6.7 Assume annealing should be completed in 1 second to permit the hot-working of the aluminum of Fig. 6–6.3 and Example 6–6.2. What temperature is required?

Answer: 685 K (or 412°C)

• *6–7.1* The following data were obtained in a creep–rupture test of Inconel "X" at 815°C (1500°F): (a) 1% strain after 10 hr, (b) 2% strain after 200 hr, (c) 4% strain after 2000 hr, (d) 6% strain after 4000 hr, (e) "neck-down" started at 5000 hr and the rupture occurred at 5500 hr. What was the creep rate?

Answer: 0.001%/hr

• *6–7.2* Design considerations would permit a pressure tube to have 3 l/o strain during a year of service. What maximum creep rate is tolerable when reported in the normal %/hr figures?

• *6–8.1* Sometimes it is assumed that if failure does not occur in a fatigue test in 10^8 cycles, the stress is below the endurance limit. A test machine is connected directly to a 1740-rpm electric motor. How long will it take to log that number of cycles?

Answer: 40 days

• *6–8.2* Examine a crankshaft of a car. Point out design features that alter the resistance to fatigue.

CHAPTER **7**

Molecular Phases

PREVIEW

"Plastics" have rapidly come into technical importance. While they have the major advantage of being lightweight, their greatest value is their ability to be processed into intricate geometric shapes with desired properties, using the minimum of labor. They constitute one of the fastest-growing areas of materials development and applications. Since they may be processed easily, it is common to find them used by designers without full knowledge of their characteristics and limitations. The purpose of this chapter is to outline the structural nature of these nonmetallic materials and to relate these structures to their properties and service behavior.

The technical name for "plastics" is *polymers*. These are large molecules, *macromolecules*, made up of many repeating units, or *mers*. We will examine these mers and how they combine into polymeric materials. We will look first at the ideal molecular arrangements. However, as with crystals, we will also look at irregularities, because they have a pronounced influence on the behavior of the polymers.

Polymers do not crystallize readily because large molecules rather than individual atoms must be arranged into an ordered structure. In addition, the molecules have relatively weak *inter*molecular bonds. Polymeric materials are insulators, because they have no free electrons.

As a result of the factors cited in the previous paragraph, the properties and service behavior of plastics differ from those of metals in many respects.

CONTENTS

STUDY OBJECTIVES

1 To become familiar with the molecules that are introductory to polymers.

2 To calculate molecular sizes, and to describe the bond changes in polymerization reactions.

3 To acquaint yourself with molecular irregularities that (a) affect the molecular alignment and crystallization of lineal polymers, and (b) lead to extended 3-D structures.

4 To interpret viscoelastic and dielectric properties in terms of molecular responses, above and below the glass temperature.

5 To anticipate degradation that is possible in the variety of environments encountered in service.

6 To be able to design products yourself in later U.G. courses, or in your future job, to communicate your engineering specifications intelligently to a design engineer so he or she will select the optimum materials.

7–1 GIANT MOLECULES

Our next principal category of materials following metals is molecular solids. These are nonmetallic compounds (Section 2–7) that commonly originate from organic raw materials. Organic substances have served as engineering materials from a time preceding the first engineer. Wood has long been a common construction material, and such natural organic substances as leather for gaskets, felt for packings, cork for insulation, fibers for binding, oils for lubrication, and resins for protective coatings are extensively used by engineers.

Early in the history of the use of organic materials, attempts were made to improve their engineering properties. For example, the properties of wood are highly directional; the strength parallel to the grain is several times greater than in the perpendicular direction. The development of plywood has helped to overcome this nonuniformity, and still better physical properties are obtained when the pores of the wood are impregnated with a thermosetting resin.

The ingenuity of technologists in working with organic materials* has not been limited to improving natural organic materials; many synthetic substances have been developed as well. For example, *plastics*, or more accurately *polymers*,[†] have given the engineer an infinitely greater variety of materials for his applications. Great strides have been, and continue to be, made in the utilization of such materials.

Whether the engineer who uses polymers is working with natural organic materials or with artificial ones, he or she is concerned primarily with the nature and characteristics of *large molecules*. In natural materials, large molecules are "built in" by nature; in artificial materials, these large *macromolecules* are built by *deliberately joining small molecules*.

Molecular sizes   The reader who has had chemistry knows that the melting temperatures of molecules in the hydrocarbon series, C_mH_{2m+2}, are related to the size of the molecules (Fig. 7–1.1). In general, those polymers containing large molecules are stronger and more resistant to thermal and mechanical stresses than are those composed of small molecules (Fig. 7–1.2). Examples of this type of relationship between molecular size and properties exist for such artificial organic polymers as polyvinyl chloride, nylon, and saran, and such natural substances as cellulose, rubber, waxes, and shellac.

These organic materials, as a class, are *polymers* because they are molecules consisting of many repeating segments or units called *mers*. Therefore, we shall use the more technical term of polymers for these molecular materials in order to distinguish the materials from their deformation behaviors.

Most of the large molecules, with which we will deal, contain many smaller repeating units called *mers* (Fig. 7–1.3). Thus, the mass of the large molecule, i.e.,

---

* Students who feel "rusty" on molecular chemistry are advised to review "polymeric molecules" in *Study Aids for Introductory Materials Courses*. This short review includes essentially all of the organic chemistry required as a prerequisite for this chapter.
† Strictly speaking, *plastic* is an adjective defining a permanently deformable material (Section 1–2), but by common usage "plastics" denote organic materials that have been shaped by plastic deformation.

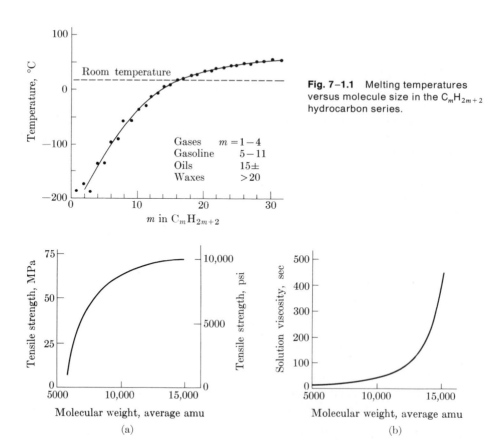

Fig. 7–1.1 Melting temperatures versus molecule size in the C_mH_{2m+2} hydrocarbon series.

Gases	$m = 1-4$
Gasoline	$5-11$
Oils	$15\pm$
Waxes	>20

Fig. 7–1.2 Polymer size versus polymer properties (copolymer of vinyl chloride and vinyl acetate). (a) Tensile strength. (b) Viscosity. (Adapted from G. O. Crume and S. D. Douglas, *Ind. Eng. Chem.*)

Fig. 7–1.3 Polymeric molecule (polyvinyl chloride). The large molecule consists of numerous repeating units (mers), in this case, $+C_2H_3Cl+_n$. (Cf. Fig. 2–3.7.)

the polymer, is simply the product of the *mer mass* times the numbers of mers in the molecule. For example, if a polyvinyl chloride molecule contains 500 mers of vinyl chloride, C_2H_3Cl, each of which has $2(12 \text{ amu}) + 3(1 \text{ amu}) + 35.5 \text{ amu} = 62.5 \text{ amu}$, the total mass of the polymer molecule would be $500(62.5 \text{ amu})$ or 31,250 amu (or 31,250 g per mole).

In addition to expressing the size of a molecule by its mass, we commonly speak of the *degree of polymerization, n*. This is the number of mers per molecule; thus,

$$n = \frac{\text{molecular mass}}{\text{mer mass}}. \qquad (7-1.1)$$

The units of the expression are

$$\frac{\text{amu/molecule}}{\text{amu/mer}} = \frac{\text{mers}}{\text{molecule}}.$$

The molecule just described, with a mass of 31,250 amu and a degree of polymerization of 500, falls in the range of commercial polymers, since they commonly possess 75 to 1000 mers. These values differ by several orders of magnitude from that of other molecules (Fig. 2–3.1). However, as large as these polymer molecules appear to be based on mass, they are smaller than the resolving power of an optical microscope, and only under certain circumstances can it be resolved even by an electron microscope. Consequently, molecular-size determinations are usually made indirectly by such physical means of measurement of viscosity, osmotic pressure, or light-scattering, all of which are affected by the number, size, or shape of molecules in a suspension or in a solution (Fig. 7–1.2b).

Molecular-size distribution   When a material like polyethylene (Fig. 7–1.4) or polyvinyl chloride is formed from small molecules, not all the resulting large molecules are identical in size. As might be expected, some grow larger than others. As a result, polymers contain a range of molecular sizes, somewhat analogous to the mixture of propane, hexane, octane, and other paraffin hydrocarbons in crude oils. Hence it is necessary to calculate the *average* degree of polymerization if a single index of size is desired.

Fig. 7–1.4   Linear polymer (polyethylene). The molecular length varies from molecule to molecule. Typically they contain hundreds of carbon atoms.

One procedure for calculating average molecular sizes utilizes the weight fraction of the molecules that is in each of several size intervals (Fig. 7–1.5a). The *"mass-average" molecular size* $\bar{M}_m$ is calculated as follows:

$$\bar{M}_m = \sum(W_i M_i), \qquad\qquad (7\text{–}1.2)^*$$

where W_i is the mass fraction in each size interval and M_i is the representative value (middle) of each size interval. The "mass-average" molecular size is particularly significant in the analysis of properties such as viscosity, where the *mass* of the individual molecules is important.

---

* Equations (7–1.2) and (7–1.3) are simply special cases of the general mathematical formula for an average:

$$\text{Average} = \sum(\text{fraction}_i)(\text{value}_i).$$

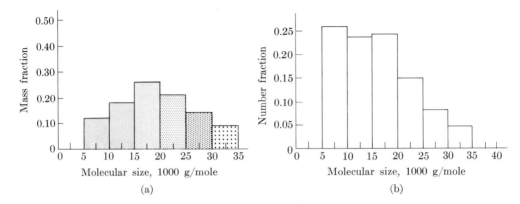

Fig. 7–1.5 Polymer size distributions (a) based on mass, (b) based on numbers. These two presentations are for the same material. In (a), the data indicate that 12% of the total *mass* is accounted for by molecules between 5,000 amu and 10,000 amu (mid-value = 7,500 amu). Other percentages of the total mass are in other size intervals. In (b), we observe that 26% of the total *number* of molecules fall in the 5,000-amu to 10,000-amu range. These values of 26% for (b) versus 12% for (a) differ because it takes many of the small molecules to account for 12% of the mass. In contrast, 4% of the molecules account for 9% of the mass that is found in the 30,000–35,000 interval. (See Example 7–1.3.)

Another procedure for determining average molecular size utilizes the *number fraction* of the molecules (Fig. 7–1.5b) that is in each of the several size intervals. Properties such as strength (Fig. 7–1.2a) are more sensitive to the *numbers* of large molecules than to the actual mass. Hence, a *"number-average"* molecular size $\bar{M}_n$ has significance:

$$\bar{M}_n = \sum (X_i M_i). \tag{7–1.3}*$$

The value X_i is the numerical fraction of molecules in each size interval (Fig. 7–1.5b). As before, M_i is the representative value (middle) of each size interval. (See Example 7–1.3.)

Molecular length   We could make a calculation from Table 2–3.1 that the average length of a polyvinyl chloride molecule with degree of polymerization of 307 mers (Example 7–1.3) is 95 nm, because each C—C bond is 0.154 nm and there are twice as many of these bonds as there are mers. However, this calculation needs a correction because the sketch of Fig. 7–1.3 is overly simplified.[†] The C—C—C bond angles are not 180°, but 109.5° (Fig. 7–1.4). Thus a "sawtooth" length would be (95 nm) sin 54.7°, or 77 nm. Even more important, however, is the fact that the single bonds of the carbon chain are free to rotate (Fig. 7–1.6). With only three bonds, the end-to-end length can vary from less than 0.3 nm, *a–d'* to practically 0.4 nm, *a–d*. The length varies randomly between these limits as the thermal agitation rotates the bond angles. With the above molecule of 307 mers, the end-to-end length could be as long as 77 nm if it were like Fig. 7–1.4, or it could be appreciably less than a nanometer if the two ends

---

*   See footnote on page 239.
[†]   We will still make use of this presentation, however, since it meets many of our needs.

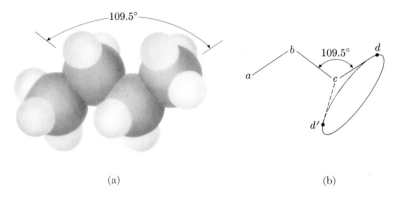

(a)   (b)

Fig. 7–1.6 Bond rotation (butane). Although there is a constant 109.5° bond angle, the end-to-end distance can vary from a–d to a–d'. Large molecules will have much more variation.

of the molecule happen to be adjacent to each other. Actually, thermal agitation keeps the molecule continuously *kinked* and changing (Fig. 7–1.7). The statistically most probable, or *root-mean-square*, length $\bar{L}$ is

$$\bar{L} = l\sqrt{m}. \qquad (7\text{–}1.4)^*$$

In this equation, l is the bond length, that is, 0.154 nm for C—C (Table 2–3.1), and m is the number of bonds. For polyvinyl chloride (Fig. 7–1.3), there are two bonds per mer; therefore $m = 2n$, and according to Eq. 7–1.4, $\bar{L}$ becomes 3.8 nm for a molecule with 307 mers.

Fig. 7–1.7 Kinked conformation. Since each C—C bond can rotate (Fig. 7–1.6), a long molecule is normally kinked and has a relatively short mean length, $\bar{L}$ (Eq. 7–1.4).

The kinked *conformation of* Fig. 7–1.7 becomes very important to us because it is the basis for the stretching and contraction of rubbers. Also, it explains why *stretched* rubber has the abnormality of a negative coefficient of thermal expansion (Section 7–5).

Example 7–1.1 The composition of "teflon" is identical to the polyethylene of Fig. 2–3.7(b) except that fluorine atoms replace hydrogen atoms. Therefore, its mer is $+C_2F_4+$. (a) What is the degree of polymerization if the mass of the molecule is 33,000 amu (or 33,000 g/mole)? (b) How many molecules per gram?

Solution: (a) The mer mass is $(4)(19) + (2)(12) = 100$ amu;

$$(33{,}000 \text{ amu/molecule})/(100 \text{ amu/mer}) = 330 \text{ mers/molecule}.$$

---

* The mathematics required to derive this equation is beyond the scope of this book. It utilizes the statistics of a "random-walk" process.

b) $(0.6 \times 10^{24} \text{ molecules})/(33,000 \text{ g}) = 1.8 \times 10^{19}/\text{g}$.

Comment. As with ethylene in Example 2–3.1, we consider the mer to be C_2F_4 rather than CF_2. This group of three atoms cannot form a monomer because only two of the four carbon bonds are used if only two fluorine atoms are present. By forming a double bond, between carbons, C_2F_4 can meet the bond requirements. ◀

Example 7–1.2 A solution contains 15 g of water, 4 g of ethanol (C_2H_5OH), and 1 g of sugar ($C_6H_{12}O_6$). (a) What is the mass fraction of each molecular component? (b) What is the number fraction of each molecular component?

Solution

a)

		W_i
H_2O	15 g/20 g	$= 0.75$
C_2H_5OH	4 g/20 g	$= 0.20$
$C_6H_{12}O_6$	1 g/20 g	$= 0.05.$

b)   Basis: 20 amu

		molecules	X_i
H_2O	15 amu/18 amu	$= 0.833$	$= 0.900$
C_2H_5OH	4 amu/46 amu	$= 0.087$	$= 0.094$
$C_6H_{12}O_6$	1 amu/180 amu	$= 0.006$	$= 0.006.$ ◀
		$\overline{0.926}$	

Example 7–1.3 It has been determined that polyvinyl chloride (PVC) has the molecular size distribution shown in Figs. 7–1.5(a) and (b)—for mass fraction and number fraction, respectively. (a) What is the "mass-average" molecular size? (b) What is the "number-average" molecular size? (c) What is the degree of polymerization n based on M_m?

Solution: (a) and (b)

Molecular size interval, amu	$(M)_i$, mid-value, amu	W_i, Mass fraction (Fig. 7–1.5a)	$(W_i)(M_i)$, amu	X_i, Number fraction (Fig. 7–1.5b)	$(X_i)(M_i)$, amu
5–10,000	7,500	0.12	900	0.26	1,950
10–15,000	12,500	0.18	2,250	0.23	2,875
15–20,000	17,500	0.26	4,550	0.24	4,200
20–25,000	22,500	0.21	4,725	0.15	3,375
25–30,000	27,500	0.14	3,850	0.08	2,200
30–35,000	32,500	0.09	2,925	0.04	1,300
			$\sum = 19,200$ amu/molecule (a) M_m		$\sum = 15,900$ amu/molecule (b) M_n

c)   Amu/mer of PVC (Fig. 7–1.3):

$$(C_2H_3Cl) = 24 + 3 + 35.5 = 62.5 \text{ amu/mer,}$$

$$\text{Degree of polymerization} = \frac{(19{,}200 \text{ amu/molecule})}{(62.5 \text{ amu/mer})} = 307 \text{ mers/molecule} \quad \text{(based on } \bar{M}_m\text{).}$$

Comments. Whenever there is a distribution of sizes, the "number-average" molecular size is always less than the "mass-average" value, because of the large number of smaller molecules per gram in the smaller size intervals. The two averages diverge more as the range of the size distribution increases. ◀

Example 7–1.4 A polyethylene $(C_2H_4)_n$ molecule (see Fig. 7–1.4), with a molecular mass of 22,400 amu, is dissolved in a liquid solvent. (a) What is the longest possible end-to-end distance of the polyethylene (without altering the 109.5° C—C—C bond angle)? (b) The shortest? (c) The most probable?

Solution

$$\text{Degree of polymerization} = (22{,}400 \text{ amu/molecule})/(28 \text{ amu/mer})$$
$$= 800 \text{ mers/molecule.}$$

Therefore,
$$\text{bonds} = 1600.$$
From Table 2–3.1,
$$l = 0.154 \text{ nm.}$$

a)   "Sawtooth length" $= (1600)(0.154 \text{ nm})(\sin 109.5°/2)$
$$= 200 \text{ nm.}$$
b)   Shortest distance $= \sim 0.3$ nm with ends in contact.
c)   Eq. (7–1.4): $L = 0.154 \text{ nm} \sqrt{1600} = 6.2$ nm.

Comments. Since the molecule is under continuous thermal agitation, the probability of its attaining a length of 200 nm would be extremely remote. A force would have to be used to stretch it out to that length. Furthermore the required force would become greater at higher temperatures because the kinking becomes more persistent with increased thermal agitation. ◀

7–2   LINEAR POLYMERS

Large one-dimensional molecules make up the basis of many of our common polymers. Examples include the *polyvinyls*, which have the structure

$$\left[\begin{array}{cc} H & H \\ | & | \\ C & C \\ | & | \\ H & R \end{array} \right]_n \qquad\qquad (7\text{–}2.1)$$

as a mer. (See Table 7–2.1 where **R** is one of a variety of side groups.) Many readers have also heard the terms *polyesters, polyurethanes,* and *polyamides,* which have linear molecules. Finally, *rubbers* have linear (and very highly kinked) molecules. Three of

Table 7–2.1
Vinyl-type molecules

Vinyl compounds
$$\begin{bmatrix} \overset{\displaystyle H}{\underset{\displaystyle H}{C}} = \overset{\displaystyle H}{\underset{\displaystyle R}{C}} \end{bmatrix}$$

	R
Ethylene	—H
Vinyl chloride	—Cl
Vinyl alcohol	—OH
Propylene	—CH$_3$
Vinyl acetate	—OCOCH$_3$
Acrylonitrile	—C≡N
Styrene (vinyl benzene)	—⬡

Vinylidene compounds
$$\begin{bmatrix} \overset{\displaystyle H}{\underset{\displaystyle H}{C}} = \overset{\displaystyle R''}{\underset{\displaystyle R'}{C}} \end{bmatrix}$$

	R′	**R″**
Isobutylene	—CH$_3$	—CH$_3$
Vinylidene chloride	—Cl	—Cl
Methyl methacrylate	—CH$_3$	—COOCH$_3$

Tetrafluoroethylene
$$\begin{bmatrix} \overset{\displaystyle F}{\underset{\displaystyle F}{C}} = \overset{\displaystyle F}{\underset{\displaystyle F}{C}} \end{bmatrix}$$

Trifluorochloroethylene
$$\begin{bmatrix} \overset{\displaystyle F}{\underset{\displaystyle F}{C}} = \overset{\displaystyle Cl}{\underset{\displaystyle F}{C}} \end{bmatrix}$$

Table 7–2.2
Butadiene-type molecules

$$\begin{bmatrix} \overset{\displaystyle H}{\underset{\displaystyle H}{C}} = \overset{\displaystyle R}{C} - \overset{\displaystyle H}{C} = \overset{\displaystyle H}{\underset{\displaystyle H}{C}} \end{bmatrix}$$

	R
Butadiene	—H
Chloroprene	—Cl
Isoprene	—CH$_3$

the simpler rubber mers are listed in Table 7–2.2. We shall consider the two main categories of polymerization reactions; *addition* and *condensation*. We will observe that the mers along linear molecular chains are not perfectly regular, but have various *configurations*.

Addition polymerization   The monomer of a vinyl is C_2H_3R, that is,

$$
\begin{array}{cc}
\text{H} & \text{H} \\
| & | \\
\text{C} & = \text{C}. \\
| & | \\
\text{H} & \textbf{R}
\end{array}
\tag{7–2.2}
$$

Vinyls are *bifunctional* because they have two reaction sites, forward and rear. For polyvinyl chloride, where —$\mathbf{R}$ of Eq. (7–2.2) is —Cl, this reaction is

$$
n\left[\begin{array}{cc}
\text{H} & \text{H} \\
| & | \\
\text{C} & = \text{C} \\
| & | \\
\text{H} & \text{Cl}
\end{array}\right]
\longrightarrow
\left[\begin{array}{cc}
\text{H} & \text{H} \\
| & | \\
\text{C} & - \text{C} \\
| & | \\
\text{H} & \text{Cl}
\end{array}\right]_n.
\tag{7–2.3}
$$

By the simplest analogy, the growth of the polymer is comparable to coupling railroad cars. The process, however, is complex, because simply placing monomers closely together does not automatically produce an addition polymerization reaction. The reaction must be *initiated*, followed by *propagation*; finally, it is *terminated*.

An initiator is commonly a free radical such as a split H_2O_2 molecule:

$$
\text{H—O—O—H} \to 2\,\text{HO}\bullet,
\tag{7–2.4a}
$$

where the reactive dot • is an unfilled orbital of the HO pair.* Therefore it will react readily. For example, when it encounters a vinyl molecule,

$$
\text{H—O}\bullet +
\begin{array}{cc}
\text{H} & \text{H} \\
| & | \\
\text{C} & = \text{C} \\
| & | \\
\text{H} & \textbf{R}
\end{array}
\longrightarrow
\text{H—O—}
\begin{array}{cc}
\text{H} & \text{H} \\
| & | \\
\text{C} & - \text{C}\bullet \\
| & | \\
\text{H} & \textbf{R}
\end{array}
\tag{7–2.6}
$$

We still have a free radical (admittedly longer), so it can react with another;

$$
\text{H—O—}
\begin{array}{cc}
\text{H} & \text{H} \\
| & | \\
\text{C} & - \text{C}\bullet \\
| & | \\
\text{H} & \textbf{R}
\end{array}
+
\begin{array}{cc}
\text{H} & \text{H} \\
| & | \\
\text{C} & = \text{C} \\
| & | \\
\text{H} & \textbf{R}
\end{array}
\longrightarrow
\text{H—O—}
\begin{array}{cccc}
\text{H} & \text{H} & \text{H} & \text{H} \\
| & | & | & | \\
\text{C} & - \text{C} & - \text{C} & - \text{C}\bullet \\
| & | & | & | \\
\text{H} & \textbf{R} & \text{H} & \textbf{R}
\end{array}
\tag{7–2.7}
$$

A chain reaction continues to propagate, to produce molecules with many units, i.e., polymers.

---

* This may be illustrated as

$$
\text{H}:\ddot{\text{O}}:\ddot{\text{O}}:\text{H} \to \text{H}:\ddot{\text{O}}\bullet + \bullet\ddot{\text{O}}:\text{H},
\tag{7–2.4b}
$$

where neither •OH radical is charged. In contrast, an OH$^-$ ion carries the charge of an excess electron:

$$
:\ddot{\text{O}}:\text{H}.
\tag{7–2.5}
$$

The driving force for Eq. (7–2.6) and continuing reactions is the change in energy. For example the left side of Eq. (7–2.7) contains a double C=C bond, whereas the right side contains an additional *pair* of single C—C bonds. All other bonds are unchanged. Table 2–3.1 indicates that the C=C bonds have 680 kilojoules per mole of 0.6×10^{24} bonds. Each C—C bond has 370 kJ per mole. By eliminating the double bond of Eq. (7–2.7), we must supply $+1.1 \times 10^{-18}$ joules; by forming *two* single bonds, $-(2)(0.6 \times 10^{-18}$ joules) are released. Thus 0.1×10^{-18} joules of energy are given off $(-)$ in Eq. (7–2.7), or 60 kJ/mole. The resulting polymer compounds are more stable than the original monomers because they have lost energy.*

The propagation of Eqs. (7–2.6) and (7–2.7) terminates when (1) the supply of monomers is exhausted, or (2) the ends of the two growing chains encounter and join. A range of molecular sizes is produced, since different molecules terminate their growth at different times (Fig. 7–1.5).

Copolymers   Each of the vinyl polymers considered so far included only one type of mer. Polyethylene has only (C_2H_4) mers; polyvinyl chloride has only (C_2H_3Cl) mers, and polyvinyl alcohol has only (C_2H_3OH) mers. A marked advance in the technology of producing polymers occurred when scientists learned that addition polymers containing mixtures of two or more different mers frequently have improved physical and mechanical properties.

It is possible, for example, to have a polymer chain composed of mers of vinyl chloride and vinyl acetate (Fig. 7–2.1). The resulting molecule is a *copolymer*. A copolymer may have properties quite different from those of either component member. Table 7–2.3 shows the variety of properties and applications of vinyl chloride–vinyl acetate mixtures with different degrees of copolymerization. The range is striking. It means that engineers may tailor their materials to a wide variety of requirements.

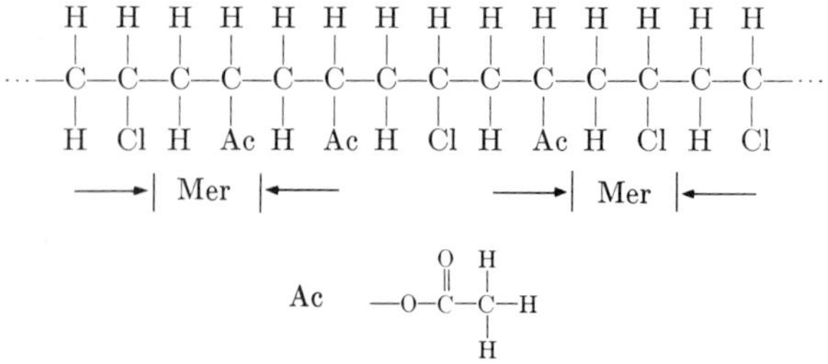

Fig. 7–2.1 Copolymerization of vinyl chloride and vinyl acetate. This is comparable to a solid solution in metallic and ceramic crystals.

The ABS polymers are triple copolymers of **a**crylonitrile, **b**utadiene, and **s**tyrene (Tables 7–2.1 and 7–2.2). Copolymerization is used extensively in producing of artificial rubbers. For example, the *buna-S* rubbers are copolymers of butadiene and styrene (Fig. 7–2.2). In principle, copolymers may be compared with solid–solution alloys of metals, e.g., brass and bronze, because they contain more than one component within the basic structure.

---

* Another factor, that of entropy, must also be considered; however, it does not alter the basic picture.

Table 7–2.3

Vinyl chloride-acetate copolymers: Correlation between composition, molecular weight, and applications*

Item	w/o of vinyl chloride	No. of chloride mers per acetate mer	Range of average mol. wts.	Typical applications
Straight polyvinyl acetate	0	0	4,800–15,000	Limited chiefly to adhesives.
Chloride-acetate copolymers	85–87	8–9	8,500– 9,500	Lacquer for lining food cans; sufficiently soluble in ketone solvents for surface-coating purposes.
	85–87	8–9	9,500–10,500	Plastics of good strength and solvent resistance; molded by injection.
	88–90	10–13	16,000–23,000	Synthetic fibers made by dry spinning; excellent solvent and salt resistance.
	95	26	20,000–22,000	Substitute rubber for electrical-wire coating; must be plasticized; extrusion-molded.
Straight polyvinyl chloride	100	—	—	Limited, if any, commerical applications *per se.*

* Adapted from A. Schmidt and C. A. Marlies, *Principles of High Polymer Theory and Practice.* New York: McGraw-Hill.

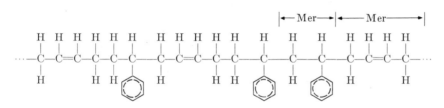

Fig. 7–2.2 Copolymerization of butadiene and styrene. This is the basis for many of our artificial rubbers. (Hydrogens are not shown on the benzene ring.)

Condensation polymerization   In contrast to addition reactions, which are primarily a summation of individual molecules into a polymer, *condensation reactions* form a second, nonpolymerizable molecule as a by-product. Usually the by-product is water or some other simple molecule such as HCl or CH_3OH. Nylon is a familiar condensation polymer. Its polymerization reaction is indicated below.

$$
H\text{---}N\text{---}\underset{\underset{H}{|}}{\overset{\overset{H}{|}}{C}}\text{---}\underset{\underset{H}{|}}{\overset{\overset{H}{|}}{C}}\text{---}\underset{\underset{H}{|}}{\overset{\overset{H}{|}}{C}}\text{---}\underset{\underset{H}{|}}{\overset{\overset{H}{|}}{C}}\text{---}\underset{\underset{H}{|}}{\overset{\overset{H}{|}}{C}}\text{---}\overset{\overset{O}{\|}}{C}\text{---}OH + H\text{---}N\text{---}\underset{\underset{H}{|}}{\overset{\overset{H}{|}}{C}}\text{---}\underset{\underset{H}{|}}{\overset{\overset{H}{|}}{C}}\text{---}\underset{\underset{H}{|}}{\overset{\overset{H}{|}}{C}}\text{---}\underset{\underset{H}{|}}{\overset{\overset{H}{|}}{C}}\text{---}\underset{\underset{H}{|}}{\overset{\overset{H}{|}}{C}}\text{---}\overset{\overset{O}{\|}}{C}\text{---}OH + \cdots \longrightarrow
$$

$$
\left[N\text{---}\underset{\underset{H}{|}}{\overset{\overset{H}{|}}{C}}\text{---}\underset{\underset{H}{|}}{\overset{\overset{H}{|}}{C}}\text{---}\underset{\underset{H}{|}}{\overset{\overset{H}{|}}{C}}\text{---}\underset{\underset{H}{|}}{\overset{\overset{H}{|}}{C}}\text{---}\underset{\underset{H}{|}}{\overset{\overset{H}{|}}{C}}\text{---}\overset{\overset{O}{\|}}{C} \right]_n + n\text{-}H_2O. \tag{7-2.8}
$$

Hydrogen and OH are removed from the ends of the molecule, and small H_2O molecules are formed as a by-product. The water can be removed from the plastic, leaving the linear polymer with several —CH_2— units plus a

$$
\overset{\overset{O}{\|}}{\underset{\underset{H}{|}}{\text{---}C\text{---}N\text{---}}}
$$

connection. One of the familiar polyesters is a product of two reactants.

$$
H\text{---}\underset{\underset{H}{|}}{\overset{\overset{H}{|}}{C}}\text{---}O\text{---}\overset{\overset{O}{\|}}{C}\text{---}\langle\bigcirc\rangle\text{---}\overset{\overset{O}{\|}}{C}\text{---}O\text{---}\underset{\underset{H}{|}}{\overset{\overset{H}{|}}{C}}\text{---}H + HO\text{---}\underset{\underset{H}{|}}{\overset{\overset{H}{|}}{C}}\text{---}\underset{\underset{H}{|}}{\overset{\overset{H}{|}}{C}}\text{---}OH \longrightarrow
$$

$$
\cdots \left[O\text{---}\overset{\overset{O}{\|}}{C}\text{---}\langle\bigcirc\rangle\text{---}\overset{\overset{O}{\|}}{C}\text{---}O\text{---}\underset{\underset{H}{|}}{\overset{\overset{H}{|}}{C}}\text{---}\underset{\underset{H}{|}}{\overset{\overset{H}{|}}{C}} \right] \cdots + 2n CH_3OH. \tag{7-2.9}
$$

In this case, the small by-product molecule is methyl alcohol. The polymer can continue to grow to produce a large linear molecule.[*][†] The condensation reactions of Eqs. (7–2.8) and (7–2.9) release energy just as does an addition reaction (Example 7–2.3). This is the driving force that activates the process.

---

[*] The product of Eq. (7–2.8) is commonly called Nylon-6 because each mer has six carbons. There are other nylons, because other sizes of mers can be used. The product of Eq. (7–2.9) goes under various trade names such as Dacron, Mylar, Terylene, Tergal, or Tervira, depending on the producing company and its physical form such as fiber or film.

[†] The product of Eq. (7–2.9) is a *polyester* since it contains a

$$
\overset{\overset{O}{\|}}{\text{---}C\text{---}O\text{---}}
$$

(Continued)

There is no absolute termination in the growth of condensation polymers. Both ends of the growing molecule remain functional. This is unlike a growing molecule in the addition reaction of Eq. (7–2.7), where the chain reactions can proceed in only one direction, thus terminating the reaction completely if two chains join. This difference affects the molecular size distributions that are eventually achieved. A condensation molecule can always join with yet another molecule for continued growth.

Example 7–2.1 Polypropylene $+H_2C=CHCH_3+_n$ is a vinyl where **R** is $-CH_3$. How much energy is released during polymerization per gram of product?

Solution: Basis: 1 mer, where one double bond forms two single bonds.

$$\text{Grams} = (36 + 6 \text{ g})/0.6 \times 10^{24}.$$

$$\text{Energy required} = 680{,}000 \text{ J}/0.6 \times 10^{24}.$$

$$\text{Energy released} = -2(370{,}000 \text{ J})/0.6 \times 10^{24}.$$

$$\text{Energy/g} = [+680{,}000 - 2(370{,}000)]/42 \text{ g}$$
$$= -1430 \text{ J}.$$

Comment. All vinyls release 60 kJ of energy per mole during polymerization because they all involve the same change from one double bond to two single bonds. ◀

Example 7–2.2 A copolymer contains 92 w/o vinyl chloride and 8 w/o vinyl acetate. What is the mer fraction of VAc?

Solution: From Table 7–2.1,

$$VC = +C_2H_3Cl+ = 24 + 3 + 35.5 = 62.5 \text{ amu/mer};$$

$$VAc = +C_2H_3OCOCH_3+ = 48 + 6 + 32 = 86 \text{ amu/mer}.$$

Basis: 100 amu

$$VC: \quad 92/62.5 = 1.472 \text{ mers} = 0.941$$
$$VAc: \quad 8/86 \quad = 0.093 \text{ mers} = 0.059.$$
$$\overline{ 1.565 \text{ mers}}$$

Comment. This copolymer has ~ 16 chloride mers per acetate mer. Possible uses are indicated in Table 7–2.3. ◀

---

linkage; comparably, the

$$\begin{array}{c} O \\ \parallel \\ -C-N- \\ | \\ H \end{array}$$

linkage of Eq. (7–2.8) identifies it as one of the *polyamides*. The *polyurethanes* possess

$$\begin{array}{c} O \\ \parallel \\ -O-C-N- \\ | \\ H \end{array}$$

groups along their molecules.

Example 7–2.3 Nylon-66 is a condensation polymer of

$$
\begin{matrix}
\text{O} & \quad & \text{O} \\
\| & & \| \\
\end{matrix}
$$

$$\text{HO—C(CH}_2)_4\text{C—OH} \quad \text{and} \quad \text{H}_2\text{N(CH}_2)_6\text{NH}_2.$$

a)   Sketch the structure of these two molecules.
b)   Show how condensation polymerization can occur.
c)   What is the amount of energy released per mole of H_2O formed?

Solution

a) and b)

c)   A mole (0.6×10^{24}) of water requires elimination of that number of C—O and H—N bonds, and the formation of equal numbers of C—N and H—O bonds.
From Table 2–3.1:

Bonds removed		Bonds formed	
C—O	+ 360 kJ/mole	C—N	− 305 kJ/mole
H—N	+ 430	H—O	− 500
	+ 790		− 805

The net energy difference is − 15 kJ/mole, or − 3.5 kcal/mole. ◀

7–3   MOLECULAR IRREGULARITIES

A lineal polymer seldom, if ever, has the geometric regularity depicted in Fig. 7–1.4. Unless the polymer chain is constrained, thermal agitation will introduce kinks (Fig. 7–1.7). Also, we have observed that it is possible to produce copolymers of more than one type of mer. We shall also see that other irregularities arise because there may be several permutations for joining monomers into a polymeric chain. Still other irregularities, such as polar groups, are inherent in many chains. Each of these irregularities has an effect on the way adjacent molecules pack together and on the mechanical behavior of the resulting products.

Crystallization is difficult in polymers because the molecules are long and cumbersome, and the intermolecular attractions are weak. We will find that any molecular irregularity adds still further interference. Therefore we will want to look closely at polymer imperfections.

Stereoisomers   The molecular chains of Fig. 7–3.1(a) show a high degree of regularity along the polymer. Not only is there an addition sequence of monomers that

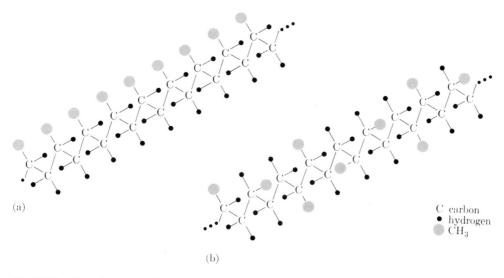

Fig. 7–3.1   Stereoisomers (polypropylene). (a) Isotactic. (b) Atactic.

HAS IMPERFECTIONS

form a linear polymer, but there is also identical ordering of the propylene mers so that the side groups are always at the corresponding position within the mer. Such an ordering is called *isotactic*, as contrasted to the *atactic* arrangement in Fig. 7–3.1(b). Adjacent molecules can pack together more efficiently if they are isotactic than when atactic.

A second example of polymer arrangement is found in rubbers that are made of butadiene-type molecules (Table 7–2.2). Natural rubber has polymerized isoprene with the structure

$$(7\text{–}3.1)$$

as a mer. Two double-bonded carbons have a CH_3 group and a hydrogen on the same side of the chain. This *cis* arrangement has important consequences in the chain behavior, because it arcs the mer (Eq. 7–3.1). Another modification has the CH_3 group and the single hydrogen on opposite sides of the chain to give a *trans* arrangement:

$$(7\text{–}3.2)$$

Although identical in compositions, these two isomers of isoprene have different structures and therefore different properties. Natural rubber, with its cis-type structure, has a very highly kinked chain, as a result of the arc within the mer. The polymer of the trans-type called *gutta percha* has a bond-angle pattern that is more typical of linear polymers (Fig. 7–1.4). In effect, the unsaturated positions balance each other across the double bond. As a result, the properties differ markedly.

Polar groups   Adjacent polyethylene chains have relatively little attraction to one another. In effect, they are very long paraffins and have the same waxy characteristics. The molecules readily slide by one another when a shear stress is applied. Other molecules exhibit greater intermolecular attractions, particularly if they contain polar groups. One such polar group is the C=O in Eqs. (7–2.8) and (7–2.9):

$$(7-3.3)$$

It possesses an electric dipole with the oxygen at the negative end. The two lone-pairs of electrons with the oxygen (cf. Fig. 2–3.4b) strongly attract hydrogen atoms in adjacent molecules because the hydrogen atoms are simply exposed protons (+) on the end of a covalent bond. The presence of these polar groups significantly affects the properties of a polymer. For example, they must be credited with the difference between nylon and polyethylene, since otherwise the two types of molecules contain the same CH_2 units.

Molecular crystals   Since each molecule contains more than one atom, we must look at a more complex packing arrangement in molecular crystals than in metal crystals. Methane is among the more easily visualized because the CH_4 molecules have a relatively symmetrical, almost sphere-like shape (Fig. 7–3.2). Iodine is not complex since the molecule contains only two atoms, I_2. However, that 2-atom

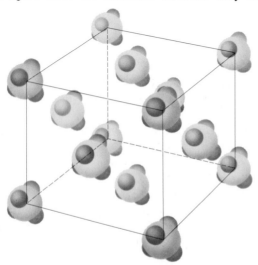

Fig. 7–3.2   FCC compound (methane, CH_4). Each fcc lattice point contains a molecule of five atoms. Methane solidifies at −183°C (90K). Between 20K and 90K, the molecules can rotate in their lattice sites. Below 20 K, the molecules have identical alignments as shown here.

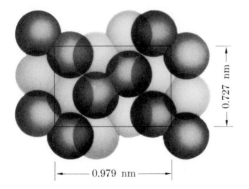

Fig. 7–3.3 Molecular crystal (iodine). The molecule of I_2 acts as a unit in the repetitive crystal structure. This lattice is *simple orthorhombic* because $a \neq b \neq c$, and the face-centered positions are *not* identical to the corner positions. (The molecules are oriented differently.) The unit cell axes have 90° angles.

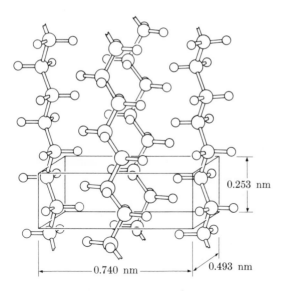

Fig. 7–3.4 Molecular crystal (polyethylene). The chains are aligned longitudinally. The unit cell is orthorhombic with 90° angles. (M. Gordon, *High Polymers*, Iliffe, and Addison-Wesley. After C. W. Bunn, *Chemical Crystallography*, Oxford.)

molecule is not spherical. Therefore, crystalline iodine does not readily form a cubic unit cell. Rather, it is orthorhombic ($a \neq b \neq c$) as shown in Fig. 7–3.3.

More complex situations arise when polyethylene crystallizes (Fig. 7–3.4) because the molecules extend beyond the unit cell. However, from this figure we can determine that there are two mers per unit cell of polyethylene. In Example 7–3.1 we will calculate the density of an ideal polyethylene crystal and obtain an answer of 1.01 g/cm^3. Densities measured from commercial grades of polyethylene are normally 5% to 10% less than this value (Table 7–3.1), because the long polyethylene molecules seldom have the perfect "zipper-like" alignment shown in Fig. 7–3.4. If they are not perfectly aligned, their packing factor decreases to the extent that there may be as much as 10 percent additional "free space" among the atoms.

Table 7–3.1
Characteristics of polyethylenes

	Low-density polyethylene (LDPE)	High-density polyethylene (HDPE)
Density, Mg/m^3 (= g/cm^3)	0.92	0.96
Crystallinity, v/o	Near zero	~ 50
Thermal expansion, $°C^{-1}$	180×10^{-6}	120×10^{-6}
Thermal conductivity (watt/m^2)/($°C$/m)	0.34	0.52
Heat resistance for continuous use, $°C$	55–80	80–120
10-min. temp. exposure, $°C$	80–85	120–125

Crystals form with difficulty in polyethylene* because the molecules extend through hundreds of unit cells. Over these distances it is highly unlikely that adjacent chains will remain perfectly aligned, particularly when they are not of the same length and are subject to thermal vibration and folding. Furthermore, only weak van der Waals bonds hold them together. Diffraction analyses (Section 3–8) reveal negligible evidence of crystallinity in the low-density polyethylenes ($\rho_{LDPE} \sim 0.92$ g/cm^3). The higher-density polyethylenes ($\rho_{HDPE} \sim 0.96$ g/cm^3) have local regions that are well crystallized, but still have intervening amorphous areas. Neither full density nor complete crystallization is observed in practice.

In view of the above complications, it is not surprising that the majority of polymers may be cooled past their melting temperatures and on down to room temperature without molecules aligning themselves into a nice crystalline pattern. Supercooling is common among polymers, especially if there are structural irregularities along the molecular chain. For example, the atactic chain shown in Fig. 7–3.1(b) crystallizes less readily than the isotactic chain of Fig. 7–3.1(a).

Glass transition temperature in polymers   Even though the molecules of the polymer are entangled, there is a continuous rearrangement within a polymer liquid due to the thermal agitation of the molecule. Because of this, there has to be "free space" in the liquid. As the temperature decreases, the thermal agitation lessens, and there is a decrease in both the "free space" and the vibrational amplitude. The resulting decrease in volume continues below the freezing point into the supercooled-liquid range (Fig. 4–5.3). The liquid structure is retained. As with liquid at a higher temperature, flow occurs; however, flow is more difficult because the viscosity increases as the temperature drops, and because there is a decrease in the amount of "free space" between molecules.

---

* But not as unlikely as in the majority of other polymers.

Those polymers that are cooled without crystallizing eventually reach a point at which the thermal agitation is not sufficient to allow for the rearrangement of the molecules into more efficient packing, nor can they respond readily to externally applied forces. Thus, the polymer becomes markedly more rigid and brittle. Likewise, its thermal expansion coefficient is limited to changes in the amplitude of thermal vibrations and does not reflect changes in the "free space" among the molecules. Thus, with this increased rigidity during cooling, there is a discontinuity in the slope of the volume-versus-temperature curve (Fig. 4–5.3). This point of change in the slope is called the *glass point*, or the *glass transition temperature*, because this phenomenon is typical of all glasses. In fact, below this temperature, just as with normal silicate glasses, a noncrystalline polymer *is* a glass (although admittedly an organic one). Conversely, we shall observe in Chapter 8 that a normal glass is an inorganic polymer.

The glass transition temperature or, more briefly, the glass temperature, T_g, is as important to polymers as the melting (or freezing) temperature, T_m, is. The glass temperature of polystyrene (Table 7–2.1) is at approximately 100°C; therefore it is glassy and brittle at room temperature. In contrast, a rubber whose T_g is at $-73°C$ is flexible even in the most severe winter temperatures.

Example 7–3.1 From Fig. 7–3.4, calculate the density of fully crystalline polyethylene.

Solution: A $(C_2H_4)_n$ mer is parallel to the two ends of the rectangular cell, for an equivalent of one mer per unit cell. Likewise, a mer is parallel to the two sides. Together, there are a total of two mers per unit cell.

$$\rho = \frac{2(24 + 4 \text{ amu})/(0.602 \times 10^{24} \text{ amu/g})}{(0.253 \times 0.740 \times 0.493)(10^{-27} \text{ m}^3)} = 1.01 \times 10^6 \text{ g/m}^3 \qquad (\text{or } 1.01 \text{ g/cm}^3).$$

Comments. Densities normally lie in the 0.92 to 0.96 g/cm³ range, depending on the degree of crystallinity (see Study Problem 7–3.3 and Table 7–3.1). As a result, there is additional space present. The chemist calls this "free space." ◀

7–4   THREE-DIMENSIONAL POLYMERS

Polymers can develop a *network* structure if the reacting monomers are *polyfunctional*, i.e., if they can connect to three or more adjacent molecules.

A familiar network polymer, which goes by various trade names, is formed from formaldehyde (CH_2O) and phenol (C_6H_5OH). The atom arrangements within these molecules (Fig. 2–3.1) are shown in Fig. 7–4.1(a). At room temperature formaldehyde is a gas; phenol is a low-melting solid. The polymerization that results from the interaction of these two compounds is shown in Fig. 7–4.1(b). The formaldehyde has supplied a CH_2 unit that serves as a bridge between the benzene rings in two phenols. Stripping two hydrogens from the benzene rings and one oxygen from the formaldehyde (to permit the connection) forms water, which can volatilize and leave the system.

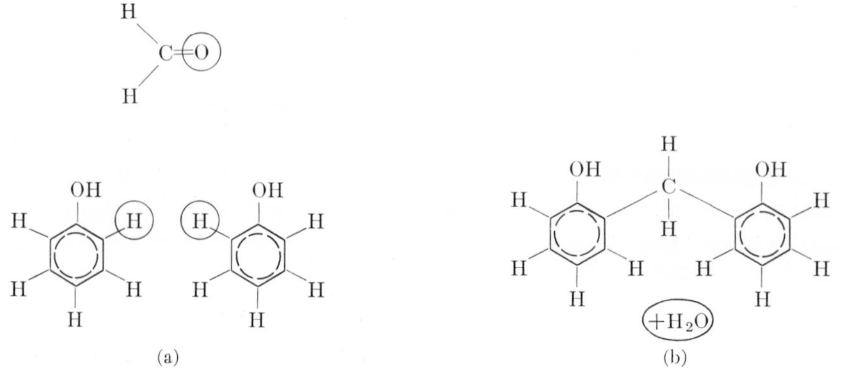

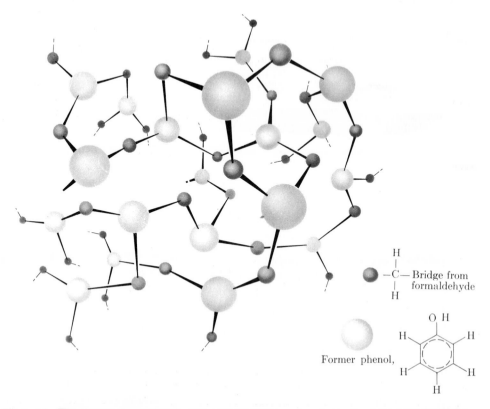

Fig. 7-4.1   Phenol-formaldehyde reaction. The phenols (C_6H_5OH) contribute hydrogen and the formaldehyde (CH_2O) contributes oxygen to produce water as a by-product. The two rings are joined by a —CH_2— bridge.

Fig. 7-4.2   Network structure of polyfunctional units. Deformation does not occur as readily as in linear polymers, which are composed of bifunctional units.

The reaction of Fig. 7–4.1 can occur at several points around the phenol molecule.*
As a result of this polyfunctionality, a molecular network is formed, rather than a
simple linear chain (Fig. 7–4.2).

Thermosetting and thermoplastic polymers   The polymer just described is phenol-
formaldehyde (PF). It was one of the first synthetic polymers. It has the attribute of
not softening when heated because its three-dimensional structure keeps it rigid. Of
course, this means that it cannot be molded in the same manner as linear polymers. In
order to mold phenol-formaldehyde, it is necessary to start with a mixture that is only
partially polymerized. At this stage, the combination is solid but averages less than
three CH_2 bridges between adjacent phenols. Thus, it can be deformed under pressure.
As it is held at 200°C–300°C, it completes its polymerization into a *more* rigid three-
dimensional structure. It is thermal setting. The manufacturer calls polymers of this
type *thermosets*. Once set, the product can be removed from the mold without waiting
for cooling to occur.

The thermosetting polymers are in contrast to *thermoplastic polymers* that are
linear and can be injected into a mold when warm because they soften at higher temperatures. The thermoplastic materials must be cooled before they can leave the mold,
or they will lose their shape. There is no further polymerization in the molding process.
The difference between thermosets and thermoplasts is mainly that the former has
three-dimensional and the latter *linear* molecular structure. This difference has considerable technical implications to both processing and engineering applications
because of effects on molding and on high-temperature uses.

Cross-linking   Some linear molecules, by virtue of their structure, can be tied
together in three dimensions. Consider the molecule of Fig. 7–4.3(a) and its polymerized combination in Fig. 7–4.3(b). Intentional additives (divinyl benzene in this

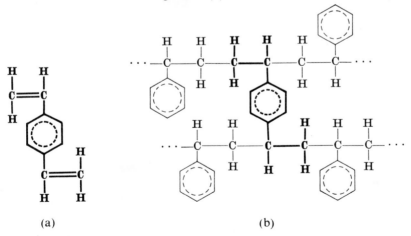

(a)   (b)

Fig. 7–4.3   Cross-linking of polystyrene. The divinyl benzene (a) becomes part of two
adjacent chains because it is tetrafunctional; i.e., it has four reaction points.

---

* Three is the normal maximum, because there simply is not space to attach more than three CH_2 bridges.
The number is limited by *stereohindrance*.

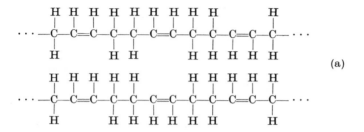

(a)

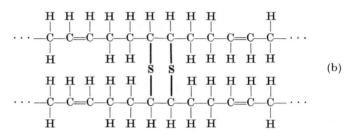

(b)

Fig. 7–4.4 Vulcanization (butadiene-type rubbers). Sulfur atoms cross-link adjacent chains. In this representation of vulcanization, two sulfur atoms are required for each pair of mers. (Other cross-linking arrangements are also possible.)

case, but we do not have to remember the name) tie together two chains of polystyrene. This causes restrictions with respect to the plastic deformation of polymers.

The *vulcanization* of rubber is a result of cross-linking by sulfur, as shown schematically in Fig. 7–4.4. The effect is pronounced. Without sulfur, rubber is a soft, even sticky, material that, when it is near room temperature, flows by viscous deformation. It could not be used in automobile tires because the service temperature would make it possible for molecules to slide by their neighbors, particularly at the pressures encountered. However, cross-linking by sulfur at about 10% of the possible sites gives the rubber mechanical stability under the above conditions, but still enables it to retain the flexibility that is obviously required. Hard rubber has a much larger percentage of sulfur and appreciably more cross-links. You can appreciate the effect of the addition of greater amounts of sulfur on the properties of rubber when you examine a hard-rubber product such as a pocket comb.

Branching Ideally, linear polymer molecules such as we have studied to date are two-ended chains. There are cases, however, in which polymer chains branch. We can indicate this schematically as in Fig. 7–4.5. Although a branch is unusual, once formed it is stable because each carbon atom has its complement of four bonds and each hydrogen atom has one bond. The significance of branching lies in the three-dimensional entanglements that can interfere with plastic deformation. Think of a pile of tree branches compared with a bundle of sticks; it is more difficult to move a branch with respect to its neighbors, than to move the individual sticks.

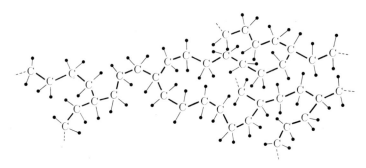

Fig. 7–4.5 Branching (polyethylene). The linear molecule of Fig. 7–1.4 branches. One mechanism for branching is shown in Fig. 7–7.3. (The hydrogen atoms are represented by dots.)

Example 7–4.1 How many grams of sulfur are required per 100 g of final rubber product to completely cross-link butadiene (C_4H_6) rubber with one sulfur per connection?

Solution: With one atom of sulfur (32) required per mer of butadiene:

$$(4)(12) + (6)(1) = 54.$$

$$\text{Fraction sulfur} = \frac{32}{32 + 54} = 0.37 \quad \text{or} \quad 37 \text{ g S/100 g product.} \quad \blacktriangleleft$$

Example 7–4.2 What fraction of butadiene (C_4H_6) is cross-linked if the product contains 18.5% sulfur? (Assume all the sulfur is utilized in cross-linking and only one sulfur per connection.)

Solution: Determine the number x of atoms of sulfur ($= 32$) per mer of butadiene $[=(4)(12) + (6)(1) = 54]$ from the weight ratio of butadiene to sulfur:

$$\frac{\text{g butadiene}}{\text{g sulfur}} = \frac{1 - 0.185}{0.185} = \frac{1(54)}{x(32)};$$

$$x = 0.383.$$

Comment. The answer cannot be obtained from Example 7–4.1 as 18.5/37. $\blacktriangleleft$

7–5 DEFORMATION OF POLYMERIC MATERIALS

Polymers undergo both elastic and plastic deformation, just as metals do. As a class, polymers have lower elastic moduli than metals. (Cf. Parts 1 and 3 of Appendix C.*) Often we use this characteristic to advantage, because the low stress per unit strain is a design feature of rubber. Also, as a class, polymers can be plastically deformed more readily than the average metal. The everyday term *plastic* emphasizes this processing advantage of polymers. Our class comparisons are not to rate one higher

---

* There are, of course, exceptions. Lead and other low-melting metals have Young's moduli of less than 70,000 MPa (10×10^6 psi). Certain polymers with aliphatic rings in their backbones have moduli of higher figures. However, both of these examples are extremes.

than the other but to give us an index for making technical specifications. By knowing what makes a rubber visibly elastic, and what accounts for the viscous flow of polystyrene into a mold, we can better select polymers and know their limitations.

Elastomers   Polymers that have large elastic deformations are called elastomers. The basis of the large strain is the kinked conformation (discussed in Section 7–1). As an example, the molecule of Example 7–1.4 had a mean end-to-end length of 6.2 nm. It can be extended many nanometers when a stress is applied that unkinks the molecular chain from the conformation of Fig. 7–1.7. Of course, we never realize the full extension to 200 nm, even in nonvulcanized rubbers, but large percentages of elastic strain are common. The stretched conformation of rubber is not natural; therefore, the molecules quickly rekink by bond rotation after the stress is removed.

In order for this unkinking and rekinking to occur, the elastomer or rubber must be above its *glass temperature*, T_g. This permits the rearrangements between adjacent molecules. Below the T_g, the elastomer is brittle and hard.*

The elastic portion of the stress-strain curve for a rubber is unlike that for metals. The first strain occurs easily with little stress, because the molecules are simply unkinking. This results in a low elastic modulus. As the molecules become straightened and are aligned together, the stress requirements increase for each additional increment of strain (Fig. 7–5.1). Thus the modulus of elasticity increases. The data for rubbers in Appendix C are for the initial stress-strain ratios, since that is the range in which elastomers are usually used.

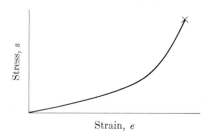

Fig. 7–5.1   Stress-strain curve (rubber). The slope (elastic modulus) of the *s-e* curve increases as the molecules are unkinked. (Cf. Fig. 1–2.1 for metals.)

The molecules of rubber are stretched by restricting the bond rotation that causes kinking. Now, if a stretched rubber band is heated, the thermal agitation for rotation and the retractive forces increase; the band will *contract* if the external load is not changed. Thus, a stretched rubber band has a *negative* coefficient of thermal expansion, and Young's modulus *increases* with temperature (Fig. 7–5.2).[†]

---

* An interesting demonstration is to place an *unstretched* rubber band into liquid nitrogen ($-196°C$, or 77 K) and then break it like glass. The glass temperature T_g of common rubbers is about $-75°C$ or 200 K.
[†] This sentence applies to *stretched* rubber only.

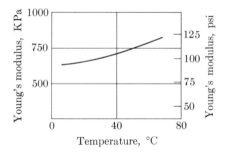

Fig. 7-5.2 Young's modulus versus temperature (stretched rubber). The retractive forces of kinking are greater with more thermal agitation. Therefore, the elastic modulus increases with temperature. (Cf. Fig. 6-3.3.)

Finally, partial *crystallization* occurs more readily in any polymer that has been deformed so that its molecules are aligned. For example, a rubber is normally completely amorphous and noncrystalline. However, a tensile force will pull the molecules into a parallel orientation. X-ray diffraction (Fig. 7-5.3) reveals that this stretching permits the adjacent oriented molecules to mesh together into a crystalline array. The crystallographer can relate the diffraction spots in Fig. 7-5.3(b) to molecular alignment. When the stress is removed and the rubber contracts, the diffraction spots disappear.

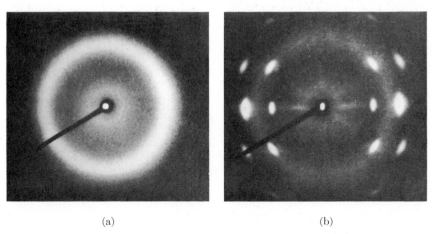

(a)   (b)

Fig. 7-5.3 Deformation crystallization of natural rubber (polyisoprene) revealed by x-ray diffraction. (a) Unstretched. (b) Stretched. (S. D. Gehman, *Chemical Reviews*, vol. 26, 1940, page 203.)

Viscous deformation   Shear stresses make a liquid flow. The rate of flow depends inversely upon the viscosity. Many polymers are supercooled liquids between their melting temperature T_m and their glass temperature T_g (Fig. 7-5.4b) and, therefore, are subject to viscous flow in addition to elastic deformation. The rate may be slow because the viscosity η is commonly high. However, even at a slow rate, this permits deformation when a polymer carries a long-term load. Likewise, it permits processing by deforming the polymer within a mold at high temperatures.

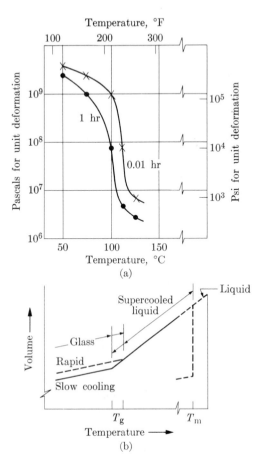

Fig. 7–5.4 Viscoelastic modulus (ordinate) versus temperature (polymethyl methacrylate, PMM). (a) With more time at a given temperature, less stress is required for unit deformation. There is a major decrease in the modulus, M_{ve}, at the glass temperature. (b) The glass temperature is lower with slower cooling, because more time permits molecular adjustments to the stress.

Figure 7–5.4(a) shows the shear stress τ required to produce unit deformation (100%) in a polymethyl methacrylate (PMM*) as a function of temperature. It is obvious that a marked change occurs slightly above 100°C. This temperature corresponds to the glass temperature, T_g, of Fig. 7–5.4(b). Recall that the molecules have the freedom to kink and turn by thermal agitation above the glass temperature. Below that temperature, there is insufficient thermal agitation to permit rearrangements of molecules into closer packing. Thus this represents a discontinuity in the thermal behavior of the material. Return to Fig. 7–5.4(a) and observe that the stress required for a deformation changes by more than two orders of magnitude at the glass temperature. Obviously the glass temperature is a temperature important for polymer behavior.

The two curves of Fig. 7–5.4(a) indicate that less stress is required when the time of stressing is increased from 36 sec (0.01 hr) to 1 hr. The two curves also indicate that the glass temperature drops $\sim 10°C$ as the time is increased from 0.01 to 1.0 hr. This change is reflected in Fig. 7–5.4(b) as a drop in the glass transition temperature with

---

* Lucite is one trade name. The composition is given in Table 7–2.1.

slower cooling. With slower cooling rates, or longer times, the molecules can be rearranged at somewhat lower temperatures.

We may compare different molecular structures and their effect on deformation. In Fig. 7–5.5, the ordinate shows the *viscoelastic modulus* M_{ve} where

$$M_{ve} = \tau/(\gamma_e + \gamma_f). \tag{7–5.1}$$

As in Chapter 6, τ is shear stress and γ is shear deformation, γ_e being elastic deformation and γ_f being displacement by viscous flow. The abscissa has been generalized. The right end includes higher temperatures and/or longer times, both of which introduce more deformation (and therefore lower values for M_{ve}). At the left end of Fig. 7–5.5(a) and below the glass temperature T_g, where only elastic deformation can occur, the material is comparatively *rigid*; a clear plastic triangle used by a draftsman is an example. In the range of the glass temperature, the material is *leathery*; it can be deformed and even folded, but it does not spring back quickly to its original shape. In the *rubbery plateau*, polymers deform readily but quickly regain their previous shape if the stress is removed. A rubber ball and a polyethylene "squeeze" bottle serve as excellent examples of this behavior because they are soft and quickly elastic. At still higher temperatures, or under sustained loads, the polymer deforms extensively by *viscous flow*.

The second part of Fig. 7–5.5 compares the deformation behavior for the different structural variants cited earlier with the amorphous polymer just described. A highly

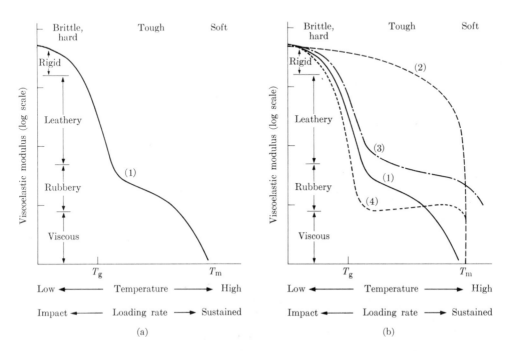

Fig. 7–5.5 Viscoelastic modulus versus structure. (1) Amorphous linear polymer. (2) Crystalline polymer. (3) Cross-linked polymer. (4) Elastomer (rubber).

crystalline polymeric material (curve 2) does not have a glass temperature. Therefore it softens more gradually as the temperature increases until the melting temperature is approached, at which point fluid flow becomes significant. The higher-density polyethylenes (Table 7–3.1) lie between curves (1) and (2) of Fig. 7–5.5(b) because they possess approximately 50% crystallinity.

The behavior of *cross-linked* polymers is represented by curve (3) of Fig. 7–5.5. A vulcanized rubber, for example, is harder than a nonvulcanized one. Curve (3) is raised more and more as a larger fraction of the possible cross-links are connected. Note that the effects of cross-linking carry beyond the melting point into the true liquid. In this respect, a network polymer like phenol-formaldehyde (Fig. 7–4.2) may be considered as an extreme example of cross-linking, which gains its thermoset characteristics by the fact that the three-dimensional amorphous structure carries well beyond an imaginable melting temperature.

Once the glass temperature is exceeded, *elastomeric* molecules can be rotated and unkinked to produce considerable strain. If the stress is removed, the molecules quickly snap back to their kinked conformations (Fig. 7–1.7). This rekinking tendency increases with the greater thermal agitation at higher temperatures. Therefore the behavior curve (4) increases slightly to the right across the rubbery plateau (Fig. 7–5.5b). Of course, the elastomer finally reaches the temperature at which it becomes a true liquid, and then flow proceeds rapidly.

- **Stress relaxation**   In the discussion of Fig. 7–5.5, we assumed constant stress and increasing strain. In some situations, strain is constant. Since viscous flow proceeds, the stress is reduced. The reader has undoubtedly observed such a phenomenon if he or she has removed a stretched rubber band from a book or bundle of papers after a period of time. The rubber band does not return to its original length. For this reason, it was not holding the papers as tightly. Some of the stress had disappeared.

The stress decreases in a viscoelastic material that is under constant strain because the molecules can gradually flow by one another. The rate of stress decrease $(-ds/dt)$ is proportional to the stress level:

$$(-ds/dt)\tau = s. \tag{7–5.2}$$

Rearranging and integrating,

$$ds/s = -dt/\tau; \tag{7–5.3a}$$
$$\ln s/s_0 = -t/\tau,$$

or

$$s = s_0 e^{-t/\tau}. \tag{7–5.3b}$$

The stress ratio, s/s_0 is between the stress at time t and the original stress s_0 at t_0. The proportionality constant τ of Eq. (7–5.2) must have the units of time; it is called the *relaxation time*. When $t = \tau$, $s/s_0 = 1/e = 0.37$.

Stress relaxation is a result of molecular movements; therefore, we find that temperature affects stress relaxation in much the same manner as it affects diffusion. Since the relaxation time is the reciprocal of a rate,

$$1/\tau \propto e^{-E/kT}, \tag{7–5.4}$$

or, as in Eq. (4–7.4),

$$\ln 1/\tau = \ln 1/\tau_0 - E/(13.8 \times 10^{-24} \text{ J/K})(T).\qquad(7\text{–}5.5)$$

In these equations, E, k, and T have the same meanings and units as in Eq. (4–7.4); $1/\tau$, like D, contains $\sec^{-1}$ in its units, since both involve rates.

Example 7–5.1 Detect evidence of orientation of rubber molecules by sensing a temperature change.

Answer: For this simple experiment, use a heavy but easily deformable rubber band. Your lip can serve as a sensitive detector of temperature changes. Place the band in contact with your lower lip. Stretch it rapidly, then quickly (without snapping) return it to its original length; repeat this cycle several times. A little care will permit one to detect a temperature increase on stretching and a temperature decrease on release. These temperature changes occur because a heat of fusion is released from the rubber band to your lip during orientation. Energy is absorbed (as entropy) during deorientation when the stress is removed.

• **Example 7–5.2** A stress of 11 MPa (1600 psi) is required to stretch a 100-mm rubber band to 140 mm. After 42 days at 20°C in the same stretched position, the band exerts a stress of only 5.5 MPa (800 psi). (a) What is the relaxation time? (b) What stress would be exerted by the band in the same stretched position after 90 days?

Solution: (a) From Eq. (7–5.3a),

$$\ln \frac{5.5}{11} = -\frac{42}{\tau},$$

$$\tau = 61 \text{ days};$$

b)   Eq. (7–5.3b),

$$s_{90} = 11e^{-90/61}$$

$$= 2.5 \text{ MPa}\qquad(\text{or } 360 \text{ psi}).$$

Alternative answer for (b), with 48 *additional* days:

$$s_{48} = 5.5e^{-48/61} = 2.5 \text{ MPa.}\quad\blacktriangleleft$$

• **Example 7–5.3** The relaxation time for 25°C is 50 days for the rubber band in Example 7–5.2. What will be the stress ratio, s/s_0, after 38 days at 30°C?

Solution: Use Eq. (7–5.5), and recall $\tau = 61$ days at 20°C.

At 20°C,

$$\ln 1/\tau_0 = \ln 1/61 + E/(13.8 \times 10^{-24} \text{ J/K})(293 \text{ K}).$$

At 25°C,

$$\ln 1/\tau_0 = \ln 1/50 + E/(13.8 \times 10^{-24} \text{ J/K})(298 \text{ K}).$$

Solving simultaneously,

$$E = 4.8 \times 10^{-20} \text{ J,}$$

$$\ln 1/\tau_0 = 7.76.$$

At 30°C,

$$\ln 1/\tau = 7.76 - 4.8 \times 10^{-20} \text{ J}/(13.8 \times 10^{-24} \text{ J/K})(303 \text{ K});$$

$$\tau = 41 \text{ days},$$

$$s_{38} = s_0 e^{-38/41},$$

$$s_{38}/s_0 = 0.4.$$

Comments. The relaxation time shortens at higher temperatures.

A rubber is more subject to oxidation when it is under stress. When this occurs, the structure is modified, and we observe other changes such as hardening and/or cracking. ◀

7-6 ELECTRICAL BEHAVIOR OF POLYMERS

Polymers are widely used for electrical insulation. Polymers have obvious advantages. They may be either rigid or flexible; they may be made as a thin film; and they may even be applied as a fluid and polymerized in place, e.g., around a wire during processing, or as a potting compound after the device has been assembled. Of primary importance is the fact that the predominantly covalent bonds of all polymers generally limit electrical conduction.

• **Conduction**   Although polymers are inherently insulators, their compositions can be adjusted to permit some conductivity. Conductivity is achieved in specialized rubbers through the addition of extremely finely powdered graphite, which provides a path for electron movements. Thus this conductivity does not arise from the polymer per se, but results from the inclusion of a second, conducting phase.

Nature also produces electrical signals, presumably by conductivity, through nerve cells, which of course are molecular in structure. One may speak of delocalized electrons that move through the molecular paths in nerve cells. However, as yet, the scientist has not analyzed this mechanism sufficiently so that it may be synthesized. This topic remains as another of the challenging areas of materials science and engineering.

Dielectric constant   Polymers contain electrical charges in the form of atomic nuclei, electrons and polar groups. These respond to electric fields. Although they cannot leave their parent molecules, electrons will shift their centers of motion a distance, d, into the positive direction of an electric field. The protons of hydrogen atoms shift their centers of vibration toward the negative electrode. Polar groups (Section 7-3) and polar molecules (Section 2-4) align with the electric field. As a result, polymers have *relative dielectric constants* κ that are greater than unity (Table 7-6.1).

The adjustments of the charges Q that have just been described produce *polarization* $\mathscr{P}$ within the material. Polarization $\mathscr{P}$ is the sum of all the dipole moments, $\sum Qd$ per unit volume V:

$$\mathscr{P} = (\textstyle\sum Qd)/V. \tag{7-6.1}$$

The amount of polarization depends upon the structure of the material and affects the *charge density* $\mathscr{D}$ (coul/cm^2) that a capacitor can carry. Thus, in Fig. 1-4.1(a), the

Table 7–6.1
Relative dielectric constants of polymers
(20°C, unless otherwise stated)

	At 60 Hz*	At 10^6 Hz
Nylon 6/6	4.0	3.5
Polyethylene, PE	2.3	2.3
Polytetrafluorethylene, PTFE	2.1	2.1
Polystyrene, PS	2.5	2.5
Polyvinyl chloride		
plasticized $(T_g \approx 0°C)^\dagger$	7.0	3.4
rigid $(T_g = 85°C)$	3.4	3.4
Rubber (12 w/o S, $T_g \approx 0°C$)		
$-25°C$	2.6	2.6
$+25°C$	4.0	2.7
$+50°C$	3.8	3.2

* Hz = cycles per sec.
† T_g = glass temperature.

capacitor builds up a charge density $\mathcal{D}_0$ that is proportional to the electric field $\mathcal{E}$ when a vacuum is present (Eq. 1–4.5). With a material present, the charge density is increased to $\mathcal{D}_m(=\kappa\mathcal{D}_0)$. This increased charge density is a second way of defining the *polarization* $\mathcal{P}$ of a material:

$$\mathcal{P} = \mathcal{D}_m - \mathcal{D}_0, \tag{7–6.2}$$

or, from Eqs. (1–4.5) and (1–4.6),

$$\mathcal{P} = (\kappa - 1)\epsilon_0\mathcal{E}, \tag{7–6.3}$$

where κ is the relative *dielectric constant* and ϵ_0 is 8.85×10^{-12} C/V·m, the proportionality constant called the *permittivity* of free space. The electric field $\mathcal{E}$ is the voltage gradient, E/d.

At frequencies in the 10^{12} Hz to 10^{15} Hz range, the polarization $\mathcal{P}$ arises solely from the displacement of the electrons, because only the electrons can respond that rapidly to the electric field, $\mathcal{E}$. At lower frequencies, $\sim 10^{10}$ Hz, we find responses in a material such as nylon, first from the protons of the side hydrogen atoms, and then from the $>C{=}0$ polar groups. The latter dipole becomes oriented with the field because the oxygen carries an excess negative charge in the form of lone-pair electrons. Both the protons and the oxygens require the movements of atoms within the material for dipole orientations. Of course, this is possible only above the glass temperature T_g. Figure 7–6.1 shows schematically the effect of polarization $\mathcal{P}$ on the relative dielectric constant, κ, as a function of frequency. Figure 7–6.2 shows data for two common organic solids.

Below T_g, molecular movements cannot contribute to the dielectric constant. As soon as T_g is exceeded, the dielectric constant increases markedly as shown in Fig. 7–6.3. However, at still higher temperatures, we see a drop in the dielectric

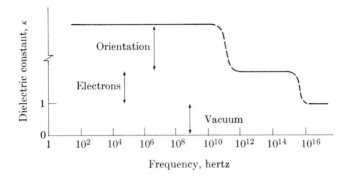

Fig. 7-6.1   Relative dielectric constant versus frequency (schematic). Electrons respond to the alternating electric field below $\sim 10^{15}$ Hz. Molecular dipoles can respond below about 10^{10} Hz.

Fig. 7-6.2   Dielectric constant versus frequency. Generally speaking, higher frequencies reduce the dielectric constant, because there is insufficient time for dipole alignment. (Ph.-For. = phenol-formaldehyde; C_2Cl_3F = polytrichlorofluoroethylene.)

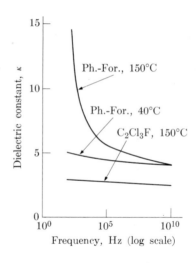

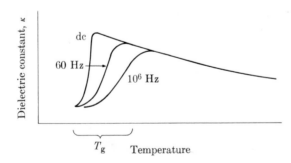

Fig. 7-6.3   Dielectric constant versus temperature. Below the glass temperature, T_g, the molecular dipoles cannot respond to the alternating electric fields. (The glass temperature is lower with low frequencies—Fig. 7-5.4b.) Above T_g, thermal agitation destroys the polarization and therefore reduces the dielectric constant κ.

constant, because the more intense thermal agitation destroys the orientation of the molecular dipoles.

The curves of Fig. 7–6.3 also illustrate another point. The glass temperature T_g varies with frequency. Just as in Fig. 7–5.4, longer times (in this case, milliseconds for 60 Hz versus microseconds for 10^6 Hz) permit molecular movements at a somewhat lower temperature. A dc field has a still lower T_g because extended time is available for molecular orientation.

Example 7–6.1 Polystyrene has a relative dielectric constant of 2.5 when subjected to a dc field. What is the polarization within the PS when a 0.5-mm sheet separates 100 volts?

Solution: The field is 2×10^5 V/m. From Eq. (7–6.3),

$$\mathscr{E} = \frac{100}{.0005}$$

$$\mathscr{P} = \underbrace{(2.5 - 1)}_{(K-1)}\underbrace{(8.85 \times 10^{-12} \text{ C/V·m})}_{\epsilon_0}\underbrace{(2 \times 10^5 \text{ V/m})}_{\mathscr{E}}$$

$$= 2.7 \times 10^{-6} \text{ C/m}^2.$$

Comment. Since the electron charge is 0.16×10^{-18} C, this polarization is equivalent to 1.7×10^7 el/mm². ◀

Example 7–6.2 The capacitance of a parallel-plate capacitor can be calculated from Eq. (1–4.7). The relative dielectric constants for polyvinyl chloride (PVC) and polytetrafluoroethylene (PTFE) are as follows:

$$C = K \underbrace{(8.85 \times 10^{-12} \text{ C/V·m})}(A/D)$$

Frequency, hertz	PVC K	PTFE K ϵ_0
10^2	6.5	2.1
10^3	5.6	2.1
10^4	4.7	2.1
10^5	3.9	2.1
10^6	3.3	2.1
10^7	2.9	2.1
10^8	2.8	2.1
10^9	2.6	2.1
10^{10}	2.6	2.1

a) Plot the capacitance-versus-frequency curves for three capacitors with 3.1 cm × 102 cm effective area separated by 0.025 mm of (1) vacuum, (2) PVC, and (3) PTFE.
b) Account for the decrease in the relative dielectric constant of PVC with increased frequency, and the constancy in the relative dielectric constant of PTFE.

Solution: (a) Sample calculations at 10^2 cycles per second:

$$C_{vac} = (1)(8.85 \times 10^{-12} \text{ C/V·m})(0.031 \text{ m})(1.02 \text{ m})/(25 \times 10^{-6} \text{ m}) = 0.0112 \ \mu f;$$

$$C_{pvc} = (6.5)(0.0112 \ \mu f) = 0.073 \ \mu f.$$

See Fig. 7–6.4(a) for the remainder of the results.

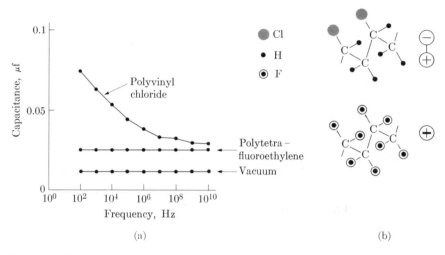

Fig. 7-6.4 Capacitance versus frequency. (a) See Example 7-6.2. (b) Symmetry of polyvinyl chloride and polytetrafluorethylene mers.

b)   The relative dielectric constant of PVC is high at low frequencies because PVC has an asymmetric mer with a large dipole moment (Fig. 7-6.4b). At high frequencies the dipoles cannot maintain alignment with the alternating field. On the other hand, PTFE has a symmetric mer, and therefore its polarization is only electronic. Although the dipoles in PTFE are weaker, they can be oscillated at the frequencies of Fig. 7-6.4. ◀

•7-7   STABILITY OF POLYMERS

Polymers deform by viscoelastic deformation (Section 7-5). Although deformation is accelerated as the temperature is raised above T_g, this softening does not break the primary covalent bonds within the molecule. Under more severe conditions, however, these bonds may be ruptured. Of course, any resulting change in structure affects the properties. Excessive heat can degrade the polymer by breaking bonds. Oxidizing environments can bring about chemical changes. Furthermore, radiation can induce scission, and in certain cases branching or cross-linking.

Degradation   The most obvious degradation of polymers is charring. If the side groups or the hydrogen atoms of a vinyl polymer (Table 7-2.1) are literally torn loose by thermal agitation, only the backbone of carbon atoms remains. We also see this occur with the carbohydrates of our morning toast or with charred wood. It is generally to be avoided in a commercial product.* Carbonization is accelerated in the presence of air, because oxygen reacts with the hydrogen atoms along the side of the polymer chain.

---

* Under controlled conditions, a graphite fiber can be formed from a polymer fiber. Such fibers hold considerable promise as a high-temperature reinforcement for composites.

Oxidation The degradation just described is accentuated by oxygen. In addition, oxygen can have other effects. For example, many rubbers are vulcanized with only 5–20 percent of the possible positions anchored by sulfur cross-links. This permits the rubber to remain deformable and "elastic." Over a period of time, the rubber may undergo further cross-linking by oxygen of the air. The result is identical to Fig. 7–4.4, except that oxygen is the connecting link rather than sulfur. Naturally the rubber becomes harder and less deformable with an increased number of cross-links.

Several factors accelerate the oxidation reaction just described.

1. Oxygen in the form of ozone, O_3, is much more reactive than normal O_2.
2. The radiation of ultraviolet light can provide the energy to break existing bonds so that the oxidation reaction can proceed.
3. The existing bonds are broken more readily when the molecules are stressed.

Because of these features, tires commonly contain carbon black or similar light absorbers to decrease the oxidation rate. Applying the same principle, accelerated testing procedures for product stability commonly expose the polymer to ozone and/or ultraviolet light.

Scission Radiation by ultraviolet light as just described, and by neutrons, can readily break a C—C bond of a vinyl-type polymer (Fig. 7–7.1). This process, which produces smaller molecules, is called scission. This, of course, affects properties such as strength and viscosity. One such example is given in Fig. 7–7.2 where the intrinsic viscosity of a polymer melt decreases with a mild exposure of 8×10^{17} neutrons/mm^2.

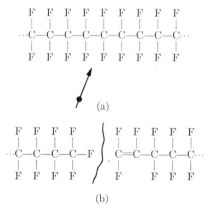

(a)

(b)

Fig. 7–7.1 Degradation by irradiation (scission of polytetrafluorethylene). Most polymers react in this manner rather than as shown in Fig. 7–7.3. As a result, most polymers lose strength through radiation damage.

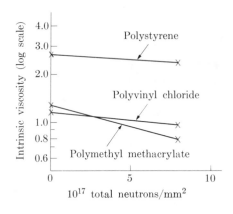

Fig. 7–7.2 Degradation by neutron exposure. The intrinsic viscosity is lowered because the polymers are ruptured into smaller molecules. (Adapted from L. A. Wall and M. Magot, "Effects of Atomic Radiation on Polymers," *Modern Plastics*.)

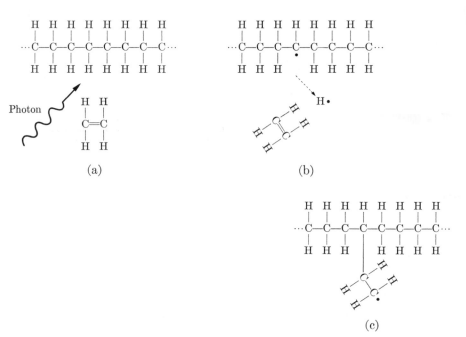

Fig. 7–7.3 Branching by irradiation during polymerization. A photon can supply the activation energy necessary to cause branching. A neutron can produce the same effect.

In certain cases (e.g., polyethylene), radiation can induce branching. If this occurs before the polymerization is complete, it can produce a structure represented by Figs. 7–4.5 and 7–7.3. The consequent effect on properties can be significant, as shown by the two pairs of squeeze bottles in Fig. 7–7.4. It is not easy to predict whether radiation will produce the desired effects of Fig. 7–7.4 or the scission of Fig. 7–7.1.

Fig. 7–7.4 Irradiation of polyethylene. All four squeeze bottles have been heated to 120°C(250°F) for 20 minutes. The two rear bottles had been previously exposed to gamma irradiation to induce branching (Fig. 7–7.3); the slumped bottles had not. Excessive radiation, however, may reverse the effect by introducing scission (Fig. 7–7.1.) More recent technology favors high-density polyethylene (HDPE) as a means of achieving boiling-water stability for sterilization (Table 7–3.1). (Courtesy of General Electric Co.)

Example 7–7.1 An isoprene rubber is 10% cross-linked with sulfur. During service in an oxidizing atmosphere, it is further cross-linked with oxygen. In doing so, it gains 2% weight. What percent of the possible cross-links have been made?

Solution: Basis: 68 g Isop. (=1 mole); therefore, 32 g sulfur at 100%.

$$\text{Amount of sulfur} = 0.1(32 \text{ g}) = 3.2 \text{ g}.$$

$$\text{Oxygen pickup} = 0.02(68 + 3.2 \text{ g}) = 1.42 \text{ g}.$$

$$\text{Moles oxygen} = (1.42 \text{ g})/(16 \text{ g}) = 0.09 \text{ moles}.$$

$$(0.09 \text{ moles})/(\text{mole Isop.}) = 0.09 \quad (\text{or } 9\% \text{ possible cross-links}).$$

$$\text{Total cross-links} = 10\% \text{ (by S)} + 9\% \text{ (by O)} = 19\%. \quad \blacktriangleleft$$

Example 7–7.2 Assume that all the energy required to remove the hydrogen atoms of Fig. 7–7.3 comes from a photon (and that none of the energy is thermal). (a) What is the maximum wavelength which can be used? (b) How many eV are involved?

Solution: (a) From Table 2–3.1,

$$\text{C—H} = 435,000 \text{ J}/(0.6 \times 10^{24} \text{ bonds})$$
$$= 0.725 \times 10^{-18} \text{ J/bond}.$$

From Eq. (2–2.1), $E = hv = hc/\lambda$,

$$\lambda = \frac{hc}{E}$$

$$\lambda = \frac{(6.62 \times 10^{-34} \text{ J} \cdot \text{sec})(3 \times 10^8 \text{ m/sec})}{0.725 \times 10^{-18} \text{ J}}$$

$$= 2.740 \times 10^{-7} \text{ m} = 274.0 \text{ nm.}$$

b) $eV = (0.725 \times 10^{-18} \text{ J/bond})(6.24 \times 10^{18} \text{ eV/J})$

$$= 4.5 \text{ eV/bond.}$$

Comments. This is in the ultraviolet region. Shorter wavelengths will supply more energetic photons, which, of course, can also break the bonds. ◀

REVIEW AND STUDY

Plastics are molecular solids called polymers. The bonding, which was considered in Section 2–3, therefore provides the basis of their structure. The response of molecules to thermal, mechanical, and electrical conditions account for the properties and behaviors of plastics.

SUMMARY: MOLECULAR STRUCTURES

The molecules in polymers are very large. They contain many units; each unit, or mer, contains the atomic features present in the total molecular structure. These mers are repeated throughout the polymeric molecule. Since not all molecules are identical, we must calculate average molecular sizes to characterize the polymer. We may use either a mass-based average or a number-based average.

The polyvinyls,

$$\left(\begin{array}{cc} H & H \\ | & | \\ -C-C- \\ | & | \\ H & R \end{array}\right)_n,$$

are a major category of addition-type polymers. They form without producing a by-product; the **R** may be any of several side groups. The polyesters are a common category of condensation-type polymers. They react to give a small by-product molecule such as water, and connecting links of

$$\begin{array}{c} O \\ \| \\ -C-O- \end{array}$$

in the polymer. Both of the above categories produce lineal polymers if the reacting molecules are bifunctional. Polyfunctional monomers (i.e., small molecules that can connect to three or more other molecules) produce network polymers.

A copolymer includes more than one type of mer within the molecule. Structural variations include stereoisomers (isotactic, atactic), and *cis* and *trans* modifications in rubbers. These arrangements, plus the natural tendency for long molecules to kink by thermal agitation, suppress crystallization.

SUMMARY: MOLECULAR BEHAVIOR

Molecules, particularly giant molecules (macromolecules), behave differently from metallic atoms. Most basic in this respect is the fact that many macromolecular solids have the structure of supercooled liquids and therefore are amorphous. At best, they form highly imperfect crystals. A supercooled liquid, however, has a transition temperature called a glass temperature, T_g. Below T_g, molecular rearrangements are precluded; above T_g the molecules can respond to mechanical stresses and forces of electric fields. Below T_g the plastic is rigid and brittle; it has a low dielectric constant. Above T_g the plastic is flexible; and when provided opportunity with extended times or a higher temperature, the plastic deforms readily. It is this latter feature that facilitates processing; it also places limits on service environments. The dielectric constant increases as the T_g is exceeded because the polar groups are more free to respond to the electric fields.

Lineal polymers are thermoplastic and deform as viscous liquids at temperatures in excess of T_g. Network polymers are less temperature-sensitive and will in fact become stiffer (take on a "set") at elevated temperatures if polymerization, cross-linking, or branching can continue.

Polymers can be degraded by excessive thermal agitation, by oxidation, and by exposure to radiation.

KEY TERMS AND CONCEPTS

Atactic	Mer
Bifunctional	Molecular crystals
Branching	Molecular length
Butadiene-type compounds, $C_4H_5\mathbf{R}$	Molecular size
Cis-	mass-average, $\bar{M}_m$
Conformation	number-average, $\bar{M}_n$
kinked	Molecules
stretched	lineal
Copolymer	network
Cross-linking	Monomer
Crystallization of polymers	Plastics
Degradation	Polar groups
Degree of polymerization, n	Polarization, $\mathscr{P}$
Dipole moment, Qd	Polyester
Elastomer	Polymer
Glass temperature, T_g	Polymerization
Initiation	addition
Isotactic	condensation

Propagation

Relative dielectric constant, κ

• Relaxation time, τ

Root-mean-square length, $\bar{L}$

Scission

Stereoisomers

• Stress relaxation

Termination

Thermoplasts

Thermosets

Trans-

Trifunctional

Vinyl-type compounds, C_2H_3R

Viscoelastic modulus

Vulcanization

FOR CLASS DISCUSSION

A_7 Explain why the "number-average" molecular size is always smaller than the "mass-average" molecular size.

B_7 Obtain 10 toothpicks. Lay them on your desk end to end, but with a 120° angle at each contact point. Flip a coin before adding each toothpick: heads, to the right; tails, to the left. (a) What is the final end-to-end distance? (b) Your results will differ from the results of Eq. (7–1.4). Why? (This may also be performed with triaxial graph paper.)

C_7 Why are polyfunctional mers necessary for network molecules?

D_7 Sketch the structure of three vinyl monomers. Repeat for the mers of the same three vinyl polymers.

E_7 Show how bonds are altered for the addition polymerization of propylene (CH_2CHCH_3).

F_7 Show how bonds are altered for the addition polymerization of butadiene $[CH_2(CH)_2CH_2]$.

G_7 Hydrogen peroxide (H_2O_2) promotes polymerization. "Twice as much is better yet." Discuss.

H_7 Heat removal requires attention by the engineers who design polymerization processes. Why?

I_7 Compare the amounts of heat evolved during the polymerization of polyethylene, of polystyrene, and of a 50–50 copolymer of the two.

J_7 Rewrite Eq. (7–2.8) so that NH_3 is a by-product, rather than H_2O. Suggest why H_2O is the normal by-product, and not NH_3.

K_7 Suggest the compositions of some rubbers other than those shown in Table 7–2.2.

L_7 By a sketch show why the $-\overset{\overset{\displaystyle Cl}{|}}{\underset{\underset{\displaystyle H}{|}}{C}}-$ portion of polyvinyl chloride is a polar group.

M_7 Why is it that metals are seldom glassy but amorphous polymers are common?

N_7 The molecule of polyethylene is "sawtoothed" but otherwise straight in its crystal (Fig. 7–3.4). When it is dissolved in cyclohexane (a liquid solvent), it has the kinked conformation of Fig. 7–1.7. Both forms are stable. Why?

O_7 Explain why the thermal-expansion coefficient (slope of the curve in Fig. 7–5.4b) is greater above T_g than below.

P_7 The exact value of T_g depends on the cooling rate. Why?

Q_7 Why does polyvinyl chloride seldom crystallize?

R_7 A copolymer of two vinyls may be *random*, or they may be in *blocks* with the mer groups clustered along the chain (e.g., —BBAAAAAABBBAAAAA-BBBBBBBA—). The latter crystallizes more readily than the random copolymer (—BAABABBBABAABAAABABBABAB—). Explain.

S_7 The glass temperature of a copolymer is lower than the average glass temperature of the two component polymers. Explain.

T_7 What is the by-product when phenol and formaldehyde are polymerized? When urea and formaldehyde are polymerized (Fig. 2–3.1)?

U_7 Will scrap from thermoplasts or from thermosets be more amenable to recycling? Why?

V_7 It is suggested that butadiene (Table 7–2.2) be used to cross-link polyethylene. Discuss pros and cons.

W_7 Explain why rubber is brittle at liquid-nitrogen temperatures (77 K).

X_7 Play again with the "silly putty" you had as a child. Explain its behavior.

Y_7 Explain the curves of Fig. 7–5.5(b) to someone who is not taking this course.

Z_7 Discuss the relationships between the viscoelastic modulus, glass temperature, and dielectric constant.

STUDY PROBLEMS

7–1.1 A common polymer has $C_2H_2Cl_2$ as a mer. (It is similar to polyvinyl chloride (Fig. 7–1.3) except that each mer has a second chlorine (and one less hydrogen).) It has an average mass of 60,000 amu per molecule. (a) What is its mer mass? (b) What is its degree of polymerization?

Answer: (a) 97 amu (or 97 g/mole)   (b) 620

7–1.2 There are 10^{20} molecules per gram of polyvinyl chloride. (a) What is the average molecular size? (b) What is the degree of polymerization?

7–1.3 The melting temperature, T_m, in K of the paraffins (C_xH_{2x+2}) of Fig. 7–1.1 is sometimes given by the empirical equation

$$T_m = [0.0024 + 0.017/x]^{-1}. \qquad (7\text{–SP.1})$$

What melting point would polyethylene have with (a) a degree of polymerization n of 10? (b) of 100? (c) of 1000? (*Note:* Mer = C_2H_4.)

Answer: (a) 35°C   (b) 130°C   (c) 142°C

7–1.4 Two grams of dextrose $(C_6H_{12}O_6)$ are dissolved in 14 grams of water. What is the number fraction of each type of molecule?

7–1.5 (a) What is the "mass-average" molecular size of the molecules in Example 7–1.2? (b) The "number-average" molecular size?

Answer: (a) 31.7 amu   (b) 21.6 amu

7–1.6 The following data were obtained in a determination of the average molecular size of a polymer:

Molecular size, interval midvalue	Mass
30,000 amu	3.0 g
20,000	5.0
10,000	2.5

Compute the "mass-average" molecular size.

7–1.7 (a) What is the number-fraction of molecules in each of the three size categories of Study Problem 7–1.6? (b) What is the "number-average" molecular size?

Answer: (a) 0.167, 0.417, 0.417   (b) 17,500 amu

7–1.8 Polyethylene contains equal numbers of molecules with 100 mers, 200 mers, 300 mers, 400 mers and 500 mers. What is the "mass-average" molecular size?

7–1.9 Polyvinyl chloride $+C_2H_3Cl+_n$ is dissolved in an organic solvent. (a) What is the mean square length of a molecule with a molecular mass of 28,500 g/mole? (b) What would be the molecular mass of a molecule with one half the mean square length as in part (a)?

Answer: (a) 4.65 nm   (b) 7125 g/mole

7–1.10 Determine the degree of polymerization and the mean square length of the average molecule in Study Problem 7–1.7, if the polymer is polyvinyl alcohol (number–average).

7–1.11 A polymeric material contains polyvinyl chloride molecules that have an average of 900 mers/molecule. What is the theoretical maximum strain this polymer could undergo if every molecule could be unkinked into a straight molecule (except for the 109.5° bond angles)?

Answer: 3300% strain

7–2.1 Utilizing the data of Chapter 2, what is the net energy change as 7 grams of ethylene polymerizes to polyethylene?

Answer: −15,000 J

7–2.2 (a) How many C=C bonds are eliminated per mer during the polymerization of butadiene? (b) How many additional C—C bonds are formed?

7–2.3 Refer to Study Problem 7–2.2. How much energy is released by polymerizing 7 grams of butadiene?

Answer: −7800 J

7–2.4 A copolymer has equal masses of styrene and butadiene. What is the mer fraction of each?

7–2.5 A triple copolymer, ABS, has equal mass fractions of each. What is the mer fraction of each?

Answer: acrylonitrile, 0.4; butadiene, 0.4; styrene, 0.2

7–2.6 How much energy is released when seven grams of the ABS copolymer of Study Problem 7–2.5 are polymerized?

7–2.7 Refer to Example 7–2.3. How much water is released per gram of product?

Answer: 0.16 g

7–2.8 Refer to Example 7–2.3. Assume the reaction formed NH_3 as a by-product (rather than H_2O). (a) Show which bonds would have to be broken. (b) What is the amount of energy change per mole of NH_3 formed? (c) Why is this reaction not expected?

7–2.9 H_2O_2 is added to 280 kg (616 lb) of ethylene prior to polymerization. The average degree of polymerization obtained is 1000. Assuming all the H_2O_2 was used to form terminal groups for polymer molecules, how many grams of H_2O_2 were added? (Mer = C_2H_4.)

Answer: 340 g (0.75 lb)

7–2.10 Two tenths of one percent by weight of H_2O_2 was added to ethylene prior to polymerization. What would the average degree of polymerization be if all the H_2O_2 were used as terminals for the molecules? (Mer = C_2H_4.)

7–2.11 What is the ratio of ethylene (C_2H_4) mers to vinyl chloride (C_2H_3Cl) mers in a copolymer of the two in which there is 30 w/o chlorine?

Answer: 2 to 1

7–2.12 The mer ratio of styrene/butadiene is $\frac{1}{3}$. What is the w/o of carbon?

7–3.1 Solid methane has a density of 0.54 g/cm^3 (=0.54 Mg/m^3). (a) What is the lattice constant of its crystal that is fcc (Fig. 7–3.2)? (b) What is the "radius" of the CH_4 molecule?

Answer: (a) 0.582 nm   (b) 0.206 nm

7–3.2 Crystalline iodine has a density of 4.93 Mg/m^3 (=4.93 g/cm^3) and the structure of Fig. 7–3.3. What is the third dimension of the unit cell?

7–3.3 A polyethylene with no evidence of crystallinity has a density of 0.90 Mg/m^3. Commercial grades of low-density polyethylene (LDPE) have 0.92 Mg/m^3, while HDPE has a density of 0.96 Mg/m^3. Estimate the volume fraction of crystallinity in each case.

Answer: C_{LDPE} = 0.18; C_{HDPE} = 0.55

7–3.4 From Example 7–3.1 and Study Problem 7–3.3, calculate the amount of "free space" in LDPE; in HDPE.

7–3.5 Sketch the three orthogonal views of the atoms in the polyethylene unit cell (Fig. 7–3.4).

7–4.1 What percent sulfur would be present if it were used as a cross-link at every possible point (a) in polyisoprene? (b) in polychloroprene?

Answer: (a) 32% S   (b) 26.5% S

7–4.2 A rubber contains 91% polymerized chloroprene and 9% sulfur. What fraction of the possible cross-links are joined by vulcanization? (Assume that all the sulfur is used for cross-links of the type shown in Fig. 7–4.4.)

7–4.3 A rubber contains 54% butadiene, 34% isoprene, 9% sulfur, and 3% carbon black. What fraction of the possible cross-links are joined by vulcanization, assuming that all the sulfur is used in cross-linking?

Answer: 0.188

7–4.4 Rubber A (200 g) contains 168 g of butadiene $-(C_4H_6)_n$ and 32 g sulfur. (a) What fraction of the cross-links are utilized if all of the sulfur forms those links?

Rubber B (217 g) contains 168 g of isoprene $-(C_5H_8)_n$ and 49 g of selenium. Although the use of selenium has some major disadvantages, it can cross-link rubber. (It lies immediately below sulfur in the periodic table.) (b) Which rubber, A or B, is more highly cross-linked?

7–4.5 What is the percent weight loss if phenol polymerizes trifunctionally with formaldehyde and all of the by-product water evaporates?

Answer: 19%

7–4.6 Company X buys a partially polymerized resin of phenol and formaldehyde. A sample (10.3 g) was pressed into a hot mold. It was initially thermoplastic, but it set after a few minutes. After the sample was thoroughly dried, it weighed 9.6 g. What was the number of $-CH_2-$ connections which the average phenol had initially if we assume the final product has 3 $-CH_2-$ bridges?

7–4.7 One kg of divinyl benzene (Fig. 7–4.3) is added to 50 kg of styrene. What is the maximum number of cross-links per gram of product?

Answer: 9×10^{19}/g

7–4.8 How much sulfur must be added to 100 g of chloroprene rubber to cross-link 10% of the mers? (Assume all of the available sulfur is used, comparable to that in Fig. 7–4.4 for butadiene.)

7–5.1 A stress relaxes from 0.7 MPa to 0.5 MPa in 123 days. (a) What is the relaxation time? (b) How long would it take to relax to 0.3 MPa?

Answer: (a) 366 days (b) an additional 187 days (total of 310 days)

• **7–5.2** An initial stress of 10.4 MPa (1500 psi) is required to strain a piece of rubber 50%. After the strain has been maintained constant for 40 days, the stress required is only 5.2 MPa (750 psi). What would be the stress required to maintain the strain after 80 days? Solve without a calculator.

• **7–5.3** The relaxation time for a polymer is known to be 45 days and the modulus of elasticity is 70 MPa (both at 100°C). The polymer is compressed 5% and held at 100°C. What is the stress (a) initially? (b) after 1 day? (c) after 1 month? (d) after 1 year?

Answer: (a) 3.5 MPa (b) 3.4 MPa (c) 1.8 MPa (d) 1000 Pa

• **7–5.4** The relaxation time for a nylon thread is reduced from 4000 min to 3000 min if the temperature is increased from 20°C to 30°C. (a) Determine the activation energy for relaxation. (b) At what temperature is the relaxation time 2000 minutes?

7–5.5 Lucite (PMM of Fig. 7–5.4) is loaded at 125°C for 1 hr. (a) How much would the load have to be increased to give the same strain in 36 sec? (b) Repeat for 100°C.

Answer: (a) $s_{0.01}/s_1 = 2.5$ (b) $14(s_1)$

• **7–6.1** The relative dielectric constant κ of an isoprene rubber at 0°C is 3.0. What voltage difference (dc) is required to develop a charge density of 10^{-5} C/m^2 in a metal coating covering a 0.1 mm film?

Answer: 37.7 V

• **7–6.2** The polarization of the rubber in Study Problem 7–6.1 contributed how many electrons per mm^2 in the metal coating?

7–6.3 (a) What is the polarization in the polyvinyl chloride of Example 7–6.2 at 100 Hz when 100 volts is applied to the capacitor? (b) What will the electron density be on the capacitor plate?
Answer: (a) 0.0002 C/m^2 (b) 1.4 × 10^{15} electrons/m^2

7–6.4 A plate capacitor must have a capacitance of 0.25 μf. What should its area be if the 0.0005-in. (0.013-mm) mylar film which is used as a spacer has a dielectric constant of 3.0?

7–7.1 Raw polyisoprene (i.e., nonvulcanized natural rubber) gains 2.3 w/o by being cross-linked by oxygen of the air. What fraction of the possible cross-links are established?
Answer: 0.10

7–7.2 A chloroprene rubber gains 3 w/o by oxidation. Assume that this oxygen produced cross-linkage. What fraction of the possible anchor points (cross-linkages) contain oxygen atoms?

7–7.3 A rubber containing 47 w/o isoprene, 38 w/o butadiene, and 15 w/o inert material is exposed to ozone, gaining 0.6 g per 100 g of initial product. Assume that the gain in weight is a consequence of cross-linking. What fraction of possible cross-links are completed?
Answer: 2.7%

7–7.4 (a) What frequency and wavelength must a photon have to supply the energy necessary to break an average C—H bond in polyethylene? (b) Why can some bonds be broken with slightly longer electromagnetic waves?

7–7.5 The average energy of a C—Cl bond is 340 kJ/mole (81 kcal/mole) according to Table 2–3.1. Will visible light [400 nm (violet) to 700 nm (red)] have enough energy to break the average C—Cl bond?
Answer: 350 nm required

7–7.6 Based on the data of Table 2–3.1, what is the net energy requirement to produce the change from Fig. 7–7.3(a) to 7–7.3(c)?

7–7.7 What are the (a) mean and (b) median of the following density data for a polyethylene?

0.926,   0.930,   0.924,   0.923   0.921,   0.926,

0.927,   0.929,   0.922,   0.928 Mg/m^3 (=0.928 g/cm^2).

• c) What is the standard deviation?
Answer: (a) 0.9256 Mg/m^3 (b) 0.926 Mg/m^3 (=0.926 g/cm^3)   (c) 0.0030 Mg/m^3

7–7.8 Determine the (a) mean, (b) median, and • (c) standard deviation of the following set of strength data for a polystyrene:

11.8 MPa	12.8	12.2	11.7	12.1	12.0	11.5	12.1
12.3	12.7	12.2	11.9	11.6	12.4	12.0	
12.1	12.9	12.3	11.7	13.0	12.3	12.2	
12.2	13.1	11.8	12.7	12.7	12.1	11.9	
12.5	12.4	11.4	11.9	11.5	12.1	12.4	
12.9	12.5	12.2	12.5	11.0	12.3	12.2	

CHAPTER 8

Ceramic Materials

PREVIEW

Ceramic materials contain phases that are *compounds of metallic and nonmetallic elements*. We can cite many of these compounds ranging from Al_2O_3 to inorganic glasses, to clay products, and on to sophisticated piezoelectrics such as $Pb(Zr,Ti)O_3$.

In general, we will find that ceramic compounds are more stable with respect to thermal and chemical environments than their components; for example, Al_2O_3 as a compound versus aluminum and oxygen separately. Since compounds inherently involve more complex atomic coordinations than their corresponding components, we will find that there is more resistance to slip, so that ceramics are generally harder, and always less ductile than their metallic or polymeric counterparts. The dielectric, semiconductive, and magnetic characteristics of selected ceramics are especially valuable to the scientists and engineers who design or utilize devices for electronic circuits.

CONTENTS

STUDY OBJECTIVES

1 To visualize the atomic arrangements in close-packed AX structures where CN = 8, CN = 6, and CN = 4.

2 To calculate density of AX compounds as a means of understanding their structures.

3 To use the previous structures as a basis for understanding several additional structures that are widely encountered in engineering applications (CaF_2, Al_2O_3, $BaTiO_3$, and ferrites).

4 To understand the source of piezoelectricity in ceramic crystals, and to be able to calculate polarization.

• **5** To understand the source of ferrimagnetism in ceramic crystals, and to be able to calculate magnetization.

6 To recognize (a) why ceramic materials (like other 3-D compounds) are strong in compression, weak in tension, and brittle, and (b) that these materials must utilize compressive stresses for optimum performance.

7 To associate atom movements and displacements with the processes of viscous forming, sintering, and tempering.

8-1 CERAMIC PHASES

The term *ceramic* is most familiar as an adjective describing artware. For the engineer, however, ceramics include a wide variety of substances such as glass, brick, stone, concrete, abrasives, porcelain enamels, dielectric insulators, nonmetallic magnetic materials, high-temperature refractories, and many others (Fig. 8–1.1). The char-

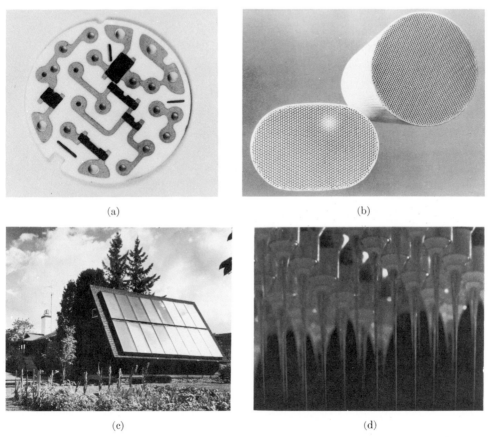

(a)   (b)

(c)   (d)

Fig. 8–1.1 Examples of technical ceramic products. (a) Printed circuit. The substrate is Al_2O_3 because its strong ionic bonds preclude electrical transport. The printed components are powders of semiconducting oxides and metals that provide the functional circuitry. (Courtesy AC Division—General Motors Corp.) (b) Catalyst supports. These honeycombed, ceramic parts are placed in automotive exhaust systems. There, they support the catalysts that eliminate the unburned hydrocarbons through oxidation. Here again, Al_2O_3 is used, because it resists elevated temperatures and may be made strong enough to withstand the vibrational punishment in service. (Alumina ceramic parts by Corning Glass Works and photo by Aluminum Company of America.) (c) Solar panels. Glass is an important component in solar heaters because the panel is transparent to radiant energy. Once converted to heat, the energy must be trapped and retained. Of course, long service lives and negligible maintenance are important. (Courtesy Libbey-Owens-Ford Glass.) (d) Glass fibers. These glass filaments are being drawn in the laboratory as continuous fibers through platinum orifices. The fiber dimensions are controlled by the orifice temperature, and by the velocity at which the glass is pulled away from the hot orifice. Commercial textile production involves hundreds of orifices and speeds of kilometers per minute to produce strands that contain fibers with diameters in the micrometer range. (Courtesy Owens-Corning Fiberglas Co.)

acteristic feature all these materials have in common is that they are compounds of metals and nonmetals. These compounds are held together with ionic and/or covalent bonds. Thus, their properties differ from metals. They are generally insulators, commonly transparent (or translucent), typically nondeformable, and unusually stable under severe environments (Fig. 8–1.1).

The compound MgO is representative of a simple ceramic material with a one-to-one ratio of metallic and nonmetallic atoms. It is used extensively as a refractory because it can withstand exceedingly high temperatures (1500°C–2500°C, or 3000°F–4500°F) without dissociation or melting. Clay is also a common ceramic material but is more complex than MgO. The simplest clay is $Al_2Si_2O_5(OH)_4$. Its crystal structure has the four different units: Al, Si, O, and the (OH) radical. Although ceramic materials are not as simple as metals, they may also be classified and understood in terms of their internal structures. Furthermore, we may use the structure of simple metals as our point of departure.

Comparison of ceramic and nonceramic phases   Most ceramic phases, like metals, are crystalline. Unlike metals, however, their structures do not contain large numbers of free electrons. Either the electrons are shared covalently with adjacent atoms, or they are transferred from one atom to another to produce an ionic bond, in which case the atoms become ionized and carry a charge.

Ionic bonds give ceramic materials relatively high stability. As a class, they have a much higher melting temperature, on the average, than do metals or organic materials. Generally speaking, they are also harder and more resistant to chemical alteration. Like the organic materials, solid ceramic minerals are usually insulators. At elevated temperatures where they have more thermal energy, ceramics do conduct electricity, but poorly as compared with metals. Due to the absence of free electrons, most ceramic materials are transparent, at least in thin sections, and are poor thermal conductors.

Crystalline characteristics may be observed in many ceramic materials. Mica, for example, has cleavage planes that permit easy splitting. Plastic deformation, similar to slip in metals, may occur in some of the simpler crystals, such as MgO. Crystal outlines can form during growth, as exemplified by the cubic outline of ordinary table salt. In asbestos, the crystals have a marked tendency toward linearity; in micas and clays, the crystals form two-dimensional sheet structures. The stronger, more stable ceramic materials commonly possess three-dimensional network structures with equally strong bonding in all three directions.

As compared with those of metals, the crystal structures of ceramic materials are relatively complex. This complexity and the greater strength of the bonds holding the atoms together make ceramic reactions sluggish. For example, at normal cooling rates glass does not have time to rearrange itself into a complicated crystalline structure, and therefore at room temperature it remains as a supercooled liquid for a long time.

The structures and properties of compounds such as the refractory carbides and nitrides fall somewhere between those of ceramic and metallic materials. These include such compounds as TiC, SiC, BN, and ZrN, which contain semimetallic elements and whose structures comprise a combination of metallic and covalent bonds. Ferromagnetic spinels are another example. Because they lack free electrons, they are not good conductors of electricity; however, the atoms can be oriented within the crystal structure so as to possess the magnetic properties normally associated with iron and related metals.

The structure and properties of silicones fall somewhere between those of ceramic and organic materials. They are often called inorganic polymers. Finally, there are close structural relationships between amorphous polymers and commercial glasses, a fact we have already encountered through the examination of the glass temperature T_g in Chapter 7.

8-2 CERAMIC CRYSTALS (AX)

The simplest ceramic compounds possess equal numbers of metal and nonmetal atoms. These may be ionic like MgO, where two electrons have been transferred from the metallic to the nonmetallic atoms and have produced cations (Mg^{2+}) and anions (O^{2-}). AX compounds may also be covalent, with a large degree of sharing of the valence electrons. Zinc sulfide (ZnS) provides an example of this type of compound.

The characteristic feature of any AX compound is that the A atoms are coordinated with only X atoms as immediate neighbors, and the X atoms have only A atoms as *first-neighbors*. Thus the A and X atoms or ions are highly *ordered*. There are three principal ways in which AX compounds can form so that the two types of atoms are in equal numbers and possess the ordered coordination just described. The prototypes are

$$CsCl \text{ with CN} = 8;$$

$$NaCl \text{ with CN} = 6;$$

$$ZnS \text{ with CN} = 4.$$

CsCl has a simple cubic array of atoms. NaCl and ZnS have an fcc array.

AX structures (CsCl-type) Each A atom of Fig. 8–2.1 has eight X neighbors (and an extension of the structure would show each X atom has eight A neighbors). This structure is *not* bcc because the 0,0,0 and $\frac{1}{2},\frac{1}{2},\frac{1}{2}$ locations are not identical. Rather it is simple cubic with equivalent sites related only by integer translations of a. Since the locations occupied by the A atoms, or ions, in Fig. 8–2.1 are in sites surrounded by eight neighbors, we can speak of them as being in 8-*fold interstitial sites* (or more simply 8-f sites) within a simple cubic lattice.

This same structure was previously noted in Section 3–5 and Fig. 3–5.4 as that of β'-brass. However, in general it is not common among ceramics, because cations, stripped of electrons, tend to be considerably smaller than anions with excess electrons. Thus, few ionic compounds meet the r/R ratio of 0.73 or more, which is required for CN = 8 (Table 2–6.1).

The lattice constant a of a CsCl-type compound is directly related to the ionic radii because $(r + R)$ is half of the body diagonal:

$$(r + R)_{CsCl} = a\sqrt{3}/2. \tag{8–2.1}$$

Of course, since CN = 8, an appropriate correction must be made to the ionic radii that are listed in Appendix B. Recall that $R_{CN=8} \approx (R_{CN=6})/0.97$.

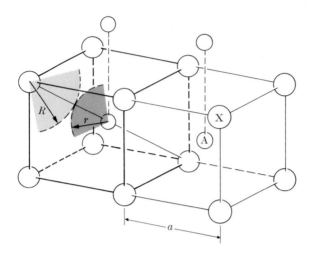

Fig. 8–2.1 AX structure (CsCl-type). The A atom, or ion, sits in the interstitial site among eight X atoms (8-f sites). These $\frac{1}{2}, \frac{1}{2}, \frac{1}{2}$ locations are occupied in all unit cells. Also note that X atoms sit among eight A atoms. The lattice constant a equals $2(r + R)/\sqrt{3}$.

AX structures (NaCl-type)   The NaCl-type structure has an fcc lattice of anions with positive ions located in 6-*fold interstitial sites* (6-f sites). One such site is identified in Fig. 8–2.2. The NaCl structure was initially shown in Fig. 3–1.1. A composite of these two figures has been resketched in Fig. 8–2.3. The fcc sites are normally positioned at

$$0,0,0 \qquad \frac{1}{2},\frac{1}{2},0 \qquad \frac{1}{2},0,\frac{1}{2} \qquad 0,\frac{1}{2},\frac{1}{2}; \qquad (8-2.2)$$

and the 6-f sites are

$$\frac{1}{2},\frac{1}{2},\frac{1}{2} \qquad 0,0,\frac{1}{2} \qquad 0,\frac{1}{2},0 \qquad \frac{1}{2},0,0. \qquad (8-2.3)$$

There are equal numbers of each; thus all 6-f sites are occupied in an AX compound. Here, as with the CsCl-type structure, it makes no difference whether all of the A ions are assigned to the fcc locations, or conversely, to the 6-f sites.

There are several hundred compounds with an NaCl-type structure; for example, MgO, NiO, FeO (Fig. 4–3.2), and MnS are among those more commonly encountered

Fig. 8–2.2 Interstitial sites in fcc lattices. The 6-f site is among six neighboring atoms. There are four such sites per unit cell. The 4-f site is among four neighboring atoms. There are eight of these sites per unit cell.

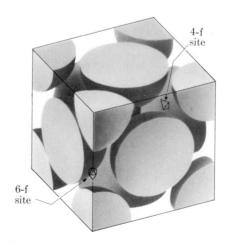

4-f site

6-f site

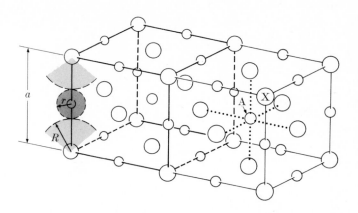

Fig. 8–2.3 AX structure (NaCl-type). Compare this figure with Figs. 3–1.1 and 8–2.2. Each A atom is among six X atoms (and each X atom is among six A atoms). There are four of these 6-f interstitial sites per unit cell. All are occupied. The lattice constant a equals $(2r + 2R)$.

in technology. We have already observed in Eq. (3–2.4) that the lattice constant a is equal to twice the sum of the two radii, $2(r + R)$, since unlike ions "touch" and like ions do not. The minimum radius ratio, r/R, for this structure is 0.41 (Table 2–6.1). There is no maximum ratio; however, those ionic compounds with $r/R > 0.73$ commonly are more stable with the CsCl-type structures.

AX structures (ZnS-type) One of the 4-f sites of an fcc lattice of atoms is also shown in Fig. 8–2.2. Compounds may assume a structure of X atoms at

$$0,0,0 \qquad \frac{1}{2},\frac{1}{2},0 \qquad \frac{1}{2},0,\frac{1}{2} \qquad 0,\frac{1}{2},\frac{1}{2}; \qquad (8-2.2)$$

and A atoms at

$$\frac{3}{4},\frac{3}{4},\frac{3}{4} \qquad \frac{1}{4},\frac{1}{4},\frac{3}{4} \qquad \frac{1}{4},\frac{3}{4},\frac{1}{4} \qquad \frac{3}{4},\frac{1}{4},\frac{1}{4} \overset{*}{.} \qquad (8-2.4a)$$

We encountered this ZnS-type structure initially in Chapter 5 for compound semiconductors (Fig. 5–4.3b). It was favored by them because covalent bonding for those compounds required that each atom have CN = 4. In fact, most of the ZnS-type compounds possess this structure to meet covalency requirements, rather than because of a small r/R ion ratio.

Figures 5–4.3(b) and 8–2.2 serve as the basis for Fig. 8–2.4. From it, one may relate the sum of the radii $(r + R)$ to the lattice constant, a, since the 4-f sites are at quarter positions of the unit cell:

$$(r + R)^2 = \left(\frac{a}{4}\right)^2 + \left(\frac{a}{4}\right)^2 + \left(\frac{a}{4}\right)^2,$$

or

$$a = 4(r + R)/\sqrt{3}. \qquad (8-2.5)$$

---

* Or alternatively at

$$\frac{1}{4},\frac{1}{4},\frac{1}{4} \qquad \frac{3}{4},\frac{3}{4},\frac{1}{4} \qquad \frac{3}{4},\frac{1}{4},\frac{3}{4} \qquad \frac{1}{4},\frac{3}{4},\frac{3}{4}. \qquad (8-2.4b)$$

There are eight 4-f sites per fcc unit cell. Thus, only half of these are occupied by an AX compound, because there are only four fcc positions per unit cell.

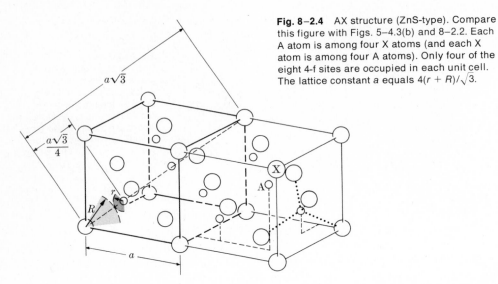

Fig. 8–2.4 AX structure (ZnS-type). Compare this figure with Figs. 5–4.3(b) and 8–2.2. Each A atom is among four X atoms (and each X atom is among four A atoms). Only four of the eight 4-f sites are occupied in each unit cell. The lattice constant a equals $4(r + R)/\sqrt{3}$.

- **AX structures (noncubic)** Not all AX compounds are cubic. Two examples are those which have hcp arrays of X atoms. In the NiAs-type structure, the A atoms are in all of the 6-f interstitial sites. In the ZnO-type structure, the A atoms are in half of the 4-f sites (Table 8–2.1). One may recall from Chapter 3 that fcc and hcp *metals* have equal packing factors (0.74) and the same coordination number (12). The only difference arises in the sequence of stacking of planes, above one another. By like token, the structures of NaCl and NiAs are related; and the structures of ZnS and ZnO are related. However, in each case the symmetry of the hexagonal version is lower than of the cubic version. Thus, we find FeS (with the NiAs-type structure) to be very hard and brittle; in contrast, MnS (with the NaCl-type structure) is among the more ductile AX compounds.

Table 8–2.1
Packing of AX compounds

		A atoms		
Structure	X atoms, lattice	Interstitial locations	Fraction filled	Other examples
CsCl	sc	8-f sites	all	
NaCl	fcc	6-f sites	all	MgO, MnS, LiF
ZnS*	fcc	4-f sites	$\frac{1}{2}$	β–SiC, CdS, AlP
NiAs	hcp	6-f sites	all	FeS, MnTe
ZnO	hcp	4-f sites	$\frac{1}{2}$	BeO, ZnS*, AlN

* ZnS, like many ceramic compounds, has more than one polymorph. The cubic polymorph is called *sphalerite* (or *blende*), after the most common zinc ore mineral; the hexagonal polymorph has the name *wurtzite*.

Example 8–2.1 The compound CsBr has the same structure as CsCl. The centers of the two unlike ions are separated 0.37 nm. (a) What is the density of CsBr? (b) What is the radius of the Br^- ions in this structure?

Solution

a) $$a = 2(0.37 \text{ nm})/\sqrt{3} = 0.427 \text{ nm}.$$

Atomic masses from Fig. 2–1.1:

$$\rho = \frac{(132.9 + 79.9 \text{ g})/(0.602 \times 10^{24})}{(0.427 \times 10^{-9} \text{ m})^3} = 4.5 \times 10^6 \text{ g/m}^3$$

$$= 4.5 \text{ Mg/m}^3 \qquad \text{(or 4.5 g/cm}^3\text{)}.$$

b) From Appendix B, $r_{Cs^+} = 0.167$ nm when CN = 6;

$$R_{Br^-} = 0.37 \text{ nm} - (0.167 \text{ nm})/0.97 = 0.20 \text{ nm}.$$

Comment. This radius is for CN = 8. R_{Br^-} is approximately 0.19 nm in CN = 6 structures. ◀

Example 8–2.2 Calculate the density of FeO which has an NaCl-type structure. (Assume equal numbers of Fe and O ions.)

Solution: From Table 2–5.1, or Appendix B,

$$r_{Fe^{2+}} = 0.074 \text{ nm} \quad \text{and} \quad R_{O^{2-}} = 0.140 \text{ nm}$$

when CN = 6. There are 4 Fe^{2+} and 4 O^{2-} per unit cell. (See Fig. 8–2.3.) Then

$$V = a^3 = [2(0.074 + 0.140) \times 10^{-9} \text{ m}]^3$$
$$= 78.4 \times 10^{-30} \text{ m}^3;$$

$$m = 4(55.8 + 16.0 \text{ amu})/(0.6 \times 10^{24} \text{ amu/g})$$
$$= 479 \times 10^{-24} \text{ g};$$

$$\rho = m/V = \frac{479 \times 10^{-24} \text{ g}}{78.4 \times 10^{-30} \text{ m}^3} = 6.1 \times 10^6 \text{ g/m}^3 \qquad \text{(or 6.1 g/cm}^3\text{)}.$$

Comments. The measured density is about 5.7 g/cm³ because the structure contains cation vacancies (Fig. 4–3.2 and Example 4–3.1). ◀

Example 8–2.3 The compound CdS has the same structure as ZnS. The centers of the two unlike ions are separated 0.25 nm. (a) What is the volume of the unit cell? (b) What is the density?

Solution

a) $$a = 4(r + R)/\sqrt{3}$$
$$= 4(0.25)/\sqrt{3} = 0.577 \text{ nm};$$
$$a^3 = 0.192 \text{ nm}^3.$$

b) Since Cd = 112.4 amu, and S = 32.1 amu (Fig. 2–1.1),

$$\rho = 4(144.5 \text{ g}/0.6 \times 10^{24})/(0.577 \times 10^{-9} \text{ m})^3 = 5.0 \text{ Mg/m}^3 \qquad \text{(or 5.0 g/cm}^3\text{)}. ◀$$

Example 8-2.4 MnS has three polymorphs. Two of these are (a) the NaCl-type structure (Fig. 8-2.3), and (b) the ZnS-type structure (Fig. 8-2.4). What percent volume change occurs when the second type (ZnS) changes to the first type (NaCl)? (See Appendix B for radii when CN = 6, and the attached footnote when CN ≠ 6.)

Solution: Each unit cell (Figs. 8-2.3 and 8-2.4) has 4 Mn^{2+} ions and 4 S^{2-} ions. Therefore, we can consider one unit cell of each.

a)
$$V = a^3 = [2(0.080 + 0.184 \text{ nm})]^3 = 0.147 \text{ nm}^3;$$

b)
$$V = a^3 = [4(0.073 + 0.167 \text{ nm})/\sqrt{3}]^3 = 0.170 \text{ nm}^3;$$

$$(\Delta V/V)_{b \to a} = -14 \text{ v/o}.$$

Comment. Only the NaCl-type is stable; however, the other two polymorphs can be formed with appropriate starting materials, and crystallization procedures. ◀

8-3 CERAMIC CRYSTALS ($A_m X_p$)

Not all binary compounds have equal numbers of A and X atoms (or ions). We shall consider only two cases out of many to illustrate the point, the CaF_2 structure and the Al_2O_3 structure. Fluorite (CaF_2) is the basic structure for UO_2, which is used in nuclear fuel elements, and provides the pattern for one of the polymorphs of ZrO_2, which is a useful high-temperature oxide. Corundum, Al_2O_3 is one of the most widely used ceramics for technical purposes. We have already encountered Al_2O_3 in a sparkplug (Fig. 2-7.3). Other uses range from emery grinding wheels to acid pumps, to substrates for printed circuits (Fig. 8-1.1a) and on to high-temperature materials for catalyst supports in exhaust systems (Fig. 8-1.1b).

The CaF_2 structure has cations, Ca^{2+}, at fcc locations: $0,0,0$; $\frac{1}{2},\frac{1}{2},0$; $\frac{1}{2},0,\frac{1}{2}$; and $0,\frac{1}{2},\frac{1}{2}$. The anions, F^-, are in 4-f interstitial positions among the cations and include *both* sets listed in "Eqs." (8-2.4a and b). This 4-to-8 ratio of Ca^{2+} to F^- accommodates the 1-to-2 ratio of *m*-to-*p* in $A_m X_p$. A sketch is shown in Fig. 8-3.1.

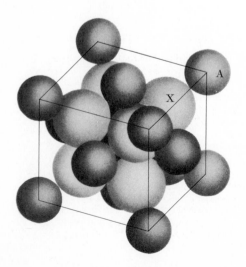

Fig. 8-3.1 AX_2 structure (CaF_2-type). There is an fcc lattice of A atoms, with X atoms occupying all of the interstitial sites among 4 atoms. (There is no atom at the center of the the unit cell.) Observe in Fig. 8-3.2 that closest F^- ions are not in contact.

The structure just described is fcc with F^- ions in 4-f sites among Ca^{2+} ions. In addition, the calcium ions are among eight fluorine ions. However, note that not all such 8-f sites are occupied; e.g., those at the center of unit cells (Fig. 8–3.1), and those at the mid-points of the edges of the unit cells. These vacant sites are important to the UO_2 of nuclear fuel elements, which has this crystal structure, because these holes provide space for fission products to reside.*

The structure of Al_2O_3 will be the final binary compound to receive consideration. It has an hcp array of O^{2-} ions. Two thirds of the 6-f interstitial sites are occupied by Al^{3+} ions.[†] With an interatomic distance of only 0.191 nm separating the 3^+ and 2^- charges of the aluminum ion and its six oxygen neighbors, it is not surprising that the bonding energies are high.[‡] This is reflected in the melting temperature ($> 2000°C$), hardness (Mohs = 9), and the resistance of Al_2O_3 to a large number of chemicals. Furthermore, its combination of low electrical conductivity and a relatively high thermal conductivity (Appendix C) makes Al_2O_3 useful for many electrical applications.

Example 8–3.1 The lattice constant of CaF_2 is 0.547 nm. (a) Sketch the arrangement of ions on the (110) plane of CaF_2. (b) What is the sum of the two radii ($r_{Ca^{2+}} + R_{F^-}$)? ●(c) What is the linear density of equivalent sites in the $[1\bar{1}2]$ direction?

Solution: (a) Figure 8–3.2 is a (110) plane sketched from Fig. 8–3.1.

b)
$$r + R = \sqrt{(a/4)^2 + (-a/4)^2 + (a/4)^2} = (a\sqrt{3})/4$$
$$= (0.547 \text{ nm})\sqrt{3}/4 = 0.237 \text{ nm}.$$

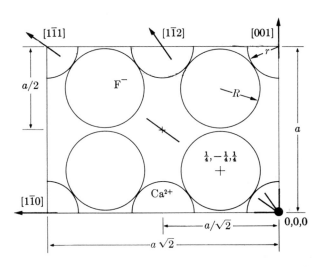

Fig. 8–3.2   CaF$_2$ structure. [(110) plane.] (See Example 8–3.1.)

---

* If uranium metal were used as a fuel element, it would expand excessively when atoms are split into two or more daughter atoms, because the metallic crystal does not have comparable vacant sites.
† This oversimplifies the structure, since it does not take into account a distortion that develops as a consequence of the vacant 6-f sites.
‡ Recall from Eq. (2–5.1) that the coulombic attractive forces are proportional to $-Z_1Z_2/a_1^2$ where Z_1 and Z_2 are the valences.

● c)   Since CaF_2 is fcc, the repetition distance t will be from 0,0,0 to $\frac{1}{2}, -\frac{1}{2}, 1$, or

$$\sqrt{a^2 + (-a/2)^2 + (a/2)^2}.$$

Therefore:

$$t = a\sqrt{1.5} = 0.547 \text{ nm } (1.225) = 0.67 \text{ nm},$$

and

$$\text{linear density} = 1.49/\text{nm} \quad (\text{or } 1.5 \times 10^6/\text{mm}).$$

Comments. The F^- ions are separated by about 0.03 nm (Fig. 8–3.2).

The data from Appendix B would indicate $a = 0.515$ nm; however, this does not take into account the mutual repulsion of the 8 F^- ions around the vacant 8-f sites. This repulsion expands the structure.

This is the same structure as UO_2. The unoccupied 8-f sites provide space for fission products to reside within the crystal. ◄

● **Example 8–3.2** Uranium dioxide has the CaF_2 structure. As such, it has a large interstitial site at the center of the unit cell. (See Fig. 8–3.2.) (a) How many such holes are there per unit cell? (b) Using adjusted data from Appendix B, what is the radius ratio, r/R, for UO_2?

Solution: (a) The unit cell is fcc, therefore, there will be four equivalent vacant sites. These are $\frac{1}{2}, \frac{1}{2}, \frac{1}{2}$ plus the permutations of $\pm\frac{a}{2}, 0, \pm\frac{a}{2}$ (Section 3–5). These include the center vacant site plus the midpoints of all 12 unit cell edges (which in turn are associated with 4 unit cells) to give $1 + \frac{12}{4} = 4$. (b) Since the coordination numbers are 8 and 4, respectively,

$$r_{U^{4+}} = 0.097 \text{ nm}/0.97 = 0.100 \text{ nm} \quad \text{and} \quad R_{O^{2-}} = 0.140 \text{ nm}/1.1 = 0.127 \text{ nm}.$$

$$r/R = 0.79.$$

Comment. A structure with $CN = 6$ (and therefore 0.097 nm/0.140 nm = 0.69) could be proposed for UO_2. However, that coordination does not permit a 1-to-2 ratio of the two atoms. ◄

8-4 MULTIPLE COMPOUNDS

$A_mB_nX_p$**-type structures**   Although the presence of three types of atoms lends additional complexity, several $A_mB_nX_p$ compounds are of sufficient interest to warrant our attention. First among these is $BaTiO_3$, the prototype for the ceramic materials used in applications such as cartridges for record players. Above 120°C, $BaTiO_3$ has a cubic unit cell with Ba^{2+} ions at the corners, O^{2-} ions at the center of the faces, and a Ti^{4+} ion at the center of the cell (Fig. 8–4.1). This structure is altered very slightly below 120°C. With the change, it becomes a useful piezoelectric material (Section 8–6).

Nonmetallic magnets may also be $A_mB_nX_p$ compounds, the most common being a ferrospinel (often called a ferrite) with the composition of MFe_2O_4, where the M are divalent cations with radii of 0.075 ± 0.01 nm. These *spinel* structures have a close-packed (fcc) array of O^{2-} ions with cations located in one half of the 6-f sites and in one eighth of the 4-f *interstitial sites* (Fig. 8–2.2). The magnetic characteristics of these materials are influenced by the cation occupants (Section 8–6).

● **Solid solutions**   Brief mention was given in Section 4–3 to solid solutions in ionic compounds. Two chief requirements were cited as necessary for solid solutions to

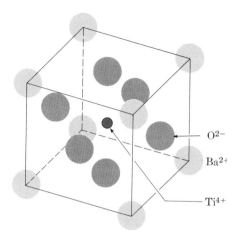

Fig. 8–4.1 Cubic $BaTiO_3$. This structure is stable above 120°C, where it has a Ti^{4+} ion in the center of the cube, Ba^{2+} ions at the corners, and O^{2-} ions at the center of each face.

occur: (1) compatibility in size and (2) balance in charge. These limitations are not as rigid as might be surmised, because compensation may be made in charge. For example, Li^+ ions may replace Mg^{2+} ions in MgO *if* F^- is simultaneously present to replace O^{2-} ions. Conversely, MgO may be dissolved in LiF. We may also find Mg^{2+} dissolved in LiF without comparable O^{2-} ions; however, in this case cation vacancies must be included. As a result, 2 Li^+ are replaced by ($Mg^{2+} + \square$).

Ceramists depend heavily on solid solutions in the previously mentioned magnetic spinels, because optimum magnetic characteristics exist when part of the divalent metal ions are zinc ($r = 0.074$ nm) and the balance of the divalent ions are ferromagnetic; for example, Ni^{2+} ($r = 0.069$ nm). In this case it is a simple matter of direct substitution; however, for certain applications it is desirable to replace $2M^{2+}$ with an Li^+Fe^{3+} pair, or to replace $2Fe^{3+}$ with an $Mg^{2+}Ti^{4+}$ pair.

Study aid (ceramic compounds)   The coordination relationships of ceramic compounds are presented with visual sketches in *Study Aids for Introductory Materials Courses*, Topic IX. This is recommended for the reader who has difficulty in viewing the 3-D patterns of ceramic compounds.

Example 8–4.1 A ferrospinel has a lattice of 32 oxygen ions, 16 ferric ions, and 8 divalent ions. (The unit cell contains 8 times as many oxygen ions as MgO does, so that the repeating pattern can be developed.) If the divalent ions are Zn^{2+} and Ni^{2+} in a 3:5 ratio, what weight fraction of ZnO, NiO, and Fe_2O_3 must be mixed for processing?

Solution: Basis: 8 mol wt of Fe_2O_3.

$$5\ NiO + 3\ ZnO + 8\ Fe_2O_3 \rightarrow (Zn_3, Ni_5)Fe_{16}O_{32}.$$

Wt fraction

$$
\begin{array}{llll}
5\ NiO = 5(58.71 + 16.00) & = & 373.5 & = 0.197 \\
3\ ZnO = 3(65.37 + 16.00) & = & 244.1 & = 0.129 \\
8\ Fe_2O_3 = 8[2(55.85) + 3(16.00)] & = & 1277.6 & = 0.674 \\
\hline
& & 1895.2 & 1.000 \quad \blacktriangleleft
\end{array}
$$

• **Example 8–4.2** A solid solution of MgO and LiF contains 11 w/o oxygen, 52 F, 20 Mg, and 17 Li. How many vacancies are present per 200 unit cells?

Solution: The atom ratios may be calculated from atomic weights.

$$
\begin{aligned}
\text{Oxygen:} &\quad 11/16 = 0.688 \\
\text{Fluorine:} &\quad 52/19 = 2.737 \\
\text{Magnesium:} &\quad 20/24.31 = 0.823 \\
\text{Lithium:} &\quad 17/6.94 = 2.450
\end{aligned}
$$

The structure is the NaCl-type; therefore, 200 unit cells have 800 anions (and 800 cation *sites*).

$$
\left.
\begin{aligned}
\text{Oxygen:} &\quad 800(0.688)/(0.688 + 2.737) = 160.7 \\
\text{Fluorine:} &\quad 800(2.737)/(0.688 + 2.737) = 639.3
\end{aligned}
\right\} \ 800 \text{ anions}
$$

Based on fluorine

$$
\left.
\begin{aligned}
\text{Magnesium:} &\quad 0.823(639.3/2.737) = 192.2 \\
\text{Lithium:} &\quad 2.450(639.3/2.737) = 572.3
\end{aligned}
\right\} \ 764.5 \text{ cations}
$$

Cation vacancies = $800 - 764.5 = 35.5/200$ unit cells.

Comment. Cation diffusion occurs much more rapidly in this solid solution than in either LiF or MgO. ◀

8–5 SILICATES

Many ceramic materials contain *silicates*, partly because silicates are plentiful and cheap and partly because they have certain distinct properties that are useful in engineering applications. Probably the most widely known silicate is portland cement, which has the very definite advantage of being able to bond rock aggregates into a monolithic material. Many other construction materials, such as brick, tile, glass (Fig. 8–1.1), and vitreous enamel, are also made of silicates. Other engineering applications of silicates include electrical insulators, chemical ware, and reinforcing glass fibers.

(a)   (b)

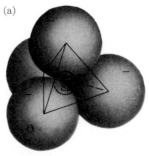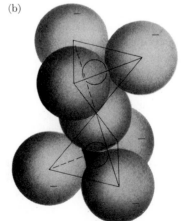

Fig. 8–5.1 (a) Tetrahedral arrangement of SiO_4^{4-}. Compare with Fig. 2–5.3(b). This SiO_4^{4-} ion has obtained four electrons from other sources. (b) Double tetrahedral unit $(Si_2O_7)^{6-}$. The center oxygen atom is shared by each tetrahedral unit. Thus, it becomes a bridging oxygen.

Silicate tetrahedral units   The primary structural unit of silicates is the "SiO_4" *tetrahedron* (Fig. 8–5.1a), in which one silicon atom is coordinated interstitially among four oxygen atoms. The forces holding these tetrahedra together involve both ionic and covalent bonds; consequently the tetrahedra are tightly bonded. However, with either the ionic or covalent bonding mechanism, each oxygen of the tetrahedra has only seven electrons rather than eight in its outer shell.

Two methods are available to overcome this deficiency of electrons for the oxygen ions: (1) An electron may be obtained from metal atoms. In this case SiO_4^{4-} ions and Metal$^+$ ions are developed. (2) Each oxygen may share an electron pair with a second silicon. In this case multiple tetrahedral coordination groups are formed (Fig. 8–5.1b). The shared oxygen is the *bridging* oxygen.

Network silicates   With pure SiO_2 there are no metal ions, and every oxygen atom is a *bridging* atom between two silicon atoms (and every silicon atom is among four oxygen atoms, as shown in Fig. 8–5.2). This gives a network-like structure. Silica (SiO_2) can have several different crystal structures, just as carbon can be in the form of graphite or diamond. The structure shown in Fig. 8–5.2 is a high-temperature form. A more common structure of silica is quartz, the predominant material found in the sands of many beaches. Just as SiO_2 in Fig. 8–5.2 contains SiO_4 tetrahedra, so does quartz, but with a more complex lattice.

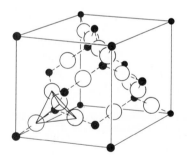

Fig. 8–5.2   Network structure (SiO_2). Each silicon atom is surrounded by four oxygen atoms, and each oxygen atom is part of two SiO_4 tetrahedra. (Cf. Figs. 8–2.4 and 5–4.3.) This polymorph of silica is called cristobalite.

Another natural silicate is feldspar. The pink or tan mineral of granite is $KAlSi_3O_8$, which may be visualized as a network silicate with one of every four silicons replaced by an Al^{3+} ion. The latter, however, has only three charges as compared to four for silicon. Thus K$^+$ is present to balance out the charges. The K$^+$ ions are interstitial ions. However, these network structures are quite open, so that there is space for extra ions to be present (cf. Fig. 8–5.2).

Glasses   The principal commercial glasses are silicates. They have the SiO_4 tetrahedra described in the preceding section, plus some modifying ions. They are *amorphous*, i.e., noncrystalline. Figure 8–5.3 contrasts the structure of a crystal with that of an amorphous solid of the same composition. Atoms in both have the same first neighbors, i.e., each oxygen atom is between two silicon atoms and *bridges* two tetrahedra. Each silicon is among four oxygens (the fourth is envisioned as being

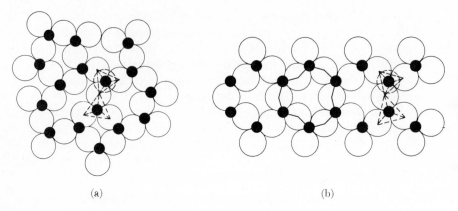

(a)   (b)

Fig. 8–5.3 Two-dimensional representations of (a) silica glass and (b) crystalline silica at room temperature. Each has a short-range network structure. Only the crystalline silica has a long-range network order. (The fourth oxygen above or below the silicon is not shown.) Oxygen serves as a bridge between adjacent tetrahedra.

above *or* below the plane of the paper). However, in glass, differences arise in more-distant neighbors because the glass, unlike the crystalline silica, does not have a regular long-range pattern.

When only silica (SiO_2) is present, and every oxygen atom serves as a bridge, the glass is very rigid. Fused silica, for example, is extremely viscous even when it gets into the temperature range in which it is a true liquid. In polymer terms, its units are polyfunctional and its *network* structure is highly cross-linked. Fused silica is very useful in some applications because it has a low thermal expansion. However, its high viscosity makes it extremely difficult to shape.

Network modifiers Most silicate glasses contain *network modifiers*. These are oxides such as CaO and Na_2O that supply cations (positive ions) to the structure. As shown in Fig. 8–5.4, the addition of Na_2O to a glass introduces two Na^+ ions, and produces two *nonbridging oxygens*, each with a single negative charge. These oxygens are attached to only one silicon. Likewise, the introduction of one Ca^{2+} ion also leads to two nonbridging oxygens. The presence of these nonbridging oxygens reduces the activation energy required for the atom in movements that permit flow in molten glass. This is shown in Fig. 8–5.5 for glasses with increasing Na_2O content (and, therefore, an increasing fraction of nonbridging oxygens).

In commercial glasses that contain soda and lime, less than three fourths of the oxygen atoms serve as a bridge between SiO_4 tetrahedra. Thus, some of the units are directly connected to only two other tetrahedra. As a result, the glass is thermoplastic at elevated temperatures and can be shaped into products such as light bulbs, window glass, and fibers that become rigid during cooling.

There are thousands of commercial glasses, because there are more than a dozen possible components that can be used in varying amounts to produce specific properties, such as index of refraction, color dispersion, or viscosity. The more prominent categories of glasses are shown in Table 8–5.1.

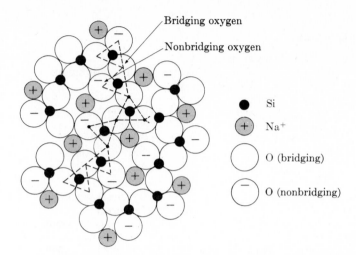

Bridging oxygen

Nonbridging oxygen

- Si
- Na⁺ ($+$)
- O (bridging)
- O (nonbridging)

Fig. 8–5.4 Structure of soda-silica glass. The addition of Na_2O to a silica glass decreases the number of bridging oxygen atoms. Each nonbridging oxygen carries a single negative charge. (As in Fig. 8–5.3, the fourth oxygen of each tetrahedron has been removed for clarity.) This glass is less viscous at high temperatures than the silica glass of Fig. 8–5.3(a).

Fig. 8–5.5 Energy for viscous flow versus bridging oxygens (Na_2O-SiO_2 glasses). The activation energy required for viscous flow increases as a larger percentage of the oxygen atoms share adjacent tetrahedra. (Adapted from Stevels, *Encylopedia of Physics*.)

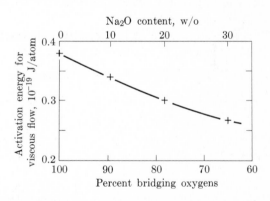

Example 8–5.1 Quartz (SiO_2) has a density of 2.65 Mg/m³ ($=2.65$ g/cm³). (a) How many silicon atoms (and oxygen atoms) are there per m³? (b) What is the packing factor, given that the radii of silicon and oxygen are 0.038 nm and 0.114 nm, respectively?

Solution

a)
$$SiO_2/m^3 = \frac{2.65 \times 10^6 \text{ g/m}^3}{(28.1 + 32.0) \text{ g/}(0.6 \times 10^{24} \text{ SiO}_2)}$$

$$= 2.645 \times 10^{28} \text{ SiO}_2/m^3$$
$$= 2.645 \times 10^{28} \text{ Si/m}^3 = 5.29 \times 10^{28} \text{ O/m}^3.$$

b)
$$V_{Si}/m^3 = (2.645 \times 10^{28}/m^3)(4\pi/3)(0.038 \times 10^{-9} \text{ m})^3 = 0.006 \text{ m}^3/m^3$$
$$V_o/m^3 = (5.29 \times 10^{28}/m^3)(4\pi/3)(0.114 \times 10^{-9} \text{ m})^3 = 0.328 \text{ m}^3/m^3$$
$$\text{Packing factor} = \overline{0.33}.$$

Table 8–5.1
Common commercial glass types

Type	Major components, weight percent						Requirements
	SiO_2	Na_2O	CaO	Al_2O_3	B_2O_3	MgO	
Window	72	14	10	1		2	High durability
Plate (Arch.)	73	13	13	1			High durability
Container	74	15	5	1		4	Easy workability, chemical resistant
Lamp bulbs	74	16	5	1		2	Easy workability
Fiber (Elect.)	54		16	14	10	4	Low alkali
Pyrex	81	4		2	12		Low thermal expansion, low ion exchange
Fused silica	99						Very low thermal expansion

Comments. Although there is considerable open space within this structure, most single atoms (except for helium) must diffuse through SiO_2 as ions. Thus their charges prohibit measurable movements at ambient temperatures. ◀

Example 8–5.2 A glass contains 80 w/o SiO_2 and 20 w/o Na_2O. What fraction of the oxygens is nonbridging?

Solution: Basis: 100 g = 80 g SiO_2 + 20 g Na_2O.

$$80 \text{ g } SiO_2/[(28.1 + (2 \times 16.0)) \text{ g/mole}] = 1.33 \text{ mole } SiO_2 \quad \text{or} \quad 80.6 \text{ m/o } SiO_2;$$

$$20 \text{ g } Na_2O/[((2 \times 23.0) + 16.0) \text{ g/mole}] = 0.32 \text{ mole } Na_2O \quad \text{or} \quad 19.4 \text{ m/o } Na_2O.$$

New basis: 100 moles.

$$80.6 \text{ } SiO_2 = 80.6 \text{ Si} + 161.2 \text{ O}$$

$$19.4 \text{ } Na_2O = \frac{19.4 \text{ O} + 38.8 \text{ Na}^+}{80.6 \text{ Si} + 180.6 \text{ O} + 38.8 \text{ Na}^+}$$

Note (from Fig. 8–5.4) that there is one nonbridging oxygen for each Na^+ added. Thus, there are 38.8 nonbridging oxygens (and 141.8 bridging oxygens):

$$\text{Fraction nonbridging oxygens} = 38.8/180.6 = 0.215.$$

Comments. A commercial soda-lime glass used for containers, etc. (Table 8–5.1), also has CaO additions. Each Ca^{2+} ion produces two nonbridging oxygens, both of which carry a single negative charge. (Cf. Fig. 8–5.4.) ◀

• **Example 8–5.3** Refer to Fig. 8–5.2 and compare with Figs. 2–3.3(a) and 2–6.1(b). The O–Si–O bond angle is 109.5°. (a) What is the closest approach (center-to-center) of the oxygens? (b) Of the silicons? (With CN = 2, $R_{O^{2-}} \approx 0.114$ nm.)

Solution

a) $$\frac{[\text{O-to-O}]/2}{r + R} = \sin(109.5°/2) = 0.82;$$

$$\text{O-to-O} = 2[(0.042 \text{ nm}/1.1) + 0.114 \text{ nm}][0.82] = 0.25 \text{ nm}.$$

b)   Assume the Si–O–Si bond angle is 180° (Fig. 8–5.2).

$$\text{Si-to-Si} = 2[(0.042 \text{ nm}/1.1) + 0.114 \text{ nm}] = 0.30 \text{ nm}.$$

Comment. In some silica polymorphs, the Si–O–Si bond bends to less than 180° to accommodate the lone-pair electrons. ◄

8–6   ELECTROMAGNETIC BEHAVIOR OF CERAMICS

Ceramics are well known for their use as electrical insulators (Fig. 8–6.1). In addition, various ceramics may serve functional roles in electromagnetic circuits as a result of their high dielectric constants, their piezoelectric responses, or their dielectric behavior. Finally, oxides of most transition elements are semiconductors.

Fig. 8–6.1   Surface breakdown. Adsorbed moisture and contaminants provide a surface path for electrical shorting. (R. Russell, *Brick and Clay Record.*)

Ceramic dielectrics   Many oxides serve well as insulators because the valence electrons of the metal atoms are transferred permanently to the oxygen atoms, forming O^{2-} ions. Let us refer once again to the sparkplug insulator of Fig. 2–7.3. The Al^{3+} ions of Al_2O_3 have been stripped of the valence electrons that would carry charge in metallic aluminum. Those electrons are now held firmly by the oxygen ions. In other insulating materials, Mg^{2+} ions lose their electrons to O^{2-} ions in MgO; and silicon and oxygen rigidly share electrons within the SiO_4 tetrahedron (Fig. 8–5.1). As a result, compositions of $MgO-Al_2O_3-SiO_2$ form some of our best insulators.

However, materials that are commonly considered to be insulators can break down under very high electrical voltages. Usually the breakdown is a *surface* phenomenon. For example, the spark plugs of an automobile may short out on a damp morning because condensed moisture on the surface of the ceramic insulators permits the current to short-circuit the spark gap. Insulators are designed with lengthened surface paths (see Fig. 8–6.1) to decrease the possibility of surface shorting, and since internal pores and cracks provide opportunity for additional "surface" failure, the insulators are usually glazed to make them nonabsorbent. *Volume breakdown* occurs only when extremely high voltage gradients are encountered. A very strong electric field can be sufficient to disrupt the induced dipoles in the insulator, and discharge occurs when the strength of the field exceeds the strength of the bonds.

Dielectric properties are listed in Table 8–6.1 for selected ceramic insulators. The *dielectric strength* is the voltage gradient that produces electrical breakdown *through* the insulator. The *relative dielectric constant* κ may be compared with similar data for plastics in Table 7–6.1. In general, these values are slightly higher in ceramic materials, since ions, rather than molecular dipoles, respond to the electric field. As with polymeric insulators, the dielectric constant is sensitive to frequency. However, in the normal temperature range, there is less variation among ceramic insulators. The electrical engineer also considers the *loss factor*, tan δ, when designing electrical circuits. The product of the relative dielectric constant and the loss factor, κ tan δ,

Table 8–6.1
Properties of ceramic dielectrics

Material	Resistivity (volume), ohm·m	Dielectric strength, volts/mm	Relative dielectric constant, κ		Loss factor, tan δ	
			60 Hz	10^6 Hz	60 Hz	10^6 Hz
Electrical porcelain	10^{11}–10^{13}	2–8 × 10^4	6	—	0.010	—
Steatite insulators	>10^{12}	8–15 × 10^4	6	6	0.005	0.003
Zircon insulators	~10^{13}	10–15 × 10^4	9	8	0.035	0.001
Alumina insulators	>10^{12}	10^4	—	9	—	<0.0005
Soda-lime glass	10^{12}	10^4	7	7	0.1	0.01
E-glass	>10^{15}	—	—	4	—	0.0006
Fused silica (SiO_2)	~10^{18}	10^4	4	3.8	0.001	0.0001

is a measure of the electrical energy consumed by a capacitor in an a.c. circuit. Thus either alumina or the glasses designed for electrical purposes (E-glasses) are selected for use in the megahertz range.*

Ceramic semiconductors   Although ceramic compounds are nominally insulators, they may become semiconductors if they contain multivalent transition elements. This was illustrated in Fig. 5–5.4(a), where electron holes carry charges by moving from one nickel ion to another. Magnetite (Fe_3O_4 or $Fe^{2+}Fe_2^{3+}O_4$) is a ceramic semiconductor with a resistivity of 10^{-2} ohm·cm, which is comparable to graphite and tin (Fig. 5–1.1). The origin of the conductivity is identical to that of NiO in Fig. 5–5.4(a); however, the number of electron holes for carrying the charge is much higher because the fraction of Fe^{3+} ions is greater.

The resistivity can be increased by solid solutions in which the multivalent iron ions are replaced by other ions. This is shown in Table 8–6.2 for solid solutions of $MgCr_2O_4$ and $FeFe_2O_4$. Neither Mg^{2+} nor Cr^{3+} ions can react with the electrons or electron holes. Thus the resistivity can be adjusted to selected levels. The temperature coefficient of resistivity of these semiconductors is equally interesting to the engineer. As noted in Table 8–6.2, the resistance change is more than 1%/°C, and in other solid solutions may be as high as 4%/°C. This sensitivity is sufficient for accurate temperature measurement and has led to devices called *thermistors* that are used for thermometric purposes. Because thermistors usually have a negative temperature coefficient of resistivity, they may also be used to compensate for positive resistance changes in the metallic components of a circuit.

Table 8–6.2
Resistivities of ceramic semiconductors*

Composition, mole percent		Resistivity, ohm·m		$\Delta\rho/\Delta T$
$FeFe_2O_4$	$MgCr_2O_4$	25°C	60°C	%/°C
100	0	0.00005	0.000045	−0.3
75	25	0.007	0.0045	−1.0
50	50	1.8	0.75	−1.6
25	75	420.	120.	−2.0
0	100	$>10^{10}$	$>10^{10}$	—

* Adapted from E. J. Verwey, P. W. Haayman, and F. C. Romeijn, *J. Chem. Phys.*

Piezoelectric ceramics   Certain ceramic crystals lack a center of symmetry. We will not consider their crystallographic features but will note an important consequence when ionic crystals are involved. The centers of positive and negative charges are not identical. As a result each unit cell acts as a small electric dipole with a positive and

* Fused silica could be very desirable—except it is difficult to process—because of its high viscosity (Section 8–5).

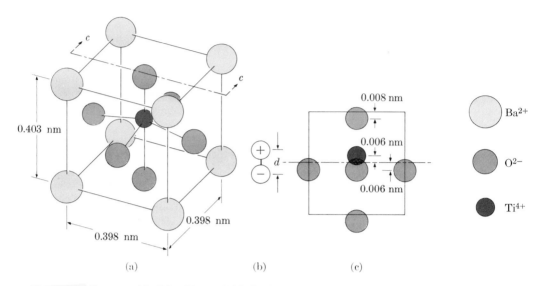

0.008 nm

0.006 nm

0.006 nm

0.403 nm

0.398 nm

0.398 nm

d

+

−

(a)

(b)

(c)

Ba^{2+}

O^{2-}

Ti^{4+}

Fig. 8–6.2   Tetragonal BaTiO$_3$. Above 120°C (250°F), BaTiO$_3$ is cubic (Fig. 8–4.1).
Below that temperature, the ions shift with respect to the corner Ba^{2+} ions. Since the Ti^{4+}
and the O^{2-} ions shift in opposite directions, the centers of positive and negative charges
are not identical. The unit cell becomes noncubic.

negative end. This is illustrated by BaTiO$_3$. Recall from Fig. 8–4.1 that BaTiO$_3$ is
cubic—above 120°C. Below that temperature, called the *Curie point*, there is a slight
but important shift in the ions. The central Ti^{4+} ion shifts about 0.006 nm with respect
to the corner Ba^{2+} ions. The O^{2-} ions shift in the opposite direction, as indicated in
Fig. 8–6.2.* The center of positive charge and the center of negative charge are
separated by the dipole length d.

A material such as BaTiO$_3$ changes its dimensions in an electric field because
the negative charges will be pulled toward the positive electrode, and the positive
charges will be pulled toward the negative electrode, thus increasing the dipole
length, d. This also increases the dipole moment Qd, and the polarization $\mathscr{P}$, since
the latter is the total of the dipole moments $\sum Qd$ per unit volume V (Eq. 7–6.1).

The above sequence of effects provides a means of changing mechanical energy
into electrical energy and vice versa. To understand this, refer to Fig. 8–6.3(a). The
cooperative alignment of the dipole moments of the many unit cells gives a polarization
that collects positive charges at one end of the crystal and negative charges at the
other end. Now consider two alternatives as shown in Figs. 8–6.3(b) and 8–6.3(d).
(1) Compress (or pull) the crystal with a stress s. There is a strain e, which is dictated
by the elastic modulus. This strain changes the dipole length d and directly affects
the polarization ($=\sum Qd/V$) since Q and V remain essentially constant. With a smaller

---

*   The shift of the Ti^{4+} in Fig. 8–6.2 is upward as drawn. Actually, the shift could be in any one of the six
coordinate directions. In any event, the O^{2-} ions shift in the opposite direction.

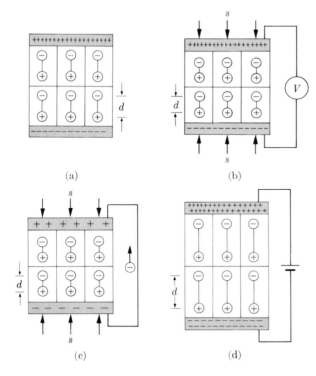

Fig. 8–6.3 Piezoelectric material. An electric field induces dimensional changes, shown in part (d). Conversely, strain from pressure induces a voltage change (b).

polarization (from compression), there is excess of charge density on the two ends of the crystal. If the two ends are isolated, a voltage differential develops (Fig. 8–6.3b). If in electrical contact, electrons will flow from one end to the other (Fig. 8–6.3c). (2) No pressure is applied in Fig. 8–6.3(d); rather a voltage is applied, which increases the charge density at the two ends. The negative charges within the $BaTiO_3$ are pulled one way and the positive charges the opposite, thus changing not only the dipole length d, but also that dimension of the crystal.

These two situations indicate how mechanical forces and dimensions can be interchanged with electrical charges and/or voltages. Devices with these capabilities are called *transducers*. Materials with the above characteristics are called *piezoelectric*, i.e., pressure-electric. Crystals of the $BaTiO_3$ type* are used for pressure gages, for phonograph cartridges, and for high-frequency sound generators. Let us examine a ceramic phonograph cartridge as an example. The stylus, or needle, follows the groove on the record. A small transducer is in contact with the stylus, which detects the vibrational pattern recorded in the groove. Both the frequency and the amplitude can be sensed as voltage changes (Fig. 8–6.3b). Although the voltage signal is small, it can be amplified through electronic circuitry until it is capable of driving a speaker.

---

* The most widely used piezoelectric ceramics are $PbZrO_3$–$PbTiO_3$ solid solutions called PZT's. They have the structure of $BaTiO_3$ in Fig. 8–6.2. However, their Curie points are higher than that of $BaTiO_3$, and they have a greater range of design applications.

Fig. 8–6.4 Piezoelectric crystals (quartz). These crystals were made artificially. They will be cut into wafers to be used as resonators for frequency control. As the quartz vibrates elastically, the electric field responds at the same frequency, and vice versa. (Courtesy of Western Electric Co.)

Quartz crystals (SiO_2) are also piezoelectric. They are produced (Fig. 8–6.4) for special applications in circuits requiring frequency control. Once cut to a selected geometry, the elastic vibrational frequency is constant to one part in 10^8! Thus, in resonance, the crystal can control the frequency of an electronic circuit to that same accuracy. Fine watch circuits and control circuits for radio broadcasting are but two applications which utilize quartz crystals for their piezoelectric characteristics.

• **Magnetic ceramics** Metallic magnets possess a major disadvantage in UHF and VHF circuits. The high frequency induces currents and therefore introduces power losses within the metallic cores. In fact the metallurgist capitalizes on this to melt high-quality alloys. Such power loss into heat cannot be tolerated in an electronic circuit.* Fortunately, ceramic compounds containing magnetic atoms can be magnetic, *and* nonconducting.

If we consider the individual iron atom (Eq. 2–1.1), we observe that there are six $3d$ electrons. The same is true for a ferrous (Fe^{2+}) ion (Eq. 2–1.2). A ferric (Fe^{3+}) ion has five $3d$ electrons (Eq. 2–1.3). These electrons have electron spin alignments in their five $3d$ orbitals as follows:

$$Fe^{2+}: \quad \underline{\uparrow\downarrow} \quad \underline{\uparrow} \quad \underline{\uparrow} \quad \underline{\uparrow} \quad \underline{\uparrow}; \tag{8–6.1}$$

$$Fe^{3+}: \quad \underline{\uparrow} \quad \underline{\uparrow} \quad \underline{\uparrow} \quad \underline{\uparrow} \quad \underline{\uparrow}. \tag{8–6.2}$$

---

* The electrical engineer overcomes this trouble in 60-Hz circuits by using laminates rather than solid cores in transformers, armatures, etc. This increases the effective resistance. In a high-frequency circuit, however, the laminates would have to be microscopically thin.

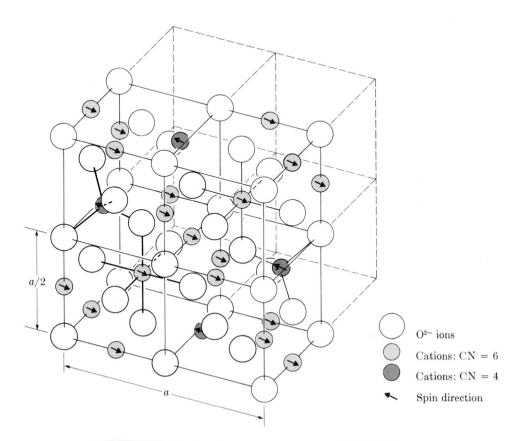

Fig. 8–6.5 Ferrospinel structure. The O^{2-} ions form an fcc array, with cations in the indicated interstitial sites. (Only the front half of the unit cell is sketched.) The cell dimensions are double those of the AX structures (Fig. 8–2.2). The magnetic moment of these cations in 6-f sites and those in 4-f sites have opposing alignments. Since the magnetic moments of the two spin directions do not balance, each unit cell has a net magnetic moment. (See the text for elaboration.)

This means that each Fe^{2+} has a magnetic imbalance of four unpaired electron spins (called four *Bohr magnetons*). Since the magnetic moment of each electron spin is 9.27×10^{-24} amp·m^2, each Fe^{2+} ion has a total magnetic moment of 37×10^{-24} amp·m^2. Each Fe^{3+} ion has an imbalance of five Bohr magnetons and magnetic moment of 46.3×10^{-24} amp·m^2.

Many ceramic magnetic phases fall in the $A_m B_n X_p$ category of Section 8–4. Examples are $Ni_8 Fe_{16} O_{32}$ and magnetite ($Fe_3 O_4$, or $Fe_8^{2+} Fe_{16}^{3+} O_{32}$). Their unit cell is shown in Fig. 8–6.5. The structure is complex, but it may be simplified by noting that the O^{2-} ions assume an fcc arrangement. The unit cell has dimensions that are double of what we considered previously in Figs. 3–1.1 and 8–2.2. As a result, there are 32 oxygen ions, 16 ferric ions, and 8 ferrous ions in each unit cell of $[Fe^{2+} Fe_2^{3+} O_4]_8$. The iron cations reside in both 4-fold and 6-fold interstitial sites

(Fig. 8–2.2). Those in the 4-f sites (CN = 4) point their magnetic moment in one $\langle 111 \rangle$ direction; those in 6-f sites (CN = 6) point their magnetic moment in the *opposite* $\langle 111 \rangle$ direction. Detailed analyses reveal that the 16 Fe^{3+} ions per unit cell are almost equally divided between 4-f and 6-f sites. Therefore there is no *net* magnetic moment from the Fe^{3+} ions:

$$\text{(8 } Fe^{3+} \text{ ions)}(5\uparrow \text{ per ion}) + \text{(8 } Fe^{3+} \text{ ions)}(5\downarrow \text{ per ion}) = 0. \qquad (8\text{–}6.3)$$

However, almost all of the Fe^{2+} ions reside in the 6-f sites. Since each of the 8 Fe^{2+} ions has four Bohr magnetons β of 9.27×10^{-24} amp·m² each, the unit cell has a magnetic moment of 32β, or about 0.3×10^{-21} amp·m². Inasmuch as the unit cell has a lattice constant of 0.837 nm, and *saturation magnetization* M_s is equal to the maximum magnetic moment per unit volume, we can calculate the saturation magnetization of Fe_3O_4 as:

$$M_s = (0.3 \times 10^{-21} \text{ amp·m}^2)/(0.837 \times 10^{-9} \text{ m})^3 = 0.5 \times 10^6 \text{ amp/m}.$$

This compares with 0.53×10^6 amp/m by experimental measurement.

- **Soft and hard magnets**   Most of the magnetization of Fe_3O_4 disappears as soon as the magnetic field is removed. This disappearance occurs because the cooperative magnetic orientations of the individual atoms are not retained. This type of "soft" behavior is desired in a transformer or in a magnetic yoke for a TV tube, because it lets the magnetization respond to an alternating current—be it a 60 Hz in a power line, or in the MHz range of a TV set.

    The magnetization is retained in a "hard" or permanent magnet after the magnetic field has been removed. There are numerous situations where this retention is required. These include not only the familiar small permanent magnets we use in shop and home but also parts for millions of electric motors, the magnetic fillers that the engineer adds to the door gasket of a refrigerator, and the coatings on recording tapes for audio- and video-cassettes, to name but a few.

    The concept of a domain is required to understand the difference between "soft" and "hard" magnets. A *domain* is a part of a crystal in which all of the magnetic moments are cooperatively aligned. That is, this small volume within a grain is a small magnet with its own magnetic orientation. The favorably oriented domains grow and the others shrink when a magnetic field is applied. The growth is by a movement of the domain wall. In a "soft" magnet, the walls move back after the field is removed to randomize the magnetic orientations. No *net* magnetization remains. In a "hard" magnet, it is necessary to apply a reverse magnetic field to introduce oppositely oriented domains.* As before, any change in orientation results from movements of the walls between adjacent domains.

    Structural features determine the amount of reverse field that is necessary to move the domain walls. They cannot move across a grain boundary. Their movements are restrained by dislocations and other imperfections. Thus a strain-hardened steel

---

* The *coercive force* H_c is the amount of reverse field required to return the net magnetization to zero.

is also a magnetically "hard" steel, i.e., a permanent magnet unless exposed to a strong coercive field H_c. An Alnico magnet contains a very fine "two-phase" micro-structure. The domain walls cannot move across the "phase" boundaries; hence, an Alnico magnet is permanent. A magnet that contains extremely fine particles of iron is permanent, because each minute particle is an individual domain; therefore, a reverse field must be strong enough to nucleate new domains within the particle rather than just move boundaries.

Example 8–6.1   The surface of silicon is oxidized to give a 25-μm silica (SiO_2) glass surface layer, and is then coated with aluminum. How does the capacitance of this layer compare at 10^6 Hz with that produced when a 2-mil (50-μm) film of polyethylene is vapor-coated with aluminum?

Solution: Using Eq. (1–4.7), and data from Tables 7–6.1 and 8–6.1,

$$\frac{C_{PE}}{C_{SiO_2}} = \frac{2.3(8.85 \times 10^{-12} \text{ C/V·m})A/(50 \times 10^{-6} \text{ m})}{3.8(8.85 \times 10^{-12} \text{ C/V·m})A/(25 \times 10^{-6} \text{ m})}$$

$$= 0.3. \quad \blacktriangleleft$$

Example 8–6.2   Nickel oxide (Fig. 5–5.4a) normally has only a limited number of Ni^{3+} ions ($Ni^{3+}/Ni^{2+} < 10^{-4}$). When Li_2O is added to NiO, more Ni^{2+} ions oxidize to Ni^{3+}, and an Li^+Ni^{3+} pair replaces two previous Ni^{2+} ions. With 1 wt % Li_2O, how many charge carriers are there per m^3? (NiO has the same structure as NaCl.)

Solution: Use 100 amu as a basis.

$$Li_2O \quad\quad 1 \text{ amu}/(13.88 + 16) = 0.0335 \quad \text{(or } 0.067 \text{ Li}^+\text{)}$$
$$NiO \quad\quad \underline{99 \text{ amu}/(58.71 + 16) = 1.3251}$$
$$\text{Oxygen ions}/100 \text{ amu} = 1.36.$$

With the NaCl-type structure, NiO has four oxygen ions/unit cell. Therefore,

$$\text{unit cells}/100 \text{ amu} = 1.36/4 = 0.34.$$

Since there are 0.067 Li^+/100 amu, there will be 0.067 Ni^{3+}, and hence 0.067 *p*-type carriers/100 amu:

$$\frac{(0.067 \text{ carriers}/100 \text{ amu})}{(0.34 \text{ unit cells}/100 \text{ amu})} = 0.197 \text{ carriers/unit cell},$$

and since $r = \sim 0.069$ nm,

$$(0.197 \text{ carriers})/[2(0.069 + 0.140)10^{-9} \text{ m}]^3 = 2.7 \times 10^{27}/\text{m}^3.$$

Comment.   One may visualize the conductivity of NiO occurring by electrons *hopping* from one nickel ion to another. Thus, as an electron hole moves from an Ni^{3+} to an Ni^{2+}, the latter atom becomes Ni^{3+} and the former Ni^{2+}. $\blacktriangleleft$

Example 8–6.3   Calculate the polarization $\mathscr{P}$ of $BaTiO_3$, based on Fig. 8–6.2 and the information of Section 7–6.

Solution: With reference to the *center* of a cell cornered on the Ba_2^+ ions:

Ion		Q, coul	d, m	Qd, coul·m
Ba^{2+}		$+2(0.16 \times 10^{-18})$	0	0
Ti^{4+}		$+4(0.16 \times 10^{-18})$	$+0.06(10^{-10})$	3.84×10^{-30}
$2\,O^{2-}$	(side of cell)	$-4(0.16 \times 10^{-18})$	$-0.06(10^{-10})$	3.84×10^{-30}
O^{2-}	(top and bottom)	$-2(0.16 \times 10^{-18})$	$-0.08(10^{-10})$	2.56×10^{-30}
				$\sum = 10.24 \times 10^{-30}$

From Eq. (7-6.1),

$$\mathscr{P} = \sum Qd/V$$
$$= (10.24 \times 10^{-30} \text{ C·m})/(4.03 \times 3.98^2 \times 10^{-30} \text{ m}^3)$$
$$= 0.16 \text{ coul/m}^2.$$

Comment. This means that polarized $BaTiO_3$ can possess a charge density of 0.16 coul/m², equivalent to 10^{12} electrons/mm². ◀

● **Example 8-6.4** By substituting $(4\,Li^+ + 4\,Fe^{3+})$ for $8\,Fe^{2+}$, magnetite is altered from (a) $Fe_8^{2+}Fe_{16}^{3+}O_{32}$ to (b) $Li_4^+ Fe_{20}^{3+}O_{32}$. Assume that there is negligible change in unit-cell size and that the Li^+ ions enter the 4-f sites and 6-f sites equally. What is the percent change in saturation magnetization?

Solution: The ferrospinel structure has 16 cations in 6-f sites and 8 cations in 4-f sites.

	6-f↑	4-f↓	fcc array
Mag:	$8\,Fe^{3+} + 8\,Fe^{2+}$	$8\,Fe^{3+}$	$32\,O^{2-}$
LiMag:	$14\,Fe^{3+} + 2\,Li^+$	$6\,Fe^{3+} + 2\,Li^+$	$32\,O^{2-}$

(O^{2-} and Li^+ are nonmagnetic; Fe^{2+} has 4β/ion; Fe^{3+} has 5β/ion)

$$\beta_{\text{Mag}} = (8)(5\uparrow) + (8)(4\uparrow) + (8)(5\downarrow) = 32 \ \beta/\text{unit cell};$$
$$\beta_{\text{LiMag}} = (14)(5\uparrow) \qquad + (6)(5\downarrow) = 40 \ \beta/\text{unit cell}.$$

With the volume of the unit cell as V,

$$\Delta M_s = [(40 - 32)(9.27 \times 10^{-24})/V]/[32(9.27 \times 10^{-24})/V] = +25\% \ ◀$$

8-7   MECHANICAL BEHAVIOR OF CERAMICS

With the exception of a few materials such as plasticized clay, ceramic materials are characterized by high shear strengths; thus, they are not ductile. We will see that this leads to high hardnesses and high compressive strengths, combined with notch sensitivity and low fracture strengths.

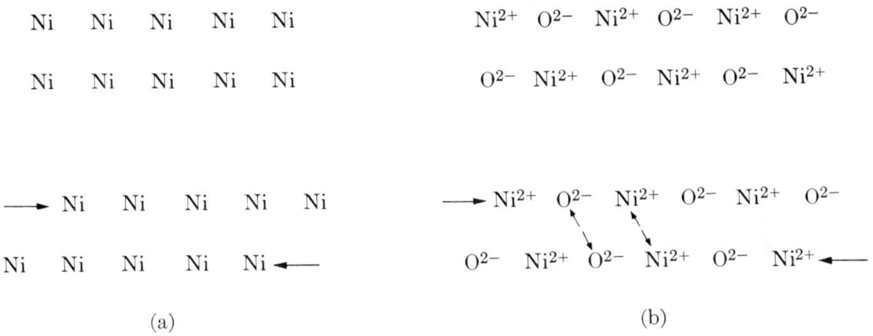

```
Ni    Ni    Ni    Ni    Ni              Ni²⁺  O²⁻  Ni²⁺  O²⁻  Ni²⁺  O²⁻

Ni    Ni    Ni    Ni    Ni              O²⁻  Ni²⁺  O²⁻  Ni²⁺  O²⁻  Ni²⁺



——▶ Ni    Ni    Ni    Ni    Ni         ——▶ Ni²⁺  O²⁻  Ni²⁺  O²⁻  Ni²⁺  O²⁻

Ni    Ni    Ni    Ni    Ni ◀——          O²⁻  Ni²⁺  O²⁻  Ni²⁺  O²⁻  Ni²⁺ ◀——

         (a)                                        (b)
```

Fig. 8–7.1 Comparison of slip processes (metallic nickel and nickel oxide). More force is required to displace the ions in NiO than the atoms in nickel. The strong repulsive forces between like ions becomes significant. Nickel also has more slip systems than does nickel oxide.

Compounds have ordered arrangements of dissimilar atoms (or ions). This was discussed for intermetallic compounds as an optional topic in Section 6–4. In ceramic phases, cations are coordinated with anions; and anions are coordinated with cations (Section 8–2). This is natural, since unlike ions attract and like ions repel. However this interferes with slip, as shown in Fig. 8–7.1. The slip process in the horizontal direction (as drawn) is precluded in NiO, MgO, and other NaCl-type crystals. The displaced arrangement would have to be achieved by first breaking strong positive-to-negative bonds of the Ni^{2+} and O^{2-} ions. Next, like charges would have to pass adjacently as the Ni^{2+} ion moves to join the next O^{2-} ions. This consideration does not exist in the metal.

In NiO, slip is possible in certain other directions without the above restriction, e.g., at a 45° direction. This would be slip in a $\langle 1\bar{1}0 \rangle$ direction on a $\{110\}$ plane; however, there are only six slip systems possible. This compares with twelve slip systems for $\{111\}\langle 1\bar{1}0 \rangle$ slip in fcc nickel (Table 6–4.1). Furthermore the slip vector **b** in NiO is $(r + R)\sqrt{2}$, or 0.3 nm, which is greater than the 0.25 nm in nickel. Everything considered—(1) coulombic repulsion, (2) fewer slip systems, and (3) longer displacement distances*—adds up to a much greater resistance to plastic deformation in crystalline ceramics than in their metallic counterparts. This difference is accentuated even further when more complex ceramic phases are considered, e.g., silicates, the spinel of Fig. 8–6.5, or portland cement.†

Hardness Ceramic phases are hard because they generally cannot undergo plastic deformation. As a result, abrasive materials such as emery consist of Al_2O_3 (usually with some Fe_2O_3 and/or TiO_2 in solid solution). Silicon carbide (SiC) and TiC are equally important for grinding and cutting metals, and similar manufacturing

---

* The energy for dislocation movements increases with the square of the slip vector, **b** (Eq. 6–4.2).
† Some chemists consider the chief hydrated phase of portland cement to be

$$Ca_2[SiO_2(OH)_2]_2 \cdot (CaO)_{y-1} \cdot xH_2O!!$$

(See Section 13–2.)

processes. One form of SiC has the ZnS-type structure (Fig. 8–2.4); TiC has the NaCl-type structure (Fig. 8–2.3).

Exceptions to the general rule that ceramics are relatively hard can be related directly to structures. Talc, clays, and mica are soft because they have sheet structures. They are strong within the sheets; however, the sheets are held together by weak secondary bonds (Section 2–4).

Notch sensitivity   A notch or crack is a *stress-raiser*. The effect is much greater than would be anticipated simply on the basis of the decreased cross-sectional area. The true stress, σ, at the tip of the notch or flaw (Fig. 8–7.2) is related to the nominal tensile stress, s, the depth of the crack, c, and the radius of curvature, r, at the tip of the notch by

$$\sigma = s2\sqrt{c/r}. \qquad (8\text{–}7.1)$$

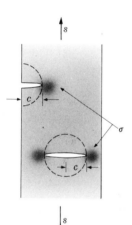

Fig. 8–7.2   Stress concentration. The concentration factor depends upon the crack depth and the radius of curvature at the tip of the crack (Eq. 8–7.1). This example is for a flat plate.

If σ exceeds the yield strength in a *ductile* material, the tip of the notch will deform to a larger radius, and the stress concentration will be reduced. If the notch is a crack in a *nonductile* material, the radius of curvature may be of atomic dimensions (say 0.1 nm). Thus a crack that reaches only 0.1 μm to 10 μm (10^{-4} mm to 10^{-2} mm) into the surface will give a *stress concentration factor*, σ/s, of 10^2 or 10^3. Deeper cracks are more severe. Even though ceramics are strong in shear, we find stress concentrations σ that exceed the bond strength between the atoms. Thus the crack may propagate. This increases the crack depth c, and accentuates the stress concentration of Eq. (8–7.1) still further, until catastrophic failure occurs.

Ceramic materials (and intermetallic compounds, Section 6–4) are generally weak in tension because of their resistance to shear at the crack tip. The same factor, shear strength, makes them strong in compression. A compressive load can be supported across a microcrack without extending the flaw.

Use of nonductile materials   This relationship of high compressive strength and lower tensile strength is important to the design engineer. Concrete, brick, and other ceramics are used primarily in compressive locations (Fig. 8–7.3). When it is necessary

Fig. 8–7.3 Reinforced concrete beam. This beam uses the nonductile material in the compressive positions.

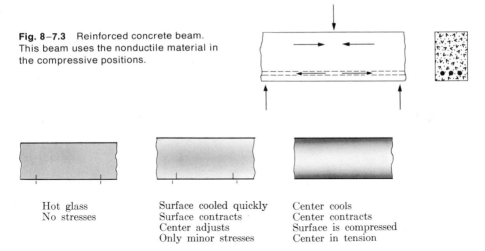

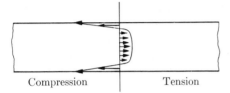

Hot glass
No stresses

Surface cooled quickly
Surface contracts
Center adjusts
Only minor stresses

Center cools
Center contracts
Surface is compressed
Center in tension

Fig. 8–7.4 Dimensional changes in "tempered" glass.

to subject such materials as glass to bending (and therefore tensile loading), it is usually necessary to increase dimensions. For example, the viewing glass of a television picture tube may be as much as 15 mm thick.

Since ceramic materials are stronger in compression than in tension, "tempered" glass is used for glass doors, rear windows of cars, and similar high-strength applications. To produce tempered glass, the glass plate is heated to a temperature high enough to permit adjustments to stresses among the atoms and is then quickly cooled by an air blast or oil quench (Fig. 8–7.4). The surface contracts because of the drop in temperature and becomes rigid, while the center is still hot and can adjust its dimensions to the surface contractions. When the center cools and contracts slightly later, compressive stresses are produced at the surface (and tensile stresses in the center). The stresses that remain in the cross section of the glass are diagrammed in Fig. 8–7.5. A considerable deflection must be applied to the glass before tensile stresses can be developed in the surface of the glass where cracks start.* In effect, since the compressive stresses must be overcome first, the overall strength of the glass is greatly enhanced.

Compression   Tension

Fig. 8–7.5 Surface compression of "tempered" glass. These compressive stresses must be overcome before the surface can be broken in tension.

---

* If a crack penetrates through the compressive skin (e.g., by scratching) into the tension zone shown in Fig. 8–7.5, the crack may become rapidly self-propagating. The aftermath of this effect can be observed in a broken rear window of a car, where the crack pattern is a mosaic, rather than spear-like shards.

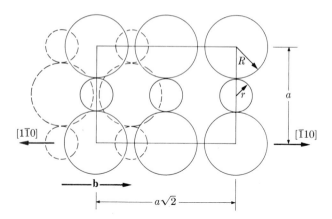

Fig. 8–7.6 The (110) plane of an NaCl-type structure. The slip system is $[1\bar{1}0](110)$ and the slip vector **b** is equal to $a/\sqrt{2}$. (The dashed circles indicate the locations of atoms in the underlying (110) plane.) (See Example 8–7.1.)

Example 8–7.1 (a) Identify the six $\langle 1\bar{1}0\rangle\{110\}$ slip systems for MnS (which has an NaCl structure).
● (b) How long is the slip vector in these slip systems?

Solution: (a) Figure 8–7.6 shows the (110) plane of the NaCl-type structure. Each plane has only one set of slip directions, in this case $[1\bar{1}0]$ or its negative $[\bar{1}10]$. Since there are six planes in the $\{110\}$, the six slip systems are

$$[1\bar{1}0](110); \quad [\bar{1}01](101); \quad [0\bar{1}1](011);$$

$$[110](1\bar{1}0); \quad [101](10\bar{1}); \quad [011](01\bar{1}).$$

b)   Since $r_{Mn^{2+}} = 0.080$ nm   and $R_{S^{2-}} = 0.184$ nm,

$a = 0.528$ nm.

$\mathbf{b} = (0.528 \text{ nm})\sqrt{2}/2 = 0.373$ nm.

Example 8–7.2 Glass has a theoretical strength in excess of 7000 MPa (10^6 psi). Strengths approaching this value have been reached for pristine glass. A piece of plate glass fails with a bending tension of 60 MPa (8,600 psi). What crack depth would be responsible for this low-stress fracture, if we assume the tip of the crack has the dimension of an oxygen ion?

Solution: From Eq. (8–7.1), and assuming $r = 0.14$ nm,

$$c = \frac{r(\sigma/s)^2}{4} = \frac{(0.14 \text{ nm})(7000/60)^2}{4} = 480 \text{ nm}.$$

Comments. Cracks of this size are observed in the surfaces of glass that is exposed to normal environments; however, special techniques are required to detect them, since that dimension is near the limit of resolution by visible light.

Tempered plate glass can support an added load, since the compressive stress of the skin (Fig. 8–7.5) would have to be exceeded before stresses start to concentrate from tension. ◀

• 8–8   PROCESSING OF CERAMIC MATERIALS   ℛℰᴀᴅ ᴏɴʟʏ

The majority of ceramic products are made by one of two general processes—by viscous forming or by sintering. *Viscous forming* involves melting and shaping of a viscous liquid. The *sintering* process starts with finely divided particles that are agglomerated into the desired shape; this is generally followed by *firing* to produce a bond between the particles. A third process with extensive, but specific, use is that of chemical bonding, e.g., like hydration of portland cement. This section will consider only the first two, viscous forming and sintering. Cement hydration is outlined later with concrete (Section 13–2). No attention will be given in this text to single crystal growth and other specialized processes; nor to raw material preparation, such as purification or spray-drying.

Viscous forming   Commercial glasses are manufactured by this process. When heated, the glass becomes sufficiently thermoplastic to be shaped into the final product. Prior to the final shaping, however, it is necessary to completely melt the component oxides in order to assume a homogeneous composition and to remove entrapped gases. This removal is particularly important because the soda and lime contents of glasses (Table 8–5.1) are obtained from Na_2CO_3 and $CaCO_3$, both of which release CO_2. A gas bubble that remains in the glass is, of course, a defect. The final shaping process may be pressing (for structural glass block), or sagging (for many car windows), or blowing (for light bulbs), or drawing (for glass fibers—Fig. 8–1.1d), etc.

Ceramic glasses have many of the characteristics of thermoplastic polymers. In fact, the initial recognition of the *glass temperature* T_g was for silicous glasses (Fig. 4–5.3). The *viscosity* η of a glass is temperature-sensitive:

$$\log_{10} \eta = C' + \frac{B'}{T},   \tag{8–8.1}$$

where C' and B' are constants* for a glass.

Viscosity control is important in glass manufacture. First, the glass must have a low viscosity (< 30 Pa·sec, or 300 poises) in the *melting range* so that bubbles can escape and homogenization can occur as discussed above. The *working range* covers several orders of magnitude. Fast operations, such as the production of light bulbs, require a relatively fluid glass ($\sim 10^{2.5}$ Pa·sec, or $10^{3.5}$ poises); the drawing operation for heavy sheet glass requires a more viscous product ($\sim 10^6$ Pa·sec).

Two other viscosity points are important: (1) the *annealing point*, the temperature with $\eta = 10^{12}$ Pa·sec (or 10^{13} poises), and (2) the *strain point*, the temperature with

---

* The flow of glass requires the movement of atoms past other atoms and therefore depends on an activation energy E. A glass with a high *fluidity*, f (low viscosity, since $\eta = 1/f$) corresponds to a glass that has fast diffusion. Thus, as in Eqs. (4–6.6) and (4–7.4,),

$$\ln f = \ln f_0 - E/kT = -\ln \eta.   \tag{8–8.2}$$

This leads to Eq. (8–8.1). Since $\ln \eta = 2.3 \log \eta$, the terms C' and B' are, respectively, $-\ln f_0/2.3$ and $E/2.3k$. Fluidity is a diffusivity (m^2/sec) per unit force, that is, m^2/s·N. Thus, viscosity has dimensions of Pa·s. Former cgs units were "poises," where 10 poises = 1 Pa·s.

$\eta = 10^{13.5}$ Pa·sec. At the temperature of the annealing point, atoms can move sufficiently so that residual thermal stresses may be relieved within about 15 minutes, a time compatible with production schedules. Below the strain point, the glass is sufficiently rigid so that it may be handled without the generation of new residual stresses. The strain point is below the glass transition temperature.

✳Sintering processes   Most nonvitreous ceramic materials are made from very finely ground particles that are sintered (fired) into a monolithic product. The artist follows the prehistoric procedure of using finely divided clays that are *hydroplastic* when wet. Products such as brick can be shaped into the desired form, allowed to dry, and then fired so that the altered clay particles sinter together and form a strong, hard bond. Alternatively, it is possible to slip-cast ceramics by making a suspension, or *slip*, that can be poured into a porous plaster mold. The mold dewaters the slip adjacent to the mold wall. The balance of the slip is dumped; the product is then *dried* and *fired*. Considerable shrinkage accompanies this process because the packing factor for the solid particles in the original slip casting is low.

Technical ceramic products are commonly formed by a *pressing* process. A spark plug insulator, to which we have referred several times, is made by *isostatic molding* (Fig. 8–8.1). This has a major manufacturing advantage in that high pressures may be applied uniformly. Also, a negligible amount of plasticizing liquid is required that must be dried. Thus, the subsequent shrinkage is small and easily controlled. Of course, high temperatures are required for firing or sintering, since these insulators are predominantly Al_2O_3. Such oxides have strong bonds and therefore low diffusivities for atom movements.

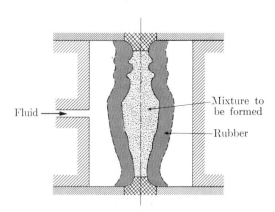

Fluid

—Mixture to be formed

—Rubber

Fig. 8–8.1   Isostatic molding (spark plug manufacture). The hydrostatic fluid provides radial pressure to the product. Maximum compaction is possible with this procedure. (Jeffery, U.S. Patent 1,863,854)

Sintering requires heating in order to agglomerate small particles into bulk materials. Sintering without the formation of a liquid requires diffusion within the solid and therefore occurs most rapidly at very high temperatures (short of melting). Many powdered metal parts and various dielectric and magnetic ceramics are produced by *solid sintering*. These ceramic materials cannot be made by melting, and often no feasible crucible or mold is available.

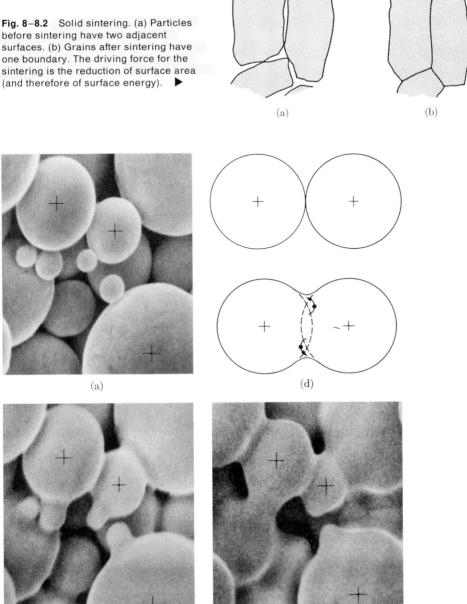

Fig. 8–8.2 Solid sintering. (a) Particles before sintering have two adjacent surfaces. (b) Grains after sintering have one boundary. The driving force for the sintering is the reduction of surface area (and therefore of surface energy). ▶

(a)

(b)

(a)

(d)

(b)

(c)

Fig. 8–8.3 Sintering (nickel powder). The initial points of contact in (a) become areas of contact in (b) and (c) while heating to 1100°C. (d) The atoms diffuse from the contact points to enlarge the contact area. (Vacancies diffuse in the opposite direction.) The particles of powder move closer together, and the amount of particle surface is reduced. (R. M. Fulrath, Inorganic Materials Research Division of the Lawrence Radiation Laboratory, University of California, Berkeley.)

The principle involved in solid sintering is illustrated in Fig. 8–8.2. As shown in part (a), there are two surfaces between any two particles before sintering. After sintering, there is a single grain boundary. The two surfaces are high-energy boundaries; the grain boundary has much less energy. Thus, this reaction occurs naturally if the temperature is high enough for a significant number of atoms to move.

The actual mechanism of sintering is shown in Fig. 8–8.3 in photographs taken through a scanning electron microscope. The points of contact between particles grow by a diffusion of atoms. The net diffusion produces a *shrinkage*, and an accompanying reduction of porosity. Figure 8–8.4 shows the progress of sintering for sodium fluoride.

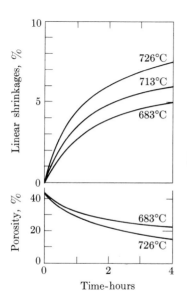

Fig. 8–8.4   Sintering shrinkage (NaF). Powdered NaF (−330 mesh) loses porosity as it shrinks during sintering. (Adapted from Allison and Murray, *Acta Metallurgica*.)

Example 8–8.1   An electrical glass has a working range of 870°C ($\eta = 10^6$ Pa·s, or 10^7 poises) to 1300°C ($\eta = 10^{2.5}$ Pa·s, or $10^{3.5}$ poises). Estimate its annealing point ($\eta = 10^{12}$ Pa·s, or 10^{13} poises).

Solution: From Eq. (8–8.1),

$$6 = C' + B'/(870 + 273),$$

$$2.5 = C' + B'/(1300 + 273).$$

Solving simultaneously,

$$B' = 14,600 \quad \text{and} \quad C' = -6.8$$

For $\eta = 10^{12}$ Pa·s,

$$12 = -6.8 + 14,600/T,$$

$$T = 775 \text{ K} \ (\simeq 500°C).$$

Comment. The annealing point corresponds approximately to the glass temperature, T_g. ◀

Example 8–8.2 A ceramic magnet has a porosity of 28 v/o before sintering and a density of 5.03 g/cm³ after sintering. The true density is 5.14 g/cm³ (=5.14 Mg/m³). (a) What is the porosity after firing (sintering)? (b) If the final dimension should be 16.3 mm, what should be the die dimension?

Solution

a)   True volume of 1 g $= 1/5.14 = 0.1946$ cm³.

                        Fired volume of 1 g $= 1/5.03 = 0.1988$ cm³.

                        Final porosity $= (0.1988 - 0.1946)/0.1988 = 2\%$.

b)   True volume $= 0.72V_0 = 0.98V_f$;

$$\frac{V_0}{V_f} = \frac{L_0^3}{L_f^3} = \frac{0.98}{0.72}.$$

$$L_0 = L_f \sqrt[3]{0.98/0.72}$$
$$= (16.3 \text{ mm})(1.108) = 18.1 \text{ mm}.$$

Comment. This problem may be set up various ways. However, a gram is convenient because it lets us compare volume changes directly.

   Note that if the die is set up for a 10% shrinkage (18.1 mm to 16.3 mm), then processing procedures must be consistent so that there is *always* a 28% porosity in the pressed (presintered) stage. Otherwise, the dimensional specifications will be missed. Processing variables must be closely controlled. ◀

Example 8–8.3 A ceramic magnet has a true density of 5.41 Mg/m³ (=5.41 g/cm³). A poorly sintered sample weighs 3.79 g dry, and 3.84 g when saturated with water. The saturated sample weighs 3.08 g when suspended in water. (a) What is its *true volume*? (b) What is its *bulk volume* (total volume)? (c) What is its *apparent (open) porosity*? (d) What is its *total porosity*?

Solution

a)   True volume $= 3.79$ g$/(5.41$ g/cm³$) = 0.70$ cm³.

b)   From Archimedes principle:

            Water displaced by bulk sample $= 3.84$ g $- 3.08$ g $= 0.76$ g;

                  $\therefore$ total volume $= 0.76$ g$/(1$ g/cm³$) = 0.76$ cm³.

   This includes material plus all pore space.

c)   Apparent porosity $= [(3.84 - 3.79 \text{ g})/(1 \text{ g/cm}^3)]/0.76$ cm³
                        $= 0.066$ (or 6.6 v/o, bulk basis).

d)   Total porosity $= (0.76 - 0.70 \text{ cm}^3)/0.76$ cm³
                        $= 0.079$ (or 7.9 v/o bulk basis).

Comment. The *closed porosity* is 1.3 v/o. The *bulk density* is 3.79 g/0.76 cm³ $= 5.0$ g/cm³. We may also speak of an *apparent density*:

                  mass/(bulk volume $-$ open pore volume).

This density is (3.79 g)/(0.76 cm³)(1 $- 0.066$) $= 5.34$ g/cm³ (or 5.34 Mg/m³). ◀

REVIEW AND STUDY

Since ceramic phases are compounds of metallic and nonmetallic elements, we find that it is not as easy to generalize about structure and properties as for metals or nonmetals alone. Metals, for example, are always electrical and thermal conductors; most ceramics are insulators but some possess semiconductivity that has technical uses. Organic polymers always transmit light in thin sections; ceramics may have the transparency of optical glass or may be opaque in magnetic spinels. Structural ceramics are very strong in compression but must be considered weak in tension. Fiberglass may have a tensile strength greater than steel and therefore can be used as a reinforcement; glass is also recognized as a brittle, friable material, so that special care must be taken when it is shipped or handled.

The examples above are just a few of those we could cite to indicate that ceramic materials have a variety of characteristics. Many of these are indispensable for current-day technology and societal needs. However, they are more complex than other materials and therefore require greater technical familiarity and understanding.

SUMMARY: STRUCTURAL

The simpler ceramic compounds may be viewed in terms of a basic lattice of anions with cations in selected interstitial sites. We used CsCl-, NaCl-, and ZnS-type structures as our prototypes. Cations reside in 8-f, 6-f, and 4-f interstitial sites, respectively. Table 8–2.1 provides a review summary of these structural types. More complex structures such as the one possessed by magnetic spinels can be analyzed by observing how cations are ordered among 4-f and 6-f interstitial sites.

Glasses involve more covalent bonding than do the majority of other ceramic compounds. Basic to their structure are the SiO_4 tetrahedral units in which a silicon is coordinated with four oxygens. With pure silica, which has only silicon and oxygen, each oxygen atom serves as a bridge between two tetrahedral units to form a rigid network. Additions of other oxides introduce nonbridging oxygens and therefore modify the network to make the glass more thermoplastic.

SUMMARY: PROPERTIES AND BEHAVIOR

Ceramic insulators are designed to be electrically inert. The constituent ions, however, can respond to electric fields. This provides a polarization inside the dielectric. The resulting dielectric constant is used advantageously in capacitors; but it produces power losses if the ion displacements lag the applied field. Permanent polarization produces piezoelectricity and leads to electromechanical transducers. Semiconducting ceramics owe their existence to atoms with multiple valence states. Magnetic ceramics require ions with unfilled subvalence shells. Each atom, therefore, acts as a small magnet; these couple together to give magnetic domains, which respond to magnetic fields.

Compounds, be they ceramic or intermetallic, strongly resist shear deformation. As a result they are notch- (or crack-)sensitive. This produces low tensile strengths. However, the resistance to shear produces high compressive strengths. Ceramic materials are thus selected for use under compression in structure and product designs. Also, ceramic materials are engineered to possess surface compression. The viscosity of glass follows the same laws as diffusion but with an applied force.

KEY TERMS AND CONCEPTS

Abrasives	• Domain	Relative dielectric constant, κ
$A_mB_nX_p$ compounds	Glass	• Silicate structures
• Annealing point	Glass temperature, T_g	network
AX structures	Insulator	• Sintering
CsCl-type	Interstitial sites	• SiO_4 tetrahedra
NaCl-type	4-f, 6-f, 8-f	• Spinel
ZnS-type	Isostatic molding	• Strain point
A_mX_p structures	Magnet	• Stress concentration factor, σ/s
CaF_2-type	• hard	Tempered glass
• Bohr magneton, β	• soft	Thermistors
Bridging oxygen	Network modifier	Transducer
• Coercive force, H_c	Nonbridging oxygen	• Viscosity, η
• Curie point	Piezoelectric	• Viscous forming
Dielectric strength	Polarization, $\mathscr{P}$	

FOR CLASS DISCUSSION

A_8 Make a list of six ceramic materials that are used in technical applications and that have not been discussed in this text.

B_8 Distinguish between CsCl-type, NaCl-type, and ZnS-type structures with regard to coordination number, minimum ion–size ratio, and the number of ions per unit cell.

C_8 Both KCl and MnSe have the same structure (NaCl-type). Although the K^+ is lighter than Mn^{2+} (39 versus 55 amu) and Cl^- is lighter than Se^{2-} (35 versus 79 amu), KCl has a larger lattice constant ($a > 0.6$ nm) than MnSe ($a < 0.6$ nm). Explain why.

D_8 Sketch the atom arrangements on the $(1\bar{1}0)$ plane of ZnS. (Cf. Fig. 8–3.2 for CaF_2.) Repeat for the (110) plane.

E_8 As shown in Fig. 8–4.1, cubic $BaTiO_3$ has an oxygen atom at the center of each face. However, it is not face-centered cubic. Why?

F_8 Figure 4–3.1 with Mg^{2+}, Fe^{2+}, and O^{2-} ions, and Fig. 8–4.1 with Ba^{2+}, Ti^{4+}, and O^{2-} ions, both have three types of ions. We call one a solid solution and the other an $A_mB_nX_p$ compound. Why the difference?

G_8 Both MnO and MnS have NaCl-type structures. A limited amount (~ 1 a/o) of solid solution occurs. Suggest the mechanism. Why is the extent of solid solution limited?

- H_8 Make a sketch of a hexagonal array of atom nuclei. Now dot in a second layer that could lie in close packing above the first layer. Indicate the locations of six-fold sites between the two layers; of four-fold sites.

- I_8 There are two polymorphs of MnTe. One has the NaCl-type structure; the other has the NiAs-type structure. In each case the Mn^{2+} sits among six Te^{2-} ions. The first is much softer and more ductile than the latter. Explain.

  J_8 Use electron dots (cf. Fig. 2–2.3a) to show covalent bonding in an SiO_4^{4-} ion; to show ionic bonding.

- K_8 Compare and contrast the structure of cristobalite (Fig. 8–5.2) with that of ZnS (Fig. 8–2.4), and with silicon, which has the structure of diamond (Fig. 2–2.4).

  L_8 Cristobalite (Fig. 8–5.2) has almost no expansion when it melts to fused silica. Suggest why.

  M_8 Distinguish between bridging oxygens and nonbridging oxygens, and their effect on glass deformation above the glass temperature.

  N_8 B_2O_3 is a good glass-former. Compare its structure and viscous behavior with fused silica (SiO_2).

  O_8 What characteristics are required of an oxide for it to be a network modifier in the structure of glass? Suggest six oxides that will fill this role.

  P_8 (a) Discuss two ways in which cracks affect an electrical insulator. (b) Why are insulators glazed before use on high-voltage lines?

  Q_8 Distinguish between dielectric constant and dielectric strength.

  R_8 Ceramic semiconductors commonly contain metals that the chemist calls transition metals (the B series of Fig. 2–1.1). Why are the oxides of these metals semiconductors?

  S_8 Explain to someone who is not taking this course how a ceramic cartridge on a phonograph provides a signal that can then be amplified.

  T_8 Suggest a concept for a piezoelectric "spark plug."

- U_8 From Appendix B indicate the number of Bohr magnetons (unpaired $3d$ electrons per atom) possessed by the elements with atomic numbers between 20 and 30.

- V_8 Normally we do not observe magnetism in MnO; yet each cation possesses five Bohr magnetons, and experiments show that MnO responds to changes in a magnetic field. Suggest a reason that we do not observe magnetization. (MnO has an NaCl-type structure.)

  W_8 Why do tensile test data for ceramics show more scatter than for metals?

  X_8 Glass can be treated to remove Na^+ ions from the surface and introduce K^+ ions in their place (called ion-exchange). When this is done below the glass transition temperature, the strength of the glass is increased significantly. This process is sometimes called "chemical tempering" because the effects on properties are similar to the tempering of Fig. 8–7.5. Account for the strengthening.

  Y_8 A ceramic (Al_2O_3) cutting tool has to be mounted in a lathe differently from a tool made of "high-speed" steel. Why?

Z_8 A glaze for dinnerware is chosen so that it has a slightly lower thermal expansion coefficient than the underlying porcelain. Explain why this makes the product stronger.

STUDY PROBLEMS

8–2.1 Calculate the density of CsCl from the data in the Appendix. (Remember that CN = 8.)

Answer: 3.94 Mg/m^3

8–2.2 CsCl has the simple cubic structure of Cl$^-$ ions with Cs$^+$ ions in the eight-fold sites. (a) The radii are 0.187 nm and 0.172 nm, respectively, for CN = 8; what is the packing factor? (b) What would this factor be if $r/R = 0.73$?

8–2.3 The intermetallic compound AlNi has the CsCl-type structure with $a = 0.2881$ nm. Calculate its density.

Answer: 6.0 Mg/m^3

8–2.4 X-ray data show that the unit-cell dimensions of cubic MgO are 0.42 nm. It has a density of 3.6 Mg/m^3. How many Mg^{2+} ions and O^{2-} ions are there per unit cell?

8–2.5 Periclase (MgO) has an fcc structure of O^{2-} ions with Mg^{2+} ions in all the six-fold sites. (a) The radii are 0.140 nm and 0.066 nm, respectively; what is the packing factor? (b) What would this factor be if $r/R = 0.41$?

Answer: (a) 0.73   (b) 0.80

8–2.6 Nickel oxide is cubic and has a density of 6.87 Mg/m^3 (= 6.87 g/cm^3). Its lattice constant is 0.417 nm. How many atoms are there per unit cell? (The nickel ion is Ni^{2+}.)

8–2.7 Lithium fluoride has a density of 2.6 Mg/m^3 and the NaCl structure. Use these data to calculate the unit cell size, a, and compare it with the value you get from the ionic radii.

Answer: 0.405 nm versus 0.402 nm

8–2.8 ZnS (Fig. 8–2.4) has a density of 4.1 Mg/m^3. Based on this, what is the spacing between the centers of the two ions?

8–2.9 Estimate the densities of the two polymorphs of MnS in Example 8–2.4.

Answer: (a) 3.9 Mg/m^3 (b) 3.4 Mg/m^3

8–2.10 The density of 1.5 g/cm^3 (or 1.5 Mg/m^3) is given for NH$_4$Cl in the Chemical Handbook. X-ray files state that there are two polymorphs for NH$_4$Cl; one has an NaCl-type structure with $a = 0.726$ nm, the other has a CsCl-type structure with $a = 0.387$ nm. The density value is for which polymorph? (The NH^{4+} ion occupies the crystal lattice as a unit.)

8–2.11 It is hypothesized that at high pressures NaCl can be forced into a CsCl-type structure. What would be the percent volume change? (See the footnote of Appendix B for differences in radii with CN = 6 and CN = 8.)

Answer: -16 v/o

8–3.1 With CN = 4, $R_{O^{2-}} = 0.127$ nm (Table 2–5.1). With CN = 8, $r_{Zr^{4+}} \simeq 0.085$ nm. Estimate the size of the unit cell of cubic ZrO$_2$, which has the CaF$_2$-type structure.

Answer: 0.49 nm (versus 0.507 nm by experiment)

8–3.2 Using the radii from the answer of Example 8–3.2, will the estimated unit cell size be larger or smaller than the experimentally measured value of $a = 0.547$ nm? (See the comment with Example 8–3.1 for an explanation.)

8–3.3 The unit cell size a of UO_2 is 0.547 nm. (a) Based on this information and Fig. 8–3.1, what is the linear density of U^{4+} ions in the $\langle 110 \rangle$ directions? Of the O^{2-} ions? (b) What is the linear density of U^{4+} ions in the $\langle 100 \rangle$ directions? Of the O^{2-} ions? (c) What is the linear density of U^{4+} ions in the $\langle 111 \rangle$ directions? Of the O^{2-} ions?

Answer: (a) 2.6×10^6/mm, 2.6×10^6/mm (b) 1.8×10^6/mm, 3.7×10^6/mm (c) 1.1×10^6/mm, 2.1×10^6/mm

8–3.4 Can MgF_2 have the same structure as CaF_2? Explain.

8–4.1 $BaTiO_3$ is made by calcining $BaCO_3$ to BaO and mixing it with TiO_2. (a) How much BaO and TiO_2 are required per 100 g of product? (b) How much $BaCO_3$ is required per gram of product? (Ba = 137.34 amu.)

Answer: (a) 65.7 g BaO and 34.3 g TiO_2 (b) 0.846 g $BaCO_3$

8–4.2 Calcium titanite ($CaTiO_3$) has the same structure as $BaTiO_3$. It is cubic and has a density of 4 Mg/m³. Calculate its lattice constant.

8–4.3 The density of cubic $PbTiO_3$ is 7.5 Mg/m³. What is the lattice dimension of the unit cell?

Answer: 0.406 nm (versus 0.397 nm by experiment)

8–4.4 Calculate the oxygen content (a) of Fe_3O_4; (b) of $NiFe_2O_4$.

• **8–4.5** A solid solution contains 30 mole percent MgO and 70 mole percent LiF. (a) What are the weight percents of Li^+, Mg^{2+}, F^-, O^{2-}? (b) What is the density?

Answer: (a) Li^+, 16 w/o; Mg^{2+}, 24 w/o; F^-, 44 w/o; O^{2-}, 16 w/o; (b) 3.0 g/cm³ (= 3.0 Mg/m³)

• **8–4.6** (a) What type of vacancies, anion or cation, must be introduced with MgF_2 in order for it to dissolve in LiF? (b) What type must be introduced with LiF for it to dissolve in MgF_2?

8–4.7 A cubic form of ZrO_2 is possible when one Ca^{2+} ion is added in a solid solution for every six Zr^{4+} ions present. Thus the cations form an fcc structure, and O^{2-} ions are located in the four-fold sites. (a) How many O^{2-} ions are there for every 100 cations? (b) What fraction of the four-fold sites is occupied?

Answer: (a) 185.7 O^{2-} ions (b) 92.9%

8–4.8 (a) What is the coordination number of Ba^{2+} in $BaTiO_3$? (b) Of Ti^{4+}? (c) The O^{2-} ion cannot be assumed to be spherical. Why?

8–4.9 The unit cell size a of the ferrospinel in Example 8–4.1 is 0.84 nm. (a) Calculate its density. (b) What is the distance of closest approach of O^{2-} ions (center-to-center)?

Answer: (a) 5.3 Mg/m³ (b) 0.297 nm

8–4.10 The lattice constant a of cubic $BaTiO_3$ is 0.40 nm. Consider $r_{Ti^{4+}} = 0.068$ nm (App. B), and $r_{Ba^{2+}} = 0.148$ nm when CN = 12. (a) What is the radius R of the O^{2-} ion in the $\langle 100 \rangle$ directions? (b) In the $\langle 110 \rangle$ directions?

8–4.11 One-tenth weight percent Fe_2O_3 is in solid solution with NiO. As such, 3 Ni^{2+} are replaced by ($2 Fe^{3+}$ + □) to maintain a charge balance. How many cation vacancies are there per m³?

8–5.1 Refer to Fig. 8–5.2, where $a = 0.693$ nm. How many silicon and oxygen atoms are there per m^3?

Answer: 2.40×10^{28} Si/m^3, 4.80×10^{28} O/m^3

8–5.2 What is the volume expansion as quartz (Example 8–5.1) changes to the SiO$_2$ polymorph shown in Fig. 8–5.2? (Use data from Study Problem 8–5.1 as required.)

8–5.3 What is the packing factor of SiO$_2$ in cristobalite (Fig. 8–5.2)? (If needed, use information from Study Problem 8–5.1 and Example 8–5.3.)

Answer: 0.29

• *8–5.4* From the information in Example 8–5.3, determine the unit cell size a of the SiO$_2$ in Fig. 8–5.2.

8–5.5 Coesite, the high-pressure polymorph of SiO$_2$, has a density of 2.9 Mg/m^3. What is the packing factor? ($r_{Si} = 0.038$ nm; $R_O = 0.114$ nm.)

Answer: 0.36

8–5.6 A glass contains only Na$_2$O and SiO$_2$ and has a structure in which 72% of the oxygen atoms serve as bridges between adjacent silicons. How many sodium and silicon atoms are there per 100 oxygen atoms?

8–5.7 A glass contains 75 w/o SiO$_2$ and 25 w/o CaO. What fraction of the oxygens serve as bridges between pairs of silicon atoms?

Answer: 70 a/o

8–5.8 A window glass is a soda-lime glass with a weight ratio of 14 Na$_2$O–14 CaO–72 SiO$_2$. What fraction of the oxygen atoms are nonbridging?

8–6.1 In Example 8–6.2, it states that Ni^{3+}/Ni^{2+} is normally $\langle 10^{-4}$. If pure nickel oxide has that ratio, how many charge carriers are there per m^3? (*Note:* (2 Ni^{3+} + $\square$) replace 3 Ni^{2+}.)

Answer: 5.5×10^{24}/m^3

8–6.2 Some TiO$_2$ is not stoichiometric but has a Ti^{3+}/Ti^{4+} ratio of 0.001. (a) Is this oxide *n*-type or *p*-type? (b) The density of TiO$_2$ is 4.26 Mg/m^3. How many carriers are there per m^3?

8–6.3 Potassium niobate (KNbO$_3$) has the same structure as BaTiO$_3$ but with K$^+$ replacing Ba^{2+}, and Nb^{5+} replacing Ti^{4+}. Assume the dimensions of the KNbO$_3$ unit cell are approximately the same as described for BaTiO$_3$ in Fig. 8–6.2. (a) What is the distance between the centers of positive and negative charges in the unit cell? (b) What is the dipole moment, $\mu = Qd$, of the unit cell? (c) What polarization is possible?

Answer: (a) 0.0117 nm   (b) 11.2×10^{-30} C·m   (c) 0.175 C/m^2

8–6.4 Refer to Example 8–6.3. A ten-millimeter cube of BaTiO$_3$ is compressed 1%. The two ends receiving pressure are connected electrically. How many electrons travel from the negative end to the positive end?

8–6.5 A piezoelectric crystal has a Young's modulus of 80,000 MPa (11,600,000 psi). What stress must be applied to reduce its polarization from 800 to 788 coul · m/m^3?

Answer: 1200 MPa (175,000 psi)

8–6.6 Refer to Example 8–6.3. What is the distance between the centers of positive and negative charges of each unit cell?

• *8–6.7* A ceramic magnetic material *nickel ferrite* has eight $[NiFe_2O_4]$'s per unit cell, which is cubic with $a = 0.834$ nm. Assume all the unit cells have the same magnetic orientation; what is the saturation magnetization? (Each Ni^{2+} ion has two Bohr magnetons.)

Answer: 250,000 amp/m

• *8–6.8* Manganese ferrite $[MnFe_2O_4]_8$ has the same structure as magnetite $[Fe^{2+}Fe_2^{3+}O_4]_8$, with approximately the same lattice constant (0.84 nm). What is the saturation magnetization? (Each Mn^{2+} ion has five Bohr magnetons.)

• *8–6.9* Refer to Study Problem 8–6.7. Assume that the N–S polarity is 180° reversed in 40% of the unit cells when the external field is partially removed. How much magnetization remains?

Answer: 50,000 amp/m

• *8–6.10* From Section 2–1, determine the number of Bohr magnetons β possessed by the divalent ions of the transition metals: Ti^{2+}, Cr^{2+}, Mn^{2+}, Fe^{2+}, Co^{2+}, Ni^{2+}, and Cu^{2+}. By the trivalent cations: Ti^{3+}, Cr^{3+}, Mn^{3+}, Fe^{3+}, and Co^{3+}.

• *8–6.11* Refer to Example 8–4.1. In this particular ferrospinel, the Zn^{2+} ions preferentially choose four-fold sites and force a corresponding number of Fe^{3+} ions into six-fold sites. (a) How many Bohr magnetons does this produce per unit cell of $(Zn_3, Ni_5)Fe_{16}O_{32}$? (b) How many are there in $[NiFe_2O_4]_8$ where all of the divalent ions are in six-fold sites?

Answer: (a) 40   (b) 16

8–7.1 Under certain conditions, the $\langle 011 \rangle \{100\}$ slip systems can operate in NaCl-type structures in addition to those listed in Example 8–7.1. (a) How many slip systems belong to this set? List them. • (b) What is the slip vector in PbS, which has the NaCl-type structure and the above slip system?

Answer: (a) 6   (b) 0.43 nm

8–7.2 A round hole (dia. $= 12.5$ mm) is cut through a glass plate which is 2.5 mm thick. Assume the surface of the hole has been made flaw-free and that a stress of 10 MPa (1450 psi) is applied in the direction shown in Fig. 8–7.2. What is the concentrated stress at the surface of the hole?

8–7.3 Microcracks that are 1 μm deep and that have atomic dimensions at their tips form at the surface of the hole in Study Problem 8–7.2. Estimate the stresses at their tips.

Answer: ~ 3400 MPa (or 500,000 psi) $\sigma = 2\sigma\,(2)\,\sqrt{10^{-6}/14\times10^{-9}}$ ← Boﬂ LENGTH

8–7.4 A large glass plate contains a hole 10 mm in diameter. (a) What is the stress-concentration factor adjacent to the hole from a tensile stress that is applied parallel to the plate? (b) When the hole was drilled, the side developed 10-μm cracks. What total stress concentration results if we assume, as we did in Example 8–7.2, that this type of crack has the dimensions of an oxygen ion?

8–7.5 A 1-mm round rod with the composition of plate glass (Appendix C) is coated with 0.1 mm of borosilicate glass, so that the rod is now 1.2 mm in diameter. Assuming no initial stresses at 200°C, what longitudinal stress is developed when the composite rod is cooled to 20°C?

Answer: $\sigma_{boro} = -55$ MPa $(-8000$ psi), $\sigma_{pl} = +25$ MPa (3500 psi)

8–7.6 Repeat the previous problem, but interchange the locations of the two glasses. (a) Which will have the higher stresses, plate or borosilicate? (b) Comment on the strength of this composite glass rod.

8–7.7 The theoretical strengths of amorphous materials are calculated to be between $G/6$ and $G/4$, where G is the shear modulus. Estimate the theoretical strength of glass based on its elastic properties, including $v = 0.25$.

Answer: 5000 MPa–7000 MPa (or 700,000 psi–1,000,000 psi)

• *8–8.1* Assume that the flow rate of a glass is inversely proportional to the viscosity η. How much faster will a molten electrical glass (Example 8–8.1) flow from a furnace at 920°C than at 900°C?

Answer: $\eta_{920}/\eta_{900} = 0.618$; $\dot{F}_{920} = 1.6\ \dot{F}_{900}$

• *8–8.2* The viscosity of window glass drops from 10^6 Pa · s at 680°C to 10^3 Pa · s at 1035°C. What is its viscosity at 900°C?

8–8.3 A ceramic insulator will have 1 v/o porosity after sintering and should have a length of 13.7 mm. During manufacturing, the powders can be pressed to contain 24 v/o porosity. What should the die dimensions be?

Answer: 15.0 mm

8–8.4 A magnetic ferrite for an oscilloscope component is to have a final dimension of 15.8 mm (0.621 in.). Its volume shrinkage during sintering is 33.1% (unsintered basis). What initial dimension should the powdered compact have?

8–8.5 A powdered metal part has a porosity of 23% after compacting and before sintering. What linear shrinkage allowance should be made if the total porosity after sintering is expected to be 2%?

Answer: 8% (sintered basis)

8–8.6 A brick weighs 3.3 kg when dry, 3.45 kg when saturated with water, and 1.9 kg when suspended in water. (a) What is the apparent porosity? (b) Its bulk density? (c) Its apparent density?

8–8.7 A piece of ceramic wall tile 5 mm × 200 mm × 200 mm absorbs 2.5 g of water. What is the apparent porosity of the tile?

Answer: 1.25 v/o

8–8.8 An insulating brick weighs 1.77 kg (3.9 lb) dry, 2.25 kg (4.95 lb) when saturated with water, and 1.05 kg (2.3 lb) when suspended in water. (a) What is the apparent porosity? (b) Its bulk density? (c) Its apparent density?

8–8.9 An electrical porcelain product has only 1 v/o porosity as sold. It had 27 v/o porosity (total) after pressing and drying. How much linear shrinkage occurred during sintering?

Answer: 9.7 l/o

8–8.10 Refer to Fig. 8–8.3. (a) How much linear shrinkage occurred during the sintering of this powder—from part (a) to part (c)? (b) How much volume shrinkage?

CHAPTER **9**

Multiphase Materials: Phase Diagrams

PREVIEW

We now shift from single-phase materials to materials with two or more phases. We direct our prime attention in Chapter 9 to *phase diagrams*, because they give us the basis for Chapters 10 and 11.

We can learn to use the phase diagrams (1) to predict *what phases* are in equilibrium for selected alloy compositions at desired temperatures; (2) to determine the *chemical composition* of each phase; and (3) to calculate the *quantity* of each phase that is present. Phase diagrams are powerful tools in the hands of scientists and engineers who design materials for specific applications, as well as others who must anticipate the stability of specific materials when designing products for service environments.

Throughout this chapter, we shall assume that *equilibrium* is attained; i.e., no further reaction is possible between the phases that are present.

CONTENTS

STUDY OBJECTIVES

1 To view a phase diagram as curves for solubility limits, and thus be able to conclude what phase(s) will be present for a specified alloy composition and at specified temperatures.

2 To determine the chemical composition of a phase at its limit of solubility.

3 To calculate by interpolation the relative amounts of two phases when the solubility limits are exceeded.

4 To take commercial alloys, such as brass, bronze, solder, sterling silver, and aluminum, under equilibrium conditions and predict phase relationships.

5 To be familiar with the structure and compositional characteristics of ferrite (α), austenite (γ), and carbide (Fe$_3$C).

6 To know the Fe–Fe$_3$C phase diagram sufficiently so that you can quickly visualize it in later discussions regarding the heat treatment of steels. This involves (a) the eutectoid temperature, composition, and reaction, and (b) the solubility limits of carbon in α and γ.

7 To handle quantitatively, the phase changes encountered during austenite decomposition in either (a) plain-carbon steels (Fe–C) or (b) simple low-alloy steels (Fe–C–X). This requires a familiarity with the AISI–SAE nomenclature pattern.

9–1 QUALITATIVE PHASE RELATIONSHIPS

The three preceding chapters have been concerned successively with metallic, organic, and ceramic phases, and with the dependence of properties on their structure. In each chapter only single-phase materials were considered. Many useful engineering materials consist predominantly of one phase;* however, a greater number are *mixtures of phases*; for example, steel, solder, portland cement, grinding wheels, paints, and glass-reinforced plastics. The mixture of two or more phases in one material permits interaction among the phases, and the resulting properties are usually different from the properties of individual phases. It is frequently possible, also, to modify these properties by changing either the shape or the distribution of the phases (see Chapter 10).

Solutions versus mixtures   Different *components* can be combined into a single material by means of *solutions* or of *mixtures*.† Solid solutions have already been discussed in Sections 4–2, 6–4, and 8–4, and we are all familiar with liquid solutions. The composition of solutions can vary widely, because (1) one atom may be substituted for another at lattice points of the phase structure, or (2) atoms may be placed in the interstices of the structure. The solute does not change the structural pattern of the solvent. A mixture, on the other hand, contains more than one phase (structural pattern). Sand and water, rubber with a carbon filler, and tungsten carbide with a cobalt binder are three examples of mixtures. In each of these examples there are two different phases, each with its own atomic arrangement.

It is, of course, possible to have a mixture of two different solutions. For example, in a lead–tin solder, one phase is a solid solution in which tin has replaced some of the lead in the fcc structure, and the other phase has the structure of tin (body-centered tetragonal). At elevated temperatures, lead atoms may replace a limited number of tin atoms in the bct structure. Thus, just after an ordinary 60–40 solder (60% Sn–40% Pb) solidifies, it contains two structures, each a solid solution.

Liquid solubility limits   Figure 9–1.1(a) shows the *solubility limit* of ordinary sugar in water; the curve is a *solubility curve*. All compositions shown to the left of the curve will form only one phase, because all the sugar can be dissolved in the liquid phase—*syrup*. With the higher percentages of sugar shown to the right of the curve, however, it is impossible to dissolve the sugar completely, with the result that we have a mixture of two phases, solid sugar and liquid syrup. This example shows the change of solubility with temperature and also demonstrates a simple method for plotting temperature (or any other variable) as a function of composition. From left to right, the abscissa of Fig. 9–1.1(a) indicates the percentage of sugar. The percentage of water may be read directly from right to left, since the total of the components must, of course, equal 100%.

Figure 9–1.1(b) shows the solubility limit of NaCl in water. As in Fig. 9–1.1(a), there is a region that contains a liquid solution; this particular liquid is called a

---

*   See the end of Section 4–5 for our definition of phases.
†   A *solution* is a phase with more than one component; a *mixture* is a material with more than one phase.

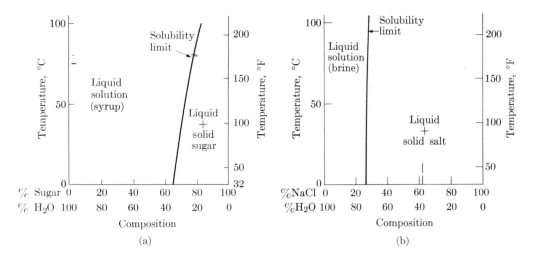

Fig. 9–1.1   (a) Solubility of sugar in water. The limit of sugar solubility in water is shown by the solubility curve. Note that the sum of the sugar and water content at any point on the abscissa is 100%. (b) Solubility of NaCl in H_2O.

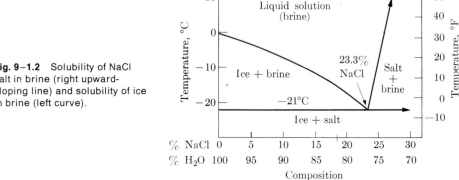

Fig. 9–1.2   Solubility of NaCl salt in brine (right upward-sloping line) and solubility of ice in brine (left curve).

brine. There is a solubility limit that increases with temperature. Beyond the solubility limit there is a region of liquid plus solid salt. Figure 9–1.2 reveals additional features of the H_2O–NaCl *system.* Here the extremes of the abscissa are 100% H_2O (0% NaCl) and 30% NaCl (70% H_2O). Note from the figure that (1) the solubility limit of NaCl in a brine solution decreases with decreasing temperature, (2) the solubility limit of H_2O in a brine solution also decreases with decreasing temperature, and (3) intermediate compositions have melting temperatures lower than that of either pure ice (0°C or 32°F), or pure salt (800.4°C). Facts (1) and (3) are well known and fact (2), the less familiar limited solubility of ice in the aqueous liquid, can be verified by a simple experiment. A salt and water solution, e.g., sea water with 1.7% NaCl, can be cooled to less than 0°C (<32°F) and, according to Fig. 9–1.2, it will still be entirely liquid until minus 1°C (30.2°F) is reached. This is in agreement with observations of any arctic saline sea.* When such a salty liquid is cooled below $-1°C$, ice crystals will form and, because the solution cannot contain more than 98.3% H_2O at that temperature, these ice crystals must separate from the liquid. At minus 20°C ($-4°F$), the maximum H_2O content possible in a brine solution is 77% (23% NaCl) as can be verified by making a slush at that temperature and separating the ice from this liquid; the ice will be nearly pure H_2O and the remaining liquid will be saltier (i.e., lower in H_2O) than the original brine solution.

Eutectic temperatures and compositions   The two solubility curves of Fig. 9–1.2 cross at $-21°C$ ($-6°F$), and at 76.7 H_2O–23.3 NaCl. This is the lowest temperature at which the brine solution remains liquid. It is called the *eutectic temperature*. The composition of this low-melting liquid is the *eutectic composition*.

Observe in Fig. 9–1.2 that the brine coexists with a solid phase below each solubility curve. On the H_2O-rich side, the mixture is ice and brine; on the other side of the eutectic, the mixture is solid NaCl and brine. A 2-phase mixture of solid H_2O (ice) and solid NaCl (salt) exists below the eutectic temperature.

Eutectic reactions   A brine of eutectic composition (~ 77 H_2O–23 NaCl) changes from a single liquid solution into two solid phases as it is cooled through the eutectic temperature:

$$\text{Liquid (23\% NaCl)} \xrightarrow[\text{Cooling}]{-21°C} \text{Ice (0\% NaCl)} + \text{Salt (100\% NaCl)}. \qquad (9\text{–}1.1)$$

Heating reverses the reaction. This eutectic reaction may be generalized to

$$L_2 \xrightleftharpoons[\text{Heating}]{\text{Cooling}} S_1 + S_3, \qquad (9\text{–}1.2)$$

where the subscripts imply progressively higher amounts of one of the components.

Eutectic alloys   It was discovered early in historic times that low-melting alloys could be made by selecting appropriate mixtures of two or more metals. One such alloy that is most familiar to the reader is an ordinary 60–40 *solder* (60 Sn–40 Pb). As seen in Fig. 9–1.3, this alloy closely approximates the eutectic composition that melts at 183°C. This solder has wide use in electrical circuitry because metallic connections may be made with a minimum of heating. If the solder contained more lead (say, 80 Pb–20 Sn), the liquid would become saturated with lead at 280°C during cooling. There would then be a 100°C range (280°C to 183°C) in which there would be a mixture of liquid and solid.

Solid solubility limits   Figure 9–1.3 differs from Fig. 9–1.2 in that the lead-rich solid can dissolve tin atoms into its structure. (Ice cannot dissolve measurable amounts of NaCl into its crystalline structure.) The amount varies from 29% tin atoms (19 w/o) in the lead-rich solid at 183°C to 10 a/o (6 w/o) tin at 300°C. Also, the tin-rich solid may contain 2.5 w/o (1.5 a/o) lead atoms at 183°C. By convention, these two solid phases have been labeled α and β, respectively.

---

* There can be a slight variation if the salt content is not exactly 1.7%.

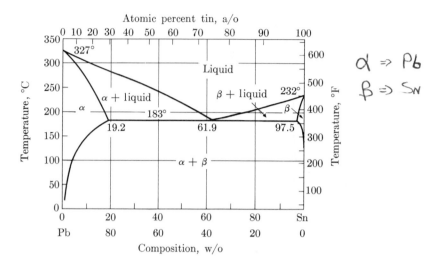

Fig. 9–1.3 Pb–Sn diagram. Such a diagram indicates the phase compositions and permits the calculation of phase quantities for any lead–tin mixture at any temperature. (Adapted from *Metals Handbook*, American Society for Metals.)

The solid solubility limit has a maximum at the eutectic temperature. Both above and below 183°C, the amount of tin that is soluble in the lead-rich, fcc α decreases. Likewise, a maximum of 2.5 w/o lead is soluble in the tin-rich, bct β at 183°C.

Phase names and/or labels In Chapters 4 and 6, we gave names to certain solid-solution phases. *Brass* is an fcc phase of zinc in copper, and *sterling silver* is a solid-solution phase of copper in fcc silver. Later we will give the name *ferrite* to the phase of bcc iron when it contains solutes; likewise, *austenite* is the common name for fcc iron when it contains added elements.

Greek-letter labels are used more commonly than names, for simplicity's sake. Thus, α and β are *phases* in the Pb–Sn *system*, while lead and tin are *components* of the Pb–Sn system.* Although the α phase has the structure of pure lead, α is not necessarily pure lead.

Example 9–1.1 According to Fig. 9–1.1(a), a syrup may contain only 67% sugar (33% water) at 20°C (68°F), but 83% sugar at 100°C (212°F). One hundred grams of sugar and 25 grams of water are mixed and boiled until all the sugar is dissolved. During cooling, the solubility limit is exceeded, so that (with time) excess sugar separates from the syrup. If equilibrium is attained, what is the weight ratio of syrup to excess sugar at 20°C?

---

* Admittedly, the label α appears in almost all systems as the first-named phase. However, this does not prove to be a problem. (The reader is aware that x is repeatedly used in algebraic calculations without special labeling confusion.)

Solution: Basis: 100 g sugar + 25 g H_2O = 125 g.
At 20°C:

$$\text{sugar in syrup} + \text{excess} = 0.67(125 - x) + x = 100 \text{ g}$$

$$x = 49.25 \text{ g excess};$$

and

$$125 \text{ g} - 49.25 = 75.75 \text{ g syrup.}$$

$$\text{Syrup/(excess sugar)} = 75.75 \text{ g}/49.25 \text{ g} = 1.54.$$

Comments. All the sugar is dissolved at the higher temperature, to give a single phase of syrup (Fig. 9–1.1a). When the mixture cools, the *solubility limit* for an 80 sugar–20 water syrup is reached at 87°C (189°F). Typically, however, some *supercooling* is encountered before the excess phase (in this case, sugar) starts to separate. In fact, in this case, supercooling can proceed to room temperature, so that the start of separation may be delayed considerably. Similar supersaturation commonly occurs in metals and other materials. ◀

Example 9–1.2 A brine contains 9% NaCl (91% H_2O by weight). How many grams of water (per 100 g brine) must be evaporated before the solution becomes saturated at 50°C?

Solution: Basis: 100 g brine = 9 g NaCl + 91 g H_2O.
From Fig. 9–1.1(b), solubility of NaCl in brine = 27%.

$$\text{Grams saturated brine} = 9 \text{ g}/0.27 = x/0.73;$$

$$x = 24.3 \text{ g } H_2O \text{ for saturation};$$

$$91 \text{ g} - 24.3 \text{ g} = 66.7 \text{ g to be evaporated.}$$

Alternatively, the salt is equal to 27% of the brine remaining after evaporation:

$$(0.09)(100 \text{ g}) = 0.27(100 - y);$$

$$y = 66.7 \text{ g.}$$

Comment. By convention, compositions of liquids and solids are reported in weight percent, unless stated otherwise. ◀

Example 9–1.3 A brine with 40 g H_2O and 10 g NaCl is cooled to -10°C (14°F). (a) This brine is placed in a beaker. How much salt can be dissolved in it? (b) An identical 40–10 brine is placed in a second beaker. How much ice can be added to it without exceeding the solubility limit?

Solution: Use the solubility curves of Fig. 9–1.2. (a) NaCl solubility limit at -10°C = 25%, leaving 75% H_2O.

$$\text{g NaCl} = 0.25(50 + x) = 10 + x;$$

$$x = 3.33 \text{ g NaCl added.}$$

b) H_2O solubility limit at -10°C = 86% H_2O, leaving 14% NaCl.

$$\text{g } H_2O = 0.86(50 + y) = 40 + y;$$

$$y = 21.4 \text{ g } H_2O \text{ added.} ◀$$

Example 9–1.4 Salt is spread on the street when the temperature is −10°C (14°F). Part of the ice melts, and the brine that forms becomes saturated with H_2O. Ninety grams of this brine is splattered over your car. How many grams of NaCl come with it?

Solution: From Fig. 9–1.2, the solubility limit at −10°C is 86% H_2O in brine. The balance is NaCl.

$$0.14(90 \text{ g}) = 12.6 \text{ g NaCl}.$$

Comments. It is recommended that the reader obtain a short, transparent millimeter rule. This will permit a more accurate reading of the solubility limits. Many phase diagrams are drawn to an accuracy of ±1% on a larger scale and then photo-reduced. Thus, your accuracy of calculation is generally limited by your reading of the graphs. ◀

9–2 PHASE DIAGRAMS (EQUILIBRIUM DIAGRAMS)

Figure 9–1.3 is the *phase diagram* for the lead–tin system. This diagram can be used as a "map" from which the phases present at any particular temperature or composition can be read if the alloy is at *equilibrium*, i.e., if all possible reaction has been completed.

For example, at 50% tin and 100°C, the phase diagram indicates two solid phases: α is a lead-rich solid solution with some dissolved tin; β is almost pure tin with very little dissolved lead. At 200°C, an alloy of 10% tin and 90% lead lies in an area that is entirely α phase. It is a solid solution of lead with some tin dissolved in it. At the same temperature, but for 30% tin and 70% lead, the phase diagram indicates a mixture of two phases—liquid and the α solid solution; if this latter alloy were heated to a temperature of 300°C, it would become all liquid.

The phase fields in equilibrium diagrams, of course, depend on the particular alloy systems being depicted. When copper and nickel are mixed, the phase diagram is as shown in Fig. 9–2.1. This phase diagram is comparatively simple, since only

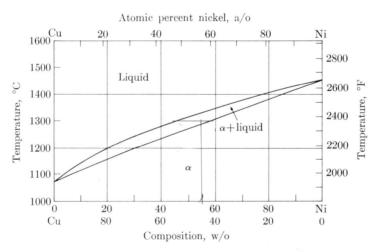

Fig. 9–2.1   Cu–Ni diagram. All solid alloys contain only one phase. This phase is fcc. (Adapted from *Metals Handbook*, American Society for Metals.)

two phases are present. In the lower part of the diagram, all alloys form only one solid solution and therefore only one crystal structure. Both the nickel and the copper have face-centered cubic structures. Since the atoms of each are nearly the same size, it is possible for nickel and copper atoms to replace each other in the crystal structure in any proportion at 1000°C. When an alloy containing 60% copper and 40% nickel is heated, the solid phase exists until the temperature of about 1235°C (2255°F) is reached. Above this temperature and up to 1275°C (2330°F) the solid and liquid solutions coexist. Above 1275°C only a liquid phase remains.

Freezing ranges   As shown in the foregoing phase diagrams, the range of temperatures over which freezing occurs varies with the composition of the alloy. This situation influences the plumber, for example, to select a high-lead alloy as a "wiping" solder when a solder is needed that will not freeze completely at one temperature. If one chooses an 80 Pb–20 Sn solder, the freezing range is from 280 to 183°C as compared with only 188 to 183°C for a 60 Sn–40 Pb solder.

The terms *liquidus*, the locus of temperatures above which all compositions are liquid, and *solidus*, the locus of temperatures below which all compositions are solid, are applied in this connection. Every phase diagram for two or more components must show a liquidus and a solidus, and an intervening freezing range. Whether the components are metals or nonmetals (Fig. 9–2.2), there are certain locations on the phase diagram where the liquidus and solidus meet. For a pure component, this point lies at the edge of the diagram. When it is heated, a pure material will remain solid until its melting point is reached and will then change entirely to liquid before it can be raised to a higher temperature.

The solidus and liquidus must also meet at the eutectic. In Fig. 9–1.3, the liquid solder composed of 61.9% tin and 38.1% lead is entirely solid below the eutectic

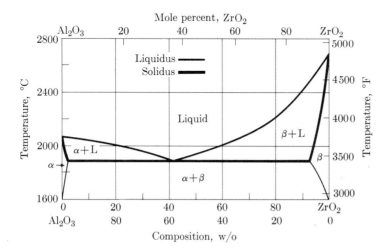

Fig. 9–2.2   The Al_2O_3–ZrO_2 diagram. Like all phase diagrams, the liquidus delineates the lowest temperatures for solely liquid. The solidus is the upper limit for completely solid. The freezing range of solid + liquid lies between the two. The two meet at the eutectic and where single phases melt. (Adapted from Alper et al, Amer. Ceram. Soc.)

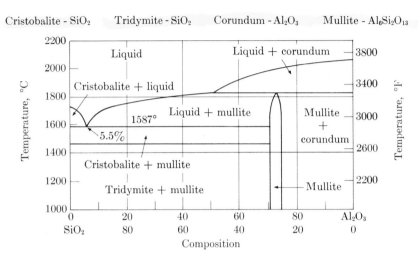

Cristobalite - SiO$_2$ Tridymite - SiO$_2$ Corundum - Al$_2$O$_3$ Mullite - Al$_6$Si$_2$O$_{13}$

Fig. 9–2.3   SiO$_2$–Al$_2$O$_3$ diagram. The phase diagrams for nonmetals are used in the same manner as those for metals. The only difference is the longer time required to establish equilibrium. (Adapted from Askay and Pask, *Science*.)

temperature and entirely liquid above it. At the eutectic temperature three phases can coexist $[(\alpha + \text{Liq} + \beta)$ in the Pb–Sn system$]$.

Isothermal cuts   A traverse across the phase diagram at a constant temperature (*isotherm*) provides a simple sequence of alternating 1- and 2-phase fields. Consider the SiO$_2$–Al$_2$O$_3$ diagram of Fig. 9–2.3 at 1650°C (3000°F), and you will see that the sequence is 1–2–1–2–1–2–1. With pure SiO$_2$, only *one* phase exists (named cristobalite). It holds negligible amounts of Al$_2$O$_3$ in solid solution.* Therefore, a second phase (liquid) appears with the addition of Al$_2$O$_3$. The *two*-phase region contains cristobalite and liquid. Between 4% Al$_2$O$_3$ (96% SiO$_2$) and 8% Al$_2$O$_3$ (92% SiO$_2$), the liquid can dissolve all of the SiO$_2$ and Al$_2$O$_3$ that is present, so just *one* phase exists. Beyond 8% Al$_2$O$_3$ (<92% SiO$_2$), the solubility limit of the liquid for Al$_2$O$_3$ is exceeded, and solid mullite precipitates. The *two* phases liquid and mullite coexist.† The solid-solution range of mullite is from 71% Al$_2$O$_3$ (29% SiO$_2$) to 75% Al$_2$O$_3$ (25% SiO$_2$). Only *one* phase is stable in this range, because it can accommodate both the SiO$_2$ and Al$_2$O$_3$ that are present. A *two*-phase field, mullite and corundum (Al$_2$O$_3$) follows and extends to within a line's width of the right side of the phase diagram. With only Al$_2$O$_3$, this *one* phase is called corundum.

---

*   There is a slight solubility, but it is so low that we are not able to show it on the phase diagram.
†   When 8% Al$_2$O$_3$ is just slightly exceeded there will be very little mullite. When the right side of this 2-phase field is approached, very little liquid remains.

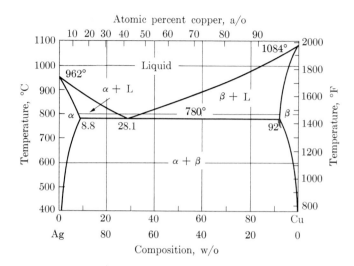

Fig. 9–2.4   Ag–Cu diagram. (Adapted from *Metals Handbook*, American Society for Metals.)

Example 9–2.1 Sterling silver, an alloy containing approximately 92.5% silver and 7.5% copper (Fig. 9–2.4), is heated slowly from room temperature to 1000°C (1830°F). What phase(s) will be present as heating progresses?

Answer

Room temperature to 760°C	$\alpha + \beta$
760°C to 800°C	Only α
800°C to 900°C	α + liquid
900°C to 1000°C	Only liquid ◀

Example 9–2.2 A combination of 90% SiO_2 and 10% Al_2O_3 is melted at 1800°C and then cooled extremely slowly to 1400°C. What phase(s) will be present in the cooling process?

Answer: (See Fig. 9–2.3.)

1800°C to 1700°C	Only liquid
1700°C to 1587°C	Liquid + mullite ($Al_6Si_2O_{13}$)
1587°C to 1470°C	Mullite + cristobalite (SiO_2)
<1470°C	Mullite + tridymite (SiO_2)

Comments. There must, of course, be three phases (liquid + mullite + cristobalite) at 1587°C as the material passes from the (Liq + Mul) field to the (Cri + Mul) field. Likewise, there are three phases at 1470°C.

The cooling will have to be extremely slow, because the process of changing the strong Si–O bonds from one structure to another is very slow.

Pure silica has three common polymorphs: cristobalite and tridymite at high temperatures, and quartz at low temperatures. ◀

Example 9–2.3 Refer to Fig. 9–2.4 for the Ag–Cu system. (a) Locate the liquidus and the solidus. (b) How many phases are present where the two meet?

Solution

a) Liquidus: 962° at 100% Ag, to 780°C at 71.9% Ag (28.1% Cu), to 1084° at 100% Cu. Solidus: 962° at 100% Ag, to 780°C at 91.2% Ag (8.8% Cu), remaining at 780°C to 92% Cu (8 % Ag), to 1084 at 100% Cu.
b) With a single component (only Ag *or* only Cu), two phases are present (solid + liquid) where the liquidus and solidus meet.

   With two components (Ag *and* Cu), three phases (α + liquid + β) are present where the liquidus and solidus meet (at the eutectic).

Comment. There are always two phases present in the temperature range between the liquidus and solidus of a 2-component, or binary, phase diagram. ◀

Example 9–2.4 Refer to the Al–Mg diagram (Fig. 9–5.4) later in the chapter. Apply the 1–2–1–2– · · · rule (a) at 500°C; (b) at 200°C.

Answer

a)

	α		$(\alpha + L)$		Liq		$(L + \epsilon)$		ϵ	
% Mg	0		11		29		76		92	100
% Al	100		89		71		24		8	0

b)

	α		$(\alpha + \beta)$		β		$(\beta + \beta')$		β'		$(\beta' + \gamma)$		γ		$(\gamma + \epsilon)$		ϵ	
% Mg	0		3		35		37		~41	~42		49		~56		96	100

Comment. The dashed lines on the phase diagram are the best estimates based on present information. ◀

9–3   CHEMICAL COMPOSITIONS OF PHASES

In addition to serving as a "map," a phase diagram shows the chemical *compositions* of the phases that are present under conditions of equilibrium after all reaction has been completed. This information, along with information on the amount of each phase (Section 9–4), constitutes very useful data for the scientists and engineers who are involved with materials development, selection, and application in product design.

One-phase areas   The determination of the chemical composition of a single phase is automatic. *It has the same composition as the alloy.* This is to be, expected; since only liquid is present in a 60 Sn–40 Pb alloy at 225°C (Fig. 9–3.1), the liquid has to have the same 60–40 composition. This also holds when the location in the phase diagram involves a single-phase solid solution.

   You will observe that we reported the chemical composition of the individual phase (and of the total alloy) in terms of the *components*—in this case, lead and tin.

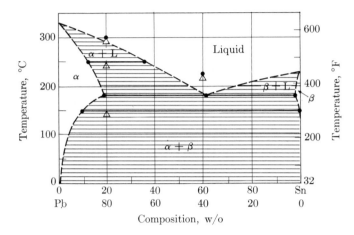

Fig. 9–3.1 Compositions of phases (Pb–Sn alloys). At 150°C, an 80Pb–20Sn alloy contains α and β. The chemical composition of α is dictated by the solubility curve. At this temperature, the solubility limit is 10% Sn (and 90% Pb) in the fcc α phase. See the text.

Two-phase areas   The determination of the chemical compositions of two phases can be handled on a rote basis. We will do this first; however, the rationale will follow in the next paragraph.

The chemical compositions of the two phases are located *at the two ends of the isotherm across the two-phase area.* To illustrate, take an 80 Pb–20 Sn solder at 150°C. As indicated in Fig. 9–3.1, α has a chemical composition of 10 w/o tin (and therefore 90 w/o lead). The composition of the β is nearly 100 w/o tin. Other isotherms on Fig. 9–3.1 permit us to read the chemical compositions of the two phases of any Pb–Sn alloy at any temperature.

The basis for the above procedure is simply that the *solubility limit* for tin in α at 150°C is 10 w/o. Our alloy exceeds this with its composition of 20 w/o tin. Therefore, α is saturated with tin and the excess tin is present in β. Conversely, the solubility limit for lead in β is <1 w/o; therefore, almost all of the lead must be in a phase other than β—specifically, in α.

Three-phase temperatures and eutectic reactions   A liquid that has the analysis of the eutectic composition (38.1 Pb–61.9 Sn, when we consider the Pb–Sn system) separates into two solid phases (α and β) at the eutectic temperature (183°C). Thus, at this temperature *only*, three phases can be in equilibrium. If this alloy is heated, the two solid phases of this solder melt into a one-phase liquid. We can write this reversible reaction as

$$L(61.9\% \text{ Sn}) \underset{}{\overset{183°C}{\rightleftharpoons}} \alpha(19.2\% \text{ Sn}) + \beta(97.5\% \text{ Sn}). \qquad (9\text{–}3.1)$$

The tin analyses are shown for the three phases that are in equilibrium at 183°C.*

---

* We could have used 38.1% Pb, 80.8% Pb, and 2.5% Pb for L, α, and β, respectively, rather than the tin values.

The reaction of Eq. (9–3.1) is called an *eutectic reaction* and involves a liquid and two solids. The more general form (presented earlier) is

$$L_2 \underset{\text{Heating}}{\overset{\text{Cooling}}{\rightleftharpoons}} S_1 + S_3, \qquad\qquad (9\text{–}1.2)$$

where the 1,2,3 subscripts refer to progressively increasing contents of one of the two components.

Study aid (phase diagrams)   Some people find it easier to introduce phase diagrams with a eutectic-free system (Cu–Ni). This pattern is followed in Topic X of *Study Aids for Introductory Materials Courses*. This is then followed by a topic (XI) that presents eutectics. This study alternative is recommended for those who initially find phase diagrams confusing.

Example 9–3.1   Consider the Cu–Ni system (Fig. 9–2.1) and at 1300°C. (a) What is the solubility limit of copper in solid α? (b) Of nickel in the liquid? (c) What are the chemical compositions of the phases in a 45 Cu–55 Ni alloy at 1300°C?

Answer: (a) 42 w/o Cu   (b) 45 w/o Ni   (c) α: (42 Cu–58 Ni); L: (55 Cu–45 Ni).

Comment. It is the solubility limit that determines the chemical compositions of the phases in a two-phase area. ◄

Example 9–3.2   An alloy of 40 Ag–60 Cu (Fig. 9–2.4) is cooled slowly from 1000°C to room temperature. (a) What phase(s) will be present as cooling progresses? (b) Indicate their compositions. (c) Write the eutectic reaction.

Answer: (a) and (b)

Temperature	α	Liquid	β
1000°C	—	40 Ag–60 Cu	—
800	—	66 Ag–34 Cu	8 Ag–92 Cu
780	91.2 Ag–8.8 Cu	71.9 Ag–28.1 Cu	8 Ag–92 Cu
600	96.5 Ag–3.5 Cu	—	2 Ag–98 Cu
400	99 Ag–1 Cu	—	Near 100 Cu
20 (extrapolated)	Near 100 Ag	—	Near 100 Cu

c)   Liquid (71.9% Ag) $\xrightarrow{780°C}$ α(91.2% Ag) + β(8% Ag).

Comments. The liquid becomes saturated with copper at about 890°C. The first β to separate at this temperature has the composition of 7 Ag–93 Cu. ◄

9-4 QUANTITIES OF PHASES

In addition to (1) identifying the stable, or equilibrium phases (Section 9–2) and (2) obtaining their chemical compositions from the phase diagrams (Section 9–3), we can (3) determine the *quantity* of each phase that is present under equilibrium conditions. This will be useful to us later when we consider the properties of multi-phase materials in Chapter 10.

One-phase areas   Again, as in Section 9–3, we have an automatic situation. With 225 g of a 80 Pb–20 Sn alloy at 300°C, only liquid is present, and its quantity is 225 g. It is equally valid to state that all (or 100%) of the alloy is liquid. In that manner we do not have to specify the exact weight of the alloy which is present.

Two-phase areas   We can use a rote basis to indicate the quantities (or the *quantity fractions*) of the phases that are present in the two-phase areas of phase diagrams. We will do this first; the rationale will follow. The quantities of the two phases are determined by *interpolating the composition of the alloy between the compositions of the two phases.* To illustrate, we will again take an 80 Pb–20 Sn solder at 150°C. As indicated in Fig. 9–4.1, the chemical composition of the alloy (80 Pb–20 Sn) is at a position which is 0.11 of the distance between the chemical composition of α (90 Pb–10 Sn) at this temperature, and the chemical composition of β (<1 Pb and ~100 Sn). Therefore, of the total amount of solder, the quantity fraction of β is 0.11 (and 0.89 of α) at 150°C.* It is equally appropriate to report this as 89% α and 11% β.

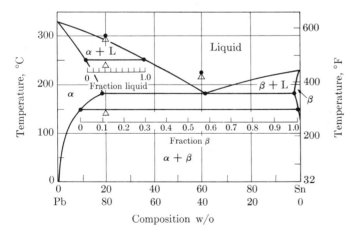

Fig. 9–4.1 Quantities of phases (Pb–Sn alloys). As observed in Fig. 9–3.1, an 80Pb–20Sn alloy contains α and β at 150°C. By interpolation, the fraction of β is 0.11. This same alloy contains 0.33 liquid at 250°C (and 0.67 α). See text.

---

*   Thus if we have 225 grams of solder, there would be ~200 grams of α and ~25 grams of β.

As another example, consider the same alloy at 250°C. On the basis of Fig. 9–4.1, we have α (88 Pb–12 Sn) and L (64 Pb–36 Sn). The chemical composition of this alloy as a whole (80 Pb–20 Sn) is $\frac{1}{3}$ of the distance between the chemical composition of α and the chemical composition of the liquid. Therefore, of the total amount of solder at 250°C, the quantity fraction of liquid is $\frac{1}{3}$ and of α is $\frac{2}{3}$.

You will observe that we have reported these quantity figures in terms of *phases*— in these cases, α and β, or α and L. This is in contrast to the chemical compositions that were reported in terms of the *components*—Pb and Sn.

To follow the above procedure, we would need to have a stretchable scale, because the length of the isotherm between the two solubility limit curves continuously varies. Interpolation by using a millimeter scale offers a simple alternative.* The alert student will also be quick to suggest interpolating on the basis of the numbers on the abscissa of the graph.

Materials balance   The rationale for the above interpolation procedure originates as a materials balance. By this we mean that the whole is equal to the sum of the parts. At 250°C, and with the above example, the lead in the total alloy is equal to the lead in the α plus the lead in liquid.

EXAMPLE   To illustrate, let us use 600 g of our same 80 Pb–20 Sn solder at 250°C. There are 480 g of Pb (and 120 g of Sn). We can consider that there are A grams of α (88 Pb–12 Sn) and L grams of liquid (64 Pb–36 Sn). Those compositions are located in Figs. 9–3.1 and 9–4.1. Of course

$$A + L = 600 \text{ g.}$$

Thus, on the basis of the chemical compositions of the two phases and the total alloy, the lead balance is

$$0.88A + 0.64L = 0.80(A + L), \tag{9–4.1}$$

or, in this case, $A = 2L$. This means that we have 200 g of liquid and 400 g of α—the same $\frac{1}{3}$ and $\frac{2}{3}$ values we calculated previously.

A generalization of Eq. (9–4.1) is

$$C_x(X) + C_y(Y) = C(X + Y), \tag{9–4.2}$$

where C_x and C_y are the chemical *compositions* of one of the components of phases x and y respectively; C is the composition of the total alloy. Likewise, X and Y are the *quantities* of the phases x and y. Algebraic rearrangement of Eq. (9–4.2) gives either

QUANTITY FRACTION OF X =
$$\frac{X}{X + Y} = \frac{C_y - C}{C_y - C_x}, \tag{9–4.3a}$$

or

QUANTITY FRACTION OF Y =
$$\frac{Y}{X + Y} = \frac{C_x - C}{C_x - C_y}. \tag{9–4.3b}$$

The value of $X/(X + Y)$ is the quantity fraction of x; and, by like token, $Y/(X + Y)$ is the quantity fraction of y. Equation (9–4.3) is the *inverse lever rule*, which is com-

---

* A small, transparent mm scale is recommended.

monly presented in other materials texts as a means for calculating the *quantity fractions of* the phases.*

• **The three-phase special case**   Lead–tin solders have three phases when they are equilibrated at 183°C, which is the eutectic temperature. In this special case, we *cannot* calculate exactly the quantity fractions of α, β, and the liquid that are present. We can calculate that a 70 Pb–30 Sn alloy contains 0.86 α at 182°C (and 14% β). We also can determine that the same solder contains $\frac{1}{4}$ liquid (and 75% α) at 184°C. At the eutectic temperature, between 182 and 184°C, the β and some of the α react on heating to give a eutectic liquid. In the process, 1000 grams of solder would have 110 g of α (that is, 860 g − 750 g) and 140 g of β consumed to form 250 g of eutectic liquid (61.9 Pb–38.1 Sn). Therefore, with three phases at 183°C, we can only indicate that the quantity of α is between 860 and 750 grams; the quantity of β, between 140 and 0 grams; and of liquid, between 0 and 250 grams.†

Example 9–4.1   Consider a 90 Cu–10 Sn bronze. (a) What phases are present at 300°C? What are their chemical compositions? What is the fraction of each? (b) Repeat for 600°C. (c) At what temperature is there $\frac{1}{3}$ liquid? $\frac{2}{3}$ liquid?

Solution: Refer to Fig. 9–5.2.
a)   α: 93 Cu–7 Sn   (37−10)/(37−7) = 0.9,   α ≠ ε FROM 7% – 37% Sn @ 300°C
      ε: 63 Cu–37 Sn   (10− 7)/(37−7) = 0.1;
b)   α: 90 Cu–10 Sn   1.0;
c)   By using an mm scale: $\frac{1}{3}$ L at 900°C, $\frac{2}{3}$ L at 975°C. ◀

Example 9–4.2   Using lead, make a materials balance that shows *A*, the quantity of α, and *B*, the quantity of β, in 75 grams of a 70 Pb–30 Sn solder at 182°C.

Solution: From Eq. (9–4.2), and using the lead analyses of Fig. 9–1.3,

$$0.808A + 0.025B = 0.70(A + B).$$

$$A + B = 75 \text{ grams.}$$

Solving simultaneously,

$$A = 64.7 \text{ g}, \qquad B = 10.3 \text{ g},$$

$$A/(A + B) = 0.86; \qquad B/(A + B) = 0.14.$$

Comment. A materials balance using tin, $0.192A + 0.975B = 0.30(A + B)$, gives the identical answer. ◀

---

* You may use this lever rule for calculating the quantity fraction of the phases, or, if simpler, interpolate according to Fig. 9–4.1. If you use the abscissa as a scale for interpolation, Eq. (9–4.3) is followed indirectly. In any event, observe that the chemical composition of the alloy is at the "center of gravity" between the compositions of the two contributing phases. This is the fulcrum of the lever; the majority phase is on the *shorter* arm of the lever.
† This special case of three phases for 2-component systems has an analog in 1-component systems. For example, with H_2O at 0°C (32°F), it is impossible to calculate the fraction that is ice and the fraction that is liquid water.

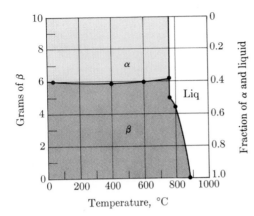

Fig. 9–4.2 Quantity of phases (10 g of 40Ag–60Cu alloy). The quantity of β was determined by interpolation (Example 9–4.3).

Example 9–4.3 As a function of temperature, plot the quantity of β in 10.0 grams of a 40 Ag–60 Cu alloy (Fig. 9–2.4).

Solution: The chemical compositions of the phases were previously determined for Example 9–3.2. At 890°C and above, all liquid; therefore, 0 grams of β. By interpolating,

At 800°C	0.45(10 g) = 4.5 g
780° +	0.50(10) = 5.0
780° −	0.615(10) = 6.15
600°	0.60(10) = 6.0
400°	0.595(10) = 5.95
20°	0.60(10) = 6.0.

This is plotted as the curve over the dark shaded area of Fig. 9–4.2.

Comments. We can also indicate the number of grams (or the quantity fraction) of α and liquid in Fig. 9–4.2, because the total must add up to 10.0 grams at all temperatures. We have done this with modified shading.

Note the discontinuity in quantities at the eutectic temperature (780°C), which is the special case where three phases are present simultaneously. ◄

9–5 COMMERCIAL ALLOYS AND CERAMICS

The reader may be relieved to learn that most commercial alloys have compositions that lie in the simpler parts of the phase diagrams. For example, the compositions of most brasses lie in the single-phase α area of Fig. 9–5.1. Likewise, the common bronzes contain less than 10% tin. There is little commercial interest in the more complex appearing central areas* of the Cu–Sn system (Fig. 9–5.2). In Chapter 11,

---

* Although more complex in appearance, all areas are either 1-phase or 2-phase. We have already considered these. Thus, for any alloy we can determine (1) *what* phase(s), (2) their chemical *compositions*, and (3) the *quantity* of each phase.

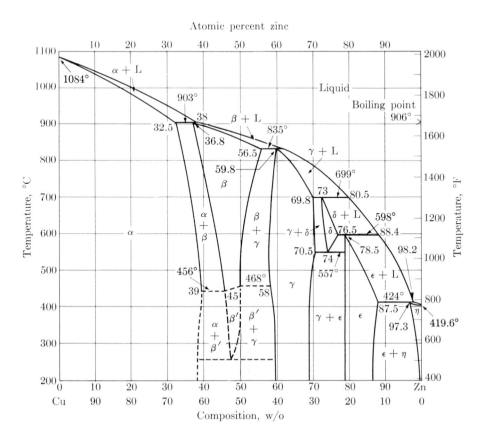

Fig. 9–5.1 Cu–Zn diagram. (Adapted from *Metals Handbook*, American Society for Metals).

we will consider alloys such as 95 Al–5 Cu, 90 Al–10 Mg, 90 Mg–10 Al (Figs. 9–5.3 and 9–5.4) because each forms one phase at elevated temperatures but crosses a solubility limit curve during cooling. By controlling the rate of separation of the second phase, the engineer can greatly increase the strength of the alloy. This, of course, is of major interest (Chapter 11). The Al–Si system (Fig. 9–5.8) reveals the commercial basis for purifying semiconducting and related materials (Example 9–5.1).

In the next two sections we shall examine the Fe–Fe₃C phase diagram for two reasons. First, steels are iron–carbon alloys that are very prominent in any technical civilization. Secondly, steels can serve as a prototype for heat-treating procedures. Suitable heat treatments and desired microstructures (and therefore desired properties) can be designed efficiently only with a knowledge of the phase diagram.

Ceramists utilize phase diagrams as do metallurgists.* However, within this text, we include only five: Al_2O_3–ZrO_2 (Fig. 9–2.2), Al_2O_3–SiO_2 (Fig. 9–2.3), Fe–O

---

* See, for example, the 3-volume compendium, *Phase Diagrams for Ceramists*, American Ceramic Society.

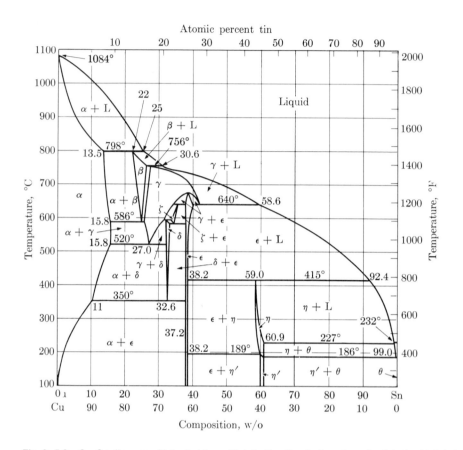

Fig. 9–5.2   Cu–Sn diagram. (Adapted from *Metals Handbook*, American Society for Metals.)

(Fig. 9–5.5), FeO–MgO (Fig. 9–5.6), and $BaTiO_3$–$CaTiO_3$ (Fig. 9–5.7).* The first involves very refractory, i.e., high-melting oxides that have a eutectic temperature just short of 2000°C. The second, Al_2O_3–SiO_2, is pertinent to clay-base ceramics since the better grade clays are approximately 40 Al_2O_3–60 SiO_2 after processing. The Fe–O diagram shows a nonstoichiometric range of $Fe_{1-x}O$ (ϵ of Fig. 9–5.5) that was discussed in Chapters 4, 5, and 8. The FeO–MgO diagram (Fig. 9–5.6) reveals a complete solid solution series below the solidus temperatures. This compares directly with the Cu–Ni system in Fig. 9–2.1. The $BaTiO_3$–$CaTiO_3$ phase diagram (Fig. 9–5.7) shows how the addition of $CaTiO_3$ affects the temperature for the change of tetragonal $BaTiO_3$ (α) to cubic $BaTiO_3$ (β) (cf. Fig. 8–6.2).

---

* Commonly, ceramic products involve three or more components, for example, MgO–Al_2O_3–SiO_2 and K_2O–Al_2O_3–SiO_2 in electrical porcelains. Although important, this text will not consider the more complex ternary phase diagrams.

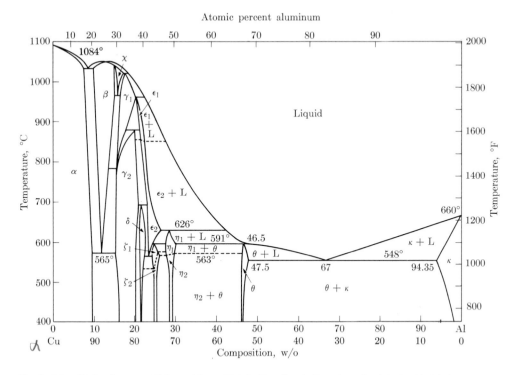

Fig. 9–5.3   Al–Cu diagram. (Adapted from *Metals Handbook*, American Society for Metals.)

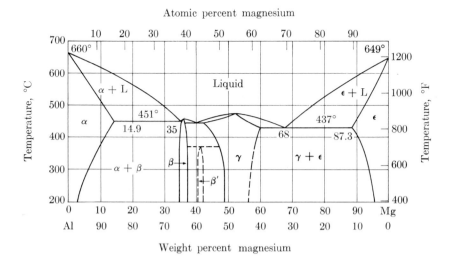

Fig. 9–5.4   Al–Mg diagram. (Adapted from *Metals Handbook*, American Society for Metals.)

Atomic percent oxygen, a/o

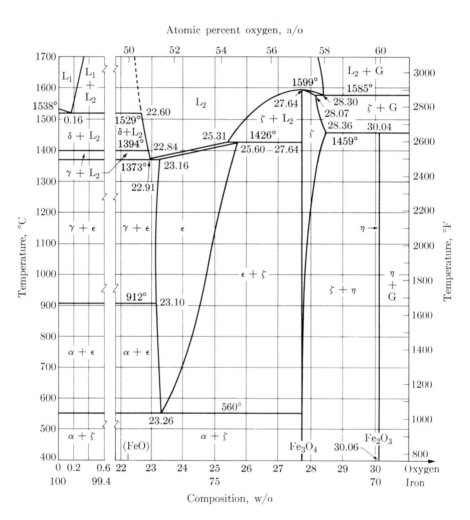

Fig. 9–5.5   Fe–O diagram. (Adapted from *Metals Handbook*, American Society for Metals.)

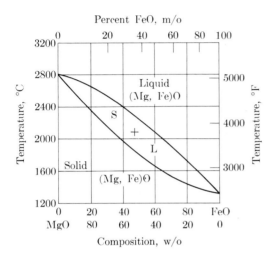

Fig. 9–5.6 MgO–FeO diagram. Comparable to the Cu–Ni diagram (Fig. 9–2.1); both the liquid and the solid phases possess full solubility. The structure of the FeO–MgO solid solution is shown in Fig. 4–3.1.

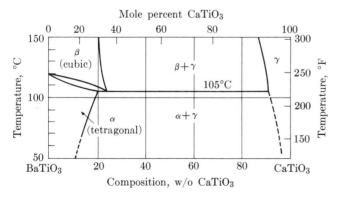

Fig. 9–5.7 $BaTiO_3$–$CaTiO_3$ diagram. (Adapted from DeVries and Roy, American Ceramic Society.)

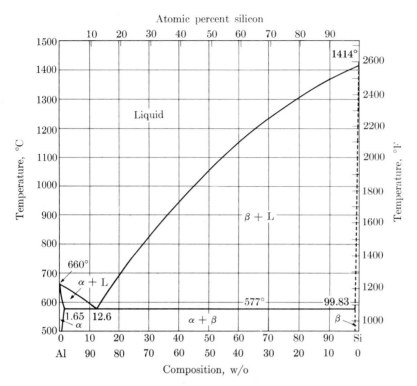

Fig. 9–5.8   Al–Si diagram. (Adapted from *Metals Handbook*, American Society for Metals.)

• **Example 9–5.1** The Al–Si diagram (Fig. 9–5.8) has the solubility limit exaggerated for Al in β because, at the eutectic temperature (577°C), β will hold a maximum of only 0.17 w/o Al (that is, 99.83% Si).

a)   Assume a linear solid solubility curve; what is the aluminum analysis of β at 1300°C?

b)   An alloy containing 98 Si–2 Al is equilibrated at 1300°C, and the liquid removed. What fraction of the aluminum is removed?

c)   The remaining solid is remelted at 1450°C. At what temperature will the new solid start to form?

d)   What is its chemical analysis?

Solution: (a) Being linear,

$$\frac{0.17\%}{(1414 - 577°C)} = \frac{C_\beta}{(1414 - 1300°C)},$$

$$C_\beta = 0.023 \text{ w/o Al.}$$

b)   From Fig. 9–5.8, $C_L = 21$ w/o Al. Using 100g,

$$L = [100 \text{ g}][(2 - 0.023)/(21 - 0.023)] = 9.4 \text{ g Liq};$$

$$\beta = 90.6 \text{ g.}$$

Fraction aluminum removed:

$$\frac{(9.4 \text{ g})(0.21)}{(100 \text{ g})(0.02)} = 0.99.$$

c) The upper end of the liquidus curve of Fig. 9–5.8 is approximately linear. Therefore,

$$\frac{21}{(1414 - 1300°C)} = \frac{0.023}{(1414 - T°C)};$$

$$T = 1414 - 0.124°C.$$

d) Repeating (a),

$$\frac{0.17\%}{(1414 - 577°C)} = \frac{C'_\beta}{0.124°C}$$

$$C'_\beta = 0.00003 \text{ w/o Al}.$$

Comment. By repeating this type of procedure several times, the metallurgist is able to produce silicon with $<1/10^9$ aluminum for controlled-purity semiconductor materials. ◀

• **Example 9–5.2** An alloy (5.8 kg) contains 6.7 Ag–93.3 Cu. It is melted and then cooled to 950°C. The first solid (β) appeared at $\sim 1050°C$ and contained $\sim 2\%$ Ag. At 950°C, the separating solid contains $\sim 6\%$ Ag; however, the cooling is sufficiently rapid so that the initial solid did not change composition. As a result the average β composition is $\sim 4\%$ Ag. (Diffusion is faster in the liquid, so its composition at 950°C is $\sim 26\%$ Ag.)
a) How much liquid is present?
b) Another step of cooling to 800°C permits solidification to continue in a comparable manner (and without changing the composition of the previous solid). What phases are present at 800°C?
c) Give the approximate amounts.

Solution

a)
$$\frac{(6.7 - \sim 4)}{(26 - \sim 4)}(5.8 \text{ kg}) = \sim 0.7 \text{ kg}.$$

b) Liquid ($\sim 65\%$ Ag); solid (2% to 8% Ag).
c) The average composition of the second generation of β is $\sim 7\%$ Ag:

$$\left[\frac{26 - \sim 7}{65 - \sim 7}\right](0.7 \text{ kg}) = \sim 0.2 \text{ kg liquid}.$$

Therefore,

$$\sim 5.6 \text{ kg } \beta.$$

Comment. This is an example of segregation (see Fig. 11–2.1). It is normal in any solidification process unless opportunity is given for diffusion.* Of course, the cooling is continuous rather than the 2-step sequence of our illustration. ◀

---

* An introduction to "delayed equilibrium" is presented in *Study Aids for Introductory Materials Courses*. This is recommended for the reader who wants to look more closely at segregation, etc.

9-6   PHASES OF THE IRON–CARBON SYSTEM

Steels, which are primarily alloys of iron and carbon, offer illustrations of the majority of reactions and microstructures available to the engineer for adjusting material properties. Also, the iron–carbon alloys are among the prominent structural engineering materials.

The versatility of the steels as engineering materials is evidenced by the many kinds of steel that are manufactured. At one extreme are the very soft steels used for deep-drawing applications such as automobile fenders and stove panels. At the other are the extremely hard and tough steels used for gears and bulldozer blades. Some steels must have abnormally high resistance to corrosion. Steels for such electrical purposes as transformer sheets must have special magnetic characteristics so that they may be magnetized and demagnetized many times each second with low power losses. Other steels must be completely nonmagnetic, for such applications as wrist watches and minesweepers. Phase diagrams can be used to help explain each characteristic described above.

Pure iron changes its crystal structure twice before it melts. As discussed in Section 3–4, iron changes from bcc to fcc at 912°C (1673°F). This change is reversed at 1394°C (2540°F) to form the bcc structure again. The bcc polymorph then remains stable until iron melts at 1538°C (2800°F).

Ferrite, or α-iron   The structural modification of pure iron at room temperature is called either α-iron or *ferrite*. Ferrite is quite soft and ductile; in the purity that is encountered commercially, its tensile strength is less than 310 MPa (45,000 psi). It is a ferromagnetic material at temperatures under 770°C (1418°F). The density of ferrite is 7.88 Mg/m^3 ($=7.88$ g/cm^3).

Since ferrite has a body-centered cubic structure, the spaces among atoms are small and pronouncedly oblate and cannot readily accommodate even a small spherical carbon atom. Therefore, solubility of carbon in ferrite is very low (<1 carbon per 1000 iron atoms). The carbon atom is too small for substitutional solid solution, and too large for extensive interstitial solid solution (Section 4–2).

Austenite, or γ-iron   The face-centered modification of iron is called *austenite*, or *γ-iron*. It is the stable form of pure iron at temperatures between 912°C and 1394°C. Making a direct comparison between the mechanical properties of austenite and ferrite is difficult because they must be compared at different temperatures. However, at its stable temperatures, austenite is soft and ductile and consequently is well suited to fabrication processes. Most steel forging and rolling operations are performed at 1100°C (2000°F) or above, when the iron is face-centered cubic. Austenite is not ferromagnetic at any temperature.

The face-centered cubic structure of iron has larger interatomic spacings than does ferrite. Even so, in the fcc structure the holes are barely large enough to crowd the carbon atoms into the interstices, and this crowding introduces strains into the structure. As a result, not all the holes can be filled at any one time ($\sim 6\%$ at 912°C). The maximum solubility is only 2.11% (9 a/o) carbon (Fig. 9–6.1). By definition, steels

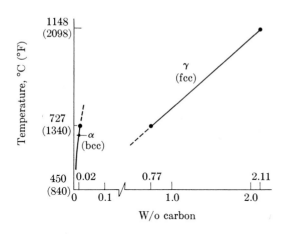

Fig. 9–6.1   Carbon solubility in iron. Negligible carbon dissolves in ferrite (α) at 20°C. The solubility increases to only 0.02 w/o at 727°C. The solubility of carbon in austenite (γ) increases from 0.77 w/o at 727°C to 2.11 w/o at 1148°C. (Cf. Fig. 9–7.1.)

contain less than 1.2% carbon; thus steels may have their carbon completely dissolved in austenite at high temperatures.

δ-iron   Above 1394°C (2540°F), austenite is no longer the most stable form of iron, since the crystal structure changes back to a body-centered-cubic phase called δ-iron. δ-iron is the same as α-iron except for its temperature range, and so it is commonly called δ-ferrite. The solubility of carbon in δ-ferrite is small, but it is measurably larger than in α-ferrite, because of the higher temperature.

Iron carbide   In iron–carbon alloys, carbon in excess of the solubility limit must form a second phase, which is most commonly iron carbide (*cementite*).* Iron carbide has the chemical composition of Fe_3C. This does not mean that iron carbide forms molecules of Fe_3C, but simply that the crystal lattice contains iron and carbon atoms in a three-to-one ratio. Fe_3C has an orthorhombic unit cell with 12 iron atoms and 4 carbon atoms per cell, and thus has a carbon content of 6.7 weight percent. Its density is 7.6 Mg/m^3 ($= 7.6$ g/cm^3).

As compared with austenite and ferrite, cementite is very hard. The presence of iron carbide with ferrite in steel greatly increases the strength of the steel (Section 10–3). However, because pure iron carbide is nonductile, it cannot adjust to stress concentrations; therefore it is relatively weak by itself. (Compare with nonductile ceramic materials in Section 8–7.)

Example 9–6.1   Carbon atoms sit in the $\frac{1}{2},\frac{1}{2},0$ positions in bcc iron.
a)   Using $r_C = 0.077$ nm and $R_\alpha = 0.124$ nm, how much must the nearest iron atoms be displaced to accommodate the carbon?
b)   A carbon atom sits in the $\frac{1}{2},\frac{1}{2},\frac{1}{2}$ position in fcc iron. Using $R_\gamma = 0.127$ nm (Table 2–5.1) for the radius of γ-iron, how much crowding is present?

---

* Graphite is the excess carbon phase in most cast irons where both silicon and carbon contents are high (Section 13–1).

Fig. 9–6.2 See Example 9–6.1. Although there is a more efficient iron packing factor in the fcc γ than in the bcc α (0.74 vs. 0.68), the interstices that are present are larger (and fewer). As a result, the carbon atoms are less crowded in the γ than in the α. This permits a higher solubility in the fcc structure (Fig. 9–6.1.)

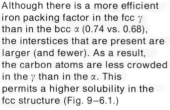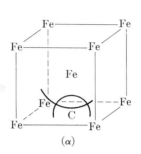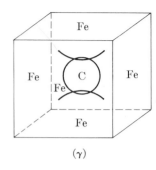

(α) (γ)

Solution: (a) Refer to Fig. 9–6.2(α). The nearest iron atoms are at $\frac{1}{2},\frac{1}{2},\frac{1}{2}$ and $\frac{1}{2},\frac{1}{2},-\frac{1}{2}$.

$$a = 4(0.124 \text{ nm})/\sqrt{3} = 0.286 \text{ nm};$$

$$\Delta = (0.124 \text{ nm} + 0.077 \text{ nm}) - (0.286 \text{ nm})/2 = 0.06 \text{ nm}.$$

This increases the center-to-center distance of iron by $2(0.06)/0.286 = \sim 40\%$.
b)   Refer to Fig. 9–6.2(γ). The nearest iron atoms are at the fcc positions.

$$a = 4(0.127 \text{ nm})/\sqrt{2} = 0.359 \text{ nm};$$

$$\Delta = (0.127 \text{ nm} + 0.077 \text{ nm}) - (0.359 \text{ nm})/2 = 0.025 \text{ nm}.$$

This increases the center-to-center distance of iron by $2(0.025)/(0.359) = \sim 14\%$.

Comments. Although bcc iron has a lower packing factor than fcc, the interstices do not accommodate a carbon atom without crowding. [The low packing factor (0.68) for bcc iron arises from six 6-f sites (related to $\frac{1}{2},\frac{1}{2},0$ and $0,0,\frac{1}{2}$ locations), plus twelve 4-f sites (related to $\frac{1}{2},\frac{1}{4},0$). The fcc iron (PF = 0.74) has only four 6-f sites and eight 4-f sites per unit cell (Fig. 8–2.2).] ◀

9–7 THE Fe–Fe$_3$C PHASE DIAGRAM

Figure 9–7.1 presents the phase diagram between iron and iron carbide, Fe$_3$C. This phase diagram is the basis for the heat-treating of the majority of our steels.

If you cover the 0%–1% carbon region of Fig. 9–7.1 by your hand, you will see a close resemblance to previous phase diagrams. The eutectic composition is at 4.3 w/o (17 a/o) carbon; the eutectic temperature is 1148°C (2100°F). *Cast irons* are based on this eutectic region, since they typically contain 2.5%–4% carbon (Section 13–1). They have relatively low melting points with certain desirable processing and mold-filling characteristics.

The iron-rich γ can dissolve up to 2.1 w/o (9 a/o) carbon. As discussed in the last section, the carbon atoms are dissolved interstitially in fcc iron. *Steels* are based on this solid-solution phase. Since steels contain less than 1.2% carbon, they can be single-phase at forging and other hot-working temperatures, which are commonly in the region of 1100 to 1250°C (2000 to 2300°F), a production advantage.

In the iron-rich region (>99% Fe and <1% C), the Fe–Fe$_3$C diagram differs from any of the diagrams we have examined previously. This difference arises from the

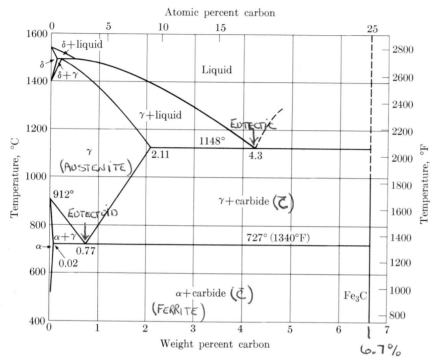

Fig. 9–7.1 Fe–Fe₃C phase diagram. The left lower corner receives prime attention in heat treating steels (Fig. 9–7.3).

fact that iron is polymorphic with bcc and fcc phases. Since we are not studying steel melting and solidification in this text, we will skip the features of this low-carbon region that are found above 1400°C. We will pay considerable attention to the features of the diagram in the 700°C–900°C (1300°F–1650°F) temperature range and the 0%–1% carbon range, because it is here that the engineer can develop within the steel those microstructures that are required for the desired properties.

The eutectoid reaction In Fig. 9–7.2 a comparison is made between the addition of carbon to austenite and the addition of common salt to water (Fig. 9–1.2). In each case, the addition of the solute lowers the stable temperature range of the solution. These two examples differ in only one respect: in the ice–salt system, a *liquid solution* exists above the eutectic temperature; in the iron–carbon system, a *solid solution* exists, so that a true eutectic reaction does not occur upon cooling. However, because of the analogy of this reaction to the eutectic reaction, it is called *eutectoid* (literally, eutectic-like).

$$\text{Eutectic:}\quad L_2 \underset{\text{Heating}}{\overset{\text{Cooling}}{\rightleftharpoons}} S_1 + S_3; \tag{9–3.2}$$

$$\text{Eutectoid:}\quad S_2 \underset{\text{Heating}}{\overset{\text{Cooling}}{\rightleftharpoons}} S_1 + S_3. \tag{9–7.1}$$

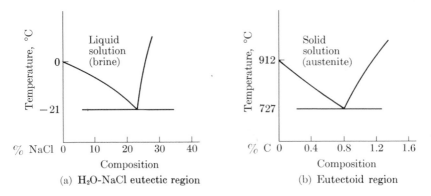

Fig. 9–7.2   Eutectic and eutectic-like (eutectoid) regions of phase diagrams.

The *eutectoid temperature* for iron–carbon alloys is 727°C (1340°F). The corresponding *eutectoid composition* is approximately 0.8% carbon. The specific eutectoid reaction for Fe–C alloys is

$$\gamma(0.77\% \text{ C}) \underset{}{\overset{727°\text{C}}{\rightleftarrows}} \alpha(0.02\% \text{ C}) + Fe_3C(6.7\% \text{ C}). \qquad (9\text{–}7.2)$$

Figure 9–7.3 shows the eutectoid region in greater detail than does Fig. 9–7.1.

Study aid (Fe–Fe₃C phase diagrams)   This short presentation in *Study Aids for Introductory Materials Courses* concisely outlines those phase relationships that are required to study steels.

Example 9–7.1   A eutectoid steel (~0.8% carbon) is heated to 800°C (1472°F) and cooled slowly through the eutectoid temperature (727°C). Calculate the number of grams of carbide that form per 100 g of steel.

Solution: At 726°C (1339°F),

$$\text{Carbide} = [(0.8 - 0.02)/(6.7 - 0.02)](100 \text{ g}) = 12 \text{ g} \qquad \text{(and 88 g } \alpha\text{)}.$$

Comment.   Careful measurements indicate that the eutectoid composition is 0.77% carbon. However, the figure, 0.8% C, is commonly used. ◀

Example 9–7.2   Plot the fraction of ferrite, austenite, and carbide in an alloy of 0.60% carbon, 99.40% iron as a function of temperature.

Solution: At 728°C (1342°F),

$$\text{Ferrite} = \frac{0.77 - 0.60}{0.77 - 0.02} = 0.23.$$

At 726°C (1339°F),

$$\text{Ferrite} = \frac{6.7 - 0.60}{6.7 - 0.02} = 0.91.$$

Results are shown in Fig. 9–7.4 for other temperatures and for other phases. ◀

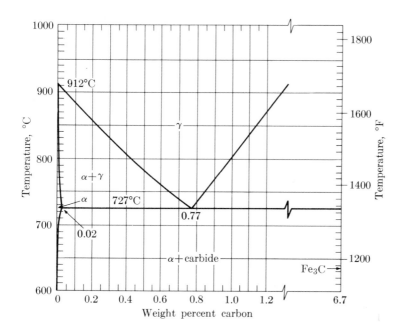

Fig. 9–7.3 The eutectoid region of the Fe–Fe$_3$C phase diagram. (Cf. Fig. 9–7.1.) Steels with ∼ 0.8% C are commonly called *eutectoid* steels. *Hypereutectoid* steels are above that value; *hypoeutectoid* steels, below.

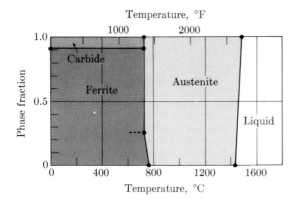

Fig. 9–7.4 Phase fractions (99.4Fe–0.6C). (See Example 9–7.2.) Approximately three fourths of this steel is austenite at 728°C (1342°F), (Since the γ has a eutectoid composition at that temperature, we will observe in the next section that the steel will contain ∼ 75% pearlite (and about one fourth proeutectoid ferrite).)

9–8 AUSTENITE DECOMPOSITION

During cooling, the Fe–C eutectoid reaction involves the simultaneous formation of ferrite α and carbide, $\bar{C}$, from the decomposition of austenite γ of eutectoid composition:

$$\gamma(\sim 0.8\% \text{ C}) \rightarrow \alpha + \bar{C}. \qquad (9\text{–}8.1)^*$$

There is nearly 12% carbide and more than 88% ferrite in the resulting mixture. Since the carbide and ferrite form simultaneously, they are intimately mixed. Characteristically, the mixture is *lamellar*; i.e., it is composed of alternate layers of ferrite and carbide (Fig. 9–8.1). The resulting microstructure, called *pearlite*, is very important in iron and steel technology, because it may be formed in almost all steels by means of suitable heat treatments.

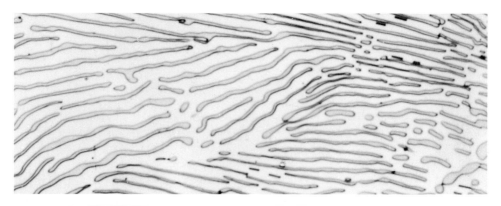

Fig. 9–8.1 Pearlite, X2500. This microstructure is a lamellar mixture of ferrite (lighter matrix) and carbide (darker). Pearlite forms from austenite of eutectoid composition. Therefore the amount and composition of pearlite is the same as the amount and composition of eutectoid austenite. (U.S. Steel Corp.)

Fig. 9–8.2 Annealed iron–carbon alloys (X500). The sequence (a) through (f) increases in carbon content; therefore there is a corresponding decrease in the amount of proeutectoid ferrite (white areas) that separated from the austenite prior to the eutectoid reaction. The gray areas are pearlite, a lamellar mixture of eutectoid ferrite and carbide. The network between the former grains of austenite in part (f) is proeutectoid carbide since the carbon content of that alloy is greater than 0.8 w/o. (U.S. Steel Corp.) ▶

---

* We will use $\bar{C}$ to denote *carbide*, to avoid confusion with C for elemental carbon. For our calculations, it will be permissible to use 0.8% C as the approximate eutectoid composition.

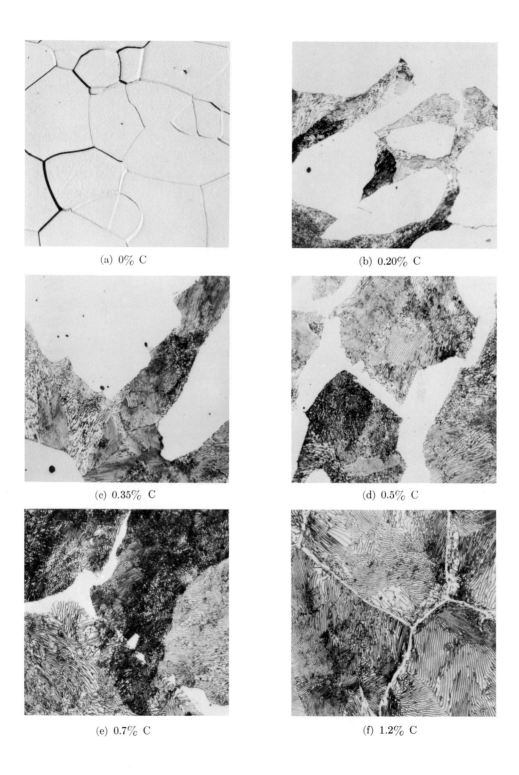

(a) 0% C

(b) 0.20% C

(c) 0.35% C

(d) 0.5% C

(e) 0.7% C

(f) 1.2% C

Pearlite is a specific mixture of two phases formed by transforming austenite of eutectoid composition to ferrite and carbide. This distinction is important, since mixtures of ferrite and carbide may be formed by other reactions as well. However, the microstructure resulting from other reactions will not be lamellar (compare Figs. 10–2.2b and 9–8.1) and consequently the properties of such mixtures will be different. (See Section 10–3.)

Since pearlite comes from austenite of eutectoid composition, the amount of pearlite present is equal to the amount of eutectoid austenite transformed (Fig. 9–8.2). We can determine this amount by measuring the fraction quantity of γ just above the eutectoid temperature. Examples 9–8.1 and 9–8.2 demonstrate this.

Study aid (austenite decomposition)   Since the reaction $\gamma \rightarrow \alpha + \bar{C}$ is central to the heat-treating of steel, it should be well understood by the reader as she or he proceeds to later sections of this text. The reader who needs further clarification on this solid reaction should consult Topic XV of *Study Aids for Introductory Materials Courses.*

Example 9–8.1 Determine the amount of pearlite in a 99.5% Fe–0.5% C alloy that is cooled slowly from 870°C (1600°F). Basis: 100 g of alloy.

Solution: From 870 to 780°C: 100 g austenite with 0.5% C.

From 780 to 727°C(+):   ferrite separates from austenite and the carbon content of the austenite increases to $\sim$0.8% C.

At 727°C(+):   composition of ferrite = 0.02% C,
amount of ferrite = 38 g;
composition of austenite = $\sim$0.8% C,
amount of austenite = 62 g.

At 727°C(−):   amount of pearlite = 62 g. (It came from, and replaced, the austenite with an eutectoid composition.)

Comments. Each of the above calculations assumes sufficient time for equilibrium to be attained. Ferrite that formed above 727°C (i.e., before the eutectoid reaction) is called *proeutectoid ferrite.* That which is part of the pearlite, having been formed from austenite with the eutectoid composition, is called *eutectoid ferrite.* (See Fig. 9–8.2b–d.) ◀

Example 9–8.2 From the results of the example above, determine the amount of ferrite and carbide present in the specified alloy (a) at 727°C(−), and (b) at room temperature. Basis: 100 g of alloy. (Some data come from Example 9–8.1.)

Solution: (a) At 727°C(−), and with the eutectoid at $\sim$0.8% C:

Amount of carbide: $62 \dfrac{0.8 - 0.02}{6.7 - 0.02} = 7.2 \dfrac{\text{g carbide}}{100 \text{ g steel}}$;

Amount of ferrite: 62 − 7.2 = 54.8 g ferrite formed with the pearlite (eutectoid),
                    $\underline{38\ \ }$ g ferrite formed before the pearlite (proeutectoid),
                    92.8 g ferrite total/100 g steel.

Alternative calculations:

Amount of carbide: $\qquad \dfrac{0.5 - 0.02}{6.7 - 0.02} = 7.2 \dfrac{\text{g carbide}}{100 \text{ g steel}};$

Amount of ferrite: $\qquad \dfrac{6.7 - 0.5}{6.7 - 0.02} = 92.8 \dfrac{\text{g ferrite}}{100 \text{ g steel}}.$

b)   At room temperature (the solubility of carbon in ferrite at room temperature may be considered zero for these calculations),

$$\frac{0.5 - 0}{6.7 - 0} = 7.5 \frac{\text{g carbide}}{100 \text{ g steel}};$$

$$\frac{6.7 - 0.5}{6.7 - 0} = 92.5 \frac{\text{g ferrite}}{100 \text{ g steel}}.$$

Comments. Additional carbide is precipitated from the ferrite below the eutectoid point because the solubility of carbon in ferrite decreases to nearly zero. This additional carbide is not part of the pearlite. (Each of these calculations assumes that equilibrium prevails.) ◄

9–9   PLAIN-CARBON AND LOW-ALLOY STEELS

An important category of steels is designed for heat-treating into the γ range (*austenitizing*), followed by cooling and decomposition of the austenite either directly (or indirectly) to ferrite plus carbide ($\alpha + \bar{C}$). If a steel contains primarily iron and carbon,* the alloy is called a *plain-carbon steel*. We use the term *low-alloy steel* if we add up to five percent of alloying elements such as nickel, chromium, molybdenum, manganese, or silicon. Alloying elements are added primarily to reduce the decomposition rate of austenite to ($\alpha + \bar{C}$) during heat treatment. The result is a much harder steel, as we will see in Chapter 10.

     The proeutectoid ferrite (Figs. 9–8.2b through e) deforms more readily than the ferrite within the pearlite since the former is more massive and is not reinforced by the harder, more rigid carbide. A metallurgist refers to *hypoeutectoid* steels (literally, lower carbon than the eutectoid composition) when speaking of steels with microstructures containing the separate ferrite areas. In a like manner, a *eutectoid steel* is predominantly pearlite if it is cooled slowly; and a *hypereutectoid steel* contains carbon *above* the eutectoid composition, thus producing proeutectoid carbide (Fig. 9–8.2f). Hypoeutectoid steels are more common than hypereutectoid steels.

Nomenclature for steels   The importance of carbon in steel has made it desirable to indicate the carbon content in the identification scheme of steel types. A four-digit numbering scheme is used, in which the last two digits designate the number of hundredths of percent of carbon content (Table 9–9.1). For example, a 1040 steel has 0.40% carbon (plus or minus a small workable range). The first two digits are simply code numbers that indicate the type of alloying element that has been added

---

* Plus approximately 0.5% Mn for processing purposes.

Table 9-9.1
Nomenclature for AISI and SAE Steels

AISI or SAE number	Composition
10xx	Plain-carbon steels*
11xx	Plain-carbon (resulfurized for machinability)
15xx	Manganese (1.0–2.0%)
40xx	Molybdenum (0.20–0.30%)
41xx	Chromium (0.40–1.20%), molybdenum (0.08–0.25%)
43xx	Nickel (1.65–2.00%), chromium (0.40–0.90%) molybdenum (0.20–0.30%)
44xx	Molybdenum (0.5%)
46xx	Nickel (1.40–2.00%), molybdenum (0.15–0.30%)
48xx	Nickel (3.25–3.75%), molybdenum (0.20–0.30%)
51xx	Chromium (0.70–1.20%)
61xx	Chromium (0.70–1.10%), vanadium (0.10%)
81xx	Nickel (0.20–0.40%), chromium (0.30–0.55%), molybdenum (0.08–0.15%)
86xx	Nickel (0.30–0.70%), chromium (0.40–0.85%), molybdenum (0.08–0.25%)
87xx	Nickel (0.40–0.70%), chromium (0.40–0.60%), molybdenum (0.20–0.30%)
92xx	Silicon (1.80–2.20%)

xx: carbon content, 0.xx%.
* All plain-carbon steels contain 0.50% ± manganese, and residual amounts (<0.05 w/o) of other elements.

to the iron and carbon. The classification (10xx) is reserved for plain-carbon steels with only a minimum amount of other alloying elements.

These designations for the steels are accepted as standard by both the American Iron and Steel Institute (AISI) and the Society of Automotive Engineers (SAE). Many commercial steels are not included in this classification scheme because of larger additions or more subtle variations in alloy contents. Usually, however, such steels have more specialized applications and may not be stocked as regular warehouse items.

• **Eutectoid shifts**   Figure 9-7.3 shows the eutectoid region for the $Fe-Fe_3C$ phase diagram, i.e., with only iron and iron carbide present. The eutectoid temperature is 727°C (1340°F); the eutectoid composition is just under 0.8% carbon.

In alloy steels, the carbon atoms and the iron atoms coordinate with atoms of other types, as well as with each other. Therefore, we should not be surprised that the eutectoid carbon content and temperature are altered from the above values when other elements are present. Figure 9-9.1(a) shows a shift in the temperature from 727° as nickel, manganese, chromium, silicon, and molybdenum are added. The first two

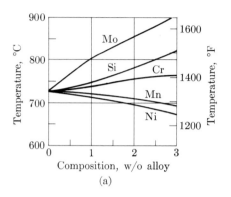
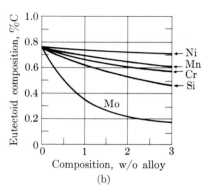

Fig. 9-9.1 Eutectoids in Fe–X–C alloys. (X is the third component—Mo, Si, Cr, Mn, or Ni.) Effect of alloy additions on (a) the temperature of the eutectoid reaction, and (b) the carbon content of the eutectoid. (Adapted from ASM data.)

elements lower the eutectoid temperature because they preferentially form fcc solid solutions with iron (austenite-formers); Cr and Mo raise the eutectoid temperature because they are bcc themselves, and are thus ferrite-formers in steels. Figure 9–9.1(b) shows that each of these alloying elements lowers the carbon content of the eutectoid composition from 0.73%. These curves are valuable because they let us predict the temperature of austenitization for heat treatment of low-alloy steels.*

• **Example 9-9.1** An SAE 1540 steel (C = 0.39%; Mn = 1.97%) is to be austenitized.
a) What temperature is required if it must be heated 30°C into the austenite range to facilitate the reaction?
b) How much pearlite will it form during cooling?

Solution: From Fig. 9–9.1, the eutectoid is shifted to C ∼ 0.65%, and T_e = 710°C. At 0.39% C, the lower side of the γ-range is ∼770°C (Fig. 9–9.2).

a) ∼770 + 30°C = 800°C (1470°F);
b) Fraction γ at 711°C = 0.6 (by interpolation between 0% and 0.65% on Fig. 9–9.2).
    Fraction pearlite below 710°C = 0.6.

Comments. For our purposes we can assume that solubility lines that outline the γ field do not change slope as small amounts of alloying elements are added. (This assumption will not hold for high alloy additions.)

In the absence of 1.97% Mn, 810°C + 30° would have been required to austenitize; and only half of the steel would be pearlite. ◀

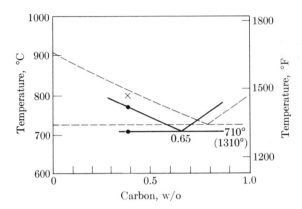

Fig. 9-9.2 Eutectoid shift (Example 9-9.1). With nearly 2% Mn, the eutectoid composition drops to 0.65% carbon, and the eutectoid temperature drops to 710°C (1310°F). Therefore, there is more γ from which pearlite forms.

REVIEW AND STUDY

The purpose of this chapter was two-fold:

1. To emphasize that one must understand how to extract necessary data from phase diagrams if one is to select, control, and modify multiphase materials.

2. To examine the Fe–C system more closely so we can use it as a prototype for solid-state microstructures in Chapter 10, and for the control of microstructures in Chapter 11.

SUMMARY: PHASE DIAGRAMS

We can obtain three types of data from a phase diagram, which is a graph of solubility-limit curves:

1. The phase diagram shows *what phase(s)* are present under equilibrium conditions for various temperatures and compositions.

2. The phase diagram also provides the *chemical compositions* for all equilibrated phases: (a) In a single-phase area the composition is simply the same as the alloy. (b) In a two-phase area, the compositions are located where the isotherm crosses the solubility-limit curves. Chemical compositions are reported in terms of the *component* percentages.

3. Finally, we can calculate the *quantity fractions* of the phases in a two-phase alloy by interpolation along the isotherm. This is the same as the "lever rule." These amounts may be reported in terms of *phase* percentages.

SUMMARY: Fe–Fe₃C DIAGRAM

The iron–carbon system has a eutectic region that is the basis for cast irons. Steels, however, are based on the fact that iron becomes fcc at elevated temperatures and can dissolve the carbon that is present. Thus steels are normally forged and hot-worked as a single-phase material.

The fcc solid solution (austenite, or γ) decomposes during cooling to bcc ferrite (α) and carbide ($\bar{C}$). The decomposition is a eutectoid reaction:

$$\gamma(\sim 0.8\% \text{ C}) \rightarrow \alpha(\sim 0.02\% \text{ C}) + \bar{C}(6.7\% \text{ C}). \qquad (9\text{--}10.1)$$

Heat treatment of steel involves forming austenite and then decomposing the austenite directly or indirectly (Chapter 10) to form a variety of microstructures. The microstructure that is obtained controls the properties of the steel (Chapters 10 and 11).

Two prominent categories of steels are (1) the plain-carbon steels, and (2) the low-alloy steels. The first are essentially Fe–Fe₃C alloys; the latter may contain up to five percent of other alloying elements, such as Mn, Ni, Cr, Mo, W, or Si.

KEY TERMS AND CONCEPTS

Austenite, γ

Austenite decomposition

Austenitization

Carbide, $\bar{C}$

Cast iron

Component

Equilibrium

Eutectic composition

Eutectic reaction

Eutectic temperature

Eutectoid composition

Eutectoid reaction

• Eutectoid shift

Eutectoid temperature

Ferrite, α

   eutectoid

   proeutectoid

Isotherm

Lever rule (inverse)

Liquidus

Materials balance

Mixture

Nomenclature, (AISI–SAE)

Pearlite, $\alpha + \bar{C}$

Phase diagram

   isothermal cut

   one-phase area

   two-phase area

   three-phase temperature

Phases

   chemical compositions of

   quantities of

Quantity fraction, $X/(X + Y)$

Solder, Pb–Sn

Solidus

Solubility limit

Solution

Steel

   eutectoid

   • hypereutectoid

   • hypoeutectoid

   low-alloy

   plain-carbon

PHASE-DIAGRAM INDEX

FOR CLASS DISCUSSION

A_9 From Fig. 9–1.2, suggest a method for purifying sea water.

B_9 Arctic oceans ($-1.5°C$) cease melting snow when the NaCl content drops to 2.5%. Why?

C_9 Explain why honey will "sugar" if it is stored in an open jar.

D_9 The solidus temperature is an important temperature limit for hot-working processes of metals. Why?

E_9 Distinguish between the chemical composition of a phase, and the quantity of a phase.

F_9 Why does the composition of an eutectic liquid, L_2, always lie between the compositions of the two solids, S_1, and S_3?

G_9 Why are the equilibrium compositions of two phases always at the limits of where the isotherm passes through the two-phase area?

H_9 We can use interpolation in the horizontal isothermal direction to get the quantity fractions of the two phases. We *cannot* obtain a valid answer by interpolating along a vertical composition line. By using a 70 Pb–30 Sn solder, show why the latter doesn't work.

I_9 Perform the algebraic rearrangement that we mentioned just before Eq. (9–4.3). (*Hint:* Solve Eq. (9–4.2) for X, to use in Eq. (9–4.3).)

J_9 Devise a "center-of-gravity" calculation to determine the quantity fraction of a phase. Show that this is identical to the "lever rule" and the "interpolation procedure."

• K_9 Explain to a classmate why the quantity of α in an 80 Ag–20 Cu alloy is indeterminate at the eutectic temperature. However, if you have 50 g of α in 100 g of alloy, the temperature would have to be 780°C, if equilibrium exists.

L$_9$ Make a sketch of the 475°C–575°C and 10%–33% Sn region of the Cu–Sn phase diagram (Fig. 9–5.2) on an enlarged scale. Prepare a problem involving phase compositions and quantities to be solved by a classmate.

M$_9$ A long rod of silicon is "zone-refined" by passing it slowly through a short heater so that a molten zone starts at one end and moves toward its other end. It solidifies as it leaves the heated zone. Explain how this could be repeated to remove impurities (cf. Example 9–5.1).

N$_9$ Assume a carbon atom is located in ferrite at $\frac{1}{2},\frac{1}{4},0$. What is its coordination number? How many such locations are there per unit cell?

O$_9$ Based on the comments of Example 9–6.1, explain why $D_{C \text{ in } \alpha} > D_{C \text{ in } \gamma}$, even though austenite can dissolve many times more carbon atoms than can ferrite.

P$_9$ The data of Fig. 9–7.3 will be used on numerous occasions during the balance of the text. They would be worth remembering so you don't have to look them up on each occasion.

Q$_9$ Pearlite is a 2-phase microstructure containing α and $\bar{C}$. Distinguish between phases and microstructures.

R$_9$ The solution to Example 9–8.2(a) mentions eutectoid ferrite and proeutectoid ferrite. Based on Fig. 9–8.2(d) and Fig. 9–7.3, explain the difference.

S$_9$ Explain the AISI–SAE nomenclature pattern to a freshman who has had chemistry but who is not in this course.

• T$_9$ Predict whether tungsten (bcc) will raise or lower the temperature of austenite decomposition.

U$_9$ Refer to Fig. 4–3.2 that shows an iron oxide with an NaCl structure. Under what conditions is this defect structure stable? This iron oxide was formed by oxidizing a piece of iron at 1000°C in air. How does the composition of the scale vary through its thickness?

V$_9$ Devise an experiment, which can be used as a class demonstration, that will show the change of iron from ferrite to austenite. (The experiment should be reversible for repetition, and visible to those in the back of the room.)

• W$_9$ The reaction,

$$S_1 + L_3 \xrightarrow{\text{Cooling}} S_2,$$

differs from the eutectic and is called a *peritectic* reaction. Although it was not discussed in this chapter, it is common, for example,

$$\gamma(69.8\% \text{ Zn}) + L(80.5\% \text{ Zn}) \xrightarrow{699°\text{C}} \delta(73\% \text{ Zn})$$

in the Cu–Zn system. Locate other peritectic reactions that appear in the phase diagrams of this chapter.

• X$_9$ Read W$_9$. A peritectoid reaction is analogous but involves solid phases only. Identify such a reaction in either the Al–Cu system or the Al–Mg system, and write an equation indicating phases and compositions.

- Y_9 The 'phase rule' as defined for a fixed pressure is

$$V = C + 1 - P,$$

where P is the number of phases present, C is the number of components, and V indicates the number of variations that are possible without changing P. Use Example 9–2.3 to illustrate the phase rule.

- Z_9 A phase diagram presents equilibrium conditions. Assume, however, that a 90 Al–10 Mg alloy is cooled rapidly from 600°C. How will this affect the compositions of the phases? From 400°C?

STUDY PROBLEMS

9–1.1 A syrup contains equal quantities of water and sugar. How much more sugar can be dissolved into 100 g of the syrup at 80°C?

Answer: 138 g added

9–1.2 A molten lead–tin solder has a eutectic composition. Fifty grams are heated to 200°C. How many grams of tin may be dissolved into this solder?

9–1.3 A 90 Pb–10 Sn* solder is to be melted at 200°C by adding more tin. What is the minimum amount of tin that must be added per 100 grams of this high-lead solder?

Answer: 104.5 g additional tin

9–1.4 One ton of salt (NaCl) is spread on the streets after a winter storm. The temperature is -15°C (5°F). How many tons of ice will be melted by the salt?

9–1.5 (a) Eight grams of salt (NaCl) are added to 32 g of ice. Above what minimum temperature will the ice completely melt? (b) Twenty *more* grams of ice are added to the above brine. Above what minimum temperature will this ice completely melt?

Answer: (a) -17°C (1.5°F)   (b) -10°C (14°F)

9–1.6 The solubility limit of zinc in α-brass is 35% at 780°C. It is desired to make 100 kg of 65–35 brass by melting some 85–15 brass and some zinc. How much of each should be melted in the crucible?

9–2.1 A 90 Cu–10 Sn* bronze is cooled slowly from 1100°C to 20°C. What phase(s) will be present as the cooling progresses?

Answer: 1100 → 1020°C, L; 1020 → 830°C, (L + α); 830°C → 340°C, α; 340 → 20°C, (α + ε)

9–2.2 A 65 Cu–35 Zn* brass is heated from 300°C to 1000°C. What phase(s) are present at each 100°C interval?

9–2.3 An alloy contains 90 Pb–10 Sn.* (a) What phases are present at 100°C? 200°C? 300°C? (b) Over what temperature range(s) will there be only one phase?

Answer: (a) 100°C, α + β; 200°C, α; 300°C, α (admittedly slight amount) + L
(b) Only α, 150–270°C; only L above 305°C

---

* The phase-diagram index precedes Topics for Class Discussion, p. 368.

9–2.4 A eutectic mixture of Al_2O_3 and ZrO_2* is molten at 2400°C. (a) At what temperature(s) will there be only one phase? (b) Two phases? (c) Three phases?

9–2.5 Locate the solidus on a Cu–Zn diagram.*

Answer: It lies immediately below all the $x + L$ fields with horizontal lines at 903°, 835°, 699°, 598°, and 424°C.

9–2.6 Locate the liquidus and the solidus as they extend from 1538°C at 100% Fe (0% O) and 1599°C for Fe_3O_4. (*Note:* There is a 2-liquid field where liquid metal separates from liquid oxide.) (See Fig. 9–5.5.)

9–2.7 Show the sequential changes in phases when the composition of an alloy is changed from 100% Cu to 100% Al (a) at 700°C, (b) at 450°C, (c) at 900°C. (See Fig. 9–5.3.)

Answer: (c) At 900°C: α, $\alpha + \beta$, β, $\beta + \gamma_1$, γ_1, $\gamma_1 + \epsilon_1$, ϵ_1, $\epsilon_1 + L$, L

9–2.8 (a) Show the sequential changes in phases when the composition of an alloy is changed from 100% Mg to 100% Al at 300°C. (b) At 400°C. (See Fig. 9–5.4.)

9–2.9 Some of the fields in the Cu–Sn diagram (Fig. 9–5.2) are not labeled. At 625°C, what field lies (a) between β and γ? (b) Between γ and ζ?

Answer: Use the 1-2-1-2-··· sequence to obtain the answer.

9–2.10 What fields lie to the right of γ_2 in the Al–Cu diagram (Fig. 9–5.3)?

9–3.1 Refer to Fig. 9–5.4. At 500°C, what is the solubility limit of magnesium (a) in solid α? (b) In liquid? (c) What are the chemical compositions of the phase(s) in a 40 Al–60 Mg alloy at 500°C? (d) a 20 Al–80 Mg alloy?

Answer: (a) 11% Mg   (b) 76% Mg   (c) Liq: 40 Al–60 Mg
(d) Liq: 24 Al–76 Mg; β: 7.5 Al–92.5 Mg

9–3.2 Refer to Fig. 9–2.2. At 2000°C, what is the solubility limit of Al_2O_3 (a) in the liquid? (b) In β? (c) What are the chemical compositions of the phase(s) in a 20 Al_2O_3–80 ZrO_2 ceramic at 1800°C? at 2000°C?

9–3.3 What are the chemical compositions of the phase(s) in Study Problem 9–2.2?

Answer: 300°, 400°, 500°, 600°, 700°: α(65 Cu–35 Zn); 800°C: α(66 Cu–34 Zn) and β(61 Cu–39 Zn); 900°C: α(67.5 Cu–32.5 Zn) and β($\sim$63 Cu–37 Zn); 1000°C: Liq(65 Cu–35 Zn)

9–3.4 What are the chemical compositions of the phase(s) in Study Problem 9–2.1?

9–3.5 (a) Write the eutectic reaction for one of the two eutectics in the Al–Cu system (Fig. 9–5.3). (b) Repeat for the second eutectic.

Answer: (a) Liq(91.5% Cu) $\overset{1035°}{\rightleftharpoons}$ α(92.5% Cu) + β(90% Cu)

9–3.6 Write the eutectic reaction for the higher-magnesium eutectic of the Al–Mg system (Fig. 9–5.4).

9–4.1 A 90 Al–10 Mg alloy is melted and cooled slowly. (a) At what temperature does the first solid appear? (b) At what temperature does it have $\frac{2}{3}$ Liq and $\frac{1}{3}\alpha$? (c) $\frac{1}{2}$ Liq and $\frac{1}{2}\alpha$? (d) 99+%α with a trace of liquid?

Answer: (a) 620°C   (b) 600°C   (c) 585°C   (d) 520°C

---

* The phase-diagram index precedes Topics for Class Discussion, p. 368.

9–4.2 Repeat Study Problem 9–4.1, but for a 10 Al–90 Mg alloy.

9–4.3 What composition of Ag and Cu will possess (a) $\frac{1}{4}\alpha$ and $\frac{3}{4}\beta$ at 600°C? (b) $\frac{1}{4}$ Liq and $\frac{3}{4}\beta$ at 800°C?

Answer: (a) 26 Ag–74 Cu   (b) 23 Ag–77 Cu

9–4.4 What composition of Al_2O_3 and ZrO_2 will possess (a) $\frac{3}{4}\alpha$ and $\frac{1}{4}\beta$ at 1800°C? (b) $\frac{3}{4}$ Liq and $\frac{1}{4}\beta$ at 1950°C?

9–4.5 The solubility of tin in solid lead at 200°C is 18% Sn. The solubility of lead in the molten metal at the same temperature is 44% Pb. What is the composition of an alloy containing 40% liquid and 60% solid α at 200°C?

Answer: 66.8% Pb, 33.2% Sn

9–4.6 (a) At what temperature will a monel alloy (70% nickel, 30% copper) contain $\frac{2}{3}$ liquid and $\frac{1}{3}$ solid, and (b) what will be the composition of the liquid and of the solid?

9–4.7 Assuming 1500 g of bronze in Study Problem 9–2.1, what are the masses of solid phase(s) at each 100°C interval?

Answer: 1100°C, 0 g; 1000°C, $\sim$150 g α; 800°C, 400°C, 1500 g α; 300°C, $\sim$1350 g α + 150 g ϵ.

9–4.8 At 175°C, how many grams of α are there in 7.1 g of eutectic Pb–Sn solder?

9–4.9 With 200 g of 65 Cu–35 Zn, how many grams of α are present at each temperature of Study Problem 9–2.2?

Answer: 300 → 700°C, 200 g α; 800°C, 180 g α; 900°C, $\sim$80 g α; 1000°C, 0 g α

9–4.10 Two kilograms of sterling silver (92.5 Ag–7.5 Cu) are cooled slowly from 1000°C to 400°C. (a) On a graph of %Ag (ordinate) versus °C (abscissa), plot the composition of α. (b) On a graph of g of α (ordinate) versus °C (abscissa), plot the amount of α. (Data on Fig. 9–2.4 are accurate to about 1 part/100. You should be correspondingly accurate.) (c) Extrapolate your curves to 20°C.

9–4.11 Plot the amount of α in 15.8 kg of a 95 Al–5 Si alloy as a function of temperature.

Answer: 0°C, 15 kg; 576°C, 15.3 kg; 578°C, 11 kg; 600°C, 8.4 kg; >630°C, 0 kg

• *9–4.12* At what temperature can an alloy of 40 Ag–60 Cu contain 55% β?

9–5.1 An alloy of 50 g Cu and 30 g Zn is melted and cooled slowly. (a) At what temperature will there be 40 g α and 40 g β? (b) 50 g α and 30 g β? (c) 30 g α and 50 g β?

Answer: (a) $\sim$780°C   (b) 750°C   (c) 800°C

9–5.2 (a) What are the chemical compositions of the phases in a 10% magnesium–90% aluminum alloy at 600°C, 400°C, 200°C? (b) What are the quantities of these phases at each of the temperatures in part (a)? (c) Make a materials balance for the distribution of the magnesium and aluminum in the above alloy at 600°C.

9–5.3 Make a materials balance for P grams of a 92 Ag–8 Cu alloy at 500°C (equilibrium conditions).

Answer: 0.938P g α, 98% Ag–2% Cu; 0.062P g β, 1% Ag–99% Cu

9–5.4 Make a materials balance for 100 g of a 90 Mg–10 Al alloy at 200°C (assume equilibrium).

9–5.5 A die-casting alloy of 95 Al–5 Si is cooled so that the metal contains primary (proeutectic) α and an eutectic mixture of $(\alpha + \beta)$. What fraction of the casting is primary α?

Answer: 0.7 primary α (i.e., proeutectic α)

9–5.6 How much mullite will be present in a 60% SiO_2–40% Al_2O_3 brick (10 kg) at the following temperatures under equilibrium conditions: (a) 1400°C? (b) 1580°C? (c) 1600°C?

9–5.7 (a) Determine the compositions of Al–Si alloys that will contain $\frac{1}{3}$ liquid and $\frac{2}{3}$ solid when brought to equilibrium at 600°C. (b) Give the chemical compositions of the liquids.

Answer: (a) 96 Al–4 Si and 29 Al–71 Si (b) 91 Al–9 Si, 86 Al–14 Si

9–5.8 Based on Figs. 9–1.3 and 9–4.2, make a graph for an alloy containing 80% Pb and 20% Sn showing (a) the fraction of liquid versus temperatures, (b) the fraction of α versus temperature, and (c) the fraction of β versus temperature.

9–5.9 (a) Plot the percent Cu in the α phase of sterling silver (92.5 Ag–7.5 Cu) versus temperature. (b) Plot the fraction of α in sterling silver versus temperature.

9–5.10 On graph paper draw the equilibrium diagram for alloys of metals A and B from the following data. Label all areas of your diagram.

Melting point of A	700°C
Melting point of B	1000°C
Eutectic temperature	500°C
Composition of liquid in equilibrium at the eutectic temperature:	30 w/o A and 70 w/o B
Solubilities at 500°C:	B in A = 15 w/o
	A in B = 20 w/o
Solubilities at 70°C:	B in A = 15 w/o
	A in B = 8 w/o

9–5.11 Sixty kg of a 90 Cu–10 Sn bronze (to be used to make a bell) are melted and then cooled very slowly to maintain equilibrium. (a) At what temperature does the first solid appear? (b) What is the composition of that solid? (c) At what temperature is there twice as much solid as liquid? (d) At what temperature does the last liquid disappear? (e) What is the composition of that liquid? (f) What phase(s) at 600°C? Give the composition(s) and amount(s) of each phase. (g) What phase(s) at 300°C? Give the composition(s) and amount(s) of each phase.

9–5.12 A kilogram of sterling silver contains 75 g copper (925 g Ag). Suggest a procedure for obtaining at least 100 grams of silver from the above that will have an analysis of <2% copper. Use melting, partial solidification, and separation steps. (You may assume the liquidus and solidus are straight lines.)

9–6.1 The maximum solubility of carbon in ferrite (α) is 0.02 w/o. How many unit cells per carbon atom?

Answer: 540

9–6.2 Iron carbide has a density of 7.6 Mg/m^3 ($= 7.6$ g/cm^3). What is the volume of the ortho-rhombic unit cell described in Section 9–6?

9–6.3 At 920°C, the maximum solubility of carbon in austenite (γ) is 1.33 w/o. How many unit cells per C atom?

Answer: 4.0

9–6.4 Based on the data of Fig. 1–3.1 and Example 3–4.1, plot an approximate volume–temperature curve for iron between 0°C and T_m. If you have to make estimates, state your assumptions.

9–7.1 Making use of the $Fe–Fe_3C$ diagram, calculate the fraction of α and the fraction of carbide present at 700°C in a metal containing 2% carbon and 98% iron.

Answer: 0.30 carbide, 0.70 ferrite

9–7.2 (a) What phases are present in a 99.8 Fe–0.2 C steel at 800°C? (b) Give their compositions. (c) What is the fraction of each?

9–7.3 Describe the phase changes that occur on heating a 0.20% carbon steel from room temperature to 1200°C. (Ferrite, α; austenite, γ; carbide, $\bar{C}$.)

Answer: RT → 727°C ($\alpha + \bar{C}$); 727°C (α + carbide + γ); 727 → 855°C ($\alpha + \gamma$) with the amount of α decreasing to zero at 855°C; 805 → 1200°C, only γ

9–7.4 Without referring to the $Fe–Fe_3C$ phase diagram, indicate the compositions of the phases in the previous problem.

9–7.5 Write the eutectoid reaction found in the $BaTiO_3–CaTiO_3$ system (Fig. 9–5.7).

Answer:

$$\beta(24\% \ CaTiO_3) \overset{105°}{\rightleftharpoons} \alpha(20\% \ CaTiO_3) + \gamma(90\% \ CaTiO_3)$$

9–7.6 (a) Locate four eutectoids in the Cu–Sn system (Fig. 9–5.2). (b) Write the reactions for two of these.

9–7.7 Refer to Fig. 9–7.4. Make a similar presentation of phase fractions for a 99 Fe–1 C steel.

9–7.8 Repeat the previous problem for a 99.8 Fe–0.2 C steel.

9–8.1 Calculate the percent ferrite, carbide, and pearlite, at room temperature, in iron–carbon alloys containing (a) 0.5% carbon, (b) 0.8% carbon, (c) 1.5% carbon.

Answer: (a) 7.5% carbide, 92.5% α, 62% pearlite (b) 12% carbide, 88% α, 100% pearlite
(c) 22.5% carbide, 77.5% α, 88% pearlite.

9–8.2 (a) Determine the phases present, the composition of each of these phases, and the relative amounts of each phase for 1.2% carbon steel at 870°C, 760°C, 700°C. (Assume equilibrium.) (b) How much pearlite is present at each of the above temperatures?

9–8.3 Assume equilibrium. For a steel of 99.5 Fe–0.5 C, determine (a) lowest temperature of 100% γ. (b) Fraction that is γ at 730°C, and its composition. (c) Fraction that is pearlite at 720°C, and its total composition. (d) Fraction that is proeutectoid ferrite at 730°C. (e) Fraction that is proeutectoid ferrite at 720°C after cooling from 730°C.

Answer: (a) 780°C (b,c) 0.6 at ~0.8% C (99.2% Fe) (d,e) 0.4 at 0.02% C

9–8.4 Substitute a 99 Fe–1 C steel for the previous problem and answer.

9–8.5 (a) Determine the amount of pearlite in a 99.1 Fe–0.9 C alloy that is cooled slowly from 870°C. Basis: 11 kg alloy. (b) From the results of part (a), determine the amount of ferrite and carbide at 725°C. (c) At 20°C.

Answer: (a) 10.8 kg   (b) 9.55 kg α, 1.45 kg $\bar{C}$ (c) 9.5 kg α, 1.5 kg $\bar{C}$

9–8.6 On the basis of this chapter, would you choose a high- or low-carbon steel for an automobile fender? Give reasons.

9–9.1 A steel has the following composition. Give it an AISI–SAE number: C 0.21, Mn 0.69, Cr 0.62, Mo 0.13, Ni 0.61.

Answer: AISI 8620

9–9.2 A steel has the following composition. Give it an AISI–SAE number: C 0.38, Mn 0.75, Cr 0.87, Mo 0.18, Ni 0.03.

9–9.3 Calculate the fractional quantities of the phase(s) at 700°C, 750°C, 800°C, and 900°C for the following steels: (a) 0.8 C–99.2 Fe;   (b) 1.2 C–98.8 Fe;   ●(c) 0.6 C–0.6 Mo–98.8 Fe.

Answer: (a) 700°C, 0.88α–0.12 $\bar{C}$; 750°C, 800°C, 900°C, 1.0γ
(b) 700°C, 0.82α–0.18 $\bar{C}$; 750°C, 0.94γ–0.06$\bar{C}$; 800°C, 0.965γ–0.035 $\bar{C}$; 900°C, 1.0γ
(c) 700°C, 750°C, 0.91α–0.09 $\bar{C}$; 800°C, ∼0.99γ–0.01 $\bar{C}$; 900°C, 1.0γ ($\bar{C}$ = carbide.)

● **9–9.4** Modify Fig. 9–7.3 for a steel containing (a) 1% Mn, (b) 1% Cr, (c) 1% Ni. (The new solubility curves remain essentially parallel to the previous ones.)

● **9–9.5** How much nickel should be added to a steel to drop the lower side of the austenite stability range in a 0.40% carbon steel as much as it is dropped by 2% Mn?

Answer: ∼2.2% Ni

CHAPTER **10**

Multiphase
Materials:
Microstructures

PREVIEW

The phase diagrams of Chapter 9 provided us with a
quantitative base for the composition and amounts of
phases. Equilibrium was assumed. Little was said about
the *time requirements* to attain that equilibrium nor the
microstructures that were obtained. This chapter
addresses those topics.

   We will first categorize solid-phase reactions on the
basis of the number of phases involved, the extent of the
atomic rearrangements, and the required compositional
modifications. Then microstructural geometry will
receive our attention. This permits us to relate properties
to microstructures.

CONTENTS

10–1 Solid–Phase Reactions:
grain growth and recrystallization, polymorphic, solution and precipitation, eutectoid, and martensitic.

10–2 Multiphase Microstructures:
phase size, phase amounts, and phase shape and distribution.

10–3 Microstructures (Mechanical Properties):
effects of phase quantities, phase size, phase shape and distribution.

• **10–4 Microstructures (Physical Properties).**

STUDY OBJECTIVES

1 To understand the principal reactions that change the internal structure of solid materials.

2 To apply the Arrhenius relationship to reactions that are limited by diffusion, and to have a qualitative understanding of why reactions are delayed when new phases must be nucleated.

3 To distinguish between conditions that lead to reactions at grain boundaries and to reactions within the grains.

4 To relate mechanical properties to microstructural variations, such as the amount, size, shape, and distribution of the phases.

5 To relate selected physical properties qualitatively to microstructural variations.

10–1   SOLID-PHASE REACTIONS

All microstructural changes within solids involve rearrangements of atoms. We initially encountered microstructural changes in Section 6–6 when we considered recrystallization. This is a very simple solid-state reaction because there is no compositional change. Likewise, the crystal structure remains the same, and every atom continues to have the same atomic coordination, for example, 12 if fcc. However, the atoms are rearranged to make more perfect crystals from the previously deformed grains. This does not occur instantaneously. The required time is a function of temperature and other variables. (Cf. Figs. 6–6.3 and 6–6.4.)

Reactions that occur in solids may be categorized in a variety of ways. We shall do this as follows:

1. *Grain growth.* Atoms move across grain boundaries ($\ll 1$ nm); no change in composition; no change in crystal structure; no new grains.

2. *Recrystallization.* New, more perfect grains; only local atom rearrangements ($\ll 1$ nm); no change in composition; no change in crystal structure.

3. *Polymorphic changes.* A new phase and, hence, new atomic coordination, by local rearrangements ($\ll 1$ nm); no change in composition.

4a. *Solution.* Disappearance of an existing phase by solution into the prime phase; major atom diffusion ($\gg 1$ nm).

4b. *Precipitation.* Separation of a new phase from a supersaturated solid solution; major atom diffusion ($\gg 1$ nm).

5. *Eutectoid reactions.* Decomposition of one phase (on cooling) into two new phases; major atom diffusion ($\gg 1$ nm).

6. *Martensitic transformation.* A polymorphic change by the shear of one or more planes of atoms with respect to adjacent planes; diffusionless ($\ll 1$ nm).

This list is not all-inclusive, but will provide a basis for understanding various commercial heat treatments (Chapter 11) and the resulting microstructures and properties.

Important features of the above listings are (1) whether a new phase is formed or not, and (2) whether the atomic movements involve local rearrangements only ($\ll 1$ nm) or a long-range diffusion over many unit cell distances ($\gg 1$ nm).

Grain growth (single-phase materials)   This reaction, which was introduced in Section 6–7, requires no nucleation time since atoms move across the boundary from one preexisting grain to another. The reaction rate is sensitive to temperature (Fig. 10–1.1), because thermal energy is required to activate the atom jumps. It is also a function of the curvature of the boundary; therefore, the rate decreases as the grains grow.

Recrystallization   Although this reaction does not lead to a new phase, a time delay occurs because new grains must be nucleated. The new grains generally form first along the slip zones in previously deformed grains (Fig. 6–6.1b). The progress

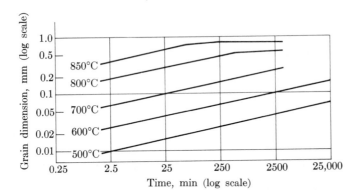

Fig. 10–1.1   Grain growth (brass). The logarithmic growth proceeds until the grains approach the dimensions of the metal sample, which was 1 mm thick. (After J. E. Burke, A.I.M.E.)

of reactions for recrystallization typically follows an S-shaped, or *sigmoidal* curve (Fig. 10–1.2). For example, aluminum (commercially pure and cold-worked 5%) shows negligible recrystallization in the first 50 hours at 310°C (Fig. 10–1.3a) but is 50% recrystallized after a total of 70 hours (Fig. 10–1.3b). The final completion requires nearly 100 hours.

Fig. 10–1.2 Isothermal recrystallization. (A. G. Guy and J. J. Hren, *Elements of Physical Metallurgy*, Addison-Wesley.)

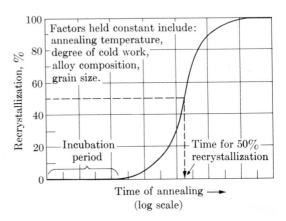

The progress of recrystallization occurs similarly at all temperatures but with a different time frame (Fig. 10–1.4a), as shown for the recrystallization of a highly cold-worked (98%) ultrapure (99.999%) copper. Observe that the 50% recrystallization time is relatively easily identified because the reaction is most rapid at that point. When we relate the time of that reaction point to the recrystallization temperature, we observe an Arrhenius behavior (Fig. 10–1.4b). This is expected because atom movements govern the reaction, and they, in turn, are dependent upon an activation energy.

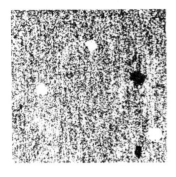

(a) Heated for 50 hours

1mm

Fig. 10–1.3 The progress of isothermal recrystallization at 310°C (590°F) in 99.95% aluminum that has been cold-worked 5%. (a) Heated for 50 hours, (b) 70 hours, (c) 80 hours, (d) 100 hours. (×7.5) (Courtesy W. A. Anderson, Alcoa.)

(b) 70 hours   (c) 80 hours   (d) 100 hours

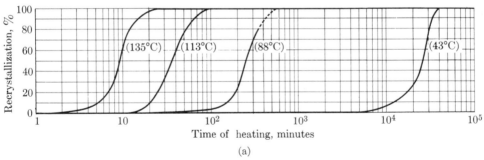

(a)

Fig. 10–1.4 Isothermal recrystallization of 99.999% pure copper. (After Decker and Harker.) (a) Isothermal recrystallization of pure copper cold-rolled 98%. (b) Plot for extrapolation from the data given by the four curves of part (a). (Adapted from A. G. Guy and J. J. Hren, *Elements of Physical Metallurgy*, Addison-Wesley.)

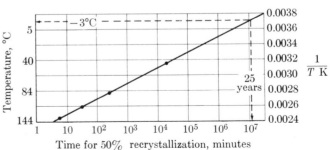

(b)

The mathematical relationship for Fig. 10–1.4(b) is

$$\ln t = C + B/T, \qquad (10\text{–}1.1)$$

where C and B are constants. We may relate this equation to Eqs. 4–7.4 and 8–8.1, if we recognize that a fast reaction rate R (a short time, since $R = 1/t$) corresponds to rapid diffusion. Thus,

$$\ln R = \ln R_0 - E/kT = -\ln t, \qquad (10\text{–}1.2)$$

where C and B of Eq. (10–1.1) are $-\ln R_0$ and E/k, respectively. (Also see Example 6–6.2.)

Polymorphic reactions   Pure materials may undergo phase transformations from one polymorphic form to another.

$$\text{Polymorphic reaction:} \quad \text{Phase } A \rightleftarrows \text{Phase } \mathscr{A}. \qquad (10\text{–}1.3)$$

Our prototype of this reaction, discussed in Section 9–6, was the $\alpha \rightleftarrows \gamma$ transformation of iron at 912°C. Other familiar polymorphic changes (out of many) include the α-to-β transformation in titanium and the quartz-to-tridymite-to-cristobalite changes in SiO_2. These transformations commonly result from temperature changes; however, a polymorphic transformation such as the one between graphite and diamond is caused by pressure variations.

Density and volume changes accompany phase transformations because the polymorphic structures do not have identical atomic packing factors. These packing factors are 0.68 and 0.74, respectively, for the bcc and fcc phases of iron; hence, there is a noticeable dimensional change at 912°C (Fig. 10–1.5a). However, the heating or cooling must be quite slow in order for the change to occur precisely at the transformation temperature without a detectable lag.

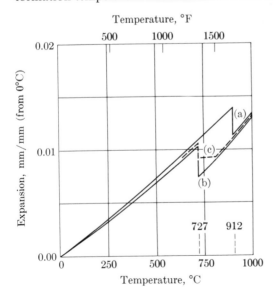

Fig. 10–1.5 Dimensional changes with phase transformations in iron and iron–carbon alloys. (a) Pure iron. (b) 99.2Fe–0.8C. (c) 99.6Fe–0.4C. Very slow changes in temperature permit an equilibrium, with transformation at the temperatures indicated by the phase diagrams (Fig. 9–7.3).

The transformation temperature can be defined as the temperature at which the two phases have identical amounts of energy available for chemical reaction. Thus, below 912°C austenite is more reactive than ferrite because it has more free energy; consequently, austenite is unstable and ferrite is stable. Immediately above 912°C, the opposite is true; then austenite is the stable phase of pure iron because the bcc iron reacts to form fcc iron.

Polymorphic reactions usually require only minor atom movements because the composition of the reactant and the product are identical (Eq. 10–1.3). However, it is still necessary to break the existing bonds and rearrange the atoms into the new structure.

Solution and precipitation within solids These two reactions are opposite to each other and may be illustrated in Fig. 10–1.6. At temperatures below 150°C, an alloy with 90% Pb and 10% Sn contains two phases, α and β. Above 150°C, all the tin may be contained in fcc α; consequently, as the metal is heated, solid β is *dissolved* in the solid α. *Solid precipitation* occurs when this alloy is cooled into the two-phase temperature range after being *solution-treated* above 150°C. Such precipitation is useful for various age-hardening alloys (Chapter 11). The following equations are appropriate for these two reactions in binary alloys. Note that only two phases are involved in each case; however, the composition of the solution changes from A to another value, $A^{\#}$, as the solubility limit is exceeded (Fig. 10–1.6).

$$\text{Solution-treating:}\quad \text{Solid } A^{\#} + \text{Solid B} \xrightarrow{\text{Heating}} \text{Solid A.} \qquad (10\text{–}1.4)$$

$$\text{Precipitation:}\quad \text{Solid A} \xrightarrow{\text{Cooling}} \text{Solid } A^{\#} + \text{Solid B.} \qquad (10\text{–}1.5)$$

The above reactions require diffusion. Consider, for example, the composition of 95.5 w/o Al and 4.5 w/o Cu, which is a widely used aluminum alloy. As shown in

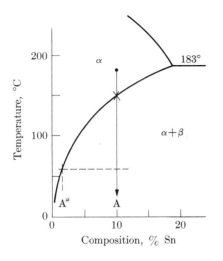

Fig. 10–1.6 Solution-treating and precipitation (90Pb–10Sn alloy). All of the tin is dissolved in α at elevated temperatures. Cooling produces a solid-precipitation reaction.

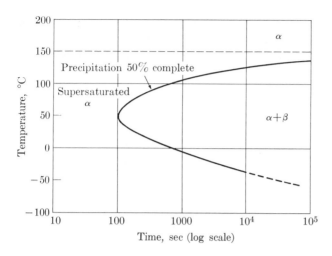

Fig. 10-1.7   Isothermal precipitation (90Pb–10Sn). Above 150°C, α can contain all of the tin. The rate of β precipitation varies with the amount of supercooling below the saturation temperature. Finally, the equilibrium of Fig. 9–1.3 is attained. (Adapted from data by H. K. Hardy and T. J. Heal, *Progress in Metal Physics* 5, Pergamon Press.)

Fig. 9–5.3, there are two phases, θ and κ, in equilibrium below 500°C (930°F). $CuAl_2$ comprises the θ phase, so that it contains an appreciably larger fraction of copper atoms than the κ phase. During solution-treatment these copper atoms must move through the κ structure to take on random substitutional positions; during precipitation, the copper atoms must be collected onto the growing $CuAl_2$ (that is, θ) particles.

Return now to the 90 Pb–10 Sn alloy of Fig. 10–1.6 so that we might consider the time requirement for the precipitation reaction to occur. Above 150°C, only α (Pb-rich) is stable; below 150°C, some β (Sn-rich) will separate from supersaturated α if time is available. Figure 10–1.7 shows the time required for *isothermal precipitation* (50% completion), where the alloy is rapidly cooled to the selected temperature and then held at that temperature for precipitation to occur. We will observe that the required time decreases as the temperature is raised from −50°C to room temperature. In this temperature range, the precipitation rate is limited by diffusion. At about 50°C, the pattern is reversed and eventually runs off the long-time end of the graph just below the solubility limit of 150°C. This reversal will need to interest us since we will find the same C-type curve in other reactions.

Above the knee of the isothermal precipitation curve, the rate is not limited by diffusion, because the atoms can diffuse relatively fast (50°C $\approx 0.6T_m$). Rather, the limitation arises from the time required to nucleate the new phase β. Nucleation occurs most readily at the boundaries of the α grains (Fig. 10–1.8b). Tin atoms move over many atomic distances to these sites.

Below the knee of the isothermal precipitation curve, the rate of precipitation is limited by the diffusion of the tin atoms. They cannot diffuse readily; furthermore, the supersaturation is great. Therefore, new β particles will nucleate wherever possible—at point imperfections such as vacancies and interstitials, along dislocation lines, and adjacent to impurities. These occur throughout the grains. Therefore, we see a different microstructure (Fig. 10–1.8c versus 10–1.8b).

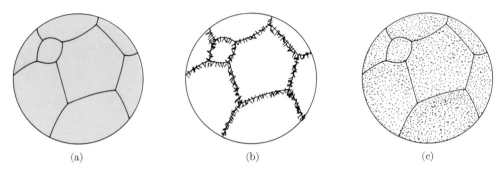

(a)   (b)   (c)

Fig. 10–1.8 Solid-phase precipitation (schematic). (a) Supersaturated solid solution, for example, 90Pb–10Sn rapidly cooled from 180°C to 20°C. (b) Grain boundary precipitation. This requires long-range diffusion to the grain boundaries, where nucleation occurs more readily. (c) Intragranular precipitation. Many minor imperfections throughout the grains nucleate the new phase; therefore diffusion distances are shorter than in structure (b). Higher precipitation temperatures favor structure (b); lower, structure (c).

Eutectoid reactions   These reactions are rather common. In fact, there are four eutectoids in the Cu–Sn system (Fig. 9–5.2), as well as being present in some of the other phase diagrams of Chapter 9. Each has the general reaction

$$\text{Solid 2} \;\underset{\text{Heating}}{\overset{\text{Cooling}}{\rightleftharpoons}}\; \text{Solid 1 + Solid 3.} \qquad (10\text{–}1.6)$$

When Solid 2 is of eutectoid composition, the eutectoid reaction occurs at a single temperature—the eutectoid temperature. On cooling, the reaction goes from a phase of intermediate composition to two phases, whose compositions are on opposite sides of the eutectoid (Eq. 9–7.2).

If Solid 2 is offset from the eutectoid composition, one of the two solubility-limit curves is encountered first, so that one phase starts to separate ahead of the other. (For example, in a 0.4 C–99.6 Fe alloy, ferrite separates ahead of carbide; in a 1 C–99 Fe alloy, carbide separates first as indicated in Figs. 9–7.3 and 9–8.2f.)

Our prototype for a eutectoid reaction is the *isothermal decomposition of austenite* (Section 9–8). The time requirement for this isothermal decomposition will be considered in Section 11–4 where we will observe a C-type curve that is somewhat similar to the isothermal precipitation curve of Fig. 10–1.7 for a Pb–Sn alloy. Currently, we will focus our attention on the mechanism of pearlite growth.

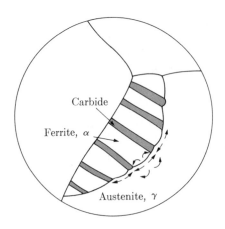

Fig. 10–1.9   Pearlite formation. Carbon must diffuse from the eutectoid austenite (~0.8%) to form carbide (6.7%). The ferrite that is formed has negligible carbon.

Pearlite growth involves the simultaneous formation of ferrite, α, and carbide, $\bar{C}$. It starts at the austenite grain boundaries. The lamellae of the two phases grow inward. As this occurs, the carbon must be segregated as shown in Fig. 10–1.9. If the cooling rate is slow, the carbon can diffuse greater distances and a *coarse pearlite* (thicker layers) is formed. If the cooling rate is accelerated, diffusion is limited to shorter distances. As a result, more (and thinner) lamellae are formed to give a *fine pearlite*.

As shown in Fig. 9–8.2, the amount of pearlite can vary from none to 100% as the carbon content increases from 0% to the eutectoid composition (~0.8% in plain-carbon steels). Concurrently, there is a decrease in the amount of proeutectoid ferrite. The amount of pearlite decreases in hypereutectoid steels because proeutectoid carbides form as a network around the original austenite grains.

Martensite formation   As indicated by Eq. (9–8.1), austenite decomposes to ferrite plus carbide ($\alpha + \bar{C}$). This assumes that there is time for the carbon to diffuse and become concentrated in the carbide phase and depleted from the ferrite. If we quench austenite very rapidly, Eq. (9–8.1) can be detoured. We can indicate this as follows:

$$\gamma \text{ (fcc) } \xrightarrow[\text{Cooling}]{\text{Slow}} \alpha \text{ (bcc)} + \text{carbide.} \qquad (10\text{–}1.7)$$

$$\begin{array}{c} | \\ \text{Quench} \\ \downarrow \\ \text{M (bct)} \end{array} \nearrow \text{Tempering}$$

This alternate route to form (α + carbide) involves a *transition phase* of *martensite*, M (Fig. 10–1.10). This polymorphic phase of iron is not stable because, given an opportunity, martensite will proceed to form ($\alpha + \bar{C}$). As a result, we do not see martensite on the Fe–Fe$_3$C diagram (Fig. 9–7.1). However, martensite is a very important phase, as we shall soon see.

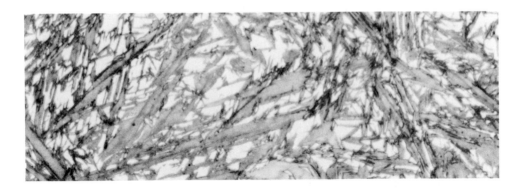

Fig. 10–1.10   Martensite (X1000). The individual grains are plate-like crystals with the same composition as that of the grains in the original austenite. (U.S. Steel Corp.)

Martensite forms at temperatures well below the eutectoid (but still above room temperature), because the fcc structure of austenite becomes sufficiently unstable so that it changes spontaneously to a body-centered structure in a special way. It does not involve diffusion but results from a shearing action. All the atoms shift in concert, and no individual atom departs more than a fractional nanometer from its previous neighbors. Being diffusionless, the change is very rapid. All of the carbon that was present remains in solid solution. The resulting body-centered structure is tetragonal (bct) rather than cubic; it is definitely different from ferrite.

Since martensite has a noncubic structure, and since carbon is trapped in the lattice, slip does not occur readily; and therefore martensite is hard, strong, and brittle. Figure 10–1.11 shows a comparison of the hardness of martensite with that of pearlite-containing steels as a function of carbon content. This enhanced hardness is of major engineering importance, since it provides a steel that is extremely resistant to abrasion and deformation.

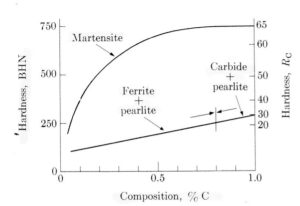

Fig. 10–1.11   Hardness of annealed iron–carbon alloys (α + carbide) and martensite versus carbon content. This difference in hardness is the reason for the quenching of steel.

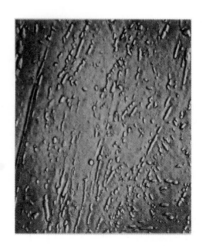

Fig. 10–1.12 Tempered martensite (eutectoid steel, X15,000). The steel was previously quenched to form martensite, which is a body-centered tetragonal (bct) phase and was 60 R_C. It was then tempered for 1 hour at 425°C (800°F). The tempered martensite is a two-phase microstructure containing carbide particles (light) in a matrix of ferrite (dark). Initially the tempered martensite was very hard and brittle; however, with the above heating, the tempered martensite is now 44 R_C, but much tougher. (A. M. Turkalo, General Electric Co.)

The existence of martensite as a *metastable* phase that contains carbon in solid solution in a bct structure does not alter the iron–carbide phase diagram (Fig. 9–7.1). With sufficient time at temperatures below the eutectoid temperature, the supersaturated solution of carbon in iron continues its progress to the more stable ferrite and carbide (Eq. 10–1.7). This process is known commercially as *tempering*:

$$\text{M} \longrightarrow \alpha + \text{carbide.} \qquad (10\text{–}1.8)$$
(Martensite)   (Tempered martensite)

The resulting $(\alpha + \bar{C})$ microstructure is not lamellar like that of the pearlite, which we previously observed, but contains many dispersed carbide particles (Fig. 10–1.12) because there are numerous nucleation sites within the martensitic steel. This *tempered martensite** is much tougher than the metastable martensite, making it a more desirable product although it may be slightly softer.

Example 10–1.1 Two identical samples of a bronze were heated to 610°C and 700°C. The grain size of the first was doubled from 0.05 mm to 0.10 mm in 30 hrs. It took the second only 30 minutes at 700°C. Estimate the time required at 750°C.

Solution: Each requires the same mass transport. Since grain growth is a diffusion process, we expect an Arrhenius relationship

$$\ln R = \ln R_0 - E/kT = -\ln t; \qquad (10\text{–}1.2)$$

or

$$\ln t = C + B/T. \qquad (10\text{–}1.1)$$

$$\ln 30 \text{ hr} = C + B/883 \text{ K} = 3.40;$$

$$\ln 0.5 \text{ hr} = C + B/973 \text{ K} = -0.70.$$

---

* Note that tempered martensite does not have the crystal structure of martensite. Rather, it is a two-phase microstructure containing *ferrite* and *carbide*. This microstructure originates by the decomposition of martensite.

Solving simultaneously,

$$B = 39{,}100 \text{ K} \quad \text{and} \quad C = -40.86;$$

$$\ln t = -40.86 + 39{,}100 \text{ K}/1023 \text{ K}.$$

$$t = 0.07 \text{ hr} \quad (\text{or } 4^{+} \text{ min}).$$

Comment. The grain size will *not* double again in a second interval of time because the growth rate decreases as the grain size increases. ◀

Example 10–1.2 An Al–Cu alloy has 2 atom percent copper in solid solution κ at 550°C. It is quenched, then reheated to 100°C, where θ precipitates (Fig. 9–5.3). The θ (CuAl$_2$) develops many *very small* particles throughout the alloy so that the average interparticle distance is only 5.0 nm. (a) Approximately how many particles form per mm^3? (b) If, by extrapolating Fig. 9–5.3, we assume that negligible copper remains in κ at 100°C, how many copper atoms are there per θ particle?

Solution: (a) Assume one particle per 5.0 nm cube.

$$\text{Count } \sim 1/(5 \times 10^{-9} \text{ m})^3 = \sim 8 \times 10^{24}/\text{m}^3, \quad (\text{or } \sim 8 \times 10^{15}/\text{mm}^3). \quad \text{↙ PARTICLES/mm}^3$$

$$a_\kappa = 4(\sim 0.143 \text{ nm})/\sqrt{2} = \sim 0.404 \text{ nm}. \quad R_{AL} \cong .143 \text{ nm}$$
$$(\text{PG } 521)$$

b)   With 4 atoms/u.c. and 2 a/o Cu,

$$\text{Cu/m}^3 = 0.02(4 \text{ atoms/u.c.})/(0.404 \times 10^{-9} \text{ m})^3$$

$$= 1.2 \times 10^{27}/\text{m}^3.$$

$$\text{Cu/particle} = (1.2 \times 10^{27}/\text{m}^3)/(\sim 8 \times 10^{24}/\text{m}^3)$$

$$= \sim 150 \text{ Cu/particle}.$$

Comment. This microstructure is approximately what we will encounter in precipitation hardening (Section 11–3). ◀

Example 10–1.3 The bct unit cell dimensions are $a = 0.2845$ nm and $c = 0.2945$ nm for martensite that contains 0.8 w/o carbon (3.6 a/o C). The lattice constant of austenite of the same composition and temperature is 0.3605 nm. How much volume change occurs when $\gamma \to M$?

Solution: Basis: 4 Fe atoms, or 2 u.c. M, and 1 u.c. γ.

$$\frac{\Delta V}{V_\gamma} = \frac{2(0.2845 \text{ nm})^2(0.2945 \text{ nm}) - (0.3605 \text{ nm})^3}{(0.3605 \text{ nm})^3}$$

$$= +0.018 \quad (\text{or } 1.8 \text{ v/o}).$$

Comments. This volume change leads to residual stresses within quenched steel. For example, the surface of a steel gear transforms to martensite first (while the center is still a hot, deformable austenite). Shortly later, the center transforms to martensite with an expansion in volume. This stresses the surface, placing the martensite in tension (and compresses the transforming austenite). Compare and contrast this with the tempering of glass (Figs. 8–7.4 and 8–7.5). ◀

10–2 MULTIPHASE MICROSTRUCTURES

We saw in Fig. 6–1.4 that a single-phase alloy could possess several geometric variants. The microstructural variables were grain *size*, *shape*, and *orientation*.

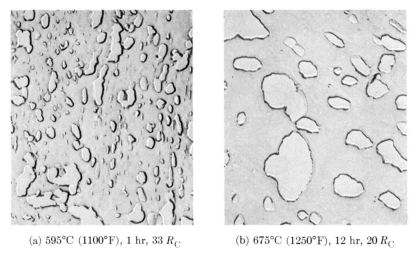

(a) 595°C (1100°F), 1 hr, 33 R_C (b) 675°C (1250°F), 12 hr, 20 R_C

Fig. 10-2.1   Electron micrographs of tempered martensite (X11,000). Each of these is eutectoid steel which was previously quenched to maximum hardness (65 R_C). (*Electron Microstructure of Steel*, American Society for Testing and Materials, and General Motors.)

The microstructures of multiphase alloys, likewise, can vary in the size, shape, and orientation of the individual phases. In addition, we encounter a range in the *amount* of each phase. Finally, we shall see a large variety of phase *distributions*.

Phase size   An example of differences in size is observed in Fig. 10-2.1. These two microstructures are from the same steel; hence, they have the same composition (0.8% C–99.2% Fe). They were both heat-treated to form austenite (austenitized) and quenched to form martensite. This was followed by heating to form tempered martensite that contains ferrite and carbide (Eq. 10-1.8). Since the two samples have the same composition, they have the same quantities of ferrite and carbide ($\alpha + \bar{C}$). The two differ in the sizes of the carbide particles. The steel with the finer carbides has been tempered for only one hour at 595°C (1100°F); the other for twelve hours at 675°C (1250°F). In the latter, there was time for the *coalesence* of carbides into larger (but fewer) particles.

Observe that the two steels of Fig. 10-2.1 differ in hardness. They also have a difference in strength. The hard carbide particles restrict the plastic deformation of the soft, ductile ferrite matrix. The amount of constraint is directly proportional to the amount of contact between the two phases, i.e., the phase boundary per unit volume between the ferrite and the carbide. It is only natural, therefore, that the finer carbides produce a harder microstructure (Example 10-2.1).

Phase amounts   The series of steels shown in Fig. 9-8.2 possesses progressively larger fractions of pearlite and, therefore, of carbide. Since the densities of ferrite and carbide are very similar (7.88 g/cm³ and 7.6 g/cm³, respectively), the volume fractions of the two phases in a steel are very similar to the weight fractions. This is not true in all microstructures. For example, a Pb/Sn solder contains two phases that have significantly different densities. Therefore, the volume fractions of α and β in Fig. 9–1.3 must be recalculated from their weight fractions (Example 10–2.2). Furthermore, we may establish a *mixture rule* for the density, ρ_m, of a phase mixture based on the volume fraction f and density of each phase:

$$\rho_m = f_1 \rho_1 + f_2 \rho_2 + \cdots. \qquad (10\text{–}2.1)$$

Such a linear relationship is applicable to density calculations if we use volume fractions, because the masses are simply additive on a volume basis.*

Phase shape and distribution Two easily encountered, but diverse, examples of phase shape and distribution can be found for steels of eutectoid composition (~ 0.8 C–99.2 Fe). As shown in Fig. 10–2.2(a), *pearlite* is a lamellar mixture of ferrite plus carbide. It can be formed only by the decomposition of austenite with a eutectoid carbon content. The two new phases ($\alpha + \bar{C}$) nucleate at the austenite grain boundaries and grow simultaneously into the grain (Fig. 10–1.9). Carbon is segregated, leaving the ferrite ($\sim 0.02\%$ C), and is concentrated into the carbide ($\sim 6.7\%$ C).

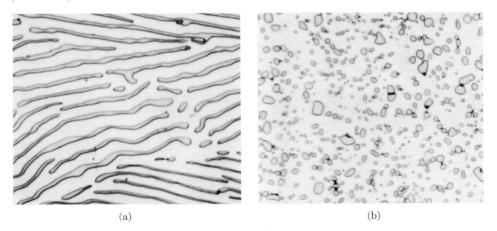

(a)   (b)

Fig. 10–2.2 Microstructures of eutectoid steels (0.80% C). (a) Pearlite ($\times 2500$) formed by transforming austenite (γ) of eutectoid composition. (b) Spheroidite ($\times 1000$) formed by tempering at 700°C (1300°F). (Courtesy U.S. Steel Corp.)

In Fig. 10–2.2(b), we have *spheroidite*. The carbide in spheroidite is a dispersion of sphere-like particles (hence the name) within a matrix of ferrite. Spheroidite can be produced by two methods. The first is through an over-tempering of martensite so that the trend started in Fig. 10–2.1 is continued to produce relatively large carbide

---

* Equation 10–2.1 is not applicable to properties such as resistivity and strength because they involve geometric factors such as paths of conduction, and surfaces between phases that do not depend on the volume fraction alone.

spheroids, and relatively large ferrite distances between the carbides. The second method is to heat pearlite for extended periods of time (18–24 hours at 700°C) so that the carbide lamellae globularize. The driving force for this change in shape to spheroids is the reduction in the amount of grain-boundary area.

A contrast in phase distribution was also shown schematically in Fig. 10–1.8. With high-temperature precipitation (or by slow cooling), the precipitating phase nucleates and grows at the grain boundaries of the principal solid solution (Fig. 10–1.8b). Thus, a grain-boundary network forms of the new phase. When this is a brittle phase, we shall see that the total material is brittle, since propagating cracks are provided with a low-energy path for fracture. By contrast, significant undercooling as a result of a rapid drop in temperature leads to a fine dispersion of preciptate particles (Fig. 10–1.8c). With a dispersion of hard particles within the ductile matrix, we obtain the favorable situation where the hard particles provide plastic constraint and therefore strengthen the matrix of the microstructure. At the same time, the ductile matrix is tough. Since a progressing crack must travel through this tough matrix phase, energy is required, giving *toughness* to the total material, as well as *strength*.

•As a further example of phase shape and distribution, we should consider Fig. 10–2.3, which shows transverse and longitudinal views of a eutectic microstructure of aluminum and Al_3Ni. The lighter phase is a continuous matrix of aluminum. The darker phase is fibers of Al_3Ni, a hard, rigid intermetallic compound of aluminum and nickel. This microstructure is geometrically comparable to glass-fiber reinforced

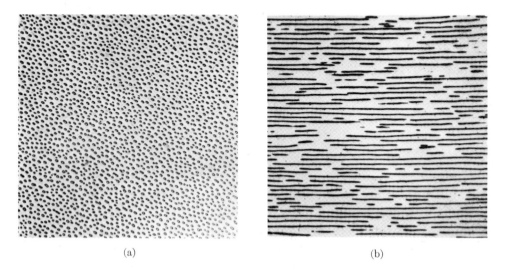

(a)   (b)

Fig. 10–2.3 Phase orientation. In each microstructure the phases have specific crystallographic relationships. (a) Transverse and (b) longitudinal eutectic microstructures of $Al–Al_3N$ (X400). (R. W. Hertzberg, F. D. Lemkey, and J. A. Ford, "Mechanical Behavior of Lamellar ($Al–CuAl_2$) and Whisker Type ($Al–Al_3Ni$) Unidirectionally Solidified Eutectic Alloys," *Trans. AIME.*)

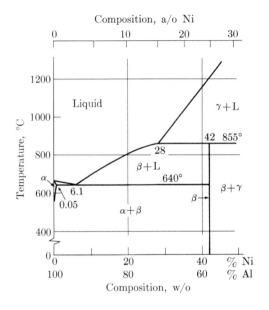

Fig. 10–2.4 Al–Ni diagram. The microstructure of Fig. 10–2.3 formed by the directional solidification of a eutectic liquid (6.1% Ni). The two phases, α and β, solidified simultaneously with β comprising only 7% of the volume (14 w/o). (Adapted from *Metals Handbook*, Amer. Soc. Metals.)

plastic. However, it was made by a markedly different procedure. In fiber-reinforced plastics, the fibers can be oriented by winding the rovings over mandrels, or by aligning them into bundles. These fibers are then impregnated with resins that can be polymerized into a bonding matrix. The microstructure of Fig. 10–2.3 was formed by directional solidification of a liquid alloy that had a eutectic composition (Fig. 10–2.4). Since the eutectic is at the intersection of the two solubility curves, the eutectic liquid simultaneously produces two solids on solidification. If the heat is extracted in one direction only, the solidification is directional with the two phases growing parallel. Since the β phase (Al_3Ni) is the minor phase, it forms a fiber-like geometry. These alloys have received specific attention because they provide a method of producing metals reinforced with rigid fibers.

Example 10–2.1 Compare the interphase boundary areas (mm^2/mm^3) in the two tempered martensites of Fig. 10–2.1 ($\times$ 11,000).

Solution: Use Eq. (4–4.1) and the perimeter of the photomicrograph. In part (b)

$$P_L = (\sim 20)/(215 \text{ mm}/11,000) = \sim 10^3/\text{mm};$$

$$S_V = \sim 2,000 \text{ mm}^2/\text{mm}^3.$$

In part (a)

$$S_V = 2(\sim 50)/(215 \text{ mm}/11,000)$$
$$= \sim 5,000 \text{ mm}^2/\text{mm}^3.$$

Comments. The tempered martensite with the greater interphase boundary area is harder. In order to make direct comparisons, such as hardness, other microstructural features, such as grain shape and phase quantities, must be comparable. ◀

Example 10-2.2 Calculate the density of a solder (eutectic) that has been equilibrated at 20°C.

Solution: Since β and α are nearly pure Sn and Pb, respectively, $\rho_\beta = 7.3$ Mg/m^3 and $\rho_\alpha = \sim 11.3$ Mg/m^3 (or 11.3 mg/mm^3). Basis: 100 mg = 61.9 mg β and 38.1 mg α.

$$\beta: \quad 61.9 \text{ mg}/(7.3 \text{ mg/mm}^3) \quad = \quad 8.48 \text{ mm}^3 \qquad f_\beta = 0.715;$$
$$\alpha: \quad 38.1 \text{ mg}/(11.3 \text{ mg/mm}^3) = \quad \underline{3.37 \text{ mm}^3} \qquad f_\alpha = 0.285.$$
$$11.85 \text{ mm}^3 \text{ total}$$

$$\rho = 0.715 \,(7.3 \text{ mg/mm}^3) + 0.285 \,(11.3 \text{ mg/mm}^3) = 8.44 \text{ mg/mm}^3, \text{(or 8.44 Mg/m}^3).$$

Alternatively, $\rho = 100$ mg/11.85 mm^3 = 8.44 mg/mm^3. ◀

10-3 MICROSTRUCTURES: MECHANICAL PROPERTIES

Properties such as hardness and strength cannot be approximated by simple mixture rules of properties of the contributing phases. For example, the steels of Fig. 10-2.1 have tensile strengths in excess of 700 MPa ($>$100,000 psi). However, the strength of the matrix phase (ferrite) is less than one third of those values. Since the ferrite is the continuous phase and therefore all of the load must be carried through it, we may conclude that "the chain is stronger than its weakest link!" While there is a mathematical rationale for this, our qualitative explanation will simply be that the rigid carbide particles inhibit slip and prevent shear of the weaker matrix.* We call this *plastic constraint.*

Effects of phase quantities   Although we have not presented a rigorous explanation for these interactive properties, we make use of them, because materials can be strengthened by the addition of "fillers." For example, the addition (1) of carbon to rubber, (2) of sand to clay, (3) of sand to tar or asphalt, (4) of wood flour to plastics, or (5) of intermetallic precipitates to soft metal increases the resistance to deformation or flow. The effect on strength in the fourth example is shown graphically in Fig. 10-3.1. Although a phenol-formaldehyde resin has considerable strength alone, it is subject to eventual shear failure under stress, and the incorporation of a second phase produces added resistance to deformation. At the other end of the composition range, the strength of wood flour (fine sawdust) alone is nil; there are no forces that hold the individual particles of cellulose into a coherent mass. The addition of the phenol-formaldehyde resin to the fine particles serves to cement the wood flour together. Maximum strength is developed with intermediate compositions, because of the strengthening that accompanies a mixture of phases.

Highway construction is another example of the use of phase mixtures. For obvious reasons, roadbeds composed solely of clay would be unsatisfactory, as would those made completely of gravel or sand. However, an appropriate combination of clay and gravel produces a practical, stabilized roadbed (Fig. 10-3.2). The clay is

---

* This can be demonstrated dramatically by using some solder ($S_t \simeq 35$ MPa, or $\sim 5,000$ psi), to join two pieces of steel together. If the soldered joint is thin, this joint can support a stress over 170 MPa ($>$25,000 psi)! Although weak, the solder does not deform because it is rigidly held by the stronger adjacent steel. With thinner and thinner joints, there is less and less chance for plastic deformation within the weak solder, so the joint will become even stronger.

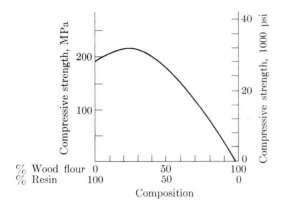

Fig. 10–3.1 Strength of mixtures (wood-flour filler in phenol-formaldehyde resin). The mixture of wood flour and resin is stronger than either alone. The wood flour prevents slip in the resin; the resin bonds the particles of wood flour.

strengthened by the hard gravel, and the gravel is bound by the clay into a coherent mass that resists concentrated loads.

Pearlite provides us with a good quantitative example of the relationship between microstructure and mechanical properties. Figures 10–3.3 and 10–3.4 plot hardness, strengths, ductilities, and toughness against the carbon content. The latter can also be expressed as percent carbide, or in terms of the microstructure as percent pearlite. Thus, the 1040 steel of Fig. 10–3.5 that contains 0.40% carbon possesses 50% pearlite with the balance being proeuctectoid ferrite. The 1080 steel possesses 100% pearlite (Fig. 10–3.5b).

The steels of Fig. 10–3.3 are strengthened because the carbide lamellae provide the plastic constraint that restricts the deformation of the ductile ferrite. Of course, a higher carbide content increases the boundary surface between the two phases and, therefore, produces a positive slope to the curves on Fig. 10–3.3. A reverse situation occurs for the curves in Fig. 10–3.4, for the same reason; specifically, less plastic deformation is possible prior to fracture with more of the nonductile carbide present. This leads not only to lower ductility but also to lower toughness.*

Fig. 10–3.2 Stabilized roadbed. The mixture of gravel and clay is more durable than either gravel or clay alone.

---

* Toughness relates to the product of stress and strain, and therefore to the product of strength and ductility. Observe from Figs. 10–3.3 and 10–3.4 that the decreases in ductility are relatively greater than the increase in strength. Therefore, there is a net decrease in toughness with added carbon.

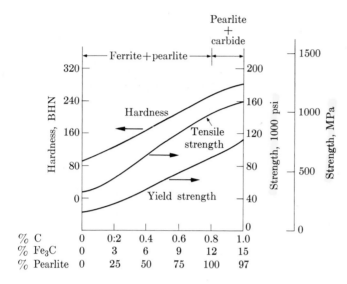

Fig. 10–3.3 Hardness and strength versus carbon content of annealed plain-carbon steels. The steels contain microstructures similar to those shown in Fig. 10–3.5. Hardness: read left. Strength: read right.

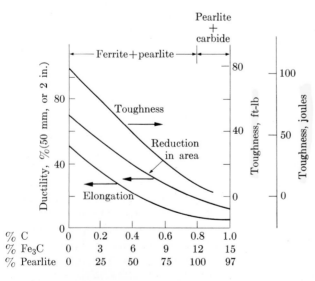

Fig. 10–3.4 Toughness and ductility versus carbon content of annealed plain-carbon steels. (Cf. microstructure of Fig. 10–3.5.) Those steels with more proeutectoid ferrite are both more ductile and tougher. Ductility: read left. Toughness: read right.

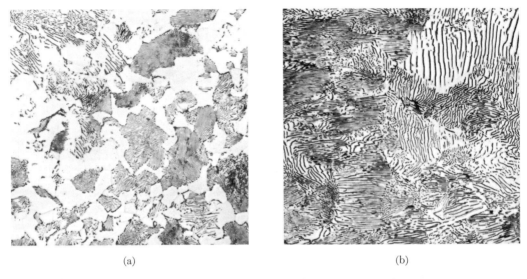

(a)   (b)

Fig. 10–3.5   Microstructure of pearlite steels (X500). (a) 0.40% carbon. (b) 0.80% carbon.
The larger (white) ferrite areas of the 0.40% carbon steel formed before the eutectoid reaction
produced the lamellar ferrite in the pearlite (gray). The 0.80% carbon steel contains only
eutectoid ferrite. (Compare with Figs. 9–7.3 and 9–8.1.) (U.S. Steel Corp.)

We may also rationalize the embrittlement (loss of toughness) of steels with
increased carbon contents on the basis of microstructures. With added pearlite
(Fig. 10–3.5b), a nearly continuous brittle path is available along carbide lamellae
that are easily followed by the progressing crack.

Effects of phase size   We have already discussed the fact that very fine sand will
strengthen asphalt more than an equal volume fraction of gravel. Likewise, finer
carbide particles strengthen ductile ferrite more than coarser carbides (Fig. 10–2.1).
The same consequences are observed in Fig. 10–3.6 for steels containing coarse and
fine pearlite; by that, we mean steels with thicker or thinner lamellae of ferrite and
carbide. Recall from our discussion of the euctectoid reaction in Section 10–1 that
accelerated cooling rates (or lower austenite decomposition temperatures) lead to
more, and thinner, lamellae because diffusion was limited to shorter distances.

A more quantitative relationship between the lamellae dimensions and yield
strength is shown in Fig. 10–3.7. The thinner layers of the fine pearlite possess a
greater grain-boundary area per unit volume. Thus, there is more plastic constraint
of the normally deformable ferrite. Figure 10–3.8 relates the hardness of a pearlitic
1080 steel to the amount of ferrite/carbide boundary area per unit volume.

Similar quantitative effects of size are observed in Fig. 10–3.9 where time and
temperature may be simultaneously considered. The initial martensite hardness
($65R_C$) is not retained by the tempered martensite. The rate of coalescence observed
in Figs. 10–2.1(a) and (b) is rapid at high temperatures. The coarsening of the carbide

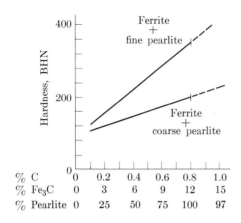

Fig. 10–3.6 Effect of microstructure dimensions on hardness of steel. The harder, finer pearlite was formed by faster cooling.

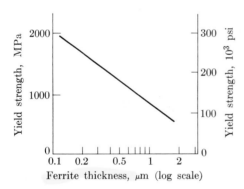

Fig. 10–3.7 Strength versus microstructure (ferrite thickness in pearlite). Thinner lamellae provide more lamellae to give plastic constraint. Therefore, the yield strength is raised.

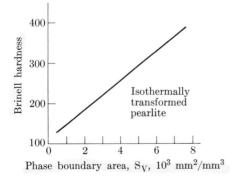

Fig. 10–3.8 Strength versus microstructure (1080 pearlitic steel). The steel is harder with more ferrite-carbide boundary area (thinner lamellae). (F. Rhines, *Metal Progress*.)

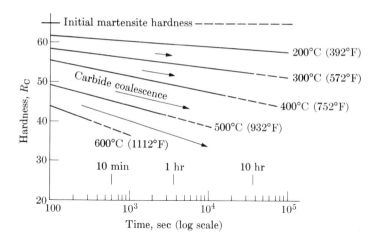

Fig. 10–3.9   Hardness of tempered martensite (1080 steel quenched to 65 R_C). Softening occurs as the carbide particles coalesce, giving greater intervening ferrite distances (Fig. 10–2.1). An Arrhenius pattern (Eq. 4–6.6) is followed in the early stages of softening.

reduces the hardness to less than $50R_C$ in a matter of seconds if the martensite is tempered at 600°C. At 400°C, the same softening takes approximately an hour. The hardness will drop to only $60R_C$ within the first hour at 200°C.*

The relationships of Fig. 10–3.9 are logarithmic because tempering involves atom movements. The required activation is thermally controlled. (See Example 10–3.2.)

Effects of phase shape and distribution   In Fig. 10–2.2, we saw two microstructures with marked contrasts in the geometry of the carbide phase. The two have the same density because they have the same volume fraction of ferrite and carbide. The two have significantly different mechanical properties (Fig. 10–3.10).

The lamellar carbides of pearlite provide greater plastic constraint to prevent deformation of the ferrite than do the spheroidal carbides of spheroidite. This contrast, which appears in Fig. 10–3.10(a) as a difference in hardness, is primarily a function of phase-boundary area because it is the adjacent ferrite that is constrained from slipping. Since the two microstructures have the same volume fraction of carbides, the nonspherical microstructure possesses greater boundary area per unit volume.

Spheroidite is softer and weaker than pearlite, but it is tougher (Fig. 10–3.10b). This is another example of the situation described earlier. A propagating crack must pass through more of the tough ferrite matrix in spheroidite than in pearlite. Thus, more energy is absorbed.

---

* At 20°C, tempering occurs almost infinitely slow. In fact, artifacts of early tools that are still martensitic after centuries have been found by archeologists.

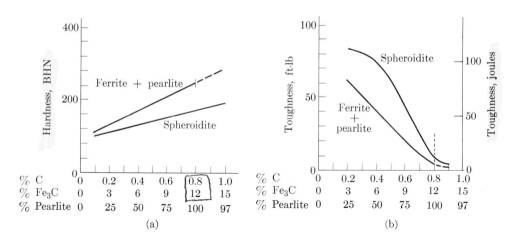

Fig. 10–3.10 Hardness and toughness versus carbide shape (annealed plain-carbon steels). The carbides in the pearlite are lamellar; those in the spheroidite are sphere-like.

Example 10–3.1 A eutectoid (1080) steel contains ferrite ($\rho = 7.88$ g/cm³) and carbide ($\rho = 7.6$ g/cm³). (a) What is the density of the steel? (b) What is the approximate ratio of ferrite lamellae thickness t_α to carbide lamellae thickness $t_{\bar{C}}$?

Solution: Basis: 100 g steel = 0.8 g carbon = 12 g Fe₃C + 88 g α   FROM GRAPH
ABOVE

$$88 \text{ g } \alpha/(7.88 \text{ g/cm}^3) = 11.17 \text{ cm}^3 \text{ } \alpha,$$

$$12 \text{ g } \bar{C}/(7.6 \text{ g/cm}^3) = \underline{\quad 1.58 \text{ cm}^3 \text{ } \bar{C}.}$$
$$12.75 \text{ cm}^3$$

a) $\rho = 100$ g/12.75 cm³ = 7.84 g/cm³ (or 7.84 Mg/m³).
b)   Since the long dimensions of the lamellae of α and $\bar{C}$ are essentially the same,

$$t_\alpha/t_{\bar{C}} = V_\alpha/V_{\bar{C}} = 11.17/1.58 = \sim 7. \blacktriangleleft$$

Example 10–3.2 The data of Fig. 10–3.9 reveal that it takes 10^4 sec to temper martensite from $65 R_C$ to $50 R_{\bar{C}}$ at 360°C and 100 sec at 490°C. Assuming an Arrhenius relationship (Eq. 10–1.1), how long will it take at 300°C?

Solution: Using $\ln t = C + B/T$,

$$\ln 10^4 = 9.2 = C + B/(360 + 273),$$

$$\ln 100 = 4.6 = C + B/(490 + 273).$$

Solving simultaneously,

$$B = 17,100 \text{ K},$$

$$C = -17.82;$$

$$\ln t = -17.82 + 17,100/573$$

$$t = 170,000 \text{ sec} \quad (\sim 2 \text{ days}).$$

Comment. This calculation is appropriate for hardness values greater than about $35 R_C$, while the carbide particles are still small and numerous. $\blacktriangleleft$

• 10–4   MICROSTRUCTURES: PHYSICAL PROPERTIES

We cannot make generalities about the physical properties of multiphase micro-structures. However, it may be advantageous to cite some relationships.

   Light transmission of nonmetallic solids is markedly affected by the presence of more than one phase. Every time a light ray crosses a boundary, the ray is refracted in accordance with Snell's law:

$$n_1 \sin \phi_1 = n_2 \sin \phi_2, \tag{10–4.1}$$

where n is the index of refraction of a phase and ϕ is the angle between the ray and the interface normal. Since two phases of a microstructure would have the same index of refraction by coincidence only, we find that ϕ_1 and ϕ_2 are usually different, with the consequence that light is refracted every time it crosses a phase boundary (Fig. 10–4.1). This leads to dispersion and translucency rather than transparency. Thus, materials with transparent phases, e.g., brick and concrete, do not transmit light because the light is scattered and eventually attenuated before it completes its path through the material.

   Heat capacities, c, of most metals and many ceramics are approximately 25 J/mole·°C (or 6 cal/mole·°C) at normal temperatures.* In calculations, we are commonly interested in the capacity on a gram basis (J/g·°C) or on a volume basis (J/cm^3·°C) because we design products in these terms rather than by moles. Multiphase materials follow a simple mixture rule:

$$c_m = f_1 c_1 + f_2 c_2 + \cdots . \tag{10–4.2}$$

In this case f can be either mass fraction or volume fraction, providing we use the corresponding heat capacity data.

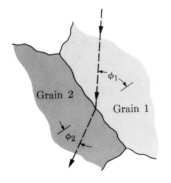

Fig. 10–4.1   Diffuse light. Microstructures with transparent phases will scatter and absorb light, because light is refracted at grain and phase boundaries. Snell's law (Eq. 10–4.1) applies; however, where there are innumerable boundaries, we do not observe transparency through the material.

---

* The general value of ~ 25 J/mole·°C pertains to materials that do not have bond bending nor bond rotations. Since most polymers and many glasses absorb energy in these forms, their heat capacities are correspondingly higher. Also, the heat capacities drop below ~ 25 J/mole·°C when the temperature drops below $\sim 0.2 T_m$.

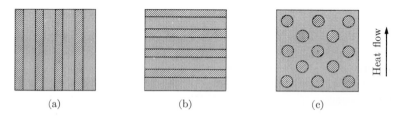

Fig. 10–4.2 Conductivity versus phase distribution (idealized). (a) Parallel conductivity (Eq. 10–4.3). (b) Series conductivity (Eq. 10–4.4). (c) Conductivity through a material with a dispersed phase (Eqs. (10–4.6) and (10–4.7), when the two phases have markedly disimilar conductivities.)

Thermal expansion coefficients of phase mixtures are also approximated by linear mixture rules. These rules may be used for estimating purposes; however, it should be appreciated that differential thermal expansion sets up internal stresses so that elastic strains are present. A precise calculation would require appropriate adjustments.

Conductivities, both electrical and thermal, are very sensitive to microstructural geometry. Two extreme situations are illustrated in Fig. 10–4.2. Consider the thermal conductivity coefficient k. In Fig. 10–4.2(a), the conduction is *parallel* to the structure; so the equation

$$k_{\|} = f_1 k_1 + f_2 k_2 + \cdots \qquad (10\text{–}4.3)$$

applies. The structure is in *series* in Fig. 10–4.2(b), so heat must be conducted perpendicularly to the layers. This leads to a reciprocal mixture rule for the conductivity of the multiphase structure.

$$1/k_{\perp} = f_1/k_1 + f_2/k_2 + \cdots. \qquad (10\text{–}4.4)^*$$

In these two equations f is the volume fraction.

The more commonly encountered microstructures involve a dispersion of two or more phases (Fig. 10–2.2b). Furthermore, even a lamellar structure such as pearlite has random orientation of the pearlite colonies. To a first approximation, the conductivities of these microstructures may be considered to vary linearly on a volume fraction basis. However, again modifications must be made if one of the two phases is continuous, c, and the other is dispersed, d. Where $k_c \ll k_d$,

$$k_m \approx k_c(1 + 2f_d)/(1 - f_d). \qquad (10\text{–}4.6)$$

Where $k_c \gg k_d$,

$$k_m \approx k_c(1 - f_d)/(1 + f_d/2). \qquad (10\text{–}4.7)$$

---

* For electrical conductivity σ, the relationship for a series sequence is

$$1/\sigma_{\perp} = f_1/\sigma_1 + f_2/\sigma_2 + \cdots, \qquad (10\text{–}4.5\text{a})$$

or for resistivity ρ, the equation is:

$$\rho_{\perp} = f_1\rho_1 + f_2\rho_2 + \cdots. \qquad (10\text{–}4.5\text{b})$$

This equation relates to the more familiar resistance R equation for series that is encountered in physics calculations.

In these equations, f_c and f_d are the volume fractions of the continuous and dispersed phases, respectively. The dominating role of the continuous phase is evident, even though it may be minor. If it has a high conductivity, it forms a route for thermal (or electrical) transport. If the continuous phase is insulative, transport is limited even though there may be major amounts of a highly conductive dispersed phase.

Example 10–4.1 Estimate the heat given off when 150 g of an 80 Pb–20 Sn solder are cooled from 180° to 20°C.

Solution: Based on 25 J/mol·°C,   ↙ MOLECULAR WEIGHT

Pb: ⟶ $(25 \text{ J}/207.2 \text{ g·°C}) = 0.121 \text{ J/g·°C};$

Sn: ⟶ $(25 \text{ J}/118.7 \text{ g·°C}) = 0.211 \text{ J/g·°C}.$

Since there are 120 g and 30 g of lead and tin, respectively,

$$\text{heat released} = [(0.121)(120) + (0.211)(30) \text{ J/°C}][-160°C]$$
$$= -3.3 \text{ kJ}.$$

Comment. The heat capacity (J/g·°C) of the two-phase solder may be calculated for 20°C as

$$c_m = (0.8)(0.121) + (0.2)(0.211)$$
$$= 0.14 \text{ J/g·°C},$$

since α and β are essentially pure Pb and Sn at that temperature. At higher temperatures there is more α, but c_α changes accordingly, since it contains more tin. ◀

Example 10–4.2 Fifty w/o SiO_2 flour (i.e., quartz powder) is added to a phenol-formaldehyde resin as a filler. (a) What is the density of the mixture? (b) What is the thermal conductivity?

Answer: From Appendix C:

$$\rho_{SiO_2} = 2.65 \text{ Mg/m}^3, \qquad \rho_{pf} = 1.3 \text{ Mg/m}^3,$$
$$k_{SiO_2} = 0.012 \text{ (W/mm}^2)/(°C/mm),$$
$$\boxed{k_{pf} = 0.00016}\text{(W/mm}^2)/(°C/mm).$$

(Basis: 100 g, and densities of 2.65 g/cm³ and 1.3 g/cm³, respectively.)

a)   50 g SiO_2 = 18.8 cm³ SiO_2; f_{SiO_2} = 0.33
           50 g   pf   = 38.4 cm³   pf; f_{pf} = 0.67
                     57.2 cm³   1.0

$$\rho_m = 100 \text{ g}/57.2 \text{ cm}^3 = 1.75 \text{ g/cm}^3,$$

or

$$\rho_m = (0.33)(2.65) + (0.67)(1.3) = 1.75 \text{ g/cm}^3.$$

b)   Since $k_{pf} \ll k_{SiO_2}$,

$$k_m \simeq k_c \left(\frac{1 + 2f_d}{1 - f_d}\right),$$

$$k_m \simeq 0.00016\left[\frac{1 + 2(0.33)}{1 - 0.33}\right] = \frac{0.0004 \text{ (W/mm}^2)}{(°C/mm)}. \quad ◀$$

REVIEW AND STUDY

SUMMARY

We depend on solid-phase reactions to change a material from one microstructure to another. Polymorphic reactions, recrystallization, and grain growth do not require extensive diffusion. Solution and precipitation involve two phases of different compositions, and therefore require major diffusion to proceed. In solution treatments, the atoms of the minor component must diffuse to become distributed throughout the solvent phase; in precipitation reactions, the excess atoms of the solute must segregate into a new phase.

When the reaction is governed by diffusion, the reaction time t (or its reciprocal, R) can be related to temperature through the Arrhenius equation, $\ln t = C + B(1/K)$. Even so, the progress of the reaction is complicated by a slow start, followed by more rapid change, and then a lingering completion (Fig. 10–1.4).

Solid-phase reactions that require the nucleation of a new phase proceed faster when the amount of supercooling is increased; however, the maximum rate (shortest time) is soon reached. At still lower temperatures, longer times are required because diffusion and compositional changes occur very slowly. This provides the typical C-curve for isothermal reactions, and gives a basis for studying (1) factors that affect reaction rates, and (2) the resulting microstructures. Above the "knee" of the C-curve, the reaction is generally nucleated at grain boundaries, with the reactants diffusing to those sites. In a temperature regime below the bend of the C-curve, the nucleation is distributed throughout the grains, and the required diffusion is decreased.

The eutectoid reaction involves three phases. During cooling, it is common to have the two product phases grow from the phase boundary into the disappearing reaction phase. This is the way pearlite is formed. Major amounts of diffusion are required.

Martensite is a transition phase that forms intermediately between austenite, and ferrite plus carbide $(\alpha + \bar{C})$. The quenching must be sufficiently fast to cool the austenite to low temperatures before normal decomposition can occur. This hard phase is of major interest in steel processing. When it finally proceeds to the more stable $(\alpha + \bar{C})$, the carbides form a very fine dispersion in the ferrite matrix, to give a microstructure called tempered martensite. It will be slightly softer than martensite, but much tougher.

Properties related primarily to the relative amounts of the phases can be described by mixture rules. These are sometimes considered to be additive. Other properties are more complex, because the behavior of one phase is constrained by its neighboring phases. We are able to observe the qualitative effects of phase amounts, phase size, and phase shape and distribution by examining the properties of various annealed steels (Fig. 10–3.3 through 10–3.10).

KEY TERMS AND CONCEPTS

C-curve

Coalescence

• Directional solidification

Dispersed phase

Isothermal precipitation

Microstructural effects

   phase distribution

   phase quantities

   phase shape

   phase size

Martensite, M

Matrix phase

Metastability

Mixture rule

Multiphase microstructures

Pearlite growth

Phase, transition

Plastic constraint

•Snell's law

Solid-phase reactions

   grain growth

   recrystallization

   polymorphic

   solution

   precipitation

   eutectoid

   martensite

Solution treatment

Spheroidite, $(\alpha + \bar{C})$

Tempered martensite, $(\alpha + \bar{C})$

Tempering

FOR CLASS DISCUSSION

A_{10} Which of the following polymorphic reactions will require the least activation energy? The most? Why? (a) Bcc Fe → fcc Fe; (b) Graphite → diamond; (c) Cubic $BaTiO_3$ → tetragonal $BaTiO_3$.

B_{10} An annealed brass is heated until the average grain dimension is 1 mm. Another sample of the same brass is cold-worked 5% and heated for the same time at the same temperature. The average grain dimension is 10 mm. A third sample is cold-worked 20%. Its grains are 3 mm after the same heat treatment. Provide a rationale for these results.

C_{10} Highly deformed copper (CW > 80%) recrystallizes at lower temperatures than slightly deformed copper (CW < 20%). Why?

D_{10} Refer to paragraph 2 of Section 10–1. Point out how reaction rates are affected by the features that are cited in each category.

E_{10} Explain solid precipitation to a student who has had college chemistry, but who has not had this course.

F_{10} Explain the three curves of Fig. 10–1.5 to a classmate.

G_{10} Why do eutectoid reactions generally require longer times than polymorphic reactions?

H_{10} Why is time required for pearlite formation from austenite?

I_{10} Austenite decomposes to $(\alpha + \bar{C})$ in less time at 500°C than at either 700°C or 300°C. Suggest reasons on the basis of Fig. 10–1.7.

J_{10} Why does martensite resist slip more than does ferrite?

K_{10} In your own words, explain why the formation of martensite is not time-dependent.

L_{10} The bct structure of martensite contains two iron atoms per unit cell and has $c = 0.291$ nm and $a = 0.285$ nm when the w/o carbon is 0.5. Point out the locations within the unit cell where the carbon atoms can reside with least strain.

M_{10} Equation 10–1.7 is irreversible as written. Under what conditions could austenite reform from $(\alpha + \bar{C})$ or from M? (*Note:* The direct reversal of $(\alpha + \bar{C})$ to M has never been observed, nor is it expected. Why?)

N_{10} Why does Fe_3C strengthen ferrite even though the carbide is so brittle that "it breaks like glass"?

O_{10} Why is fine pearlite stronger than coarse pearlite?

P_{10} Cite those properties of pearlite and spheroidite (Fig. 10–2.2) that are the same; those that are different. Explain your choices.

Q_{10} Compare and contrast (a) pearlite, (b) martensite, and (c) tempered martensite. Distinguish between a phase and a microstructure.

R_{10} The growth of the carbide particles in Fig. 10–2.1 is not directly comparable to the growth mechanisms of Sections 6–7 and 10–1. Rather, they are related to the mechanisms of solution and precipitation. Discuss.

S_{10} A hacksaw blade contains enough carbon so its hardness is $>60R_C$ after quenching. Before use, however, it is tempered. As a result, its hardness drops to $<60R_C$. How will the tempering affect the ability of the steel blade to cut other steel?

T_{10} On the basis of the information in *this* chapter, suggest a way of preparing spheroidite.

U_{10} The average width of the ferrite lamellae may be measured in Fig. 10–2.2(a). What correction must be made in that data before we estimate its yield strength from Fig. 10–3.7?

V_{10} The 95 Al–5 Cu alloy is cooled slowly so that θ forms a network at the κ grain boundaries. Based on Fig. 9–5.3 and Sections 6–4 and 8–7, predict various mechanical properties.

W_{10} Explain to a classmate why the curves of Fig. 10–3.4 have a negative slope.

X_{10} Explain to a freshman chemistry student why the curves of Fig. 10–3.9 have a negative slope.

Y_{10} Give a basis for using the Arrhenius assumption in Example 10–3.2.

Z_{10} "Oilless" bronze bearings are made by sintering bronze powders to have a porosity of 10 v/o–20 v/o. What mixture rule would best describe their density? Their thermal conductivity?

STUDY PROBLEMS

10–1.1 Refer to Example 10–1.1. At what temperature will the grain size grow from 0.05 mm to 0.10 mm in 15 hours?

Answer: 625°C.

• **10–1.2** The data of Fig. 10–1.1 obey the following relationship:

$$\ln \delta = C + n \ln t \quad (\text{or } \delta \propto t^n)$$

independently of temperature and as long as the grain dimensions δ do not approach the sample dimensions. What is the value of the grain growth exponent n for the brass in Fig. 10–1.1?

10–1.3 Solve graphically from Fig. 10–1.4. (a) The time to recrystallize (50%) this copper at 75°C (167°F). (b) The temperature required for hot-working if the recrystallization is to be 50% completed while the copper moves through the rolls (~ 0.06 sec).

Answer: (a) 1000 min (17 hrs)   (b) $1/T = \sim 0.00165$ $T = \sim 605$ K $= \sim 330$°C

10–1.4 Solve Study Problem 10–1.3 mathematically. (Cf. Example 6–6.2.)

10–1.5 An alloy of 95 Al–5 Cu is solution-treated at 550°C, then cooled rapidly to 400°C where it is held for 24 hours. During that time it produces a microstructure with 10^6 particles of θ per mm^3. The θ particles ($CuAl_2$) are nearly spherical and have twice the density of κ matrix. (a) Approximately how far apart are these particles? (b) What is the average particle dimension?

Answer: (a) ~ 10 μm   (b) 3.8 μm

10–1.6 A microstructure has spherical particles of β with an average dimension $\bar{d}$ that is 10% of the average distance $\bar{D}$ between the centers of the adjacent particles. (a) What is the volume percent of β? (b) What is the ratio $\bar{d}/\bar{D}$ with 0.5 v/o β?

10–1.7 A 90 Pb–10 Sn solder is held at 185°C until equilibrium is established. The average grain size is 10^6 μm^3. The alloy is then *rapidly* cooled to room temperature where a β precipitate forms within the initial α grains. The resulting matrix within those former α grains is 99 Pb–1 Sn ($\rho = 13.3$ Mg/m^3), and the β particles ($\sim 100\%$ Sn) are separated by an average distance of 0.1 μm. (a) How many particles are there per original grain? (b) What is the volume fraction of β? (c) Approximately how many tin atoms are there per β particle?

Answer: (a) $\sim 10^9$ (b) 15 v/o   (c) $\sim 5 \times 10^6$ Sn

10–1.8 A sterling silver (92.5 Ag–7.5 Cu) is quenched after solution treatment, then reheated to 400°C until equilibrium is attained; at equilibrium it contains a β precipitate (approximately spherical in shape) in an α matrix. (a) If the representative dimension d of the β particle is 0.1 μm, how many will there be per mm^3? If $d = 0.05$ μm? (b) What is the average distance between particles?

10–1.9 (a) Identify the eutectoid reaction in the Fe–O system (Fig. 9–5.5). (b) How much metallic iron is in the two-phase product below the eutectoid temperature?

Answer: (b) 16 w/o

10–1.10 An oxide containing 75 w/o iron and 25 w/o oxygen is solution-treated at 1300°C (2370°F). It is then cooled slowly to room temperature. Compare and contrast the reactions and possible microstructures with those found after slow cooling a 1060 steel from 900°C (1650°F).

10–1.11 Refer to Study Problem 10–1.9. What is the Fe^{3+}/Fe^{2+} ratio in the oxide of the eutectoid composition?

Answer: 0.13

10–1.12 Refer to the Cu–Sn diagram in Chapter 9. Locate and state the reactions (on cooling) for four eutectoid transformations.

10–1.13 Refer to discussion topic L_{10}. Calculate the density of the martensite that contains 0.50 w/o C.

Answer: 7.89 Mg/m^3 ($=$7.89 g/cm^3)

10–2.1 An Al–Cu alloy contains 2.5 v/o θ ($\rho = 5.5$ Mg/m^3) in a matrix of κ which is essentially pure Al. What is the density of the alloy?

Answer: 2.77 Mg/m^3 ($=$2.77 g/cm^3)

10–2.2 The tempered martensites of Fig. 10–2.1 and the spheroidite of Fig. 10–2.2(b) have the same composition and comparable microstructures (sphere-like carbide in a matrix of ferrite). How will the hardness of the spheroidite compare with the two products in Fig. 10–2.1? (See Example 10–2.1 and its comments.)

10–2.3 Silica flour (finely ground quartz, $\rho = 2.65$ Mg/m^3, or 2.65 g/cm^3) is used as a filler for polyvinyl chloride ($\rho = 1.3$ Mg/m^3). (a) What volume fraction is required to give a product density of 1.70 Mg/m^3? (b) What is the w/o SiO_2?

Answer: (a) 30 v/o SiO_2 (b) 47 w/o SiO_2

10–2.4 Calculate the density of a glass-reinforced phenol-formaldehyde rod, in which the glass content is 15 w/o. (A borosilicate glass is used for the longitudinal glass fibers.)

10–2.5 The density of a Pb–Sn alloy is 10.0 Mg/m^3 ($=$10.0 g/cm^3) after equilibration at 20°C. (a) What is the volume fraction β? (b) What is the % Sn?

Answer: (a) 33 v/o β (b) 24 w/o Sn

10–3.1 Assume that the perimeter provides a random sampling of the pearlite in Fig. 10–2.2(a). Estimate its hardness, based on the data in Fig. 10–3.8.

Answer: ~160 BHN

10–3.2 A 1045 steel is austenitized at 800°C (1470°F), quenched rapidly to form 100% martensite, and tempered at 400°C for 10 minutes, producing $\alpha + \bar{C}$. What is the density?

10–3.3 It takes ~20 secs of heating at 300°C to soften the steel of Fig. 10–3.9 to $60R_C$, and ~1500 sec at 200°C. Assume the Arrhenius relationship applies. Estimate how long it will take at 100°C.

Answer: ~10^6 sec (or ~13 days)

• *10–4.1* In going from phase A ($n = 1.60$) into phase B ($n = 1.58$), what angle is the typical light ray refracted? (We may assume that the phase boundary lies at 45° to the typical light ray.)

Answer: 0.7°

• *10–4.2* Estimate the heat capacities in J/g·°C for the pure metals (99+) in Appendix C.

• **10–4.3** CaF_2 has a density of 3.2 Mg/m^3 ($=$3.2 g/cm^3). What is its heat capacity in J/m^3·°C?

Answer: ~3 MJ/m^3·°C

• **10–4.4** A cemented carbide cutting tool contains 60 v/o TiC ($c = 0.8$ J/g·°C) and 40 v/o nickel. Utilize needed data from the appendices to estimate the heat capacity (J/cm^3·°C).

• *10–4.5* Estimate the thermal conductivity (longitudinal) of the reinforced plastic of Study Problem 10–2.4.

Answer: 0.0002 $(W/mm^2)/(°C/mm)$

10–4.6 A radiation shield is made by using lead powder (90 w/o) as a filler in polystyrene, then pressing it above the T_g of the polystyrene. (a) What is the volume fraction lead? (b) What is the shield's density? •(c) Estimate its thermal conductivity.

10–4.7 A cube (25 mm along each side) is made by laminating alternate sheets of aluminum and vulcanized rubber (0.5 mm and 0.75 mm thick, respectively). What is the thermal conductivity of the laminate (a) parallel to and (b) perpendicular to the sheets? (Use the data of Appendix C.)

Answer: (a) 0.09 $(W/mm^2)/(°C/mm)$ (b) 0.0002 $(W/mm^2)/(°C/mm)$

10–4.8 The presence of 2 v/o Al_2O_3 within aluminum as numerous ($>10^9/mm^3$) very fine particles (<0.1 μm) adds significantly to the metals strength. (a) How much (%) does the Al_2O_3 change the density? •(b) How much (%) does the Al_2O_3 change the thermal conductivity? The electrical conductivity?

CHAPTER 11

Multiphase Materials: Thermal Processing

PREVIEW

Long ago, craftsmen learned that materials were altered by heating, and by the manner they were cooled. They established names for the various heat-treating processes, although they did not know exactly what was happening inside the material. These microstructural changes are now known. As a result, it is helpful for us to relate the two—*thermal processes and micro-structures*—because we can better prescribe procedures and optimize their use. Thus, we can have better control of the final properties. An understanding of the internal structural changes has also led to new processes that give added capabilities to many materials for engineering applications.

CONTENTS

STUDY OBJECTIVES

  1 To interpret the structural changes that occur during the annealing of various materials, and to anticipate the consequent property changes.

  2 To relate the heat-treating steps of age-hardening (precipitation-hardening) and of over-aging (a) to the solubility limits of phase diagrams, and (b) to the microstructural changes of particle development and growth.

  3 To anticipate time requirements for age-hardening through the use of Arrhenius calculations.

  4 To understand isothermal transformation diagrams for steels so that you may (a) use them to predict reaction rates, (b) anticipate their alteration with changes in carbon content, alloy additions, and grain size, and (c) rationalize critical cooling rates that will produce all martensite or no martensite.

  5 To relate commercial steel-treating processes to various time–temperature–transformation sequences.

• **6** To utilize hardenability curves for predicting hardness values of selected steel products, and to understand the rationale for differences in hardenability curves.

11–1   ANNEALING PROCESSES

The term *annealing* originated from craftsmen. They found benefits in heating some materials to elevated temperatures, after which the materials were generally cooled slowly (as opposed to quenching.)

The benefits of annealing vary from material to material. Glass is annealed and slowly cooled to reduce the possibility of delayed cracking. Annealing does not change the hardness of the glass product that we use. Cold-worked brass is annealed to soften it. We now know that the brass is recrystallized during annealing (Section 6–6). Of course, this cannot happen in glass because glass is noncrystalline, both before and after annealing (Section 8–5). Annealing leads to greater ductility in cast iron, and we often conclude that softening is synonymous with ductility. Such is the case with the brass just cited; however, annealed steels and other multiphase alloys are less ductile than is possible by other heat treatments.

Annealing cannot be defined in terms of the resulting properties. Rather, it is a process by which *the material undergoes extended heating and is slowly cooled.* We must look at the individual materials to predict the properties that result from this heat treatment.

The following outline will point out the purposes and results of several common annealing treatments, together with their technical requirements. The "shop name" of the heat treatment is given in bold face type.

Anneal

Material: Glass.

Purpose: To remove residual stresses, and to avoid thermal cracking. (Initial and final product have the same hardness.)

Procedure: Varies with glass composition since the glass transition temperature T_g must be approached to allow stress relief. Glass technologists select a temperature at which the viscosity η drops to 10^{12} Pa·s (10^{13} poises). This *annealing point* permits stress relaxation within a few minutes without deformation. The glass must be cooled slowly past the *strain point* where the viscosity is $10^{13.5}$ Pa·s ($10^{14.5}$ poises). The slow cooling avoids the reintroduction of new thermal stresses. Below the temperature of the strain point where there has been a $\times 30$ increase in viscosity, cooling can be rapid since no new residual stresses can be introduced.

Microstructural changes: None.

Stress relief

Materials: Any metal, but steels in particular since they have a discontinuous volume change during transformation.

Purpose: To remove residual stresses (as in an annealing glass); however, most metals are not subject to thermal cracking. Rather, they are subject to distortion and warping, if machining removes residual stresses nonsymmetrically.

Procedure: A few minutes at ~600°C (~1100°F) for steels. Allowance must, of course, be made for the centers of large masses to attain this temperature.

Microstructural changes: None.

Recrystallization (or anneal*)

Materials: Cold-worked metals.

Purpose: To soften by removing strain hardening.

Procedure: $0.3T_m$ to $0.6T_m$, chosen to give a time compatible with the production sequences. Recrystallization is more rapid in pure metals than in alloys. It is more rapid with more highly strained metals (Section 6–6). (Steel sheet and wire should not be recrystallized above the eutectoid temperature, unless special care is taken for slow cooling. Otherwise, brittle martensite could form.)

Microstructural changes: New grains (Figs. 6–6.1 and 10–1.3).

Full anneal

Materials: Steels.

Purpose: To soften prior to machining.

Procedure: Austenization 25°C–30°C (50°F) above the stability of the last ferrite (Fig. 11–1.1). This is followed by furnace cooling, so that the austenite decomposes into a coarse pearlite. The product is sufficiently soft so that the steel may be machined. Annealed steel has less ductility than is possible with other heat treatments, but this is desirable since it leads to better chip formation

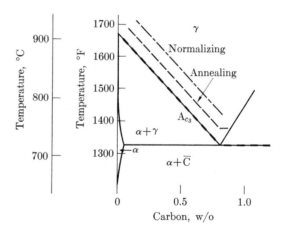

Fig. 11–1.1   Annealing and normalizing (plain-carbon steels). The heat-treating temperature varies with the carbon content. For annealing, the temperature is sufficiently high to ensure complete austenization. The steel is then cooled slowly to form a coarse pearlite and a relatively soft product. For normalizing (Section 11–2), the steel is heated somewhat higher to promote more rapid atom diffusion and microstructural uniformity. Excessive heating, however, would permit undesirable grain growth. Following austenization, the normalized steel is air-cooled to produce uniformly fine pearlite. Low hardness values are not a primary objective. (See Fig. 11–2.2a.)

---

*   Depending on the equipment used, this may be called *box anneal, continuous anneal, process anneal,* etc.

during machining. (This is a relatively expensive operation because it ties up the furnace for a considerable period of time.)

Microstructure formed: Coarse pearlite (Fig. 10–2.2a).

Spheroidization

Materials: High-carbon steels, e.g., ball bearings.

Purpose: To toughen an otherwise brittle steel. In certain steels, it is necessary to have a high carbide content for wear resistance; however, as pearlite, it would have very little toughness (Fig. 10–3.10).

Procedure: If initially pearlite, 16–24 hours just below the eutectoid tempera-ture ($\sim 700°C$). If initially martensite, 1–2 hours at that temperature (Fig. 10–2.1).

Microstructure formed: Spheroidite (Fig. 10–2.2b).

• Malleabilization

Material: Cast iron. (See Section 13–1.)

Purpose: To introduce ductility into iron castings. This is achieved by providing the necessary conditions for the reaction

$$Fe_3C \rightarrow 3\ Fe + C \text{ (graphite)}. \tag{11–1.1}$$

This reaction occurs, since the carbide is not completely stable, particularly in the presence of silicon.

Procedure: Anneal just below the eutectoid temperature if maximum ductility is desired, since

$$Fe_3C \xrightarrow{<750°C} 3\ Fe(\alpha) + C\text{(graph)}. \tag{11–1.2}$$

This produces a *ferritic malleable iron* (Fig. 13–1.6). If the annealing temperature is above the eutectoid temperature,

$$Fe_3C \xrightarrow{>750°C} 3\ Fe(\gamma) + C\text{(graph)}. \tag{11–1.3}$$

The product is a *pearlitic malleable iron* following normal cooling rates, since austenite decomposes to $(\alpha + \bar{C})$. The resulting product is stronger and less ductile than the result from Eq. (11–1.2).

Microstructural changes: Clusters of graphite are formed (Fig. 13–1.6).

Example 11–1.1   Select the heat treatment to improve the machinability of a 1030 steel.

Solution: Since a combination of low ductility and low hardness are required, use a full anneal.

> Temperature: 855°C (25°C–30°C into range of complete γ).
> Time: depends on size of product.
> Cooling: furnace cooled to produce coarse pearlite.

Comment.   Heating to a higher temperature will require extra fuel and may cause undesirable grain growth. ◀

11–2 NORMALIZATION PROCESSES

A uniform product is desirable in most cases.* Greater uniformity can sometimes be achieved through appropriate heat treatments. "Shop names" for these processes include *homogenization, soaking,* and *normalizing.*

Homogenization (soaking)

Materials: Cast metals

Purpose: To make the composition more uniform. During solidification, the first solid to form is not the same as the overall composition.† This is illustrated in Fig. 11–2.1(a) for a 96 Al–4 Cu alloy. The appropriate phase diagram (Fig. 9–5.3) reveals that κ, the very first solid to form at $\sim 650°C$ has only 1% copper; the copper content of the solid gradually increases as the temperature falls. At the eutectic temperature, the liquid contains 33% copper (and κ contains $\sim 5.5\%$ copper).

Procedure: Temperature as high as possible short of forming liquid or producing excessive grain growth. Some atoms must diffuse long distances compared to the atomic scale (~ 0.1 mm, or $\sim 10^6$ unit-cell distances). It is therefore necessary to heat the metal considerably above the recrystallization temperatures so that homogenization may occur in a reasonable length of time.

Microstructural changes: Greater homogeneity, and a closer match to the phase diagram.

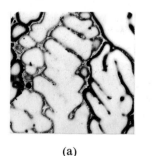   

(a)   (b)

Fig. 11–2.1 Solidification segregation and homogenization (96Al–4Cu). Segregation may arise from rapid solidification, during which there has been insufficient time for diffusion to provide equilibrium (Fig. 9–5.3). (a) The cast alloy has a copper-rich residual (dark) between the growing crystals. (b) The same alloy after "soaking." The copper has diffused until it is uniform throughout each grain. (These grains are oriented differently; therefore, they reflect light differently to give the several shades of gray. Within each grain there is general homogeneity.) (Courtesy Alcoa Research Laboratories.)

---

* Of course there are exceptions. A carburized gear is intentionally nonuniform in order to provide a hard, wear-resistant, high-carbon surface over a tougher steel core. Likewise, an asphalt road is nonuniform on a microscale since it is a mixture of particles of rock within a soft matrix that has a markedly different composition. Even so, the engineer will specify good mixing to provide "uniform heterogeneity."
† Cf. Example 9–5.2.

Normalizing

Material: Steel

Purpose: To produce a uniform, fine-grained microstructure

Procedure: Austenization, 50°C–60°C (100°F) into the range of complete austenite, followed by air-cooling (Fig. 11–1.1). The air-cooling avoids excessive proeutectoid segregation. Since low hardnesses are not required, the product is removed from the furnace and air-cooled, permitting the furnace to receive new products.

Microstructure: Fine pearlite, and absence of massive proeutectoid ferrite.

These processes as they pertain to steels are summarized in Fig. 11–2.2. The reference heat-treating temperatures vary with the steel composition (Figs. 9–7.3 and 9–9.1).

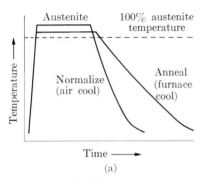
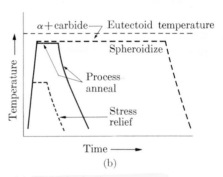

Fig. 11–2.2 Steel heat-treatment processes (schematic). (a) Austenization processes. (b) Subeutectoid processes.

Example 11–2.1 Let 100 g of a 96 Al–4 Cu alloy equilibrate at 620°C (1150°F), forming the κ and liquid, as shown by the phase diagram. The alloy is then cooled rapidly to 550°C (1020°F) with no chance for the initial solid to react. A liquid phase will still be present.
a)   What will its composition be?
b)   How many grams will there be of this final liquid?

Solution: At 620°C, and from Fig. 9–5.3,

$$\text{Liq: 88 Al–12 Cu,} \qquad \kappa: \text{98 Al–2 Cu.}$$

$$\text{Grams liquid} = 100 \text{ g} \left(\frac{98 - 96}{98 - 88} \right) = 20 \text{ g liquid.}$$

At 550°C, and with only the 20 g of liquid reacting,

$$\kappa: \text{94.4 Al–5.6 Cu.}$$

a)   Liq: 67 Al–33 Cu.

b)   Grams liquid $= 20 \text{ g} \left(\dfrac{94.4 - 88}{94.4 - 67} \right) = 4.6 \text{ g liquid.}$

Comments. In the final 4.6 g of liquid, there will be local areas with high copper segregation (33%). There will also be nearly 80 g of metal with only 2% Cu.

Cooling may be considered to take place in a series of small steps, in which the solute in the residual liquid is progressively concentrated. ◀

11–3 PRECIPITATION PROCESSES

Precipitation-hardening (age-hardening)   A very noticeable increase in hardness may develop during the *initial stages of precipitation* from a supersaturated solid solution (Section 10–1 and Eq. 10–1.5). In fact, the *start* of the precipitation in Fig. 10–1.7 can be detected by this increased hardness. (This *precipitation-hardening* is commonly called *age-hardening* because it develops with time.) The prime requirement for an alloy that is to be age-hardened is that solubility decreases with decreasing temperature, so that a supersaturated solid solution may be obtained. Numerous metal alloys have this characteristic.

The process of age-hardening involves a *solution treatment* (Eq. 10–1.4) followed by a *quench* to supersaturate the solid solution. Usually the quenching is carried to a temperature where the precipitation rate is exceedingly slow. After the quench, the alloy is reheated to an intermediate temperature* at which precipitation is initiated in a reasonable length of time. These are the two steps XA and AB in Fig. 11–3.1, and Table 11–3.1.

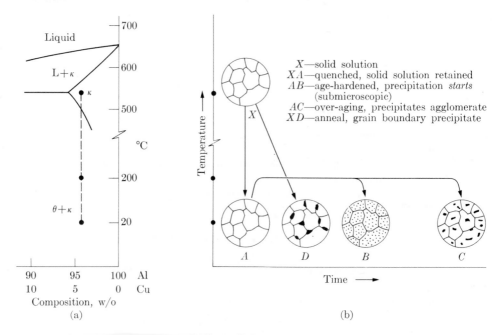

X—solid solution
XA—quenched, solid solution retained
AB—age-hardened, precipitation *starts*
      (submicroscopic)
AC—over-aging, precipitates agglomerate
XD—anneal, grain boundary precipitate

90	95	100 Al
10	5	0 Cu

Composition, w/o
(a)

Time ⟶
(b)

Fig. 11–3.1   Age-hardening process (96% Al–4% Cu alloy). See Table 11–3.1. The precipitates are still submicroscopic at the time of maximum hardness.

---

* But below the "knee" of the C-curve (Fig. 10–1.7), to produce *intra*grain precipitation (Fig. 10–1.8c).

Table 11–3.1
Properties of an age-hardenable alloy (96% Al–4% Cu)

Treatment (See Fig. 11–3.1)	Tensile strength, MPa (psi)	Yield strength, MPa (psi)	Ductility, % in 5 cm (2 in.)
XA Solution-treated (540°C) and quenched (20°C)	240 (35,000)	105 (15,000)	40
XAB Age-hardened (200°C, 1 hr)	415 (60,000)	310 (45,000)	20
XAC Overaged	~170 (25,000)	~70 (10,000)	~20
XD Annealed (540°C)	170 (25,000)	70 (10,000)	15

Observe the enhanced properties for the age-hardened alloy (XAB) in Table 11–3.1, as compared to annealing the same alloy (XD). The former has several times the yield strength of the annealed material and at the same time possesses greater ductility. As a result, the toughness is increased markedly by age-hardening. Greatest ductility is obtained when only one phase is present, i.e., after solution-treatment (plus quenching to preserve the single phase).

An interesting example of the utility of the age-hardening process is the way it is used in airplane construction. Aluminum rivets are easier to drive and fit more tightly if they are soft and ductile, but in this condition they lack the desired strength. Therefore the manufacturer selects an aluminum alloy that can be quenched as a supersaturated solution, but that will age-harden at room temperature. The rivets are inserted while they are still relatively soft and ductile, and they harden further after they have been riveted in place. Since hardening sets in fairly rapidly at room temperature, there arises the practical problem of delaying the hardening process if the rivets are not to be used almost immediately after the solution treatment. Here advantage is taken of the known effects of temperature on the reaction rate. After the solution treatment the rivets are stored in a refrigerator, where the lower temperature will delay hardening for reasonable lengths of time.

Detailed studies have produced the following interpretation of the age-hardening phenomenon. The supersaturated atoms (Cu atoms in Example 10–1.2 and Fig. 11–3.1"B") tend to accumulate along specific crystal planes in the manner indicated in Fig. 11–3.2(b). The concentration of the copper (solute) atoms in these positions lowers the concentrations in other locations, producing less supersaturation and therefore a more stable crystal structure. At this stage, the copper atoms have not formed a phase that is wholly distinct; a *coherency* of atom spacing exists across the boundary of the two structures. Dislocation movements occur with difficulty across these distorted regions, and consequently the metal becomes harder and more resistant to deformation under high stresses.

• Solute atom   ○ Solvent atom

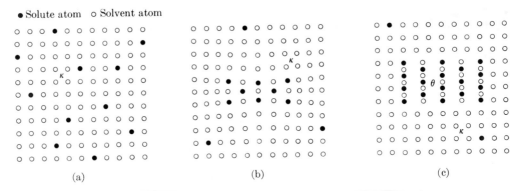

(a)   (b)   (c)

Fig. 11–3.2 Age-hardening mechanism. (a) κ solid solution. (b) Age-hardened; the θ precipitation has been initiated. Since the two structures are coherent at this stage, there is a stress field around the precipitate. (c) Over-aged. There are two distinct and noncoherent phases, κ and θ. With limited numbers of solute atoms, maximum interference to dislocation movements occurs in part (b). (A. G. Guy and J. J. Hren, *Elements of Physical Metallurgy*, Addison-Wesley.)

Over-aging   A continuation of the local segregation process over long periods of time leads to true precipitation and *over-aging*, or softening. For example, the development of a truly stable structure in an alloy of 96% aluminum and 4% copper involves an almost complete separation of the copper from the fcc aluminum at room temperature. Nearly all the copper forms $CuAl_2$ (θ in Fig. 11–3.2c). Because the growth of the second phase provides larger areas that have practically no means of slip resistance, a marked softening occurs.

   Figure 11–3.3 shows data for the aging and overaging of a commercial aluminum alloy (2014). The initial hardening is followed by softening as the resulting precipitate is agglomerated. Two effects of the aging temperature may be observed: (1) precipita-

Fig. 11–3.3 Over-aging (2014–T4 aluminum). Softening occurs as the precipitated particles grow. This proceeds more rapidly at elevated temperatures. (*Aluminum*, American Society for Metals.)

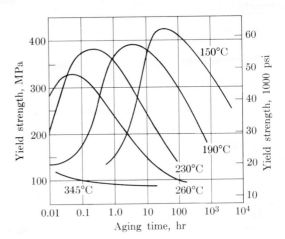

tion, and therefore hardening, starts very quickly at higher temperatures; (2) over-aging, and therefore softening, occurs more rapidly at higher temperatures. These two phenomena overlap to affect the maximum hardness that is attained. Lower temperatures permit greater increases in hardness, but longer times are required.

Combined hardening   Occasionally it is desirable to combine two methods of hardening. The cold-working of an alloy that has previously been age-hardened increases the hardness still further. However, there are some practical difficulties encountered in this process. Age-hardening increases resistance to slip and therefore increases the energy required for cold-working, and it also decreases ductility so that rupture occurs more readily during cold-working. A possible alternative is to cold-work the metal prior to the precipitation-hardening treatment. The metal is cold-worked more readily, and the age-hardening reaction occurs at a lower temperature because the dislocations serve as nuclei for the precipitation. However, the temperature of the aging process that follows cold-working may relieve some of the strain hardening and cause a slight loss in hardness. Although it does not produce hardnesses as great as those obtained from the reverse order, the final hardness is greater than that developed by using either method alone (Table 11–3.2).

Table 11–3.2
Tensile strengths of a strain- and age-hardened alloy (98% Cu–2% Be)

Annealed, 870°C	240 MPa	35,000 psi
Solution-treated, 870°C and cooled rapidly	500	72,000
Age-hardened only	1200	175,000
Cold-worked only (37%)	740	107,000
Age-hardened, then cold-worked*	1380	200,000
Cold-worked, then age-hardened	1340	195,000

* Cracked

• **High-strength, low-alloy (HSLA) steels**   Low-alloy steels have undergone some recent developments that have led to significantly higher strength products. Their development has provided the engineer with *structural* steels that have yield strengths of greater than 500 MPa ($>70,000$ psi) in contrast to earlier structural steels of only half that figure.

The HSLA steels gain their strength from a very small grain size and finely distributed precipitates within the ferrite. In the latter respect, they may be compared to precipitation-hardened aluminum alloys, which have just been discussed. The origin of the precipitate is significantly different, however. If "carbide-formers" such as vanadium or niobium are present, along with 0.05%–0.2% carbon, we can still obtain a single phase (austenite) at hot-rolling temperatures. Thus beams, pipe, etc., can be readily processed. However, as the proeutectoid ferrite forms during austenite decomposition, there is a concurrent precipitation of alloy carbides. The precipitate forms not because of the decreasing solubility curve for *one* phase but rather because these alloy carbides are less soluble in α than in γ.

Unlike quenched and tempered steels (Section 11–5), these HSLA steels have the advantage of being weldable and of maintaining their mechanical properties without subsequent heat treatment. As a result, they are attractive to those engineers who design cars, major structures, pressurized pipelines, etc.

Example 11–3.1 Use information in Fig. 11–3.3 to estimate the temperature required to reach the maximum hardness for that aluminum alloy in 10,000 hours ($\sim$14 months).

Solution: Assume that the Arrhenius relationship applies (Eq. 10–1.1), and use 150°C (423 K) and 260°C (533 K). At 150°C, with maximum hardness after 30 hrs:

$$\ln t_{150} = \ln 30 = C + B/423 = 3.4.$$

At 260°C, with maximum hardness after 3 min (or 0.05 hr):

$$\ln t_{260} = \ln 0.05 = C + B/533 = -3.0.$$

Solving simultaneously,

$$C = -27.6, \qquad B = 13,100 \text{ K}.$$

$$\ln 10^4 \text{ hr} = 9.21 = -27.6 + 13,100/T,$$

$$T = 356 \text{ K} = 83°C.$$

Comments. We may check our Arrhenius assumption at 230°C (503 K) and 190°C (463 K):

$$\ln t_{230} = -27.6 + 13,100/503 = -1.56, t = 0.2 \text{ hr};$$

$$\ln t_{190} = -27.6 + 13,100/463 = 0.69, t = 2 \text{ hr}.$$

The agreement with the experimental data in Fig. 11–3.3 is not exact, but reasonable; however, an extrapolation to still longer times or lower temperatures is an approximation. ◀

11–4 $\gamma \rightarrow (\alpha + \bar{C})$ REACTION RATES

Isothermal transformation The time requirement for austenite decomposition has been studied in considerable detail because of its practical importance. Figure 11–4.1 shows time–temperature data for this reaction in a eutectoid steel (AISI–SAE 1080). The left solid curve t_s is the time required for the decomposition to *start*. The right solid curve t_f is the time to *finish* the $\gamma \rightarrow (\alpha + \bar{C})$ reaction. These curves are commonly called *isothermal-transformation diagrams*, or *I-T diagrams*.* The data for Fig. 11–4.1 were obtained as follows. Small samples of eutectoid steel were heated into the austenite temperature range sufficiently long to assure complete transformation to austenite. These wire-size samples were then quenched to a lower temperature (e.g., 620°C) and held there for varying lengths of time before being quenched further to room temperatures (Fig. 11–4.2). The change $\gamma \rightarrow (\alpha + \bar{C})$ was not observed in samples held at 620°C for less than one second, and complete transformation to α + carbide was not observed until after more than 10 seconds had elapsed (Figs.

---

* They may also be called C-curves (because of their shape) or T–T–T curves (Temperature–Time–Transformation).

SAE 1080

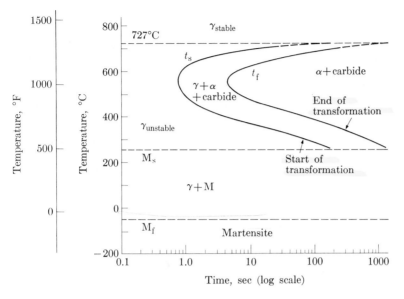

Fig. 11-4.1   Isothermal-transformation curves for austenite decomposition (SAE 1080). (Adapted from U.S. Steel Corp. data.)

PROBS
4.1 & 5.1

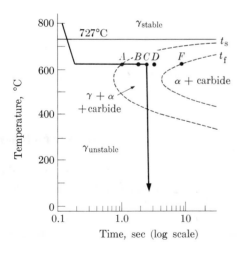

Fig. 11-4.2   Interrupted quench (eutectoid steel). This technique is used in establishing isothermal-transformation curves. The initial quench is made into a hot bath and it is held there for a prescribed time before the second quench to room temperature. (Cf. Figs. 11-4.3, 4, 5.)

11-4.3 through 11-4.5). Similar data were obtained for other temperatures until the completed diagram shown in Fig. 11-4.1 was established.

The I-T diagram shows that the austenite transformation occurs slowly, both at high temperatures (close to the eutectoid) and at low temperatures. It is slow at higher temperatures because there is not enough supercooling to readily nucleate

Fig. 11–4.3 The beginning of austenite transformation at 620°C (1150°F) ($\gamma \rightarrow \alpha$ + carbide). After 1 sec (point A of Fig. 11–4.2) the change to pearlite P is underway. M is metal that has not transformed to α + carbide.

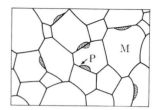

Fig. 11–4.4 Transformation 25% complete at 620°C (1150°F). (Point B in Fig. 11–4.2.)

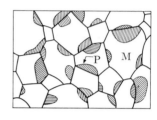

Fig. 11–4.5 Transformation 75% complete at 620°C (1150°F). (Point D in Fig. 11–4.2.)

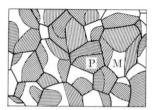

the new ferrite and carbide from the previous austenite. The austenite decomposition is slow at lower temperatures because diffusion rates are slow and therefore carbon separation from the ferrite into the carbide is slow. At intermediate temperatures, the nucleation is sufficiently rapid and the atomic migration is fast enough to quickly start and complete the reaction.

These curves may be compared with the isothermal precipitation curve of Fig. 10–1.7. That was a single curve simply because only the midpoint of the reaction was shown. A similar 50% transformation curve could be drawn on Fig. 11–4.1; it would be about halfway between the t_s and t_f curves.*

With extremely fast cooling (severe quenching), it is often possible to miss the "knee" of the curve for the beginning of the transformation, and to cool the steel to room temperature without the formation of ferrite and carbide, because the decomposition is detoured through the martensite (Eq. 10–1.7). In fact, this is the purpose of quenching steel in regular heat-treating operations.

The example just considered is the transformation of a eutectoid steel when there has been no advance separation of ferrite (or cementite) from the austenite prior to the formation of the pearlite ($\alpha + \bar{C}$). Figure 11–4.6(a) shows a similar transformation diagram for an SAE 1045 steel. Two features differ from those in Fig. 11–4.1.

---

* The interpolation is not linear, however. The reaction quartiles (25% and 75%) lie close to the 50% point because of the sigmoidal nature of the reaction curves. (See Figs. 10–1.2 and 10–1.4.) Thus, points B and D of Fig. 11–4.2, which are represented by Figs. 11–4.4 and 11–4.5, lie closer to C than to t_s and t_f.

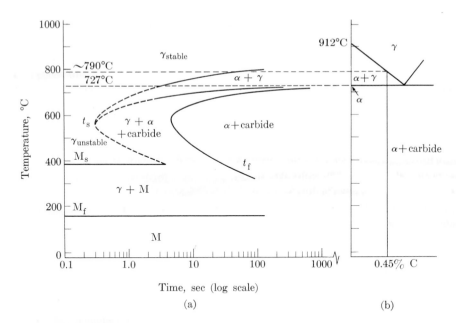

Time, sec (log scale)

(a)

(b)

Fig. 11–4.6 Isothermal-transformation diagram for <u>SAE 1045 steel</u>. The stable phases of the phase diagram (right) are not achieved immediately. However, in this steel, the austenite decomposition reaction is faster than in a eutectoid steel (Fig. 11–4.1).

1. Some ferrite may separate from the austenite above the eutectoid temperature. This could be predicted from the phase diagram shown at the right side of Fig. 11–4.6.

2. The isothermal transformation of 0.45 carbon steel occurs somewhat faster than the transformation of eutectoid steel. Comparison of the "knees" of the two curves shows this difference: the higher carbon steel starts to transform in about one second; in the 0.45% carbon steel, the reaction starts sooner. In fact, in the latter case the reaction occurs sufficiently fast so that we are not able to measure the rate at the "knee" of the curve with the interrupted-quench technique described above. A lower carbon content permits faster reaction, since part of the transformation delay is associated with the movement of the carbon atoms.

Isothermal-transformation curves show the start and finish of a reaction with *time* as a variable. Therefore, they are *not* equilibrium diagrams such as we encountered in Chapter 9. However, they show one thing in common with the equilibrium diagrams. The far right side of an I-T diagram represents extended periods of time. Therefore, the phases that are present should match the phases shown in a phase, or an equilibrium diagram. As an example, the phases eventually produced according to the I-T diagram for 1045 steel (Fig. 11–4.6a) match those found at 0.45% carbon on the Fe–Fe$_3$C diagram (Fig. 11–4.6b).

Also observe that an I-T diagram is for a particular material and a specific reaction. Thus, once $\alpha + \bar{C}$ is formed from austenite, that I-T diagram has no further use. This may be illustrated by the $\gamma \rightarrow \alpha + \bar{C}$ reaction in 1080 steel (Fig. 11–4.1). If the austenite is quenched to 550°C and held there 10 seconds, *the reaction is complete.* Now, if we were to lower the temperature to 300°C, we must *not* expect to make continued use of the curves shown in Fig. 11–4.1. Specifically, the $\alpha + \bar{C}$ *cannot go back to γ at 300°C.* That would be contrary to the phase diagram. The only way to get austenite again is to raise the temperature into the austenite field of Fig. 9–7.3.

Martensite temperatures   At the base of Figs. 11–4.1 and 11–4.6 are shown the temperatures at which the martensite reaction (Eq. 10–1.7) starts, M_s, and finishes, M_f, when the steel is quenched rapidly enough to miss the "knee" of the C-curve. Since diffusion is not involved,* the process of the martensite formation is not time-dependent. However, M_s and M_f depend on the amount of carbon that is present (Fig. 11–4.7). As a result, a water-quench does not complete the $\gamma \to M$ reaction in high-carbon alloys such as those used in tool steels. Thus, they possess *retained austenite.*

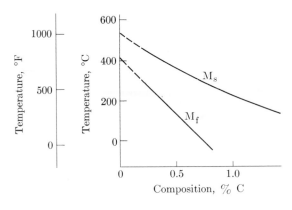

Fig. 11–4.7   Martensite transformation temperatures (plain-carbon steels). Transformation is first detected at M_s and is virtually complete at M_f. Between M_s and M_f, austenite is retained as a result of induced stresses.

Delayed austenite transformation   From the foregoing discussion, we see that temperature obviously affects the isothermal decomposition of austenite γ to ferrite plus carbide $(\alpha + \bar{C})$. Other factors modify the time requirements, too. We will consider two: (1) *austenite grain size*, and (2) *alloy retardation.*

As sketched schematically in Figs. 11–4.3 through 11–4.5, the ferrite and carbide formation from the austenite starts at the grain boundary. A fine-grained steel offers more grain-boundary area per unit volume on which decomposition can be nucleated than does a coarse-grained steel. This is demonstrated in Fig. 11–4.8 where the two curves are for fine-grained (G.S. #8) steel and coarse-grained (G.S. #2) steel. The time requirements for the *start* of transformation differs by a factor of three at almost all temperatures.

The alloy retardation effect is even more pronounced. As an example, only 0.25% molybdenum (Fig. 11–4.9) delays the start of isothermal transformation by a factor of four at temperatures below 650°C (1200°F). This delay occurs because not only the carbon but also the molybdenum must be relocated when the austenite decomposes. Most of the Mo goes into the carbide. When silicon is the alloying element,

---

* Austenite changes to martensite by shear (Section 10–1).

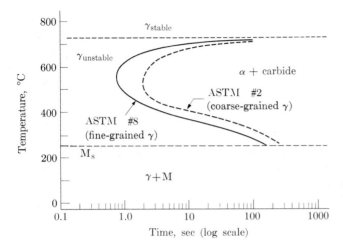

Fig. 11–4.8 Grain-boundary nucleation (eutectoid steel). The effect of austenite grain size on the *start* of austenite decomposition is shown for eutectoid steels. The steel with the fine-grained austenite has more grain-boundary area from which the $(\gamma \to \alpha + \bar{\text{C}})$ reaction can start.

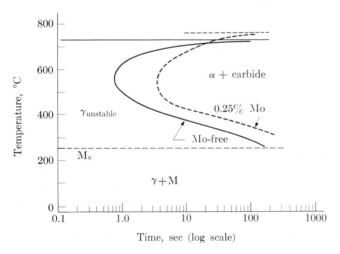

Fig. 11–4.9 Transformation retardation. Molybdenum, like other alloying elements, retards the *start* of transformation of austenite.

most of it shifts into the ferrite. Since almost all alloying elements—Cr, Ni, Si, Mn, Mo, Ti, W, etc.—diffuse more slowly in iron than carbon does (Fig. 4–7.3), we find that all low-alloy steels (Table 9–9.1) have their isothermal transformation curves shifted to the right. In fact, it is easy to cool many low-alloy steels to room temperature and miss the "knee" of the transformation curve entirely, thus forming all martensite.

Continuous-cooling transformation (CCT)   Isothermal-transformation diagrams are
convenient for interpretation, because we can hold the effect of temperature constant
during the austenite decomposition. In practice, however, we more commonly
encounter continuous cooling: (1) a hot piece of steel is removed from the furnace and
air cooled, or (2) the steel is quenched in water. In neither is there an isothermal
holding period while the $(\alpha + \bar{C})$ forms.

Let us consider the progress of austenite decomposition with different reaction
rates of continuous cooling. A severe quench will miss the "knee" of the transformation curve, with the result that the austenite is changed to martensite rather than to
pearlite (i.e., $\alpha + \bar{C}$). Slow cooling permits pearlite to form; however, the start of
decomposition occurs after a longer time (and therefore at a lower temperature) than
for isothermal transformation, simply because part of the time was spent at higher
temperatures where reactions were initiated more slowly. Thus the isothermal-
transformation curves are displaced downward and to the right for *continuous-cooling
transformation* (Fig. 11–4.10).

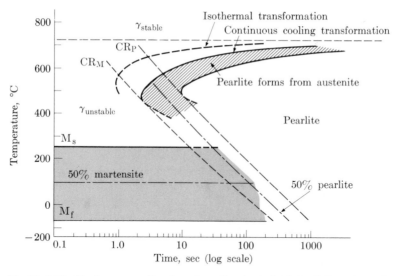

Fig. 11–4.10 Continuous-cooling transformation (eutectoid steel). Transformation
temperatures and times are displaced from the isothermal-transformation curve for the same
steel. (Cf. Fig. 11–4.1.) CR_M = minimum cooling rate for 100% martensite.
CR_P = maximum cooling rate for 100% pearlite.

There are two important *critical cooling rates* for the continuous-cooling decomposition of austenite; these are included in Fig. 11–4.10. The first is the cooling
rate, CR_M, which just misses the "knee" of the transformation curve; *more* rapid
cooling rates to the left of this curve produce *only martensite*. The second is the cooling
rate, CR_P, which produces no martensite; *less* rapid cooling rates to the right of the

CR$_P$ curve produce *only pearlite*. In a eutectoid steel (0.8 C–99.2 Fe), these two critical cooling rates are approximately 140°C/sec and 35°C/sec, respectively, through 700°C (1300°F), which is in the eutectoid temperature range. Both of these rates are significantly slower in any steel which contains alloying elements because the alloy content slows down the $\gamma \to (\alpha + \bar{C})$ reaction (Fig. 11–4.9).

Study aid (transformation rates in steels)   The isothermal transformation, or time–temperature–transformation (TTT), diagram is presented somewhat differently in *Study Aids for Introductory Materials Courses* than in this text. Some students may find that this second presentation will be helpful.

Example 11–4.1   Three AISI–SAE 1045 steel wires underwent the following thermal steps in the indicated sequences. Give the phases after each step and their approximate chemical analyses. (The symbol # means that the wire was held at that temperature until equilibrium was reached.)

		Time held
Wire (a)	1) Heated to 820°C (1510°F):	#
	2) Quenched to 560°C (1040°F):	0
	3) Held at 560°C:	1 min
	4) Reheated to 820°C:	#
Wire (b)	1) and 2) Same as wire (a), 1) and 2):	#
	3) Held at 560°C:	1 sec
	4) Quenched to 430°C (805°F):	0
Wire (c)	1) Heated to 730°C (1345°F):	#
	2) Quenched to 430°C (805°F):	0
	3) Quenched to 330°C (625°F):	10 sec
	4) Held longer at 330°C:	#

Solution and comments

Wire (a) 1) γ 0.45 C–99.55 Fe.
2) Same, but austenite is metastable (Fig. 11–4.6).
3) α Negligible carbon; $\bar{C}$ 6.7% carbon. p̄g 357
4) γ Same as (1).

Wire (b) 3) $\gamma(0.45\% \text{ C}) + [\alpha(\text{negligible carbon}) + \bar{C}(6.7\% \text{ C})]$.
4) Same as (3). $(\alpha + \bar{C})$ will *not* revert to austenite below the eutectoid temperature.

Wire (c) 1) $\alpha(0.02\% \text{ C}) + \gamma(0.8\% \text{ C})$. See the phase diagram.
2) Note that nothing is expected to happen to the ferrite as it cools. The austenite is of eutectoid composition; therefore, we should turn to Fig. 11–4.1 for its transformation. At zero time, unstable γ is present.
3) Still, $\alpha(0.02\% \text{ C}) + $ unstable $\gamma(0.8\% \text{ C})$.
4) $\alpha + \bar{C}(6.7\% \text{ C})$. ◀

Example 11–4.2   A thin (<0.5 mm), one-gram sample of 1045 steel undergoes the following steps during a heat treating process. (The symbol # means equilibrium was reached.)

	Time held
1) Heated to 730°C:	#
2) Quenched to 550°C:	10 sec
3) Quenched to 100°C:	0 sec

What phases are present after each step? What is their carbon content? Approximately how much of each phase?

Solution and comments

$$\overset{\text{AUSTENITE}}{\underset{\downarrow}{}}$$

1) 0.44 g α(0.02% C), and 0.56 g γ(0.8% C)

We now have a mixture of α and γ. Upon quenching, nothing happens to the ferrite (except getting colder). The austenite is 1080 steel on a microscopic scale. Therefore, we must use the I-T diagram for 1080 (Fig. 11–4.1), and not the 1045 diagram (Fig. 11–4.6).

2) 0.44 g α(0.02% C);
   0.56 g γ → 0.5 g α(0.02%), and 0.06 g $\bar{C}$(6.7% C).
   Total α = 0.94 g α.

We are now through with Fig. 11–4.1, since the γ-decomposition is complete.

3) 0.94 g α(0.02% C), and 0.06 g $\bar{C}$(6.7% C).

The quench to 100°C does not permit the last, small amount (0.02%) of carbon to separate from the ferrite. ◀

• **Example 11–4.3** By what ratio was the grain size changed to decrease the start of transformation in Fig. 11–4.8 from the left curve (G.S. = #8) to the right curve (G.S. = #2)? Repeat for the boundary area.

Solution: Refer to Eq. (4–4.2). With X100 and G.S. #8,

$$N = 2^{8-1} = 128/(0.0645 \text{ mm}^2);$$

With X100 and G.S. #2,

$$N = 2^{2-1} = 2/(0.0645 \text{ mm}^2);$$

$$N_8/N_2 = 128/2 = 64.$$

Let δ be a representative linear dimension of a grain, which varies inversely with the square root of the number of grains:

$$\delta_2/\delta_8 = \sqrt{N_8/N_2} = \sqrt{64} = 8.$$

For the boundary area, assume either a cubelike or a spherelike grain, where

$$\text{Boundary area} = 3/\delta. \qquad \text{(See comments.)}$$

Therefore,

$$B_2/B_8 = (3/\delta_2)/(3/\delta_8) = \tfrac{1}{8}.$$

Comments. The surface area/volume ratio of a cube is $6a^2/a^3 = 6/a$; the same ratio of a sphere is $4\pi r^2/(4/3)\pi r^3 = 6/d$. The boundary area/volume is $3/\delta$ since each boundary is shared by two adjacent grains. Of course, the grains are neither spheres nor cubes, but something in between; thus the B/V ratio remains $3/\delta$. ◀

11–5   COMMERCIAL STEEL-TREATING PROCESSES

Since austenite can decompose in several different ways, the engineer has a choice of different microstructures. These are summarized in Table 11–5.1 and Fig. 11–5.1. The resulting properties vary significantly.

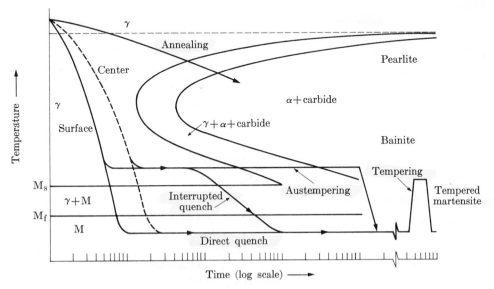

Fig. 11–5.1   Transformation processes. *Annealing:* The normal $\gamma \rightarrow \alpha$ + carbide transformation occurs. *Direct quench:* Martensite forms, first in the surface, then in the center. Severe stresses result. *Interrupted quench:* Time is available for the surface and center to transform nearly simultaneously, thus avoiding the quench-cracking found in direct quenching. *Tempering:* Both the direct and the interrupted quench must be followed by a tempering process to complete the transformation. *Austempering:* Quenching avoids pearlite formation, but the $\gamma \rightarrow \alpha$ + carbide transformation may still occur above the M_s. The resulting microstructure is bainite.

Annealing   This has already been considered in Section 11–1 (also see Fig. 11–2.2a). The resulting properties are shown in Figs. 10–3.3 and 10–3.4 for plain-carbon steels.

Quenching   Cooling rates faster than CR_M of Fig. 11–4.10 give hard (and relatively brittle) martensite, by avoiding the transformation to $(\alpha + \bar{C})$. This is the purpose of quenching steels.

Austenite is denser than martensite (and also denser than ferrite plus carbide). This presents a problem with a direct austenite-to-martensite quench, because the slower-cooling central region transforms and expands after the quickly cooled surface has formed brittle martensite. Hence, cracking can occur if the steel is larger than sheet or wire dimensions, particularly if the carbon content is greater than 0.5%. This is an added reason for alloy additions to the steel. With more time available in which to form martensite (Fig. 11–4.9), an alloy steel can be cooled more slowly so that the surface and center transform more or less concurrently; thus it is possible to avoid the sharp volume change differentials and the resulting stresses that promote cracking.

Interrupted quench   By examining the isothermal-transformation diagram, we can see an alternative to the direct quench just described. That is an interrupted quench (also called *martempering* or *marquenching*).

In this process, the steel is quenched rapidly past the "knee" of the transformation curve to avoid $(\alpha + \bar{C})$ formation, but the cooling is interrupted just above the M_s temperature. Cooling is then continued at a slow rate through the martensite range to ambient temperatures, so that the surface and the center of the steel may transform more or less simultaneously, thus avoiding quenching cracks. Slower cooling is possible at these lower temperatures because the $(\alpha + \bar{C})$ transformation is delayed, while the martensite forms directly with the drop in temperature.

This process is more complicated from the production viewpoint because the cooling rate must be shifted from a quench to a "hold," and then to a slow cooling rate. As with the earlier direct quench, martensite is the product, and this must be tempered to secure toughness.

Tempering   Martensite has the attribute of being exceptionally hard. It is also very brittle, because it is nonductile, when it contains carbon. This is not surprising since hardness and ductility are generally inversely related. (Compare Figs. 10–3.3 and 10–3.4, where we see that hardness and strength increase with carbon content while toughness and ductility decrease.) Fortunately, by tempering martensite, we can develop toughness faster than the hardness and strength decrease.*

Table 11–5.2 shows that a 4140 steel is relatively tough in both the annealed and the spheroidized conditions, but only moderately strong. Martensite, although very hard and basically strong, lacks toughness; therefore it may crack readily. However, tempered martensite can have a relatively high strength *and* good toughness if it is heat-treated appropriately. This is achieved because the brittle martensite is replaced by a fine dispersion of rigid carbide particles within a tough ferrite matrix (Fig. 10–2.1). The carbide particles arrest dislocation movements and thereby prevent slip, and they strengthen the alloy in a way similar to the precipitation hardening of Section 11–3. At the same time, the ductile ferrite can deform locally at points of stress concentration (Fig. 8–7.2) to blunt the tip of the cracks that start to form.[†]

---

[*] This fact was discovered by trial and error several centuries ago. Its governing principles have only become understood during the lifetime of the present-day metallurgist.

Tempering is certainly among the more important discoveries of nature that have affected the technological evolution of man. We can anneal steels so they can be machined. In turn, *these steels* can be hardened and tempered so they can machine *other steels*. A technological "bootstrap operation" such as this is not possible to the same degree of efficiency with other materials, because they lack the concurrent hardness and toughness. They either soften and become dull, or they break from brittleness.

[†] Metallurgists are aware that extended heating around 500°C (~900°F) will embrittle a steel because there are secondary reactions that can occur among the alloying elements. Therefore, they require that the steel be cooled rapidly through this temperature region if tempering is performed at higher temperatures. They also know that the toughness has a transition temperature (Section 6–8), that varies with the austenite grain size and other factors.

Table 11–5.1
Transformation processes for steels*

Process	Purpose	Procedure	Phase(s)
Annealing	To soften	Slow cool from γ-stable range	α + carbide
Quenching	To harden	Quench more rapidly than CR_M	Martensite[†]
Interrupted quench	To harden without cracking	Quench, followed by slow cool from M_s to M_f	Martensite[†]
Austempering	To harden without forming brittle martensite	Quench, followed by isothermal transformation above the M_s	α + carbide
Tempering	To toughen (usually with minimum softening)	Reheating of martensite	α + carbide

*   Cf. Fig. 11–5.1.
†   Steels containing martensite must be toughened by the tempering process.

Table 11–5.2
Effects of heat treatments (SAE 4140)

Microstructure		Tensile strength, MPa (psi)	Toughness, J (ft-lb)
Annealed, ($\alpha + \bar{C}$)	—	655	55
Lamellar carbides		(95,000)	(40)
Spheroidite, ($\alpha + \bar{C}$)	—	480	110
Large "spherical" carbides in ferrite matrix		(70,000)	(80)
Martensite, M	—	~ 1400	<3
		($\sim 200{,}000$)	(<2)
Tempered martensite, ($\alpha + \bar{C}$) Dispersed carbides in ferrite matrix:			
500°C (930°F), 1 hr	—	1275	55
		(185,000)	(40)
600°C (1110°F), 1 hr	—	1035	110
		(150,000)	(80)

Austempering   The final transformation process that we will consider for austenite decomposition is that of *austempering*. In this treatment, the austenite is allowed to transform *isothermally* to ferrite and carbide just above the M_s temperature. This requires a quench to avoid transformation to pearlite at higher temperatures. The advantage of austempering is that transformation occurs by a combination of shear and diffusion, to give a fine dispersion of carbides in ferrite and thus a strong, tough

Fig. 11–5.2 Bainite at X11,000. This SAE 1080 steel was austenitized, then transformed isothermally at 260°C (500°F) to form a fine dispersion of carbides in a ferrite matrix. The hardness is 57 R_C. (Courtesy General Motors and ASTM, *Electron Microstructure of Steel*.)

Fig. 11–5.3 Pearlite and bainite formation: (a) eutectoid steel, (b) 4340 steel. Pearlite forms in less time than bainite above 550°C, and bainite forms in less time than pearlite below 550°C (1000°F). ▼

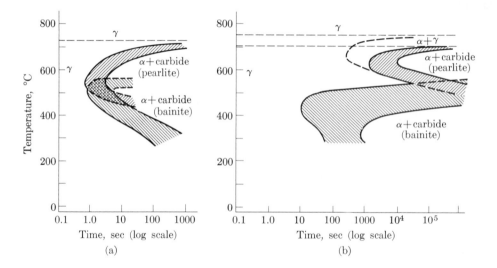

product. Also, quench-cracking is avoided because the reaction (and volume change) takes place at constant temperature. In many respects the product, *bainite* (Fig. 11–5.2), is similar to tempered martensite, and the physical properties of the two microstructures are closely related.

A close examination of the isothermal transformation processes in a 1080 steel for $\gamma \rightarrow \alpha + \bar{C}$ reveals that they are different above and below the "knee" of the I-T curve. Above that temperature, the nucleation is limited to austenite grain boundaries, and growth of $(\alpha + \bar{C})$ proceeds as pearlite into the former austenite grains. Below the "knee," the reaction is delayed by sluggish atom movements; however, the severely supercooled metal readily nucleates α and $\bar{C}$ at numerous imperfection sites within the austenite grains to produce bainite. This means that the

I-T curves of Fig. 11–4.1 are in reality two sets of curves shown in Fig. 11–5.3(a). By chance, they happen to be tangent to each other for the 1080 steel. They are not tangent in other steels, as shown in Fig. 11–5.3(b) for 4340 steel. Above about 550°C (for those two steels), pearlite forms in less time than bainite; below that temperature, bainite forms first. Both temperature ranges lead to $(\alpha + \bar{C})$.*

When the two curves are tangent (Fig. 11–5.3a), it is impossible to obtain bainite by continuous cooling, because pearlite would always form first. The 4340 steel, however, can be continuously cooled to produce bainite, because the pearlite formation is retarded more by the addition of the alloying elements than is the bainite formation (Fig. 11–5.3b).

Study aid (microstructures of steel)   The goal of heat treatments is to gain a desired microstructure. This topic in *Study Aids for Introductory Materials Courses* focuses on the ways different microstructures are developed in steels.

Example 11–5.1 (a) A small piece of 1045 steel is heated to 850°C, quenched to 650°C, held for 5 sec, and then quenched to 20°C. What phases are present after each step? (b) A small piece of 1080 steel is heated to 800°C, quenched to 100°C, reheated to 290°C, and held 1 min. What phases are present after each step?

Solution

a)   At 850°C: γ_{stable}
      After 5 sec at 650°C: $\gamma_{unstable} + \alpha + $ carbide
      After quenching to 20°C: $M + \alpha + $ carbide†.

b)   At 800°C: γ_{stable}
      After quenching to 100°C: $\gamma_{unstable} + M$
      After 1 min at 290°C: $\gamma_{unstable} + M$†.

Comment. With additional time at 290°C, both γ and M will transform to $\alpha + $ carbide. ◀

Example 11–5.2 Two drill rods (4-mm dia.) of 1080 steel receive the following sequences of treatments:

	(1)	(2)
Austenitize:	775°C	775°C
Quench to:	275°C	275°C
Held at 275°C for:	45 sec	45 min
Cooled to 30°C in:	30 sec	30 sec

a)   Indicate the phase(s) that will be present after each step.
b)   What microstructure does each product have?
c)   What process name does each have?

---

* This double pattern of microstructural development is not unique to steels. In fact, this was discussed first in connection with isothermal precipitation (Fig. 10–1.8). Thus, we could redraw that curve as two tangent curves, one for a *grain-boundary* reaction, and the other for *intragrain* precipitation. The age-hardening process (XAB of Fig. 11–3.1) is also designed to utilize the intragrain precipitation.
† The reactions of Eq. (10–1.7) are *not* reversible below the eutectoid temperature.

Answer

Rod (1) a) γ, $(\gamma)_{unstable}$, $(\gamma)_{unstable}$, martensite + (γ);
         b) martensite, plus retained austenite;
         c) interrupted quenching (or marquenching).

Rod (2) a) γ, $(\gamma)_{unstable}$, $\alpha + \bar{C}$, $\alpha + \bar{C}$;
         b) bainite;
         c) austempering.

Comments. The hardness of the martensite for Rod (1) is $65R_C$; however, it must be tempered to become toughened. The final product loses some of that hardness (Fig. 10–3.9). When bainite is formed at 275°C (525°F), its hardness is $\sim 50R_C$. It does not need to be tempered because it already contains a fine dispersion of carbides within the ferrite matrix. ◀

Example 11–5.3 Prescribe a heat-treating procedure to austemper (a) 1040 steel, (b) 4340 steel. Comment on the production considerations of each.

Procedure

a) From Fig. 11–4.6: austenize above 800°C (1475°F); quench to ~ 425°C (~ 800°F); hold ~ 1 min; cool to ambient.
b) From Fig. 11–5.3(b): austenize above 750°C (>1380°F); quench to ~ 325°C (~ 620°F); hold ~ 1 hour; cool to ambient.

Comments. The required quenching rate for 1040 to exceed CR_M is impossibly fast (except for piano wire and razor blades). The 4340 steel allows slightly more time to completely miss the start of transformations. More important, if the decomposition starts before reaching 325°C, the product that forms is bainite rather than pearlite, thus tough and relatively hard.

Steels with still higher alloy contents than 4340 could be used to give more quenching time. However, they would also require more furnace time for the final bainite formation. ◀

• **Example 11–5.4** There is still 5 v/o austenite present after quenching a 1080 steel. This retained austenite is present as small residual grains ($\delta = \sim 1\ \mu m$) within a martensite matrix. What pressure must be overcome within the metal for one of these small grains of this retained austenite to complete its transformation to martensite?

Solution: For simplicity, assume the grain changes ($\gamma \rightarrow M$), and then must be compressed back into the original space within the rigid martensite matrix.

From Example 10–1.3, $\Delta V/V = -0.018$ (minus for compression).

From Example 6–3.2, $K = 162{,}700$ MPa   (or 23,600,000 psi).

Eq. (6–3.5), $P_h = K(\Delta V/V)$

$$= (162{,}700 \text{ MPa})(-0.018)$$

$$= 2930 \text{ MPa compression} \quad \text{(or 425,000 psi)}.$$

Comments. This assumes that the adjacent metal is absolutely rigid. Actually, it relaxes some and reduces the pressure slightly. Even so, the compressive stresses are sufficient to stop the $(\gamma \rightarrow M)$ reaction short of completion. ◀

• 11–6 HARDENABILITY

It is important to distinguish between *hardness* and *hardenability: Hardness* is a measure of resistance to plastic deformation. *Hardenability* is the "ease" with which hardness may be attained.

Figure 11–6.1 shows the maximum possible *hardnesses* for increasing amounts of carbon in steels; these maximum hardnesses are obtained only when 100% martensite is formed. A steel that transforms rapidly from austenite to ferrite plus carbide has low *hardenability* because these high-temperature transformation products are formed at the expense of the martensite. Conversely, a steel that transforms very slowly from austenite to ferrite plus carbide has greater hardenability. Hardnesses nearer the maximum can be developed with less severe quenching in a steel of high hardenability, and greater hardnesses can be developed at the center of a piece of steel even though the cooling rate is slower there.

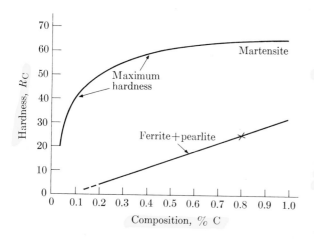

Fig. 11–6.1   Maximum hardness versus carbon content of plain-carbon steels, showing maximum hardnesses arising from martensite compared with hardness developed by pearlitic microstructures. To produce maximum hardness, the reaction $\gamma \rightarrow \alpha$ + carbide must be avoided during quenching.

Hardenability curves   For any given steel, there is a direct and consistent relationship between hardness and cooling rate. However, the relationship is highly nonlinear. Furthermore, the theoretical bases for quantitative analyses are complex.* Fortunately, it is possible to use a standardized test that lets the engineer make necessary predictions of hardnesses for many applications in a minute or two, and hardness comparisons between steels at a glance. This is the *Jominy end-quench test.* In this standardized test, a round bar of a specified size is heated to form austenite and is then end-quenched with a water stream of specified flow rate and pressure, as indicated in Fig. 11–6.2(a). Hardness values along the cooling-rate gradient are determined on a Rockwell hardness tester, and a *hardenability curve* is plotted (Fig. 11–6.2b).

---

* Variables include each and every alloying element and/or impurity, grain size, and austenitizing temperature. Also recall from the discussion of Fig. 11–4.10 that the cooling rates are measured at 700°C. This rate decreases at lower temperatures and approaches zero before cooling is complete.

The quenched end is cooled very fast and therefore has the maximum possible hardness for the particular carbon content of the steel that is being tested. The cooling rates at points behind the quenched end are slower (Fig. 11–6.3), and consequently the hardness values are lower (Fig. 11–6.2b). The cooling-rate data of Fig. 11–6.3 are generally valid for all types of plain-carbon and low-alloy steels since they have comparable values for density, heat capacity, and thermal conductivity—the three properties that affect thermal diffusivity.*

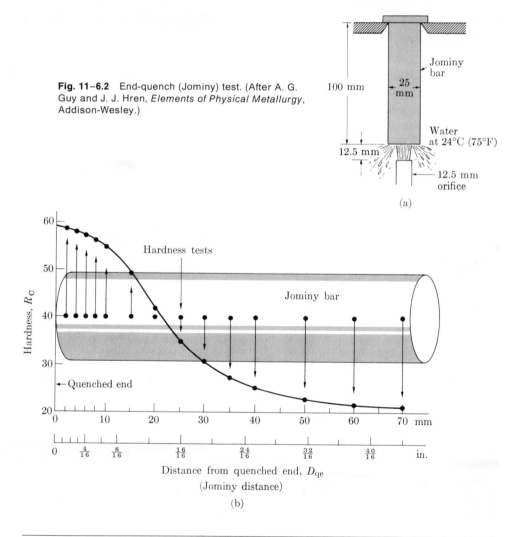

Fig. 11–6.2   End-quench (Jominy) test. (After A. G. Guy and J. J. Hren, *Elements of Physical Metallurgy*, Addison-Wesley.)

---

* Stainless-type steels do not follow the pattern shown in Fig. 11–6.3, since their high-alloy contents reduce their thermal conductivities significantly without a comparable effect on density and/or heat capacity. However, these steels are seldom quenched for hardness requirements.

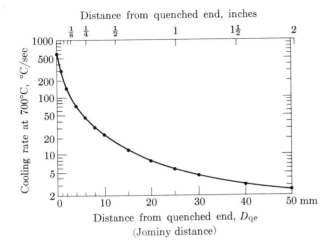

Fig. 11–6.3 Cooling rates (at 700°C) versus the distance, D_{qe}, from the quenched end of a Jominy bar. (Since the cooling rate decreases continuously as the temperature drops, 700°C is selected as a reference for comparisons. It is approximately the eutectoid temperature for most low-alloy steels and therefore critical for the $\gamma \to \alpha + \bar{C}$ reaction.)

Figure 11–6.4 shows hardenability curves for several common grades of steels. They are plots of hardness versus cooling rates. The rates are shown in °C/sec on the upper abscissa. In general, however, it is more convenient to use the "distance from the quenched-end," or D_{qe} (called the Jominy distance), because it can be plotted directly from the laboratory data. We will use this simplified procedure.*

Observe several things from Fig. 11–6.4. The low-alloy steels (4140 and 4340) have greater hardenability than the plain-carbon steels, i.e., for a *given cooling rate* their hardnesses are nearer the maximum possible. Specifically, for a 0.40% C steel, the maximum hardness is $57R_C$ as indicated in Fig. 11–6.1. At $D_{qe} = 10$ mm (where CR = 25°C/sec), the hardnesses of 4340 and 4140 are $55R_C$ and $53R_C$, respectively; the hardness of 1040 steel is only $26R_C$. Expectedly, higher-carbon steels are harder (1060 vs. 1040 vs. 1020); this is true with rapid cooling rates ($D_{qe} = 0$ mm) as well as with slow cooling rates ($D_{qe} = 30$ mm). Finally, observe the second curve for the 1060 steels. The coarser-grained steel (#2) has higher hardenability as a direct result of the slower decomposition of the austenite. (Cf. Fig. 11–4.8.) This means that given a cooling rate, coarse austenite produces more martensite than fine-grained austenite.[†]

Use of hardenability curves   End-quench hardenability curves are of great practical value because (1) if the cooling rate of a steel in any quench is known, the hardness may be read directly from the hardenability curve for that steel, and (2) if the hardness at any point can be measured, the cooling rate at that point may be obtained from the hardenability curve for that steel.

---

* This means that identical distances from the quenched-end on two different hardenability curves have the same specific cooling rates.
† The other steels of Fig. 11–6.4 also have different hardenability curves for other grain sizes. A major advantage of the Jominy end-quench test is that a test specimen can be made from the same steel as that being processed into gears, tools, etc. Therefore, the exact hardenability curve is known. This means that variables such as grain-size, and minor composition differences, are automatically considered.

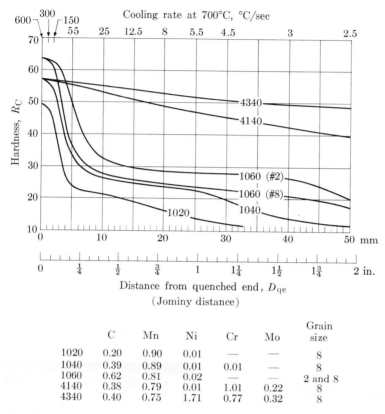

	C	Mn	Ni	Cr	Mo	Grain size
1020	0.20	0.90	0.01	—	—	8
1040	0.39	0.89	0.01	0.01	—	8
1060	0.62	0.81	0.02	—	—	2 and 8
4140	0.38	0.79	0.01	1.01	0.22	8
4340	0.40	0.75	1.71	0.77	0.32	8

Fig. 11–6.4 Hardenability curves for six steels with the indicated compositions and grain sizes. The steels were end-quenched as shown in Fig. 11–6.2(a). In commercial practice, the hardenability curve of each type of steel varies because of small variations in composition. As a result, hardenability tests are commonly made for each heat of steel that is produced for quench-and-temper applications. (Adapted from U.S. Steel data.)

Figure 11–6.4 presents the end-quench hardenability curve for an AISI–SAE 1040 steel with the grain size and composition indicated.* The quenched end has nearly maximum hardness for 0.40% carbon steel because the cooling was very rapid and only martensite was formed. However, close behind the quenched end, the cooling rate was not rapid enough to avoid some ferrite and carbide formation, and so maximum hardness was not attained at that point.

In the laboratory, it is also possible to determine the cooling rates within bars of steel. Table 11–6.1, for example, shows the cooling rates at eutectoid temperatures

---

* These data apply to this 1040 composition (and grain size). A slight variation is possible in the chemical specifications of any steel (e.g., in a 1040 steel, C = 0.37/0.44, Mn = 0.60/0.90, S = 0.05, P = 0.04, and Si = 0.15/0.25). As a result, two different 1040 steels may have slightly different hardenability curves.

Table 11–6.1
Cooling rates in a 75-mm (~3-in.)
diameter steel bar (at 700°C)

Position	Agitated water quench		Agitated oil quench	
	°C/sec	D_{qe}*	°C/sec	D_{qe}*
Surface	~100	3	~20	11
$\frac{3}{4}$-radius	27	9	9.5	18
Mid-radius	14	14	7.5	21
Center	11	17†	5.5	25

* Distance from the quenched end of a Jominy bar that has the same cooling rate at 700°C (Jominy distance).
† Observe that during a water-quench the center of a 75-mm diameter bar, which is 37 mm from the surface, cools at the same rate as when $D_{qe} = 17$ mm, and therefore much faster than the steel that is 37 mm from the quenched end of a Jominy bar (11°C/sec vs. 3.5°C/sec). Of course, heat is removed radially from the 75-mm bar, but primarily from one end of the Jominy bar.

for the surfaces, mid-radii, and centers of 75-mm (~3-in.) rounds quenched in mildly agitated water and oil. These cooling rates were determined by thermocouples embedded in the bars during the quenching operation. Similar data may be obtained for bars of other diameters. These data are summarized in Fig. 11–6.5.

By the use of the data of Fig. 11–6.5 and a hardenability curve, the *hardness traverse* that will exist in a steel after quenching may be predicted. For example, the center of the 75-mm (~3-in.) round bar quenched in oil has a cooling rate of 5.5°C per second. Since the center of this large round bar has the same cooling rate as a Jominy test bar at a point 25 mm (~1-in.) from the quenched end, the hardnesses at the two positions will be the same. Thus if the bar is 1040 steel (Fig. 11–6.4), the center hardness will be 23R_C. Figure 11–6.4 also shows that the following center hardnesses may be expected for 75-mm bars of other oil-quenched steels (cooled at 5.5°C per second):

AISI–SAE:	1040	4140	1020	4340	1060 (G.S. 8)	1060 (G.S. 2)
R_C:	23	47	14	52	24	28

Several determinations of quenched hardnesses are given in examples that follow shortly.

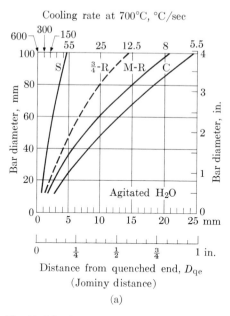
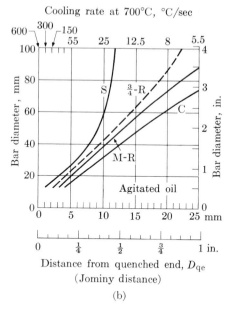

Fig. 11–6.5 Cooling rates in round steel bars quenched in (a) agitated water and (b) agitated oil. Top abscissa, cooling rates at 700°C; bottom abscissa, equivalent positions on an end-quench test bar. (C, center; M-R, mid-radius; S, surface; Dashed line, approximate curve for $\frac{3}{4}$-radius.) The high heat of vaporization of water produces a severe quench in that quenching medium.

Tempered hardness   The results of Examples 11–6.3 and 11–6.4 are hardnesses of quenched steel. As indicated in Fig. 10–3.9, the hardness decreases with continued tempering because the carbide particles coalesce. The data of that figure are for a plain-carbon eutectoid steel (1080). Alloy steels temper more slowly.* Data for tempering rates are available in metallurgical books for use by the engineer who must specify heat-treating processes for specific steels.

Study aid (hardenability)   This topic is presented somewhat differently in *Study Aids for Introductory Materials Courses* (Topic XVIII) than in this section. Therefore, the reader is referred to this paperback if more clarification is needed.

Example 11–6.1 Determine the cooling rate for the center and mid-radius of a 20-mm (0.79-in.) diameter round steel bar when quenched in agitated water. (The bar is long enough so there is no end effect.)

---

* For example, a "high-speed" tool steel contains elements such as V, Cr, W, and Mo that make very stable carbides. These carbide particles coalesce much more slowly than Fe_3C does in a plain-carbon steel. Thus, tool steels can be used at higher temperatures (higher-speed operation) before overtempering, softening, and consequent destruction of the cutting edge becomes critical.

Solution: From Fig. 11–6.5(a) and for a 20-mm round bar,

$$\text{CR}_\text{center}: \qquad D_\text{qe} = 4 \text{ mm}.$$

From Fig. 11–6.3 (or from the upper abscissa of Fig. 11–6.5a),

$$\text{cooling rate} = 75°\text{C/sec}.$$

$$\text{CR}_\text{mid-rad}: \qquad D_\text{qe} = 2.5 \text{ mm}, \quad \sim 125°\text{C/sec}.$$

Comments. Observe that the center of the 20-mm bar is 10 mm from the surface of the bar. However, its cooling rate is much faster in this location than it would be at 10 mm from the end of the Jominy bar, where the cooling rate is 25°C/sec. It is faster because heat is extracted in all radial directions and not just in the lengthwise direction. Furthermore, there is a reservoir of heat that must pass through the 20-mm point of the end-quenched bar. ◄

Example 11–6.2 What is the quenched hardness at a point 5 mm from the surface of a 40-mm diameter bar of 4140 steel that was quenched in agitated oil.

Solution: Since the radius is 20 mm, this is a $\frac{3}{4}$-radius point.

From Fig. 11–6.5, $D_\text{qe} = 9^\text{plus}$ mm   (and CR $= \sim 27°\text{C/sec}$).

From Fig. 11–6.4,   Hardness $= 53R_\text{C}$.

Comment. Data may be interpolated from Figs. 11–6.4 and 11–6.5. In doing so, it is recommended that D_qe be used rather than °C/sec, because the latter is highly nonlinear. ◄

Example 11–6.3 Sketch the *hardness traverses* for two steel rounds quenched in water; each is 38 mm (1.5 in.) in diameter, with AISI–SAE 1040 and 4140 compositions, respectively.

Solution

	From Fig. 11–6.5(a)		From Fig. 11–6.4		
Position	Approximate cooling rate at 700°C		Cooling rate at 700°C	AISI–SAE 1040	AISI–SAE 4140
	°C/sec	D_qe*	D_qe*		
Surface	200	1.5 mm	1.5 mm	$55R_\text{C}$	$57R_\text{C}$
$\frac{3}{4}$-radius	75	4 mm	4 mm	$38R_\text{C}$	$56R_\text{C}$
Mid-radius	55	5 mm	5 mm	$34R_\text{C}$	$55R_\text{C}$
Center	35	7.5 mm	7.5 mm	$28R_\text{C}$	$54R_\text{C}$

* D_qe = cooling rate, expressed as distance from the quenched end of a Jominy bar.

Comments. The hardness traverses for the two steels of Example 11–6.3 are shown in Fig. 11–6.6. Although the surface hardnesses of the two are very similar, the difference in their hardenability produces a higher center hardness for the AISI–SAE 4140 steel. As indicated, this steel has a higher alloy content, which slows down the transformation of austenite to ferrite and carbide. Consequently, more martensite can form. ◄

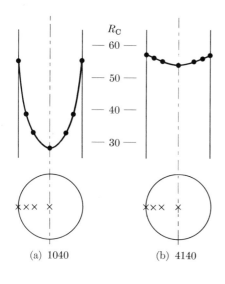

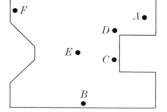

Fig. 11–6.6   Hardness traverses. See Example 11–6.3.

Fig. 11–6.7   V-bar cross section. See Example 11–6.4.

Example 11–6.4   Figure 11–6.7 shows the points in the cross section of a V-bar of AISI–SAE 1060 steel (G.S. #2) in which the following hardness readings were obtained after oil quenching: A—$40R_C$, B—$36R_C$, C—$33R_C$, D—$32R_C$, E—$31R_C$, F—$63R_C$. What hardness values would be expected for an identically shaped bar of AISI–SAE 4068 steel?

Solution: For a given steel, the hardness is dependent on the cooling rate.

| | AISI–SAE 1060 (G.S. #2) | | AISI–SAE 4068 | |
| | (From Fig. 11–6.4) | | (From Fig. 11–6.8) | |
Point	Hardness	Approximate cooling rate D_{qe}*	Cooling rate at 700°C D_{qe}*	Hardness
A	$40R_C$	7 mm	7 mm	$62R_C$
B	36	8	8	61
C	33	10	10	59
D	32	11	11	57
E	31	12	12	55
F	63	1	1	64

* D_{qe} = cooling rate, expressed as distance from the quenched end of a Jominy bar.   ◄

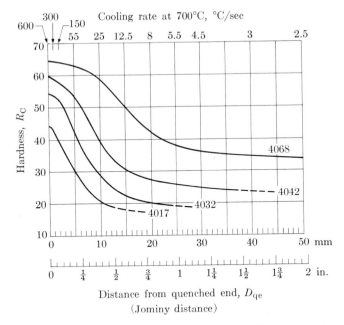

Fig. 11–6.8 Hardenability curves for 40xx steels. Except for carbon content, the composition is the same for each. Additional carbon gives harder martensite and harder $\alpha + \overline{C}$, according to Fig. 11–6.1.

Example 11–6.5 A 1020 steel rod (dia. = 32 mm, or $1\frac{1}{4}$ in.) is *carburized* to 0.62% C at the surface, to 0.35% C at 2 mm below the surface, and unaltered beyond the depth of 4 mm.
a)   Determine the hardness profile of the steel after quenching in water.
b)   What would the profile have been without carburizing?(G.S.= #8.)

Solution

	D_{qe}	(a)		(b)	
Surface	1.5 mm	0.62% C	$62R_C$	0.20% C	$47R_C$
2 mm	~2.5	0.35	~50	0.20	~40
4 mm	3	0.20	34	0.20	34
MR	4	0.20	26	0.20	26
Center	6.5	0.20	22	0.20	22 ◀

REVIEW AND STUDY

Annealing processes involve extended heating and slow cooling. This permits the material to closely match equilibrium conditions. The resulting microstructures and properties are a function of the material being annealed. In some materials, annealing produces stress-relief; in others, recrystallization; softening is common; ductility will usually increase if only one phase is present, but will decrease if the resulting microstructure provides a brittle path for crack propagation.

Normalization processes are designed to introduce more uniformity into the product—compositional and/or microstructural.

Rapid cooling thwarts equilibrium. In the simplest case, we can retain a solid solution to temperatures below the solubility limit. This permits more hardening than would be possible otherwise. It also sets the stage for the age-hardening process, a low-temperature* isothermal precipitation.

Isothermal transformation of $\gamma \rightarrow (\alpha + \bar{C})$ in steels follows the typical C-type curve. Continuous-cooling transformation is more common, but requires a measure of the cooling rates, CR. With rapid cooling (quenching), the regular eutectoid reaction is avoided and hard martensite forms. Alloy steels have an increased *hardenability* because the critical cooling rate necessary to produce martensite is slower; conversely, alloy steels more closely attain the maximum possible *hardness* for any given cooling rate than do plain-carbon steels. Hardenability curves permit one to estimate hardness profiles from standardized end-quench tests.

The examples of this chapter were primarily metals, and steels in particular, since their heat-treatments have been extensively developed and refined. However, the same concepts apply to other materials. For example, glass may first be shaped by viscous deformation and then heat-treated to give a stronger product by devitrification (crystallization).

KEY TERMS AND CONCEPTS

Alloy retardation

Annealing

Austenite grain size

Austenization

Austempering

Bainite $(\alpha + \bar{C})$

Continuous-cooling transformation, CCT

• Carburize

Critical cooling rates, CR_M and CR_P

• End-quench test (Jominy-bar)

• Hardenability

• Hardness traverse

Homogenization

Interrupted quench

Isothermal transformation, I-T

• Jominy distance, D_{qe}

Normalize

Nucleation

Over-aging

Precipitation-hardening (age-hardening)

Quenching

Rate, $R = t^{-1}$

Solution treatment

Spheroidization

Stress relief

Tempering

---

*   Below the "knee" of the C-curve.

FOR CLASS DISCUSSION

A_{11} Why do diffusion, recrystallization, stress relaxation (Section 7–5), and viscous flow (Section 8–8) have similar time–temperature relationships?

B_{11} Point out the differences between normalizing a steel and giving it a full anneal. What are the technical bases for these differences?

C_{11} How will the glass transition temperature T_g compare with the temperature of the annealing point of glass ($\eta = 10^{12}$ Pa·s), and with a temperature of the strain point ($\eta = 10^{13.5}$ Pa·s)?

D_{11} Give two reasons why superheating occurs less readily in a solution reaction than supercooling occurs in a precipitation reaction.

E_{11} Explain the data variations in Table 11–3.1.

F_{11} As a practical production matter, why is it simpler to first quench a 96 Al–4 Cu alloy to room temperature and then reheat to 100°C for aging, than to quench directly to 100°C?

G_{11} (a) Refer to the phase diagrams of Chapter 9. What alloys are candidates for precipitation-hardening? (b) Cu–Zn brass is never age-hardened. Why?

H_{11} Steel used for food cans has less than 0.1% carbon, so that it is deformable. Therefore, it is not amenable to quenching and tempering treatments; however, it hardens slightly during the food-packing process. Suggest an explanation.

I_{11} The lever rule is not applicable to an isothermal-transformation diagram. Why?

J_{11} Why does the "start" curve in Fig. 11–4.6 have two branches on its upper arm?

K_{11} The start (or finish) of any reaction is generally very difficult to determine (compared to the midpoint). Give two reasons. (*Note:* Because of this, M_s is commonly considered to be ~1% reaction; and M_f is commonly considered to be 99% completion.)

L_{11} Why does the austenite in a fine-grained steel transform to ($\alpha + \bar{C}$) faster than in a coarse-grained steel?

M_{11} Why does the austenite of a 2% Cr steel transform to ($\alpha + \bar{C}$) less rapidly than the austenite in an 0.7% Cr steel?

N_{11} Design a laboratory setup to allow 1 sec ($\pm 10\%$) for isothermal transformation to occur in steel.

O_{11} Why did we *not* use Fig. 11–4.6 for wire (c), step 3, of Example 11–4.1?

P_{11} On an I-T diagram, Fig. 11–4.6, the t_f curve ends before it meets the M_f curve. Suggest a reason why.

Q_{11} Compare and contrast: tempered martensite, pearlite, bainite, martensite, spheroidite, ferrite, carbide.

R_{11} Assume CR_M and CR_P for a steel are 75°C/sec and 15°C/sec, respectively. What phases will be present if the steel is cooled at 10°C/sec? At 20°C/sec? At 50°C/sec? At 100°C/sec?

S_{11} Assume the same CR_M and CR_P as in R_{11}. If this steel is cooled at 10°C/sec, it is harder than if it is cooled at 5°C/sec. Why? (They both contain only α and $\bar{C}$.)

T_{11} The t_s and t_f curves of isothermal transformation are "displaced downward and to the right for continuous-cooling transformation (Fig. 11–4.10)." Why do M_s and M_f remain the same in each process?

U_{11} Explain why low-alloy steels are commonly specified for steels that are to be heat-treated by quenching and tempering.

• V_{11} Bainite cannot be obtained by continuously cooling a 1080 steel. Why?

• W_{11} Explain the difference between hardness and hardenability.

• X_{11} Why should a steel (or the water) be agitated when a gear is water-quenched?

• Y_{11} Why does the center of a 25-mm (1-in.) diameter round steel bar quench to a greater hardness than the steel that is 25 mm behind the end of a Jominy end-quench bar? Which will quench faster?

• Z_{11} The temperature chosen for carburization is generally the same as that for normalizing. Why not use a temperature in the α-range since the carbon diffuses faster in that structure? Why not use a temperature further into the γ-range?

STUDY PROBLEMS

11–1.1 A copper alloy, which received 35% cold work, had recrystallized (50% completion) after 1 hour at 295°C (563°F). At 325°C (617°F), the time required was only 15 min. What minimum temperature is required for hot-working where this recrystallization must be 50% complete in 6 sec?

Answer: 466°C (871°F)

11–1.2 Select an annealing temperature for (a) 1040 steel; (b) 1080 steel; (c) 1% carbon steel.

• **11–1.3** What is the recommended annealing temperature for a steel containing 2.5% nickel and 0.5% carbon?

Answer: 760°C

11–1.4 Select a heat treatment to form spheroidite in 1090 steel.

11–2.1 (a) What temperature should be used to normalize a 1030 steel? (b) 1080 steel? (c) 1% carbon steel?

Answer: (a) 880°C

• **11–2.2** What is the normalizing temperature for a steel of Example 9–9.1?

11–2.3 Twenty kilograms of an 8 Al–92 Mg alloy are first melted and then cooled rapidly to 500°C. Since there wasn't time for diffusion in the solid, the average composition of the ϵ is 5% Al. (a) What is the composition of the liquid at 500°C? (Assume rapid diffusion in the liquid.) (b) How much liquid at 500°C? (c) At what temperature will γ appear? (d) What is the composition of the final liquid?

Answer: (a) 23 Al–77 Mg (b) 3.3 kg (c) 437°C (d) 32 Al–68 Mg

11–3.1 Maximum hardness is obtained in a metal when the aging time is 10 seconds at 380°C, or 100 seconds at 315°C. Neither of these is satisfactory for production because we cannot be certain the parts are uniformly heated. Recommend a temperature for a maximum hardness in 1000 seconds (15 min–20 min), a time compatible with production.

Answer: ~260°C

11–3.2 How long should it take for the metal in the previous problem to reach maximum hardness at 100°C?

11–3.3 Explain why a 92% copper, 8% nickel alloy can (or cannot) be age-hardened.

11–3.4 Explain why the following alloys can (or cannot) be considered for age-hardening. (a) 97% aluminum, 3% copper. (b) 97% copper, 3% zinc. (c) 97% nickel, 3% copper. (d) 97% copper, 3% nickel. (e) 97% aluminum, 3% magnesium. (f) 97% magnesium, 3% aluminum.

11–3.5 A slight amount of age-hardening is realized when a steel (99.7 w/o Fe, 0.3 w/o C) is quenched from 700°C (1300°F) and reheated for 3 hr at 100°C. Account for the hardening.

Answer: Carbon solubility in ferrite decreases.

11–3.6 An aircraft manufacturer receives a shipment of aluminum alloy rivets that have already age-hardened. Can they be salvaged? Explain.

11–3.7 (Refer to Fig. 11–3.3.) Select a plotting procedure that shows a relationship between temperature and time for peak hardness and that lets you estimate (graphically) the time for peak hardness at 100°C. What is that time for this 2014 aluminum?

Answer: ~2000 hrs.

11–3.8 Estimate graphically (as in Study Problem 11–3.7) the time that would be required to attain 200 MPa (30,000 psi) at 100°C.

11–4.1 A small piece of 1080 steel is heated to 800°C, quenched to −60°C, reheated immediately to 300°C, and held 10 sec. What phases are present at the end of this time?

Answer: Martensite (with some possible tempering to α + carbide).

11–4.2 Some 1045 steel is quickly quenched from 850°C to 400°C and held for 1 sec; for 10 sec; for 100 sec. What phase(s) will be present at each time point?

11–4.3 Some 1045 steel is quickly quenched from 850°C to 425°C and held 5 sec before quenching again to 20°C. (a) What phase(s) will be present just before the second quench? Give the composition of each. (b) What phase(s) will be present immediately after the second quench?

Answer: (a) γ(0.45% C), α(0.02% C), and $\bar{C}$(6.7% C)   (b) M(0.45% C), α, and $\bar{C}$

11–4.4 Repeat Study Problem 11–4.3, but change 20°C to 275°C.

11–4.5 A 1020 steel is equilibrated at 790°C (1450°F) and quenched rapidly to 400°C (750°F). How long must the steel be held at that temperature to reach the midpoint of the austenite decomposition?

Answer: ~8 sec

11–4.6 Sketch an isothermal-transformation diagram for a steel of 1.0 w/o C and 99.0 w/o Fe.

11–4.7 Sketch an isothermal-transformation diagram for a 1020 steel.

Answer: $M_s = 450°C$, $M_f = 290°C$; α and γ are stable between 727°C and 855°C after long times; at 550°C the curve is farther to the left than in 1045 steel.

11–4.8 Six different wire samples of an AISI–SAE 1045 steel received one of the following six heat-treating sequences. Indicate the phases that exist *immediately after* the completion of each sequence.
a) Heated to 825°C,# quenched to 550°C, held 10 sec;
b) Heated to 900°C,# quenched to 550°C, held 10 sec, quenched to 250°C;
c) Heated to 925°C,# quenched to 300°C, held;#
d) Heated to 700°C,# quenched to 250°C;
e) Heated to 250°C,# heated to 425°C, held 1 sec;
f) Heated to 750°C,# quenched to 550°C.

11–4.9 One hundred grams of an AISI–SAE 1045 steel are heated and quenched in the ways indicated below. Indicate the phase(s) and grams of each phase at the *end* of each sequence.
a) Heated to 825°C,# quenched to 120°C;
b) Heated to 750°C,# quenched to 20°C;
c) Heated to 825°C,# quenched to 550°C, held 7 sec, quenched to 250°C;
d) Heated to 825°C,# quenched to 550°C, held 10 sec, heated to 750°C.#
Answer: (a) 100 g M   (b) 70 g M, 30 g α (c) 93 g α, 7 g carbide   (d) 70 g γ, 30 g α

11–4.10 A small wire of AISI–SAE 1045 steel is subjected to the following treatments as *successive* steps:
1) heated to 875°C, held there for 1 hr;
2) quenched to 250°C, held there 2 sec;
3) quenched to 20°C, held there 100 sec;
4) reheated to 550°C, held there 1 hr;
5) quenched to 20°C and held.
Describe the phases or structures present *after each step* of this heat-treatment sequence.

11–4.11 (a) Repeat Problem 11–4.10 with steps (1), (2), (5). (b) Repeat Problem 11–4.10 with steps (1), (3), (4), (5). (c) Repeat Problem 11–4.10 with steps (1), (2), (4), (5).

• **11–4.12** Refer to Fig. 11–4.9. A steel without molybdenum is equilibrated at 740°C then quenched to 400°C and held for ~ 5 sec for γ decomposition to start. What sequence would be necessary to achieve the same results if 0.25% Mo were present? Explain.

11–5.1 A small piece of 1080 steel has its quench interrupted for 20 seconds at 300°C (570°F) before final cooling to 20°C. What phase(s) will be present?

Answer: M(0.8% C), and some retained γ(0.8% C)

11–5.2 A 1045 steel is austenitized at 825°C (1520°F), quenched to 670°C (1240°F), and held one minute before quenching to ambient. (a) What phase(s) and microstructure will be present? (b) Repeat for a 4340 steel.

This symbol indicates that equilibrium was attained at this step of the heat treatment before proceeding to the other steps in the sequence.

11–5.3 A 1080 steel is to be quenched and tempered to a hardness of $50R_C$. (a) What should the austenitizing temperature be? (b) How long should it be tempered if the tempering temperature is 400°C (750°F)?

Answer: (a) $\sim 750°C$ ($\sim 1380°F$)   (b) ~ 40 min

11–5.4 Compare and contrast: (a) martensite and tempered martensite; (b) tempered martensite and bainite; (c) bainite and spheroidite.

11–5.5 Draw temperature (ordinate) and time (abscissa) plots for the following heat treatments. Indicate the important temperatures, relative times, and reasons for drawing the curves as you do. (a) Normalizing a 1095 steel, contrasted with annealing the same steel. (b) Solution-treating a 95 Al–5 Cu alloy, contrasted with aging the same alloy. (c) Austempering a 1080 steel, contrasted with martempering (interrupted quench) the same steel. (d) Spheroidizing 1080 steel, contrasted with spheroidizing 10 · 105 steel.

11–5.6 Using the data of Example 10–1.3 and Appendix C, determine for a 1080 steel whether the volume change for tempering martensite $(M \rightarrow \alpha + \bar{C})$ is greater or less than during the quenching step $(\gamma \rightarrow M)$.

● **11–5.7** A 50-mm diameter bar of 1080 steel is quenched and forms a martensite "rim" that is 2.5 mm thick. While the austenitic center is hot, it can still adapt to the $(\gamma \rightarrow M)$ volume changes (Example 10–1.3). The more slowly cooled center changes directly to $(\alpha + \bar{C})$. What hoop stress is placed on the martensite rim? (For 1080, $\rho_\gamma = 7.99$ Mg/m³ $(= 7.99$ g/cm³$)$; $\rho_M = 7.85$ g/cm³; and $\rho_{(\alpha + \bar{C})} = 7.84$ g/cm³.)

Answer: 1170 MPa tension (170,000 psi)

● *11–6.1* How hard will the quenched end of an AISI–SAE 4620 steel test bar be?

Answer: Approximately $49R_C$

● *11–6.2* The quenched end of a Jominy bar should be $44R_C$. What must the carbon content be for the steel? Explain.

● *11–6.3* (a) What is the cooling rate at the mid-radius of a 50-mm (1.97-in.) round steel bar quenched in agitated oil and reported in °C/sec? (b) Reported as the distance from the quenched end of a Jominy bar?

Answer: (a) $\sim 17°C$/sec   (b) 13 mm

● *11–6.4* Repeat Study Problem 11–6.3 but for the $\frac{3}{4}$-radius and a water-quench.

● *11–6.5* (a) What is the quenched hardness at the mid-radius of a 50-mm (1.97-in.) round steel bar of 1040 steel quenched in agitated oil? (b) In agitated water?

Answer: (a) $25R_C$ (b) $28R_C$

● *11–6.6* Repeat Study Problem 11–6.5, but for the $\frac{3}{4}$-radius.

● *11–6.7* What hardness would you expect the center of a 50-mm (2-in.) round bar of 1040 steel to have if it were quenched in (a) agitated oil? (b) Agitated water?

Answer: (a) $24R_C$ (b) $26R_C$

● *11–6.8* A round bar of 1040 steel has a surface hardness of $41R_C$ and a center hardness of $28R_C$. How fast were the surface and center cooled through 700°C?

• *11–6.9* A 63-mm (2.5-in.) diameter round of 1040 steel is quenched in agitated oil. Estimate the hardness 25 mm (1 in.) below the surface of the round bar. (Show reasoning.)

Answer: $24R_C$

• **11–6.10** How would the hardness traverse of the 1040 steel shown in Fig. 11–6.6 vary if it were quenched in (a) still oil? (b) Still water? (c) If it had a coarser austenite grain size? Explain.

• **11–6.11** (a) A 40xx steel is to have a hardness of $40R_C$ after it is quenched at the rate of 17°C/sec. What is the required carbon content? (b) What will the hardness of this steel be if quenched twice as fast?

Answer: (a) 0.5% C   (b) 4050 at 34°C/sec ($\therefore D_{qe} = 7.5$ mm) $53R_C$

• *11–6.12* An 80-mm (3.15-in.) round 4340 steel bar is quenched in agitated oil. Plot the hardness traverse.

• *11–6.13* Plot a hardness traverse for a 1060 steel bar (diameter of 38 mm, or 1.5 in.) that is to be quenched in (a) agitated water. (b) Agitated oil. (G.S. = #2)

Answer: (a) S: $63R_C$; MR: 48; C: 37

• *11–6.14* A 40-mm (1.6-in.) bar of 1040 steel (i.e., with diameter = 40 mm, and length ≫ 40 mm) is quenched in agitated water. (a) What is the cooling rate through 700°C at the surface? At the center? (b) Plot a hardness traverse.

• **11–6.15** The center hardness of six bars of the same steel are indicated below. From these data, plot the hardenability curve for the steel. (*Hint:* There should be only *one* curve.)

Diameter	Water quench	Oil quench
25-mm (1.0-in.)	$58R_C$	$57R_C$
50-mm (2.0-in.)	$55R_C$	$47R_C$
100-mm (3.9-in.)	$34R_C$	$30R_C$

Answer: $58R_C$ at $D_{qe} = 5$ mm; $56R_C$ at 10 mm; $49R_C$ at 15 mm; $39R_C$ at 20 mm; $33R_C$ at 25 mm; $31R_C$ at 30 mm; $30R_C$ at 35 mm

• **11–6.16** The mid-radius hardnesses of various bars of the same steel are indicated below. From these data, plot the hardenability curve for the steel. (*Hint:* There should be only *one* curve.)

Diameter	Water quench	Oil quench
25-mm (1.0-in.)	$59R_C$	$58R_C$
50-mm (2.0-in.)	$56R_C$	$52R_C$
100-mm (3.9-in.)	$38R_C$	$31R_C$

• *11–6.17* A spline gear had a hardness of $45R_C$ at its center when it was made of the 4068 steel shown in Fig. 11–6.8. What hardness would you expect the same gear to have if it were made of 1040 steel?

Answer: $24R_C$

● **11–6.18** Two 76-mm (3-in.) round bars of steel were quenched, one in agitated water, the other in agitated oil. The following hardness traverses were made:

Distance below surface		Water, R_C	Oil, R_C
mm	inches		
0	0	57	40
9.5	$\frac{3}{8}$	46	33
19	$\frac{3}{4}$	36	32
28.5	$1\frac{1}{8}$	34	31
38	$1\frac{1}{2}$	33	30

Calculate and sketch the hardenability curve that would be obtained from an end-quenched test of the same steel. (*Hint:* There will be only *one* curve.)

● **11–6.19** The hardness of the surface of a round bar of 1040 steel quenched in agitated oil is $40R_C$. Determine the hardness of the center of a round bar of 4068 steel quenched in water if this bar is twice the diameter of the 1040 bar. (Indicate all steps in your solution.)

● **11–6.20** A 40-mm (1.6-in.) diameter steel rod of 4017 steel has been carburized to 0.60% carbon at the surface, 0.3% C at 2 mm below the surface, and 0.17% C at 5 mm below the surface. Determine the hardness profile of the steel bar after water quenching.

● *11–6.21* Repeat Example 11–6.5, but with oil quenching.

Answer: (a) S: $33R_C$; 2 mm: $\sim 28R_C$; $\frac{3}{4}$R: $23R_C$; MR: $22R_C$; C: $21R_C$

CHAPTER 12
Corrosion
of Metals

PREVIEW

The reader is aware that corrosion degrades metals. The importance of corrosion is accentuated by the estimate that nearly 5% of every modern country's annual income goes directly or indirectly to maintain, repair, or replace deteriorated products. Most obvious to you, the reader, is the effect of corrosion on your automobile, particularly if you live in the northern latitudes where road salt is used. In addition, there are many other serious corrosion problems ranging from bridge maintenance to household plumbing, and from prosthetic implants (Fig. 12–1.1) to transatlantic cables.

If we own cars, or similar products, and do not simply store them in garages, we must expect to encounter corrosion. Corrosion cannot be eliminated by specifying gold-quality metals, because automobile design requires materials that are cheaper, stronger, and more available. However, corrosion can be minimized if technical designers become familiar with the causes of corrosion and how corrosion can be minimized.

CONTENTS

12-1 Electroplating:
corrosion in reverse.

12-2 Galvanic Couples:
electrode potentials, cathode reactions, rust.

12-3 Types of Galvanic Cells:
composition, stress, concentration.

• **12-4 Corrosion Rates:**
current density, polarization, passivation.

12-5 Corrosion Control:
coatings, stainless steel, galvanic protection.

STUDY OBJECTIVES

1 To refresh your chemistry background so you may handle simple calculations involving electrical currents and corrosion (or electroplating, which is "corrosion in reverse.")

2 To understand the significance of electrode potentials at standard (1-molar) concentrations, and to be able to calculate the potential for other concentrations (Nernst Equation).

3 To expand your chemistry background to include cathode, as well as anode reactions that lead to galvanic cells.

• **4** To become acquainted with factors such as polarization and passivation that affect corrosion rates.

5 To recognize engineering alternatives to reduce or avoid excessive corrosion.

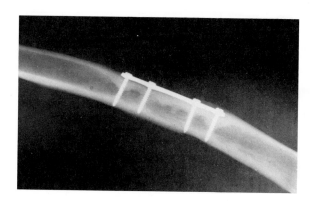

Fig. 12–1.1 Metallic prosthesis. This metal plate and screws, which are used for an internal fixation of a bone, must resist corrosion in the complex tissue environment (1) without loss of strength, and (2) without introducing undesirable corrosion products into the body. (Courtesy S. F. Hulbert, Rose–Hulman Polytechnic University.)

12–1 ELECTROPLATING: CORROSION IN REVERSE

This common process is the reverse of corrosion. Metal ions are removed from an *electrolyte* by providing each with an electron to produce metallic atoms, M:

$$M^{n+} + n\,e^- \rightarrow M^0 \tag{12–1.1}$$

Since each electron carries a charge of 0.16×10^{-18} amp·sec, or coulombs, we can easily calculate the current required to plate metal at any given rate. One mole (0.6022×10^{24}) of monovalent ions requires $(0.6022 \times 10^{24}$ electrons) $(0.1602 \times 10^{-18}$ coul/electron), or 96,500 coul. This value is called a *faraday*, $\mathscr{F}$. Thus, to plate 107.87 g of silver from an Ag^+ solution, it would be necessary to use 1 amp for 96,500 sec (or some other combination of current and time to give 96,500 amp·sec).

The requirements for electroplating include two electrodes, an electrolyte, and a source of electrons. In commercial practice, the electrons are supplied to one of the electrodes by a d.c. current source; however, they eventually come from the second electrode (Fig. 12–1.2). In this sketch the righthand electrode, or *cathode*, *receives*

Fig. 12–1.2 Electroplating. The battery provides electrons to the cathode where the plating occurs (Eq. 12–1.1). However, the ultimate source of the electrons is the metal in the anode, which undergoes corrosion by Eq. (12–1.2).

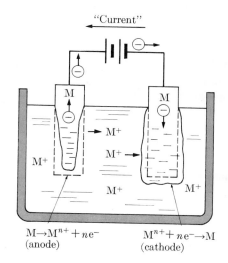

$$M \rightarrow M^{n+} + n\,e^- \qquad\qquad M^{n+} + n\,e^- \rightarrow M$$
$$\text{(anode)} \qquad\qquad\qquad \text{(cathode)}$$

*electrons from the external circuit.** Reaction (12–1.1) occurs, and plating proceeds. As this reaction continues, the left-hand electrode, or *anode, supplies the electrons to the external circuit** via the battery, which acts as a "pump." The electrons come from metal atoms of the left electrode, which are being oxidized to a higher valence level,

$$M^0 \rightarrow M^{n+} + n\,e^-. \qquad\qquad (12–1.2)$$

The *anode undergoes corrosion.*

Example 12–1.1 An electroplating process (as in Fig. 12–1.2) must plate 100 g/hr of copper on a surface containing 2500 mm². (a) What current density, i, is required if the process is fully efficient? (b) How much copper will be corroded from the anode?

Solution

a)   Moles/mm² = 100 g/[(63.54 g/mole)(2500 mm²)] = 0.00063;

Electrons required = (0.00063 moles/mm²)(0.6 × 10²⁴ Cu⁺⁺/mole)(2 electrons/Cu²⁺)

= 7.55 × 10²⁰ electrons/mm²;

Coulombs required = (7.55 × 10²⁰ electrons/mm²)(0.16 × 10⁻¹⁸ coul/electron)

= 121 coul/mm²;

Current density = (121 amp·sec/mm²)/(3600 sec)

= 0.034 amp/mm².

b)   The same amount, 100 g of copper, will be corroded from the anode and will appear in the electrolyte as Cu^{2+} ions.

Comments. Electroplating commonly has an inefficiency of a few percent because of side reactions. For example, some electrons may combine with H^+ ions adjacent to the anode to form H_2 gas. ◀

12–2 GALVANIC COUPLES (CELLS)

Although the reaction at the anode (left electrode of Fig. 12–1.2) illustrates corrosion, it is atypical because it is assisted by a battery, and the two electrodes are identical. Corrosion, as we normally encounter it, involves dissimilar electrode conditions. We encounter a *galvanic couple.* To explain this, consider Fig. 12–2.1, where we have one electrode of zinc and the other of copper. Each metal is subject to oxidation:

$$Zn^0 \longrightarrow Zn^{2+} + 2\,e^-, \qquad\qquad (12–2.1)$$

$$Cu^0 \rightarrow Cu^{2+} + 2\,e^-. \qquad\qquad (12–2.2)$$

As indicated by the arrow lengths, we can envision that the zinc reaction produces a greater push to the right than does the copper reaction. Thus, if we connect the two

---

* These italicized statements apply to anodes and cathodes in all electrical circuits, even in TV tubes that have chemically inert electrodes. In electrochemical reactions, it is always the *anode that undergoes corrosion.*

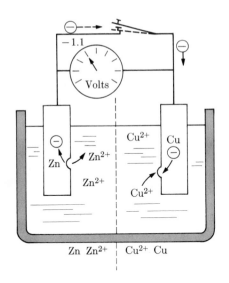

Fig. 12–2.1 Galvanic cell (Zn–Cu). With the switch closed, zinc provides electrons through the external circuit to copper. A 1.1-volt potential difference develops when the circuit is opened. (Table 12–2.1, with molar solutions.)

metals through a voltmeter only, a voltage difference is registered. This difference is −1.1 volts when we use standardized (1-molar) electrolytes at 25°C. This means that if we bypass the voltmeter, electrons will flow from the zinc electrode into the external circuit, and the copper will receive electrons from the external circuit. With electrons removed from the anode, Eq. (12–2.1) will progress to the right, and more zinc will be oxidized. The zinc is corroded. Meanwhile, electrons are supplied to the cathode in excess of the number released by Eq. (12–2.2). They will react with the copper ions that are available in the electrolyte, and Eq. (12–2.2) is *reversed* to give Eq. (12–1.1). Corrosion occurs at only one electrode of the galvanic couple, specifically the *anode*.

Electrode potentials   The production of ions and electrons in reactions such as Eq. (12–2.1) builds up a potential called an *electrode potential*, which depends on (1) the nature of the metal and (2) the nature of the solution. Not all metals oxidize to ions and electrons with equal facility; this inequality is indicated by the first two equations of this section. In addition, atoms along a grain boundary are less stably located than those in the crystal lattice (Section 4–4), so they ionize more readily. Furthermore, the reaction in Eq. (12–2.1) will produce equilibrium with a greater electrode potential if the metal ions enter a solution in which they are relatively stable (e.g., the positive zinc ions are more stable in a concentrated Cl^- solution than in a dilute Cl^- solution).

The electrode potentials of all metals (and therefore their corroding tendencies) are referenced to a standard *hydrogen electrode*. According to Eqs. (12–1.1) and (12–1.2), iron has the following electrochemical reaction:

$$Fe \rightleftharpoons Fe^{2+} + 2\,e^-. \tag{12–2.3}$$

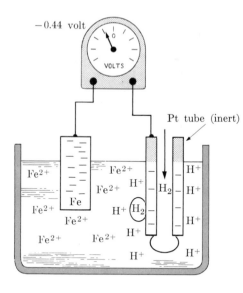

Fig. 12–2.2 Potential difference, Fe versus H_2. Iron produces a lower electron potential than does H_2 (see Table 12–2.1, center column with electrochemical notation). Therefore iron is the anode and hydrogen the cathode. (Platinum is not the cathodic element because there are no Pt^{4+} ions present to receive electrons.)

With hydrogen, equilibrium occurs by this reaction:

$$H_2 \rightleftharpoons 2\,H^+ + 2\,e^-. \tag{12–2.4}$$

The potential *difference* between the iron and hydrogen electrodes is 0.44 volt (Fig. 12–2.2).

Similar measurements for other metals yield the voltage comparisons listed in Table 12–2.1. The alkali and alkaline earth metals, which hold their outer-shell electrons rather loosely, show a greater potential difference with respect to hydrogen than does iron. Conversely, the noble metals, such as silver, platinum, and gold, release their electrons less readily than hydrogen does; therefore, they are at the other end of the electrochemical potential scale (Fig. 12–2.3 and Table 12–2.1).

Electrode potentials (dilute solutions) One-molar solutions (Table 12–2.1) are not regularly encountered. Almost always, the electrolyte is more dilute. The effect of metal-ion concentration C on the electron potential $\mathscr{E}$ is easily calculated by the Nernst equation. At 25°C, it is commonly written as

$$\mathscr{E} = \mathscr{E}_0 + (0.0257 \text{ V}/n)(\ln C). \tag{12–2.5a}*$$

---

* Equation (12–2.5a) originates from the chemical equation

$$\mathscr{E} = \mathscr{E}_0 + (kT/n) \ln \mathscr{K}, \tag{12–2.6a}$$

where k is 86.1×10^{-6} V/K, $T = 298$ K, and $\mathscr{K}$ is the equilibrium constant of the mass law. Thus, we can change Eq. (12–2.5a) to

$$\mathscr{E} = \mathscr{E}_0 + (0.0257 \text{ V}/n)(T/298 \text{ K})(\ln C) \tag{12–2.6b}$$

to accommodate temperature variations.

In terms of base-10 logarithms, Eq. (12–2.5a) becomes

$$\mathscr{E} = \mathscr{E}_0 + (0.059 \text{ V}/n) \log C. \tag{12–2.5b}$$

Table 12–2.1
Electrode potentials (25°C; 1-molar solutions)

Anode half-cell reaction (the arrows are reversed for the cathode half-cell reaction)	Electrode potential used by electrochemists and corrosion engineers,* volts		Electrode potential used by physical chemists and thermodynamists,* volts
$Au \rightarrow Au^{3+} + 3e^-$	+1.50	↑	−1.50
$2H_2O \rightarrow O_2 + 4H^+ + 4e^-$	+1.23		−1.23
$Pt \rightarrow Pt^{4+} + 4e^-$	+1.20		−1.20
$Ag \rightarrow Ag^+ + e^-$	+0.80	Cathodic (noble)	−0.80
$Fe^{2+} \rightarrow Fe^{3+} + e^-$	+0.77		−0.77
$4(OH)^- \rightarrow O_2 + 2H_2O + 4e^-$	+0.40		−0.40
$Cu \rightarrow Cu^{2+} + 2e^-$	+0.34		−0.34
$H_2 \rightarrow 2H^+ + 2e^-$	0.000	Reference	0.000
$Pb \rightarrow Pb^{2+} + 2e^-$	−0.13		+0.13
$Sn \rightarrow Sn^2 + 2e^-$	−0.14		+0.14
$Ni \rightarrow Ni^{2+} + 2e^-$	−0.25		+0.25
$Fe \rightarrow Fe^{2+} + 2e^-$	−0.44		+0.44
$Cr \rightarrow Cr^2 + 2e^-$	−0.74		+0.74
$Zn \rightarrow Zn^{2+} + 2e^-$	−0.76	Anodic (active)	+0.76
$Al \rightarrow Al^{3+} + 3e^-$	−1.66		+1.66
$Mg \rightarrow Mg^{2+} + 2e^-$	−2.36		+2.36
$Na \rightarrow Na^+ + e^-$	−2.71		+2.71
$K \rightarrow K^+ + e^-$	−2.92		+2.92
$Li \rightarrow Li^+ + e^-$	−2.96	↓	+2.96

* The choice of signs is arbitrary. Since we are concerned with corrosion, we will use the middle column.

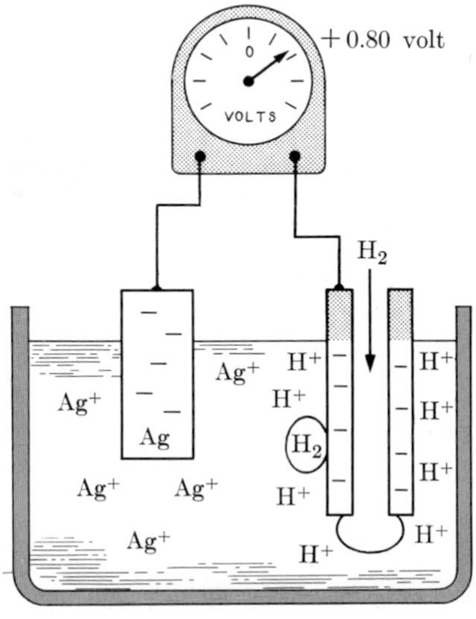

Fig. 12–2.3 Potential difference, H_2 versus Ag. H_2 produces a lower electron potential than does silver (see Table 12–2.1, center column with electrochemical notation) and is therefore the anode. Silver is the cathode.

The standard 1-molar potential, $\mathscr{E}_0$, is obtained from Table 12–2.1. The denominator, n, is the number of electrons removed per ion, for example, $n = 2$ for Zn^{2+} in Eq. (12–2.1). In dilute solutions, C is less than one mole per liter and $\mathscr{E} < \mathscr{E}_0$, since Eq. (12–2.5) is written to match the signs used in corrosion calculations (middle column of Table 12–2.1).

Cathode reactions   In our consideration of corrosion, Eq. (12–1.2) is the prime anode reaction. However, a variety of cathode reactions must receive attention, since an electron-consuming (cathodic) reaction always accompanies the anodic (electron-producing) reactions of corrosion. The major cathodic reactions are

Electroplating: $$M^{n+} + n\,e^- \rightarrow M \qquad (12\text{--}1.1)$$

Hydrogen generation: $$2\,H^+ + 2\,e^- \rightarrow H_2\uparrow \qquad (12\text{--}2.4)$$

Water decomposition: $$2\,H_2O + 2\,e^- \rightarrow H_2\uparrow + 2(OH)^- \qquad (12\text{--}2.7)$$

Hydroxyl formation: $$O_2 + 2\,H_2O + 4\,e^- \rightarrow 4(OH)^- \qquad (12\text{--}2.8)$$

Water formation: $$O_2 + 4\,H^+ + 4\,e^- \rightarrow 2\,H_2O \qquad (12\text{--}2.9)$$

Each reaction consumes electrons.

The reaction that predominates depends upon the variables of the electrolytic environment, such as temperature and concentration. Obviously, metal ions must be present for the first of these five reactions to occur. Furthermore, as a metal–ion concentration increases, those ions consume more of the electrons at the cathode. This will be important to us when we consider concentration cells shortly. Reaction (12–2.9) requires the presence of oxygen and a low pH (i.e., acidic solution). Equation (12–2.8) becomes predominant in alkaline or neutral environments if oxygen is present (Fig. 12–2.4). This reaction will be important to us when we consider oxidation cells. Reaction (12–2.7) is encountered in anaerobic conditions, particularly if sulfur or other hydrogen-consuming materials are present.

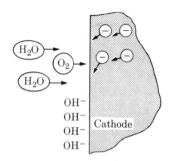

Fig. 12–2.4 Hydroxyl formation at the cathode. The rate of reaction (12–2.8) increases with increased oxygen content. It occurs at the cathode, where the electrons are consumed. If the electrons are depleted from this electrode by a reversed d.c. current, Eq. (12–2.8) is reversed and O_2 is released (Study Problem 12–2.4).

Rust   Figure 12–2.5 presents schematically the mechanism of iron rusting. Iron has an electrode potential ϕ of -0.44 volts in Table 12–2.1 when it forms Fe^{2+} (and in fact will be oxidized further to Fe^{3+} if the electrons can be consumed). Equation (12–2.8) is cathodic, thus consuming electrons; therefore, the following reactions occur in oxygen-enriched water.

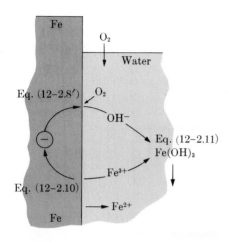

Fig. 12-2.5   Rust formation. The oxidation of iron produces iron ions and electrons. The electrons are combined with oxygen and water at the cathode (Eq. 12-2.8) to form $(OH)^-$ ions. Rust contains a combination of Fe^{3+} and $(OH)^-$ ions.

Anode: $Fe \rightarrow Fe^{3+} + 3\,e^-$ (12-2.10)

Cathode: $3\,e^- + \frac{3}{2}\,H_2O + \frac{3}{4}\,O_2 \rightarrow 3(OH)^-$ (12-2.8')

Precipitation: $Fe^{3+} + 3(OH)^- \rightarrow Fe(OH)_3\downarrow$ (12-2.11)

The hydrated form of the final product, $Fe(OH)_3$, is insoluble in water and therefore precipitates. It is *rust*; it will form on iron in air even if no more than an adsorbed layer of moisture is present as an electrolyte.

Example 12-2.1 A standard cell (1-molar, 25°C) is made in which the anode reaction is Eq. (12-2.1) and the cathode reaction is Eq. (12-2.8). What potential is established between the two electrodes?

Solution: From Table $12-2.1$ (where $H_2 \rightarrow 2\,H^+ + 2\,e = 0.0$ V), the two *anode* reactions are

$(12-2.1)$: $Zn \quad \rightarrow Zn^{2+} + 2\,e^- \qquad \mathscr{E}_0 = -0.76$ V;

$(12-2.8)$: $4(OH)^- \rightarrow O_2 + 2\,H_2O + 4\,e^- \qquad \mathscr{E}_0 = +0.40$ V.

Difference $= -1.16$ V.

Comment. Since Eq. (12-2.8) is "above" Eq. $12-2.1$ in Table 12-2.1, it becomes a *cathode* reaction and its direction is reversed to one of consuming electrons. ◀

Example 12-2.2 Determine the electrode potential (with respect to hydrogen) for a chromium electrode in a solution containing 2 g of Cr^{2+} ions per liter.

Solution: Since the atomic weight of chromium is 52 g,

$$C = (2 \text{ g/l})/(52 \text{ g/mole}) = 0.0385 \text{ M}.$$

$$\mathscr{E} = -0.74 \text{ V} + (0.0257 \text{ V}/2)(\ln 0.0385)$$

$$= -0.78 \text{ V}. \quad ◀$$

Example 12–2.3 Describe how hydrogen and oxygen may be used to operate a fuel cell.

Explanation: Electrons must be stripped from the hydrogen and returned from the external circuit to combine with oxygen. The two half-cell reactions are

Anode: $\quad\quad\quad\quad\quad$ $2\,H_2 \rightarrow 4\,H^+ + 4\,e^-$ $\quad$ (0.000 volts) Reference

Cathode: $\quad$ $O^2 + 4\,H^+ + 4\,e^- \rightarrow 2\,H_2O$ $\quad\quad$ (1.23 volts)

A cell with 1.23 volts would be expected theoretically, with standard 1-molar electrolytes. The electrode materials must be a conductive, chemically inert, porous material through which the gas can be fed. ◀

12–3   TYPES OF GALVANIC CELLS

Galvanic corrosion cells may be categorized in three different groups: (1) *composition* cells, (2) *stress* cells, and (3) *concentration* cells. Each produces corrosion because one half of the couple acts as the anode, and the other half serves as the cathode. Only the anode is corroded, and then only when it is in electrical contact with a cathode. If the anode were present alone, it would quickly come to equilibrium with its environment (see Eq. 12–2.3, where only iron is present).

Composition cells   A composition cell may be established between any two *dissimilar* metals. In each case the metal lower in the electromotive series as listed in Table 12–2.1 acts as the anode. For example, on a sheet of *galvanized* steel (Fig. 12–3.1), the zinc coating acts as an anode and protects the underlying iron even if the surface is not completely covered, because the exposed iron is the cathode and does not corrode. Any corrosion that does occur is on the anodic zinc surface. So long as zinc remains it provides protection to adjacent exposed iron.

Conversely, a *tin* coating on sheet iron or steel provides protection only so long as the surface of that metal is completely covered. However, if the surface coating is punctured, the tin becomes the cathode with respect to iron, which acts as the anode (Fig. 12–3.2). The galvanic couple that results produces corrosion of the iron.

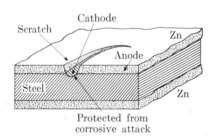

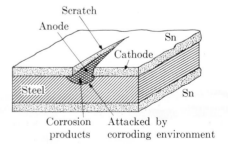

Fig. 12–3.1 Galvanized steel (cross section). Zinc serves as the anode; the iron of the steel serves as the cathode. Therefore the iron is protected even though it is exposed where the zinc is scraped off.

Fig. 12–3.2 Tinplate (cross section). The tin protects the iron while the coating is continuous. When the coating is broken, the iron of the steel becomes the anode and is subject to accelerated corrosion.

Since the small anodic area must supply electrons to a large cathode surface, very rapid localized corrosion can result.

Other examples of galvanic couples often encountered are (1) steel screws in brass marine hardware, (2) Pb–Sn solder around copper wire, (3) a steel propeller shaft in bronze bearings, and (4) steel pipe connected to copper plumbing. Each of these is a possible galvanic cell unless protected from a corrosive environment. Too many engineers fail to realize that the contact of dissimilar metals is a potential source of galvanic corrosion. Recently, in an actual engineering application, a brass bearing was used on a hydraulic steering mechanism made of steel. Even in an oil environment, the steel acted as an anode and corroded sufficiently to permit leakage of oil through the close-fitting connection.

Galvanic cells can be microscopic in dimension, because each phase has its individual composition and structure; therefore, each possesses its own electrode potential. As a result, galvanic cells can be set up in two-phase alloys when those metals are exposed to an electrolyte. For example, the pearlite of Fig. 9–8.1 reveals the carbide lamellae because the carbide was the anode in the electrolyte that was used as an etch.* Figure 12–3.3 shows the microstructure of an Al–Si casting alloy. Again we depend on corrosion to reveal the two phases. One is the anode, the other the cathode.

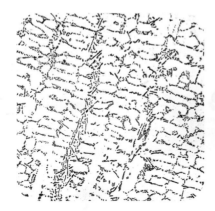

Fig. 12–3.3 Galvanic microcells (Al–Si alloy). Any two-phase alloy is more subject to corrosion than is a single-phase alloy. A two-phase alloy provides anodes and cathodes. (Alcoa Research Laboratories.)

Heat treatment may affect the corrosion rate by altering the microstructure of the metal. Figure 12–3.4 shows the effect of tempering on the corrosion of a previously quenched steel. Prior to tempering reactions, the steel contains a single phase, martensite. The tempering of the martensite produces many galvanic cells and grain boundaries of ferrite and carbide, and the corrosion rate is increased. At higher temperatures the coalescence of the carbides reduces the number of galvanic cells and the number of grain boundaries, which decreases the corrosion rate markedly.

---

* The etch was 4% picral. The carbides are darkened because a corrosion reaction product remains on the surface. The electrode potentials of ferrite and carbide are sufficiently close so that with other electrolytes their cathodic and anodic roles may be interchanged.

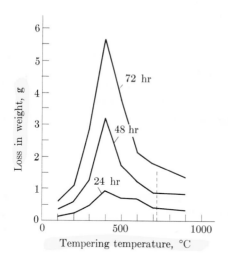

Fig. 12–3.4   Microcells and corrosion. After quenching, only martensite exists. After intermediate-temperature tempering, many small galvanic cells exist as a result of the fine (α + carbide) structure in tempered martensite. After high-temperature tempering, the carbide is agglomerated and fewer galvanic cells are present. (Adapted from F. N. Speller, *Corrosion: Causes and Prevention*, McGraw-Hill.)

Fig. 12–3.5   Age-hardening and corrosion (schematic). The single-phase, quenched alloy has a lower corrosion rate than the subsequent two-phase modifications.

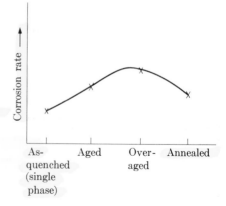

When only a single phase is present, the corrosion rate of an age-hardenable aluminum alloy is low (Fig. 12–3.5), but the corrosion rate is significantly increased with precipitation of the second phase. Still greater agglomeration of the precipitate once again decreases the rate, but never to as low a level as in the single-phase alloy. The maximum corrosion rate occurs in the over-aged alloy.

Stress cells   As shown in Fig. 4–4.9, where the grain boundaries had been etched (i.e., corroded), the atoms at the boundaries between the grains have an electrode potential different from that of the atoms within the grains; thus an anode and a cathode were developed (Fig. 12–3.6). The grain-boundary zone may be considered to be stressed, since the atoms are not at their positions of lowest energy.

The effect of internal stress on corrosion is also evident after a metal has been *cold-worked*. A very simple example is shown in Fig. 12–3.7(a), where strain-hardening

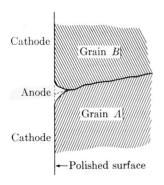

Fig. 12–3.6   Grain-boundary corrosion. The grain boundaries served as the anode because the boundary atoms have a higher energy. (Cf. Fig. 4–4.9.)

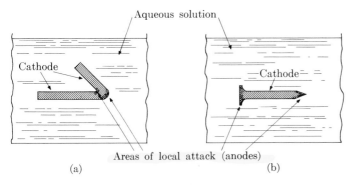

(a)   (b)

Fig. 12–3.7   Stress cells. In these two examples of strain hardening, the anodes are in the more highly cold-worked areas. The electrode potential of a deformed metal is higher than that of an annealed metal.

exists at the bend of an otherwise annealed wire. The highly cold-worked metal serves as the anode and the unchanged metal as the cathode.*

The engineering importance of the effects of stress on corrosion is plain. When engineering components must be used in a corrosive environment, the presence of stress may significantly accelerate the corrosion rate.

Concentration cells   According to the Nernst equation (12–2.5), an electrode in a dilute electrolyte is anodic with respect to a similar electrode in a concentrated electrolyte. We may view this in terms of Fig. 12–3.8 and Eq. (12–3.1):

$$Cu^0 \underset{\text{Conc.}}{\overset{\text{Dilute}}{\rightleftharpoons}} Cu^{2+} + 2\,e^-. \tag{12–3.1}$$

The metal on side (D) of the figure is in the more dilute Cu^{2+} solution. Therefore, reaction (12–3.1) readily proceeds to the right. The metal on side (C) is in a solution

---

* *Corrosion in Action*, published by the International Nickel Company, demonstrates (with illustrative experiments) the effect of cold work on galvanic corrosion.

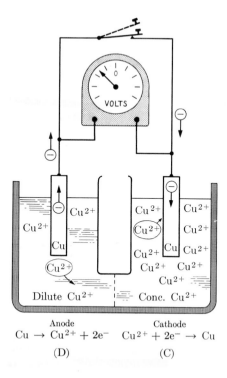

Fig. 12-3.8 Concentration cell. When the electrolyte is not homogeneous, the less concentrated area becomes the anode.

$$\text{Dilute } Cu^{2+} \qquad \text{Conc. } Cu^{2+}$$

Anode	Cathode
$Cu \rightarrow Cu^{2+} + 2e^-$	$Cu^{2+} + 2e^- \rightarrow Cu$
(D)	(C)

with a higher concentration of Cu^{2+}. Therefore, reaction (12-3.1) more readily plates copper on that electrode. The electrode in the concentrated electrolyte is protected and becomes the cathode; the electrode in the dilute electrolyte undergoes further corrosion and becomes the anode.

The concentration cell accentuates corrosion but it accentuates it where the concentration of the electrolyte is lower.

Concentration cells of the above type are frequently encountered in chemical plants, and also under certain flow-corrosion conditions. However, in general, they are of less widespread importance than are *oxidation-type concentration cells*. When oxygen in the air has access to a moist metal surface, corrosion is promoted. However, the most marked corrosion occurs in the part of the cell with an oxygen deficiency.

This apparent anomaly may be explained on the basis of the reactions at the cathode surface, where electrons are consumed. Equation (12-2.8) is restated below because it indicates the role of O_2 in promoting corrosion in oxygen-free areas:

$$2\,H_2O + O_2 + 4\,e^- \rightleftarrows 4(OH)^-.$$

Since this cathode reaction, which requires the presence of oxygen, removes electrons from the metal, more electrons must be supplied by adjacent areas that do not have as much oxygen. The areas with less oxygen thus serve as anodes.

The oxidation cell accentuates corrosion but it accentuates it where the oxygen concentration is lower. This generalization is significant. Corrosion may be accelerated

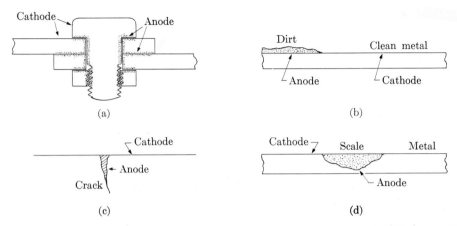

(a)   (b)

(c)   (d)

Fig. 12–3.9 Oxidation cells. Inaccessible locations with low oxygen concentrations become anodic. This situation arises because the mobility of electrons and metal ions is greater than that of oxygen or oxygen ions.

in apparently inaccessible places such as cracks or crevices, and under accumulations of dirt or other surface contaminations (Fig. 12–3.9) because the oxygen-deficient areas serve as anodes. This frequently becomes a self-aggravating situation, because accumulation of rust or scale restricts the access of oxygen and establishes an anode, to promote still greater accumulation. The result is localized *pitting* due to non-uniform corrosion (Fig. 12–3.9d), and the useful life of the product is thereby reduced to a greater extent than the weight loss would indicate.

Example 12–3.1 Copper concentrations of 0.03 M and 0.002 M occur in the electrolyte at the two ends of a copper wire. (a) What electrode potential develops between the two ends? (b) Which end will be corroded?

Solution:

a)
$$\mathscr{E}_{0.03} = +0.34 \text{ V} + (0.0257/2)(\ln 0.03) \ \ = 0.295 \text{ V}$$
$$\mathscr{E}_{0.002} = +0.34 \text{ V} + (0.0257/2)(\ln 0.002) = 0.260 \text{ V}$$
$$\Delta = \underline{35 \text{ mV.}}$$

b)   The anode is at the end with dilute electrolyte. ◀

•12–4   CORROSION RATES

The reader probably has already raised some questions about Table 12–2.1. Aluminum is more anodic than iron; why does it corrode less than iron? How can chromium impart corrosion resistance to stainless steels when its electrode potential is 0.3 volts less noble than iron? To answer these questions, we must consider factors that affect current densities.

Current density   An open circuit between an iron anode and a hydrogen cathode (Fig. 12–2.2) produces an electrode potential difference of − 0.44 volts under standard conditions (Table 12–2.1). In effect, we have a battery that will produce a current if the two electrodes are connected in a circuit. Of course, the voltage difference will decrease if a connection is made.

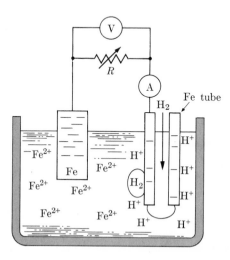

Fig. 12–4.1   Current density measurement (schematic). With direct contact ($R = 0$, and therefore, $\Delta\phi = 0$), the corrosion current density i_{co} is established. For this couple, it is 2 amp/m². The current density drops to 0.1 amp/m² when the resistance is adjusted to give $\Delta\phi = 0.2$ V (Fig. 12–4.2).

Figure 12–4.1 is another presentation of Fig. 12–2.2, but with a variable resistor included, and with an iron tube for establishing the hydrogen cathode. Experiments show that if a direct contact is made between the two electrodes ($R = 0$), then the current density i on the electrodes is 2 amp/m², i.e., the current I (read with an ammeter) per unit area. This is called the *corrosion current density*, i_{co}. Of course the potential difference, $\Delta\phi$, between the two electrodes becomes 0 V with direct contact. With an increased resistance, the potential difference increases until $\Delta\phi = -0.44$ V when $R = \infty$ as shown in Fig. 12–2.2.

The corrosion engineer uses Fig. 12–4.2 to show the relationships between current density and potential difference. Although a mathematical presentation can be made,* our interest will focus on the two curves. Curve C is for the cathode, curve A is for the anode. The curves usually cross as straight lines at 2 amp/m² (with electrodes of hydrogen and iron and under standard conditions). At $\Delta\phi = 0.2$ V (dots of Fig. 12–4.2), the current density drops to 0.1 amp/m².

Of course, other couples will be different. For example, the A-curve for zinc lies appreciably below the A-curve for iron. As a result, the corrosion current density

---

* The current density i changes according to

$$i = i_0 e^{\phi/B},\qquad(12\text{–}4.1a)$$

where i_0 is the exchange current density, ϕ is the overpotential, and B is the Tafel constant. (See a textbook on corrosion.) Alternatively,

$$\ln i = C + \phi/B.\qquad(12\text{–}4.2b)$$

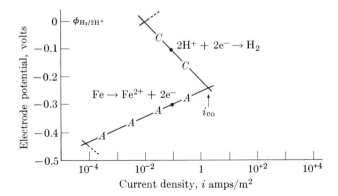

Fig. 12–4.2 Current density versus potential difference (cf. Fig. 12–4.1). There is a corrosion current, i_{co}, of 2 amp/m^2 when the anode and the cathode of Fig. 12–4.1 are in direct contact. The current density decreases when the potential difference between the anode, A, and cathode, C, is increased. (It is only 0.1 amp/m^2 when the potential differential is 0.2 volts ($C = -0.1$V, and $A = -0.3$V).)

i_{co} will be much higher if the iron of Fig. 12–4.1 is replaced by zinc; ($i_{co} = \sim 105$ amp/m^2 for Zn–H$_2$). In general, anode and cathode curves cross at higher current densities when their open-circuit electrode potential differences are greater. Thus, from Table 12–2.1, we expect a greater current density for an iron/copper couple [$\Delta\phi = -0.44$ V $- (+0.34$ V)] than for an iron/hydrogen couple [$\Delta\phi = -0.44$ V $- 0$], or an iron/nickel galvanic cell [$\Delta\phi = -0.44$ V $- (-0.25$ V)].

The above discussions assumed the anode and cathode had equal areas. Since the anode and cathode seldom have the same areas, we observe a higher current density, i, on the electrode with the smaller area. (The total current, I, must be the same for each electrode, since we are not storing charge.) This accelerates corrosion markedly when the anode is the smaller electrode. Consider the steel screw that has been used inadvisedly in some brass marine hardware, or the scratch through the "tin-plate" of Fig. 12–3.2. In either case, the total number of electrons for the corrosion current must be obtained by corroding metal in a small anodic area, so the corrosion rate is very high locally. Corrosion thus penetrates through the sheet steel in Fig. 12–3.2. extremely rapidly; also, the steel screw is destroyed in a short time when it is immersed with brass in sea water.

Cathodic polarization When a corrosion current is predicted from Fig. 12–4.2, one must assume that ions can leave or approach the electrode surfaces at a rate sufficient to match the corrosion current. This assumption is reasonably valid at the anode, where the corroded ions enter a dilute electrolyte (as is usually the case in corrosion). However, the H$^+$ ions, the dissolved O$_2$, or other dilute reactants can easily be depleted from the cathode region to cause diffusional delays. This has become known over the years as *cathodic polarization* and may be described by modifying the cathode curve, C, of Fig. 12–4.2, as shown in Fig. 12–4.3. In a stagnant situation, at

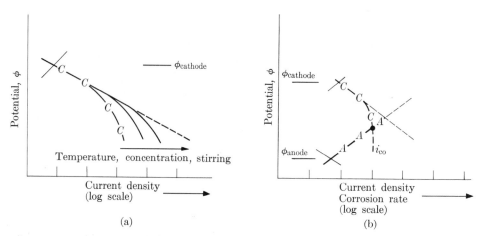

Fig. 12–4.3   Polarization. (a) The current density is normal (dashed line) only when the electrolyte concentration at the cathode and the temperature are high, and the electrolyte is not stagnant. Otherwise the current density lags. (b) When the current density lags, the corrosion current, i_{co}, is less than predicted. This polarization is eliminated if the service environment is altered to increase the availability of reactants.

low temperatures, or with dilute concentrations, the reactants are not continuously available to consume the electrons, and the current density lags behind the expected cathode curve. Thus, the resulting corrosion current density at the intersection of the two curves is lower than anticipated. Since this leads to slower corrosion, we would like to capitalize on it.* In the steam power plant this is done by deaerating the feed water. However, a complication arises from the situation shown in Fig. 12–4.3(a) because, under stagnant conditions, corrosion may be negligible; then a change of environment, such as a higher temperature or electrolyte movement, may suddenly decrease the cathodic polarization. The rate of corrosion may thus change by a couple of orders of magnitude. This accounts for some of the erratic corrosion rates that are commonly encountered in service.

Passivation   It was stated in the previous paragraph that the electrolyte is commonly dilute with respect to the anode product. Thus, anodic polarization is uncommon. However, the anode may become isolated from the electrolyte by noncorrosive reactions. These are tremendously important for corrosion control because they alter the rate of attack.

Consider that the iron of Fig. 12–4.2 is replaced by stainless steel. In the presence of excess oxygen, an oxide surface film containing chromium forms on the anode to isolate it from the electrolyte. As the oxidation potential is increased, the anode, current density curve, A, is markedly affected (Fig. 12–4.4). We call this *passivation*. Its effect on the corrosion rate is pronounced because the corrosion current density, i_{co}, may be reduced by two or more orders of magnitude, from (A) to (P).

The chromium of stainless steel produces passivation because of its strong attraction for oxygen. Thus we will find, under oxidizing conditions, that stainless steel

---

* It is undesirable in a dry cell or battery, where we utilize corrosion to produce a current.

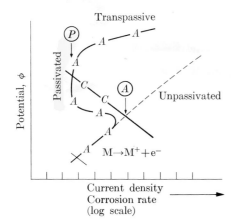

Current density
Corrosion rate
(log scale)

Fig. 12–4.4 Passivation. Metals such as aluminum, titanium, and stainless steel form a persistent protective film on the anode surface in oxidizing environments. The anode current density departs from the normal relationship (dashed line). Thus, the corrosion current density (where the two curves cross) is reduced significantly. If the oxide film is destroyed, the corrosion rate increases quickly from the passivated corrosion rate ⓟ to the activated corrosion rate Ⓐ.

does not have a significant corrosion rate (but it is not zero). Likewise we find that many metals are only slowly attacked by HNO_3 because of its oxidizing nature. The same metals will be rapidly attacked by HCl, which does not have a supply of oxygen.

Equally significant is the fact that aluminum and titanium will form a protective film of Al_2O_3 (and TiO_2) on their surfaces. This film is so protective that the corrosion current density is almost nil, and we can make boats of aluminum!* The commercial process that produces a protective oxide coating on aluminum is called *anodizing*.

Example 12–4.1 The average corrosion current density for iron is 2 amp/m².
a)   How many Fe^{2+} ions are formed per second from each sq. mm of surface?
b)   What is the monthly corrosion penetration if the anodic area is limited to 10% of the surface?

Solution

a) $(2 \text{ amp/m}^2)/(2 \text{ el./Fe}^{2+})(0.16 \times 10^{-18} \text{ amp·sec/el.}) = 6 \times 10^{18} \text{ Fe}^{2+}/\text{sec·m}^2$

$$= 6 \times 10^{12} \text{ Fe}^{2+}/\text{sec·mm}^2.$$

b)   Local corrosion rate $= (6 \times 10^{18} \text{ Fe/sec·m}^2)/(0.10)$

$$= 6 \times 10^{19} \text{ Fe/sec·m}^2 \text{ locally};$$

$$\frac{(6 \times 10^{19} \text{ Fe/sec·m}^2)(2.6 \times 10^6 \text{ sec/mo})(10^3 \text{ mm/m})}{(0.6 \times 10^{24} \text{ Fe/55.85 g})(7.86 \times 10^6 \text{ g/m}^3)} = \sim 2 \text{ mm/mo.}$$

Comment. Any factor that localizes corrosion (see Fig. 12–3.9d) greatly affects the life of a product because it may fail although much of the metal has been unaffected. Examine the rust spots on the fender of a car. ◄

---

*   This is another example of why the casual student can be confused by the corrosion mechanism. However, to summarize: (1) oxygen accelerates corrosion; (2) in an oxidation cell, the corrosion is accelerated where the oxygen *ain't*, because the oxygen contributes to the cathodic side of the cell and drains the electrons from the anode where the corrosion occurs; (3) finally, some metals (e.g., aluminum and stainless steels) can be passivated because they react with oxygen to form protective surface films. An electrically isolated metal cannot corrode.

Example 12–4.2 The Al_2O_3 layer that forms on aluminum passivates the metal. How thick must the layer be to limit the current density to 10^{-4} amp/m^2 in a cell that develops 2 volts?

Solution: Basis: 1 cm^2 = 10^{-4} m^2; therefore, 10^{-8} amp. Since $E = IR$,

$$R = 2V/10^{-8} \text{ amp} = 2 \times 10^8 \text{ ohm.}$$

From $\rho = RA/L$ and Appendix C,

$$L = (2 \times 10^8 \text{ ohm})(10^{-4} \text{ m}^2)/(>10^{12} \text{ ohm·m})$$

$$= <2 \times 10^{-8} \text{ m, or 20 nm.}$$

Comment. This is well below the wavelength of light (~ 500 nm). As a result we do not see it. Even so, it lowers the current density and the corrosion rate to insignificant levels. Of course, if the aluminum encounters an environment such as an alkaline solution that dissolves the Al_2O_3 film, the passivation is lost. ◀

12–5 CORROSION CONTROL

EXCEPT FOR HEAT, CORROSION (OR TORCH, ETC.)
Only in the absence of an electrolyte can corrosion be completely avoided. Even this is difficult. As is well known, tools hanging in a workshop can rust from adsorbed moisture films. The tool manufacturer is aware of this and therefore coats his products with an oil or grease film before shipping them to the retailer or customer.

Since corrosion is a galvanic action, there must be two kinds of metal in order for corrosion to proceed. The design engineer is normally cognizant of this and therefore will avoid the "brass bolt–steel washer" pitfall. Less familiar to many technical people is the fact that a cathode and anode may develop in a single material because of microstructure, stress concentrations, or electrolyte heterogeneities (Section 12–3). These must be considered by those seeking to control corrosion.

In addition to (1) *providing protective coatings* and (2) *avoiding of galvanic couples*, it is also possible to minimize corrosion by (3) *introducing galvanic protection*. These three procedures will be the subject of this section.

Protective coatings Protecting the surface of a metal is probably the oldest of the common procedures for corrosion control. A coat of paint, for example, isolates the underlying metal from the corroding electrolyte. The chief limitation of this method is the service behavior of the protective coating. The greasy film cited in the first paragraph of this section is obviously not very permanent. High temperatures or abrasive wear place limitations on organic coatings.

But protective coatings need not be limited to organic materials. For instance, tin can be used as an "inert" coating on a steel base. Copperplate, nickelplate, and silverplate are other examples of corrosion-resistant surfaces. Some metals may be applied as hot-dip coats, such as passing steel wire or sheet through molten zinc, a process called *galvanizing*. Inert ceramic materials can also be used for protective coatings. For example, true enamels are oxide coatings applied as powdered glass and fused to become a vitreous surface layer. A comparison of the advantages and disadvantages of the several categories of protective coatings is given in Table 12–5.1.

Passivation, discussed in the previous section, also provides a protective film, admittedly thin. Its importance, particularly for aluminum and for chromium-bearing stainless steels, is revealed in Table 12–5.2. This galvanic series differs from Table

Table 12–5.1
Comparison of inert protective coatings

Type	Example	Advantages	Disadvantages
Organic	Baked "enamel" paints	Flexible Easily applied Cheap	Oxidizes Soft (relatively) Temperature limitations
Metal	Noble metal electroplates	Deformable Insoluble in organic solutions Thermally conductive	Establishes galvanic cell if ruptured
Ceramic	Vitreous enamel oxide coatings	Temperature resistant Harder Does not produce cell with base	Brittle Thermal insulators

Table 12–5.2
Galvanic series of common alloys (ELECTROLYTE : SEAWATER)

Graphite	Cathodic ↑	Nickel—A
Silver		Tin
12% Ni, 18% Cr, 3% Mo steel—P		Lead
20% Ni, 25% Cr steel—P		Lead–tin solder
23 to 30% Cr steel—P		12% Ni, 18% Cr, 3% Mo steel—A
14% Ni, 23% Cr steel—P		20% Ni, 25% Cr steel—A
8% Ni, 18% Cr steel—P		14% Ni, 23% Cr steel—A
7% Ni, 17% Cr steel—P		8% Ni, 18% Cr steel—A
16 to 18% Cr steel—P		7% Ni, 17% Cr steel—A
12 to 14% Cr steel—P		Ni-resist
80% Ni, 20% Cr—P		23 to 30% Cr steel—A
Inconel—P		16 to 18% Cr steel—A
60% Ni, 15% Cr—P		12 to 14% Cr steel—A
Nickel—P		4 to 6% Cr steel—A
Monel metal		Cast iron
Copper–nickel		Copper steel
Nickel–silver		Carbon steel
Bronzes		Aluminum alloy 2017-T
Copper		Cadmium
Brasses		Aluminum, 1100
80% Ni, 20% Cr—A		Zinc
Inconel—A		Magnesium alloys
60% Ni, 15% Cr—A	Anodic ↓	Magnesium

* Adapted from C. A. Zapffe, *Stainless Steels*, American Society for Metals.
A–active; P–passivated.

12–2.1 in that common alloys that are widely used under corrosive conditions are included along with those elemental metals that were previously cited. Note that a number of alloys are listed twice, in both their active and passivated conditions. The presence of an oxygen-containing film on the surface shifts these alloys toward the cathodic end of the series. In fact, when passivated, these steels are less corrodable than copper, bronze, and brass. This introduces an important point. A passivated metal may have the corrosion current density indicated by point $\textcircled{P}$ in Fig. 12–4.4. If service conditions change so that the protective oxide film is destroyed, the corrosion current density can change by several orders of magnitude, to point $\textcircled{A}$. Of course, this has major implications to both the design engineer and the technical manager.

Inhibitors are compounds that are added to the electrolyte to restrict corrosion of the metallic container. We are most familiar with these as *rust inhibitors*, to decrease the corrosion in automobile radiators. They may also be used in steam boilers and similar hot-water systems. An inhibitor is effective because it contains polyatomic anions that adsorb onto the surface of the metal to give a protective, oxygen-rich, monolayer film closely related to that found with passivation. Commonly, inhibitors involve compounds containing chromates, phosphates, tungstates, or ions of other highly oxidizable transition elements.

Avoidance of galvanic couples   The simplest method of avoiding galvanic couples is to limit designs to only one metal, but this is not always feasible. In special circumstances, the cells may be avoided by electrically insulating metals of different compositions.

Other, less simple, methods are frequently warranted, and *stainless steel* provides a good specific example. There are many types of stainless steel, whose chromium content varies from 13% to 27%. The purpose of the chromium is to provide a composition that will normally develop a passive surface. Many, but not all, stainless steels also contain 8% to 10% nickel, which is more noble than iron (Table 12–2.1).

• The high alloy content of such a metal as 18–8 stainless steel (so-called because it contains 18% Cr–8% Ni) causes the formation of austenite, which is stable at ambient temperatures. Such a steel is not used primarily for applications requiring high hardness, but rather in corrosive applications. Therefore carbon, which is more soluble in austenite at high than at low temperatures (Fig. 12–5.1), is kept to a minimum. If steel containing 0.1% carbon is cooled rapidly from about 1000°C ($\sim$1800°F), a separate carbide does not form and galvanic cells are not established. On the other hand, if the same steel is cooled slowly, or held at 650 $\pm$ °C ($\sim$1200°F) for a short period of time, the carbon precipitates as a chromium carbide, usually in the form of a fine precipitate at the grain boundaries (Fig. 12–5.2). In the latter case, two effects are possible: (1) galvanic cells may be established on a microscopic scale, or (2) the carbon forms chromium carbide (more stable than Fe_3C), which depletes the grain-boundary area of chromium and removes its passivating protection locally (Fig. 12–5.3). Either of these effects accentuates corrosion at the grain boundaries and is to be avoided (Fig. 12–5.4).

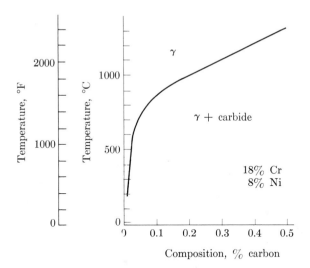

Fig. 12–5.1 Carbon solubility in austenitic stainless steel. The carbon solubility in an 18–8 type stainless steel decreases markedly with temperature. Consequently, the carbon will precipitate if cooling is not rapid. The precipitated carbide is rich in chromium. (Adapted from E. E. Thum, *Book of Stainless Steels*, American Society for Metals.)

Fig. 12–5.2 Carbide precipitation at the grain boundaries, X1500. The small carbon atom readily diffuses to the grain boundary. It will precipitate there as a chromium carbide if sufficient time is available (a few seconds at 650°C). Galvanic cells are then formed. (P. Payson, *Trans. AIME.*)

• There are several ways to inhibit intergranular corrosion; the choice, of course, depends on the service conditions:

1. *Quenching to avoid carbon precipitation.* This method is commonly used unless (a) service conditions require temperatures in the precipitation range, or (b) forming, welding, or size prevent such a quenching operation.

2. *Provision for an extremely long anneal in the carbide separation range.* This technique offers some advantage because of (a) agglomeration of the carbides, and (b) homogenization of the chromium content so that there is no deficiency at the grain boundary. However, this procedure is not common because the improvement in corrosion resistance is relatively small.

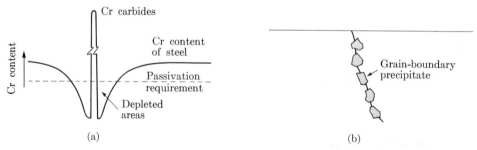

Fig. 12–5.3 Chromium depletion adjacent to the grain boundary. The carbide precipitation consumes nearly ten times as much chromium as carbon. Since the larger chromium atoms diffuse slowly, the Cr content of the adjacent areas is lowered below protection levels.

Fig. 12–5.4 Intergranular corrosion. This type of corrosion becomes severe if the steel has been heated into the carbide-precipitation range. (W. O. Binder et al., *Corrosion of Metals*, American Society for Metals, Chapter 3.)

3. *Selection of a steel with less than 0.03% carbon.* As indicated in Fig. 12–5.1, this would virtually eliminate carbide precipitation. However, such a steel is expensive because of the difficulty of removing enough of the carbon to attain this very low level.

4. *Selection of a steel with high chromium content.* A steel that contains 18% chromium corrodes less readily than a plain carbon steel. The addition of more chromium (and nickel) provides additional protection. This, too, is expensive because of the added alloy costs.

5. *Selection of a steel containing strong carbide formers.* Such elements include titanium, niobium, and tantalum. In these steels, the carbon does not precipitate at the grain boundary during cooling because it is precipitated earlier as titanium carbide, niobium carbide, or tantalum carbide, at much higher temperatures. These carbides are innocuous because they neither deplete the chromium from the steel nor localize the galvanic action to the grain boundaries. This technique is used frequently, particularly with stainless steel which must be fabricated by welding.

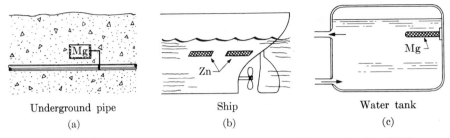

Underground pipe

(a)

Ship

(b)

Water tank

(c)

Fig. 12–5.5   Sacrificial anodes. (a) Buried magnesium plates along a pipeline. (b) Zinc plates on ship hulls. (c) Magnesium bar in an industrial hot-water tank. Each of these sacrificial anodes may be easily replaced. They cause the *equipment* to become a cathode.

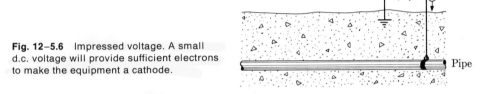

Fig. 12–5.6   Impressed voltage. A small d.c. voltage will provide sufficient electrons to make the equipment a cathode.

Pipe

Although the above examples are somewhat specific, they do indicate methods that are used to reduce the extent of corrosion in metals. The exact choice of procedure depends on the alloy and the service conditions involved.

Galvanic protection   It is possible to restrict corrosion by turning some of the mechanisms of corrosion to protective ends. A good example is the galvanized steel discussed earlier in Section 12–3. The zinc coating serves as a *sacrificial anode* that itself corrodes, instead of the underlying steel. The same method may be used in other applications. Three examples are shown in Fig. 12–5.5. An advantage of such procedures is that the spent anode can be replaced quite easily. For example, the magnesium plates in Fig. 12–5.5(a) can be replaced at a fraction of the cost of replacing the underground pipe.

A second method of galvanic protection is the use of an *impressed voltage* on the metal. This is illustrated in Fig. 12–5.6. Both the sacrificial anode and the impressed voltage methods involve the same protection principle; that is, extra electrons are supplied so that the metal becomes the cathode and the corrosion reactions do not proceed.

REVIEW AND STUDY

SUMMARY

We first considered that corrosion is the reverse of electroplating. This gave us a basis for relating the amount of corrosion to the amount of current. Two electrodes are required, plus an electrolyte.

Table 12–5.3
Summary of galvanic cells

Specific examples	Anode	Cathode
	Base metal	Noble metal
Zn versus Fe	Zn	Fe
Fe versus H_2	Fe	H_2
H_2 versus Cu	H_2	Cu
	Higher energy	Lower energy
Boundaries	Boundaries	Grain
Stresses	Cold-worked	Annealed
Stress corrosion	Stressed areas	Nonstressed areas
	Lower conc.	Higher conc.
Electrolyte	Dilute solution	Concentrated solution
Oxidation	Low O_2	High O_2
Dirt or scale	Covered areas	Clean areas

Most corrosion results from galvanic cells and the accompanying electrical currents. Two dissimilar electrodes are required; these may be provided by (1) differences in composition, (2) differences in energy level (disordered, or stressed areas), or (3) differences in electrolytic environment. They are detailed further in Table 12–5.3. The electrode that undergoes corrosion is the anode; the cathode is protected.

Oxygen accelerates the corrosion of metals; however, the corrosion is accelerated where the oxygen content is low, since the oxygen reacts on the cathode side of the galvanic cell. Oxygen forms a protective film on aluminum, titanium, and those steels that contain chromium. This phenomenon is called passivation and is the basis for the relative inertness of stainless steels.

Polarization may occur on the cathode because the reactants are depleted from the electrolyte near the electrode surface. This reduces the current density. Since the reactants (ions and oxygen) are replenished more rapidly at higher temperatures or with stirring, we find that polarization varies significantly with service conditions.

Corrosion is controlled most simply by isolating the metal surface by protective coatings. These may be organic (paint), ceramic (enamel), or other metals. Satisfactory corrosion control, however, requires that the galvanic couples be avoided. Unlike metals must be electrically separated when the corrosive environment is severe. In those severe situations where stainless steel is used, further care must be taken that galvanic cells are not developed within the microstructures (e.g., carbides versus carbon-depleted areas). In some applications, corrosion protection can be provided through sacrificial anodes, or by an impressed voltage. In both situations, the metal that is to be protected is made to be the cathode.

KEY TERMS AND CONCEPTS

Anode

Anodized

Cathode

Cathode reactions

Cells

   composition

   concentration

   galvanic

   oxidation

   stress

• Corrosion current density, i_{co}

Electrode potential, $\mathscr{E}$

Electrolyte

Electroplating

Faraday, $\mathscr{F}$

Galvanic couple

Galvanization

Hydrogen electrode (reference)

Impressed voltage

Inhibitor

Nernst equation

• Passivation

• Polarization

Rust

Sacrificial anode

Stainless steel

FOR CLASS DISCUSSION

A_{12} Distinguish between anode and cathode on the basis of electron movements. How does a cathode-ray tube fit into the definitions of Section 12–1?

B_{12} Show the origin of 96,500 coul as the value of one Faraday, $\mathscr{F}$.

C_{12} What is the source of electrons in an ordinary dry cell? (The electrolyte is a gelatinous paste containing NH_4Cl. The cathode reaction changes Mn^{4+} in MnO_2 to Mn^{2+}.)

D_{12} Why do we use the hydrogen electrode as the reference electrode rather than some other element, such as lead?

E_{12} A discarded tin-coated can is dropped in a lake. Eq. (12–2.3) develops as the anodic reaction. The cathode reaction is *not* $Sn^{2+} + 2\,e^- \rightarrow Sn$. Why? What is the probable cathode reaction?

F_{12} Under what conditions will rusting *not* occur? Why?

G_{12} It is suggested that sodium might be produced by taking an aqueous solution of NaOH ($pH = 14$) and electrolyzing it to plate-out sodium. Discuss.

H_{12} Distinguish between the type of protection given to iron by zinc and by tin.

I_{12} Plumbing codes call for an insulator such as Teflon to be placed between copper tubing and steel pipe when they are coupled in a plumbing system. Explain.

J_{12} Electrical codes call for a "jumper" to be placed around the insulated coupling just described, to provide a ground. Does this defeat the purpose of the plumbing code? Explain.

K$_{12}$ A zinc-coated nail is sheared in half and placed in an electrolyte. (a) What couples must be considered in judging where the anode will be? (b) Cite the location that will be corroded initially.

L$_{12}$ Explain why the metal in a dilute electrolyte becomes the anode to metal in a more concentrated electrolyte.

M$_{12}$ Explain why the metal in an oxygen-deficient electrolyte becomes the anode to metal in an oxygen-enriched electrolyte.

N$_{12}$ If oxygen accelerates corrosion (Eqs. 12–2.8 and 12–2.9), why does the corrosion occur in the oxygen-lean area?

O$_{12}$ How do barnacles accelerate corrosion on a ship hull?

• P$_{12}$ Why does aluminum corrode less than iron when used in kitchen utensils?

• Q$_{12}$ Why is corrosion accelerated when the anode is smaller than the cathode, but not when the situation is reversed?

• R$_{12}$ The headache of metal-can manufacturers is "pinholes" in the tinplate. Explain.

• S$_{12}$ A flashlight with an old drycell will dim during use. If left unused for a short time, it will shine more brightly again. Explain.

T$_{12}$ Steel piles corrode at the tide level much more than at lower levels. Why?

• U$_{12}$ A vitreous enamel (glass) is used to coat the inside of a home hot-water heater. How should the thermal expansion of the glass and metal compare? (Check Chapter 8.)

V$_{12}$ An enterprising mechanic suggests the use of a magnesium drain plug for the crankcase of a car as a means of combating engine corrosion. Discuss.

W$_{12}$ Undercoatings of various types are used on new cars. Under what conditions are they helpful? Under what conditions are they detrimental?

X$_{12}$ Is it possible to use an impressed voltage to protect a ship hull? Explain.

Y$_{12}$ Cite three examples of corrosion from your experience. Describe the nature of the deterioration, and account for the corrosion.

• Z$_{12}$ A stainless-steel sheet is welded into a circular duct. After a period of time, rust appears along a band that is parallel to each side of the weld, about a centimeter away. Why did this occur? Could it have been avoided?

STUDY PROBLEMS

12–1.1 How many coulombs are required to plate each gram of nickel from an Ni^{2+} electrolyte?
Answer: 3300 coul

12–1.2 What current is required if a 0.1-mm coating of nickel is to be plated onto a piece of copper (1230 mm^2) in a period of 90 minutes?

12–1.3 A metallic reflector that has 1.27 m^2 of surface is being electroplated. The current is 100 amp. (a) How many grams of Cr^{2+} must be added to the electrolyte per hour of plating? (b) What thickness will form per hour?

Answer: (a) 97 g   (b) 0.01 mm (or 10 μm)

12–1.4 How long does it take to electroplate one gram of chromium from a Cr^{2+} electrolyte with a current of 1 ampere?

12–1.5 Bubbles of hydrogen are evolved from the cathode at a rate of one per second. Each is 1 mm^3 when it emerges from the surface of the electrolyte. What electron current is involved? (1 mole of gas at STP = 0.0224 m^3.)

Answer: 8.6 mA

12–1.6 Aluminum is refined by first fluxing Al_2O_3 in a molten salt and then electroplating the metal onto the cathode. What current density, amp/mm^2, does it take to produce 1.0 grams per day per mm^2 of electrode area?

Answer: 0.124 amp/mm^2

12–2.1 The electrodes of a standard galvanic cell are nickel and magnesium. What potential difference will be established?

Answer: 2.1 volts (with nickel as the cathode)

12–2.2 In order to determine the electrode potential of cadmium, Fig. 12–2.1 was duplicated with cadmium and silver, rather than zinc and copper. The voltmeter read − 1.20 volts. What is the cadmium electrode potential with respect to hydrogen?

12–2.3 What copper concentration (g/l) is required in an electrolyte for it to have an electrode potential of 0.32 V (with respect to hydrogen)?

Answer: 13.4 g/liter

12–2.4 Describe how water may be electrolyzed to produce H_2 and O_2.

12–2.5 In Fig. 12–2.1, copper is replaced by gold and zinc by tin. Based on the half-cell reactions of Table 12–2.1, will the weight of tin corroded per hour be greater or less than the weight of gold which is plated?

Answer: 10 w/o more gold

12–3.1 The cell of Fig. 12–2.1 has electrolytes that contain 3 g of zinc per liter and 23 g of copper per liter. What is the cell voltage?

Answer: − 1.13 V

12–3.2 An iron wire lies in some mine waste water with a concentration gradient of Fe^{2+} ions from 0.09 M at one end (A) to 0.005 M at the other end (B). (a) Which end is the anode? (b) What is the potential difference between the two ends?

12–3.3 Two pieces of metal, one copper and the other zinc, are immersed in sea water and connected by a copper wire. Indicate the galvanic cell by writing the half-cell reactions (a) for the anode and (b) for the cathode; also, by indicating (c) the direction of the electron flow in the wire, and (d) the direction of the "current" flow in the electrolyte. (e) What metal might be used in place of copper so that zinc changes polarity?

12–3.4 Consider Eqs. (12–1.1), (12–2.4), (12–2.7), (12–2.8), (12–2.9), and the corrosion of iron. Which, if any, of these, provide the prevalent cathode reaction if (a) iron is in water along a seashore? (b) Iron is in a copper sulfate solution? (c) Iron is a can containing a carbonated beverage? (d) Iron is the blade of a rusty spade? (e) Iron is a scratched car fender?

12–3.5 Select the anode from each of the following pairs. (Factors other than those cited are identical for the two members of each pair.) (a) Copper vs. silver (b) Copper vs. iron (c) Zn (in 0.1–M solution) vs. Zn (in 0.2–M solution) (d) Grain of nickel vs. grain boundary of nickel (e) Product being nickel-plated vs. nickel electrode • (f) Passivated metal vs. activated metal (g) Stressed aluminum vs. stress-free aluminum.

• **12–4.1** Pit corrosion punctured the hull of an aluminum boat after 12 months in sea water. The aluminum sheet was 1.1-mm thick. The average diameter of the hole is 0.2 mm. (a) How many atoms were removed from the pit per second? (b) What was the corrosion current density?

Answer: (a) 6.6×10^{10}/sec (b) 1 A/m^2

• *12–4.2* With a current density of 0.1 A/m^2, how long will it take to corrode an average of 0.1 mm from the surface of aluminum?

• **12–4.3** An average current density of 10^2 amp/m^2 is used to build up an anodized coating on aluminum. The Al_2O_3 coating is to be 1-μm thick. What time is required?

Answer: 215 sec

•CHAPTER 13

Cast Iron, Concrete, Wood, and Composites

(Widely Used, More Complex Materials)

The title of this chapter cites four examples of widely used materials in engineering and technical design. Cast iron dates from early history and is the least expensive of metallic materials. At the same time, it has many advantages as illustrated by the fact that $>80\%$ of an automobile engine is this material. More tons of concrete are produced each year than any other engineering material (if one excludes the technical usage of air, fuel, and water). Wood is a replaceable resource in this day of materials shortages and has the advantage of easy processing with a minimum of energy. Composites represent an obvious example of "tailor-made" or "engineered" materials, sometimes called the "materials of tomorrow."

None of the above materials is simple. However, each provides an example of how properties relate to structures. This final chapter can serve to integrate our concepts of structure versus properties, and will introduce the technology of these common materials for students who will not encounter them in later, more specialized courses.

The contents of Chapter 13 are limited to the materials of the chapter heading but could have included a variety of other materials, such as soils, tool steels, brick, paper, and whitewares. Each is relatively complex but may be analyzed on the basis of its internal structure.

STUDY OBJECTIVES

Choose one or more of the above materials. If you include *cast iron* among your choices,

1 Distinguish among various types of cast iron on the basis of micro-structure.
2 Be able to detail with an Fe–C phase diagram how the ferritic and pearlitic versions of gray, nodular, and malleable irons may be produced.
3 Relate selected properties to the nature of the microstructure.

If you include *concrete* among your choices,

4 Recognize the features that characterize the aggregate and the particle packing.
5 Know the general nature of the hardening process for portland cement.
6 Be able to calculate unit mixes of concrete.
7 Identify factors that lead to concrete failures.

If you include *wood* among your choices,

8 Become familiar with the biological units that are found in the structure of wood.
9 Relate the structural anisotropy of wood to its directional properties.
10 Identify ways in which wood may be upgraded for improved properties and/or service.

If you include *composites* among your choices,

11 Acquaint yourself with types of composites other than the reinforced concrete and fiber-glass reinforced plastics that are cited, e.g., glass-coated steels, aluminum-coated mylar, and plastic-laminated glass.
12 Recognize the role of the interfacial boundary between the components in establishing a coherent composite.
13 Familiarize yourself with the mixture rules that apply to simple com-posites where the geometries of the reinforcement and the matrix may be defined.

•13–1   CAST IRONS

In principle, cast irons are eutectic alloys of iron and carbon. Thus, they are relatively low melting ($\sim 1200°C$, or $2200°F$). This is advantageous because they are easily melted, requiring less fuel and more easily operated furnaces. Also, the molten metal will easily fill intricate molds completely. These characteristics lead to an inexpensive material and considerable versatility in product design (Fig. 13–1.1).

Fig. 13–1.1   Cast-iron engine block (V-8). Not only is cast iron one of the most available and least expensive alloys, it has the advantage of castability into intricate designs, thus requiring a minimum of machining before assembling the final engine (Fig. 1–1.1). (Courtesy of the Ford Motor Company.)

Cast irons are somewhat more complex than the simple eutectic alloy just described. For example, most cast irons contain 1%–3% silicon. This is partially a consequence of the fact that silicon is retained with the iron in the production process, and additional efforts would be required to remove it. More important, however, is the role of silicon in the final product. First, silicon increases the strength of the ferrite within the cast iron. Second, with silicon, the low-melting eutectic is achieved with 2%–3.5% carbon rather than the 4.3% carbon indicated in Fig. 9–7.1. Finally, the silicon leads to a decomposition of the carbide to iron and graphite:

$$Fe_3C \xrightarrow{Si} 3\,Fe + C_{(gr)}. \qquad (13–1.1)$$

Graphitization   The above reaction produces graphite in a cast iron, because iron carbide, Fe_3C, is not truly stable. It is only metastable. Its instability is accentuated in the presence of silicon and with extended exposures to elevated temperatures.*

Since cast irons do graphitize, we must revise the phase diagram to be an Fe–C diagram rather than an Fe–Fe_3C diagram. This has been done in Fig. 13–1.2. Fortunately, the difference is very slight—with only one significant change. Graphite

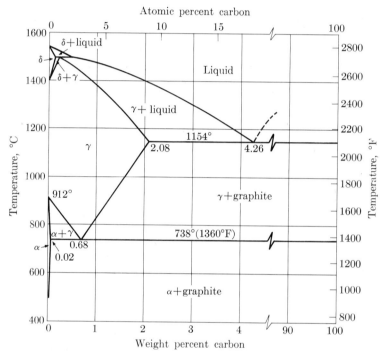

Fig. 13–1.2   Fe–C phase diagram. The Fe–Fe_3C phase diagram (Fig. 9–7.1) is not truly an equilibrium diagram, because Fe_3C eventually decomposes to iron and graphite, particularly with the higher carbon contents of cast iron. The only significant difference between this diagram and Fig. 9–7.1 is the presence of graphite as the carbon-rich phase beyond the solubility-limit curves for carbon in ferrite and for carbon in austenite.

---

* We have encountered metastable phases previously. For example, diamond is supposedly "forever"; however, given an opportunity it will alter to form graphite. Diamond is stable only when it is subjected to very high pressures and temperatures. Likewise, martensite and glass are two metastable phases. Martensite will temper to $(\alpha + \bar{C})$, and glass will crystallize if given an opportunity. Admittedly, the time requirements at 20°C may be almost infinitely long.

We used the Fe–Fe_3C diagram of Figs. 9–7.1 and 9–7.3, because under normal steel treating conditions (low Si, and short or moderate heating times) we do not observe graphitization. Thus, Fe_3C is the carbon-containing phase that is present in the microstructure of steels and influences their properties.

replaces Fe_3C! Thus, at 400°C, the amount of ferrite in cast iron containing 3.5% carbon is

$$\alpha = (100 - 3.5)/(100 - \sim 0) = 0.965,$$

and *not*

$$\alpha = (6.7 - 3.5)/(6.7 - \sim 0) = 0.48.$$

The former uses 100% C as the composition of graphite; the latter uses 6.7% C as the composition of Fe_3C.

Gray cast iron A cast iron with a high silicon content ($\sim 2\%$ Si) graphitizes so readily that Fe_3C never forms. Graphite flakes form within the metal during solidification. This is shown in Fig. 13–1.3 where we see the traces of graphite flakes in the polished 2-dimensional section of the metal. When this metal is fractured in tension, the fracture path progresses from flake to flake because the mica-like graphite is very weak in tension. Thus, a very high percentage of the fractured surface is graphite and is gray in appearance. Hence, the name, *gray iron.*

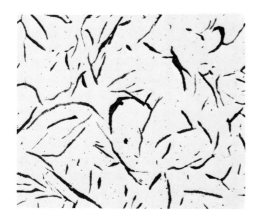

Fig. 13–1.3 Gray cast iron, X100. Carbon is present as graphite *flakes.* The high carbon content gives a eutectic composition that permits easy casting. However, the strength and ductility are less than for plain-carbon steels (J. E. Rehder, Canada Iron Foundries.)

Gray iron has almost negligible ductility because the graphite flakes are present; however, it has the merit of being an inexpensive metal. Furthermore, with these graphite flakes present, a gray iron is an excellent absorber of vibrational energy. In engineering terms, its *damping capacity* is high. Thus, this type of metal finds extensive use as a base for machine tools and heavy equipment (Fig. 13–1.4).

Gray iron may be heat-treated in a variety of ways to give pearlitic, ferritic, martensitic, and bainitic microstructures to the metal. For example, if a gray iron is heated to $\sim 750°C$ (1380°F), the resulting phases are austenite (0.75% C) and graphite. Moderately slow cooling produces *pearlite* from the austenite as described with Fig. 10–1.9. The carbide forms more rapidly than does the graphite because less extensive diffusion is required. Specifically, $(\alpha + \bar{C})$ can form without major iron diffusion. In contrast, the graphitic areas within the solid must be completely depleted of iron.

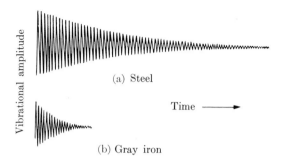

Vibrational amplitude

(a) Steel

Time ⟶

(b) Gray iron

Fig. 13–1.4   Relative damping capacity of steel and gray iron. The vibrations decay much faster in the gray iron because the graphite flakes absorb the vibrational energy. (After Wallace, *Metals Engineering Qtrly.*)

A quench from 750°C to low temperatures produces *martensite* from the austenite. Alternatively, a quench from 750°C to intermediate temperatures, followed by austempering (Section 11–5), produces *bainite*. These alternatives lead to corresponding variations in strengths and hardnesses for pearlitic gray iron, martensitic gray iron, and bainitic gray iron (Table 13–1.1), much as expected from our previous knowledge of pearlite, martensite, and bainite.

Still another microstructure is available to the engineer if the silicon is increased slightly to facilitate complete graphitization of gray iron at ~700°C (~1300°F), because the resulting microstructure is ferrite and graphite. This *ferritic* gray iron is a soft, machinable cast iron, which meets many needs at low cost. It must be cooled slowly through 700°C to avoid all carbide formation.

Nodular (ductile) cast irons   Careful laboratory work revealed that the microstructure of easily graphitized cast irons (Fig. 13–1.3) could be replaced by that shown in Fig. 13–1.5 if small additions of magnesium or cerium are introduced into the molten metal. The resulting graphite spheroids (or nodules) drastically affect the ductility of the cast iron (Table 13–1.1). Whereas a normal gray iron has essentially no ductility, this *nodular iron* may have 10%–20% elongation. Although an explana-

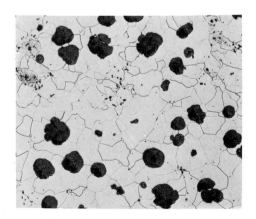

Fig. 13–1.5   Nodular cast irons, X100. Carbon is present as graphite *nodules.* Unlike gray cast iron, nodular cast iron is ductile (Table 13–1.2). The graphite forms *during* solidification in both gray cast iron and nodular cast iron. (G. A. Colligan, Dartmouth)

Table 13–1.1
Properties of unalloyed cast irons*

Cast iron	Type	Minimum mechanism properties			Typical uses
		S_t	S_y	Elong. (5 cm)	
Gray (3.2 C–2 Si)	Pearlitic	275 MPa (40,000 psi)	240 MPa (35,000 psi)	<1%	Engine blocks
	Martensitic	550 (80,000)	550 (80,000)	nil	Wearing surfaces
	Bainitic	550 (80,000)	550 (80,000)	nil	Cam shafts
	Ferritic[†]	172 (25,000)	138 (20,000)	<1%	Pipe, machine bases
Nodular[§] (3.5 C–2.5 Si)	Ferritic	413 MPa (60,000 psi)	275 MPa (40,000 psi)	18	Pipe
	Pearlitic	550 (80,000)	380 (55,000)	6	Crankshafts
	Tempered martensite	825 (120,000)	620 (90,000)	2	Special machine parts
Malleable (2.2 C–1 Si)	Ferritic	365 MPa (53,000 psi)	240 MPa (35,000 psi)	18	Hardware
	Pearlitic	450 (65,000)	310 (45,000)	10	Railroad equipment
	Tempered martensite	700 (100,000)	550 (80,000)	2	Railroad equipment
White (3.5 C–0.5 Si)	As cast (pearlitic)	275 MPa (40,000 psi)	275 MPa (40,000 psi)	nil	Wear-resistant products

* Adapted from Flinn and Trojan, *Engineering Materials and Their Applications*, and from Amer. Soc. Metals.
[†] Ferritic gray cast iron generally has $\sim 3.5\%$ C and $\sim 2.5\%$ Si.
[§] Also called *ductile* cast iron.

tion that accounts for this microstructural change is very complex, the practical effects are dramatic, because it permits a cast iron to be used in applications such as crankshafts where a brittle failure would be catastrophic.

Nodular iron, like gray iron, may be treated to be ferritic, pearlitic, or to contain tempered martensite. The former is the more ductile, the latter produces an exceptionally strong cast product (Table 13–1.1).

Malleable cast iron   We have just seen that cast irons with 3%–3.5% C and 2%–2.5% Si graphitize readily. Typical steels (<1% C and <0.25% Si) do not graphitize under normal circumstances. Expectedly, intermediate compositions behave in an intermediate fashion; however, the usefulness of the product is highly dependent upon how the engineer manipulates the composition, the cooling rates, and subsequent heat treatments. For example, a cast iron with 2.25% carbon and 1% silicon will produce a *white cast iron* if it is solidified rapidly, i.e., if it forms the metastable Fe_3C during freezing and is not given an opportunity for graphitization as a solid.

The above white cast iron* finds applications in certain wear-resistant parts, because the large amount of carbide (~ 30 v/o of the product) withstands abrasion well. However, this metal finds wider use if it is annealed to form the *malleable iron* that was initially cited in Section 11–1. By annealing a white cast iron, the Fe_3C will decompose to iron and graphite. Since the reaction occurs in the absence of a liquid, the resulting graphite is neither flake-like (Fig. 13–1.3) nor spheroidal (Fig. 13–1.5). Rather, it grows as "clusters" within the solid metal (Fig. 13–1.6). This graphite is not "crack-like" on a submicroscopic scale as are the edges of graphite flakes. Therefore, the metal retains some ductility (Table 13–1.2) and is more malleable than gray iron. Hence, the name.

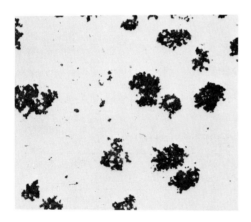

Fig. 13–1.6 Malleable cast iron, X100. Carbon is present as graphite clusters (*temper carbon*). As a result, there is significant ductility (Table 13–1.2). The graphite clusters are formed by the solid reactions of Eqs. (11–1.2) and (11–1.3). (From ASM Committee on Nodular Iron, "Nodular Cast Iron," *Metal Handbook* Supplement, 1A, American Society for Metals.)

Table 13–1.2
Ductility of ferritic alloys

Type	Ductility % elong. (5 cm)	See figure
Ferritic iron	40–50	9–8.2(a)
Ferritic nodular C-I	15–20	13–1.5
Ferritic malleable C-I	15–20	13–1.6
Ferritic gray C-I	<1	13–1.3

---

* Its fracture is metallic white, rather than gray.

As indicated with Eqs. (11–1.2) and (11–1.3), it is possible to produce either ferritic malleable iron or pearlitic malleable iron. Furthermore, appropriate quenching and tempering can lead to other relatively strong and inexpensive castings usable in railway and similar equipment.

The reader is asked to observe one limitation on designs that specify malleable iron castings. Since the solidification must be sufficiently fast to form a white iron initially, these castings cannot have thick sections. Otherwise, graphite flakes would form during the time required for solidification. Depending on the silicon content and the thermal conductivity of the mold wall, the maximum thickness of malleable iron castings is normally limited to 20 mm–30 mm (~1 in.).

Example 13–1.1 Plot the ferrite and austenite contents as a function of temperature for a cast iron containing 2.5% carbon, 1% silicon, and the balance iron.
a)   As a white cast iron.
b)   As a carbide-free, malleable cast iron.

Solution: The silicon dissolves in the ferrite and austenite. Therefore, we will approximate the composition as being 97.5 Fe–2.5 C.
a)   Use Fig. 9–7.1 for the Fe–Fe$_3$C diagram.

$$\text{At } 700°C \text{ (and below)}, \qquad \alpha = (6.7 - 2.5)/(6.7 - 0) = 0.63$$

$$\text{At } 800°C, \qquad \gamma = (6.7 - 2.5)/(6.7 - 1.0) = 0.74$$

$$\text{Etc.,} \qquad \text{See Fig. 13–1.7.}$$

b)   Use Fig. 13–1.2 for the Fe–C diagram.

$$\text{At } 700°C \text{ (and below)}, \qquad \alpha = (100 - 2.5)/(100 - 0) = 0.975$$

$$\text{At } 800°C, \qquad \gamma = (100 - 2.5)/(100 - 0.95) = 0.985$$

$$\text{Etc.,} \qquad \text{See Fig. 13–1.7.}$$

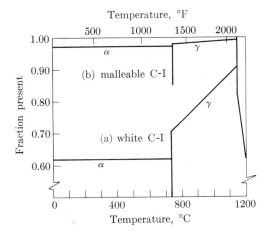

Fig. 13–1.7   Phase fractions (cast iron with 2.5% C). The second phase in white cast irons (a) is carbide below the eutectic temperature (Fig. 9–7.1). After malleabilization (b), the second phase below the eutectic temperature is graphite (Fig. 13–1.2). (See Example 13–1.1 .)

Comments. Silicon, like all elements, will shift the eutectoid to lower carbon contents (Section 9–9), This decreases the carbon solubility in austenite somewhat. Therefore, the curves in Fig. 13–1.7 for γ are slightly high. ◀

Example 13–1.2 Indicate the heat-treating sequence to produce a nodular iron (3 C–2 Si–95 Fe) that will contain tempered martensite. Rationalize any factors that receive different considerations than if the metal were a 1080 steel.

Procedure and Comments
a) Melting, $\sim 1300°C$ ($\sim 2400°F$). The metal must be innoculated with magnesium (or cerium) to produce graphite nodules during solidification. The melting temperature is lower than steel since more carbon is present.
b) Austenization, 800°C (1475°F). Since there is 2% silicon present, the eutectoid temperature and composition are $\sim 775°C$ and $\sim 0.55\%$, respectively (Fig. 9–9.1). Austenite ($\sim 0.55\%$ C) + graphite (100% C).
c) Oil-quench to ambient. The 2% Si increases the hardenability appreciably. An oil-quench is thus possible and will reduce the probability of cracking. Martensite ($\sim 0.55\%$ C) + graphite (100% C).
d) Temper as if it were a 1055 steel. Ferrite (nil C) + carbide (6.7% C) + graphite (100% C).

Example 13–1.3 Compare the densities of the white cast iron and ferritic malleable cast iron of Example 13–1.1. With 1% Si,

$$\rho_\alpha = \sim 7.8 \text{ g/cm}^3;$$

$$\rho_{Fe_3C} = 7.6 \text{ g/cm}^3;$$

$$\rho_{graph} = \sim 2.2 \text{ g/cm}^3 \qquad (\text{or } \sim 2.2 \text{ Mg/m}^3).$$

Solution: Basis: 100 g, and use g/cm³.
White cast iron: 0.63α and $0.37 \text{ Fe}_3\text{C}$ (Example 13–1.1).

$$(63 \text{ g } \alpha)/(7.8 \text{ g/cm}^3) = 8.08 \text{ cm}^3 \ \alpha;$$
$$(37 \text{ g Fe}_3\text{C})/(7.6 \text{ g/cm}^3) = 4.87 \text{ cm}^3 \ \text{Fe}_3\text{C};$$
$$\rho_{wci} = 100 \text{ g}/(8.08 + 4.87 \text{ cm}^3)$$
$$= 7.72 \text{ g/cm}^3 \ (\text{or } 7.72 \text{ Mg/m}^3).$$

Ferritic malleable: 0.975α and 0.025 graph. (Example 13–1.1).

$$(97.5 \text{ g } \alpha)/(7.8 \text{ g/cm}^3) = 12.50 \text{ cm}^3 \ \alpha;$$
$$(2.5 \text{ g graph})/(2.2 \text{ g/cm}^3) = 1.14 \text{ cm}^3 \ \text{graph};$$
$$\rho_{amci} = 100 \text{ g}/(12.50 + 1.14 \text{ cm}^3)$$
$$= 7.33 \text{ g/cm}^3 \ (\text{or } 7.33 \text{ Mg/m}^3).$$

Comment. There is a volume expansion during malleabilization that must be considered in processing control. ◀

• 13–2   CONCRETE (AND RELATED PRODUCTS)

It is common knowledge that engineering structures such as buildings, bridges, and highways, can be made out of gravel, sand, and cement. The resulting monolithic product is called *concrete*.* The characteristics of the concrete depend upon the nature of the *aggregate*, the nature of the admixed water-cement *paste*,† and, of course, upon their relative amounts. In other words, the properties are governed by the internal structure, just as for any material.

Aggregates   Of course, the sand and gravel must be durable materials, such as quartzite, crushed limestone, or basalt, having greater strength than what is expected of the final concrete. If not, and slate (or a similar friable rock) is present, failure will occur through the aggregate, even with the best formulated cement additions. Concrete codes specify aggregate quality, for example, ASTM–C 88. In addition, the engineer requires a measure of aggregate size. This is obtained by *sieve analyses.* Two series of screens are in common use for sieve analyses (Table 13–2.1) and are generally comparable. The U.S. Series of sieve sizes is most widely used for concrete codes. The Tyler Series, which is based on $\sqrt{2}$, has wider use in the ceramic industry.

Table 13–2.1
Sieve series used in particle analyses (openings)

	U.S. Series						Tyler Series			
	1.5	38	mm	1.50	in.		1.5	38	mm	1.50	in.
	1	25.4		1.00			1	26.7		1.050	
	$\frac{3}{4}$	19.1		0.75			$\frac{3}{4}$	18.9		0.742	
	$\frac{1}{2}$	12.7		0.50			$\frac{1}{2}$	13.3		0.525	
	$\frac{3}{8}$	9.5		0.375			$\frac{3}{8}$	9.4		0.371	
					No.	3	6.7		0.263	
No.	4	4.75		0.187			4	4.7		0.185	
						6	3.3		0.131	
	8	2.38		0.0937			8	2.36		0.093	
						10	1.66		0.065	
	16	1.19		0.0469			14	1.18		0.046	
						20	0.83		0.0328	
	30	0.59		0.0232			28	0.59		0.0232	
						35	0.42		0.0164	
	50	0.297		0.0117			48	0.295		0.0116	
						65	0.208		0.0082	
	100	0.150		0.0059			100	0.147		0.0058	
							*		*	

* Series continues to No. 400 with an opening of 0.037 mm (0.0015 in.); steps are on the basis of $\sqrt{2}$.

---

* The term, *cement*, should refer to the material used to bond the concrete, and not to the final product.
† When 0.4 to 0.7 units (weight) of water are mixed with a unit of cement, the product initially has the consistency of a paste. We may view concrete as an admixture of this *paste* with an aggregate. This paste slowly hardens by hydration.

An average particle size has only limited significance. For example, the distribution could be very narrow, as is typical of beach sands, or very broad, such as river sands, and still have the same *mean* size (Fig. 13–2.1). Rather, the engineer is commonly interested in what fractional quantity is retained on successive screens. Several such analyses are shown in Fig. 13–2.2.

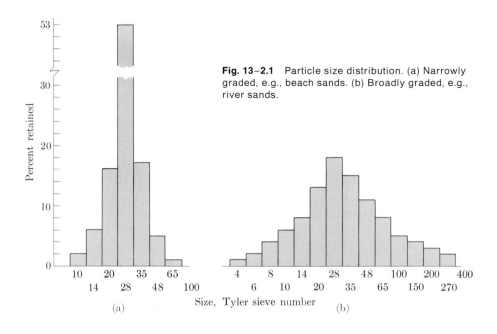

Fig. 13–2.1   Particle size distribution. (a) Narrowly graded, e.g., beach sands. (b) Broadly graded, e.g., river sands.

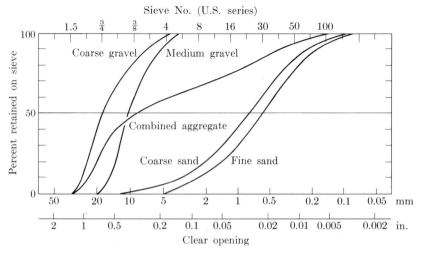

Fig. 13–2.2   Analyses of typical aggregates used for concrete mixes. The combined aggregate is a mixture of coarse gravel (60%) and sand (40%). The combination fills space more completely than gravel alone or sand alone. (Data adapted from Troxell, Davis, and Kelly, *Composition and Properties of Concrete.*)

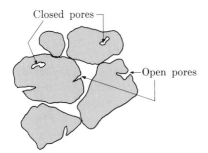

Closed pores

Open pores

Fig. 13–2.3 Pore space (schematic). Closed pores are inside pores and, therefore, impermeable. Open pores are permeable voids. Open pores include space between particles (interparticle porosity), plus small cracks and capillaries that extend into the aggregate particles.

Packing density and porosity are also important. Figure 13–2.3 shows several subdivisions in volume that affect the amount of water and/or a cement paste that can be associated with the aggregate. *Closed* (inside) pores are impermeable. These may be encountered in artificial light-weight aggregates and in some crushed basalt aggregates. *Open* pores are permeable voids. They include capillaries, fine striations, and reentrant shape irregularities of the aggregate particle. In addition, there can be major amounts of *interparticle* porosity among sand, gravel, etc., which is open pore space.

These pores lead to several identifiable volumes. The *true volume* is that volume that is inherent to the material. The *apparent volume* is the sum of the true volume and the closed-pore volume. The *bulk volume* includes not only the closed-pore volume (and the true volume) but also the open-pore volume. With this *total volume*, we must define our basis. If only the aggregate particles are considered, the open-pore volume is limited to the small cracks and capillaries. If our basis involves bulk handling, the total volume must include the relatively large amount of interparticle porosity.

These several volumes lead to two principal porosities, true (or total) and apparent (or open):

$$Total\ porosity = \frac{open\text{-}pore\ volume + closed\text{-}pore\ volume}{total\ volume}. \qquad (13\text{–}2.1)$$

$$Apparent\ porosity = \frac{open\text{-}pore\ volume}{total\ volume}. \qquad (13\text{–}2.2)$$

In either case, we should identify the basis of the total volume—either the particles of aggregate, or the volume that includes the unfilled voids between aggregate particles.

There are also three principal densities:

$$True\ density = mass/true\ volume. \qquad (13\text{–}2.3)$$

$$Apparent\ density = mass/apparent\ volume \qquad (13\text{–}2.4)$$
$$= mass/(true\ volume + closed\text{-}pore\ volume)$$
$$= mass/(total\ volume - open\text{-}pore\ volume).$$

$$Bulk\ density = mass/total\ volume. \qquad (13\text{–}2.5)$$

Here again, we should identify our basis for the total volume, as to whether we are considering only the aggregate particles or are including the interparticle space present that accompanies bulk materials.

Cement A strength of the hardened cement-water paste essentially establishes the strength of concrete, because with a strong aggregate, any failure occurs *between* the sand and/or rock particles. Therefore, in principle, a driveway that could be produced from only cement and water mixtures would be as strong as when an aggregate is present. However, it would be prohibitive on the basis of cost. Therefore, the cement–water mixture is "diluted" with the stronger, cheaper aggregate materials.

Conceptually, concrete contains gravel or crushed rock, with sand filling the interstices; then a paste of cement and water fills the space among the sand grains. In practice, some excess sand is required to ensure the complete filling of the space within the gravel (Fig. 13–2.4); likewise, approximately 10 percent more cement paste is required than the theoretical amount for the space among the sand grains. Otherwise, some local voids could remain where mixing is not perfect.

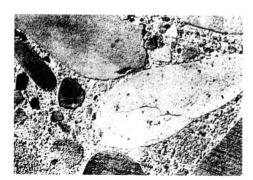

Fig. 13–2.4 Cross section of concrete (X1.5). The space among the gravel is filled with sand. The space among the sand is filled with a hydrated portland cement paste. (Portland Cement Association.)

Cement hydration The cement used in almost all concretes is called *portland cement.** It is primarily a mixture of calcium silicates and calcium aluminate that hydrate in the presence of water:

$$Ca_3Al_2O_6 + 6H_2O \rightarrow Ca_3Al_2(OH)_{12}; \tag{13–2.6}$$

$$Ca_2SiO_4 + x\,H_2O \rightarrow Ca_2SiO_4 \cdot x\,H_2O; \tag{13–2.7}^\dagger$$

$$Ca_3SiO_5 + (x + 1)H_2O \rightarrow Ca_2SiO_4 \cdot x\,H_2O + Ca(OH)_2, \tag{13–2.8}$$

---

* It was initially manufactured as early as the 18th century on the Isle of Portland.
† This reaction may be more correctly stated as

$$2\,Ca_2SiO_4 + (5 - y + x)H_2O \rightarrow Ca_2[SiO_2(OH)_2]_2 \cdot (CaO)_{y-1} \cdot x\,H_2O + (3 - y)Ca(OH)_2, \tag{13–2.9}$$

where x varies with the partial pressure of water and y is approximately 2.3. Equation (13–2.8) may also be modified accordingly.

In each case, the hydrated product is less soluble in water than was the original cement. Therefore, in the presence of water, the above reactions are ones of solution and reprecipitation.

Each of these reactions emphasizes that cement does not harden by drying, but by the chemical reaction of hydration. In fact it is necessary to keep concrete moist to ensure proper *setting*. The hydration reactions described above release heat, as shown in Fig. 13–2.5. Engineers take advantage of this in cold climates in which it may be necessary to pour concrete at temperatures that are slightly below the freezing temperature of water. Conversely, the *heat of hydration* presents a problem when concrete is poured into massive structures such as dams. In these extreme cases, special cooling is required, and a refrigerant is passed through embedded pipes that are left in place and later act as reinforcement.

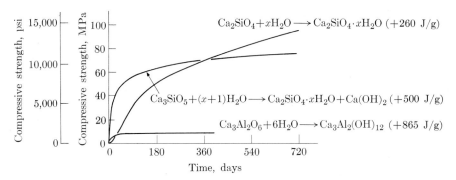

Fig. 13–2.5 Cement hydration and strength (portland cement components). For strength to develop, water must be present for hydration. Concrete does *not* set by drying. Heat is given off during hydration as indicated. Its removal may require special provisions in massive structures, such as dams.

The variables we have discussed suggest that the engineer may utilize various kinds of cement to obtain different characteristics. For example, a cement with a larger fraction of tricalcium silicate (Ca_3SiO_5) sets rapidly and thus gains strength early.* In contrast, in cases in which the heat of reaction might be a complication, and more setting time is available, a Ca_2SiO_4-rich cement is used. The American Society for Testing and Materials (ASTM) lists the following types of cement: (1) Type I, used in general concrete construction in which special properties are not required. (2) Type III, used when one wants an early-strength concrete. (3) Type IV, used when a low heat of hydration is required. There are also two other types (II and V) for special applications in which sulfate attack is a design consideration.

Concrete mixes   *Unit mixes* of concrete are commonly based on the amount of cement used. Thus, an engineer may specify a mix of 3.1 gravel–2.6 sand–1 cement–0.55 water on a weight basis (or these may be converted to appropriate volumes).

---

* A *high, early-strength cement* can also be achieved by grinding the cement more finely.

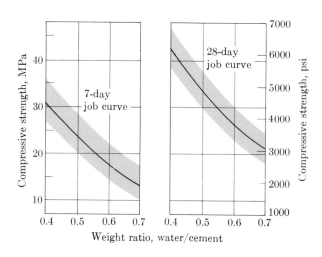

We will observe first that the cement/water ratio is critical for the strength of the concrete (Fig. 13–2.6). Since excessive amounts of water do not enter the hydration reaction, it occupies space and precludes solid-to-solid bonding. If the excess water eventually evaporates, hairline cracks remain. Therefore, the mix is designed with the water/cement ratio as low as possible. However, there is a limit. There must be enough water present to make the concrete workable, and to allow it to fill the forms completely without voids. Vibration procedures help with form filling. Also the incorporation of air bubbles (~ 1 mm dia.) into the concrete through the use of appropriate *air-entraining agents* increases the workability (and therefore lowers the required amount of water). These spherical voids do not weaken the concrete as much as the capillary voids that remain after excess water is evaporated.*

The small bubbles of entrained air also increase the resistance of the concrete to deterioration resulting from the freezing and thawing of pore water. As a result, almost all concrete road pavements of northern latitudes have air entrainment specifications.

We may calculate the volume of a concrete mix from the volumes of the aggregate, cement, and water additions. As demonstrated in Example 13–2.3, we assume that the sand will (more than) fill the space among the gravel and that the cement/water parts will more than fill the passages between the sand. Finally, an adjustment may be made for the volume of the entrained air.

The density and porosity of the concrete product may be related to the various volumes of that product—true, apparent, and total (Eqs. 13–2.1 to 13–2.5). In turn, density and porosity are significant for properties. For example, the heat losses through a wall are a function of the true volume (and of the total porosity), while a wall's permeability to water is related to the amount of apparent (open) porosity.

---

* Cf. the effect of graphite nodules and graphite flakes on the strength of cast iron (Section 13–1).

Other cements   Portland cement is made in large tonnages. Of course, other cements also exist. Gypsum plaster and lime are also hydraulic cements. For example, gypsum plaster (plaster of paris) reacts as

$$2 \, CaSO_4 \cdot H_2O + 3 \, H_2O \rightarrow 2(CaSO_4 \cdot 2 \, H_2O). \tag{13–2.10}$$

In principle, its hardening mechanism is identical to that of portland cement since the initial compound dissolves and reprecipitates in the hydrated form.

Polymeric cements involve polymerization reactions, as is the case for silicic acid:

$$x \, Si(OH)_4 \rightarrow \begin{bmatrix} OH \\ | \\ Si \\ | \\ OH \end{bmatrix}_x + x \, H_2O. \tag{13–2.11}$$

In this example the polymerization reaction is a condensation reaction. The above reaction leads to an amorphous, 3-dimensional, SiO_2 structure when it is extended to involve the remaining two —OH radicals. Other, more complex polymeric reactions occur in epoxy cements, some of which produce exceptionally strong bonds.

Bonding may also be achieved by thermoplastic materials. Asphalt falls in this category and is used in kiloton quantities for highway purposes. Typically, it is heated to provide adherence to the aggregate and to permit more efficient packing on the road bed. It hardens as it cools below its transition temperature. As described for Fig. 13–2.7, the civil engineer tailors the composition and structure of the final product to meet the varying needs of the road bed requirements. The interested reader is referred to books on highway construction materials for more elaboration.

Example 13–2.1   A sample of wet gravel is dried to remove surface water, but not the absorbed water. It weighs 8.91 kg. In this condition, its bulk volume (excluding interstitial pores) of the gravel particles was determined to be 3,570 cm³ by water displacement. After thorough drying, its weight is 8.79 kg.
a)   What is its apparent porosity?
b)   The apparent density of the gravel particles?

Solution: Bulk volume = 3,570 cm³ (or 0.00357 m³). Since it absorbs 0.12 kg of water (120 cm³) in its open pores, its apparent volume is (3,570 − 120 cm³) = 3,450 cm³ (or 0.00345 m³).

a)   Apparent porosity of the gravel = 120 cm³/3,570 cm³

$$= 3.4\%.$$

b)   Apparent density of the gravel = (8,790 g)/(3,450 cm³)

$$= 2.55 \text{ g/cm}^3 \qquad (\text{or } 2.55 \text{ Mg/m}^3).$$

Comment.   We can also calculate the bulk density of the gravel particles, 8790 g/3570 cm³ = 2.46 g/cm³ (or 2.46 Mg/m³). ◀

Example 13–2.2   The true density of a concrete to which an air-entraining agent was added is 2.80 Mg/m³ (=174.7 lb/ft³). However, a dry core [152 mm × 102 mm in diameter (6.0 in. ×

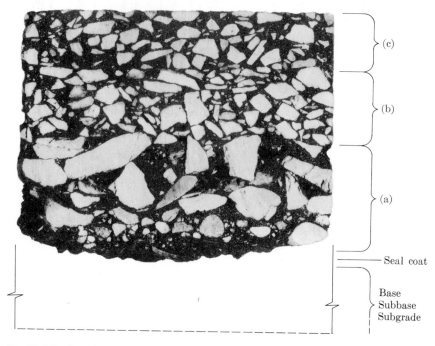

(c)

(b)

(a)

Seal coat

Base
Subbase
Subgrade

Fig. 13–2.7 Core from asphalt pavement (X0.5). Like concrete, this product is agglomerated; but viscous asphalt, rather than a hydrated silicate, serves as the bond. (a) Bonding course. Sand (30 w/o) and asphalt (5 w/o) fill the interstices among the coarse stone 65 w/o) to give a rigid support above the subgrade aggregate. (b) Leveling course. The stone is smaller, otherwise the composition is similar to the preceding course. (c) Wearing course. More asphalt (6 w/o), more sand (55 w/o), and 4–5 w/o mineral filler (fly-ash) provide an impermeable, tough surface to resist traffic wear. (Core from Michigan Highway Testing Laboratory.)

4.0 in.)] of this concrete weighs only 2.982 kg (6.56 lbs). The same core weighs 3.08 kg (6.776 lb) when saturated with water.

a)   What are the percents of the open and closed pores of the concrete product?

b)   What are the bulk and apparent densities of the concrete product?

Solution

$$\text{Bulk volume} = \pi(51 \text{ mm})^2(152 \text{ mm})(1 \text{ cm}^3/1000 \text{ mm}^3)$$
$$= 1{,}242 \text{ cm}^3 \quad (\text{or } 75.4 \text{ in.}^2).$$

$$\text{True volume} = [(2982 \text{ g})/(2.8 \times 10^6 \text{ g/m}^3)](10^6 \text{ cm}^3/\text{m}^3)$$
$$= 1{,}065 \text{ cm}^3 \quad (\text{or } 65.0 \text{ in.}^2).$$

$$\text{Total porosity} = (1242 \text{ cm}^3 - 1065 \text{ cm}^3)/(1242 \text{ cm}^3) = 14 \text{ v/o}.$$

$$\text{Volume open pores} = (3080 \text{ g} - 2982 \text{ g})/(1 \text{ g H}_2\text{O/cm}^3)$$
$$= 98 \text{ cm}^3 \text{ H}_2\text{O} \quad (\text{or } 5.98 \text{ in.}^3).$$

a)   Apparent (open) porosity = 98 cm^3/1242 cm^3 = 0.08 (or 8 v/o).

$$14 \text{ v/o} - 8 \text{ v/o} = 6 \text{ v/o closed pores.}$$

b)   Bulk density = 2982 g/1242 cm^3 = 2.4 g/cm^3

$$= 2.4 \text{ Mg/m}^3 \qquad \text{(or 150 lb/ft}^3\text{)}.$$

Apparent density = 2982 g/(1242 − 98 cm^3) = 2.6 g/cm^3

$$= 2.6 \text{ Mg/m}^3 \qquad \text{(or 163 lb/ft}^3\text{)}. \blacktriangleleft$$

Example 13–2.3 A concrete unit mix contains 136 kg (300 lbs) of gravel, 95 kg (210 lbs) of sand, 42.7 kg (94 lbs) of portland cement, and 22.5 kg (50 lbs) of water, plus 5 v/o entrained air. How much cement is required to build a low retaining wall, 28.6 m × 1.07 m × 0.305 m (94 ft × 3.5 ft × 1 ft)?

Density Mg/m^3*

Material	Apparent	Bulk†
Gravel	2.60	1.76
Sand	2.65	1.68
Cement	3.25	1.50

* Multiply by 62.4 to obtain lbs/ft^3.
† As-received basis (includes interstitial volume).

Solution: Basis: 1 unit mix.

Volume of gravel = 136,000 g/(2.60 × 10^6 g/m^3) = 0.0523 m^3

Volume of sand = 95,000 g/(2.65 × 10^6 g/m^3)   = 0.0358
(in the space among the gravel)

Volume of cement = 42,500 g/(3.25 × 10^6 g/m^3) = 0.0131
(in the space among the sand and gravel)

Volume of water = 22,500 g/(10^6 g/m^3)   = 0.0225
(in the remaining space among all solids)
                                                          ──────────
                                                          0.124 m^3

Total m^3/unit mix = 0.124 m^3/0.95 = 0.1305 m^3

Unit mixes = 9.33 m^3/0.1305 m^3 = 71.5

Cement required = (71.5)(42.7 kg) = 3050 kg.

Comment. Using lbs and ft^3,

Volume of gravel = 300 lbs/(2.60 × 62.4 lbs/ft^3) = 1.85 ft^3;

$$V_s = 1.27 \text{ ft}^3;$$

$$V_c = 0.46 \text{ ft}^3;$$

$$V_{H_2O} = 0.80 \text{ ft}^3;$$

$$\sum V = 4.38 \text{ ft}^3/0.95 = 4.6 \text{ ft}^3;$$

Unit mixes = 329 ft^3/4.6 ft^3 = 71.5.

Cement required = (71.5)(94 lbs) = 6700 lbs. $\blacktriangleleft$

Example 13–2.4 A concrete mix contains 585 kg of sand and gravel, 100 kg of cement, and 55 kg of water. Assume the cement is (a) all tricalcium aluminate, $Ca_3Al_2O_6$, (b) all tricalcium silicate, Ca_3SiO_5, (c) all dicalcium silicate, Ca_2SiO_4. What is the temperature rise during hydration, given that the average specific heat of the concrete mix is 0.8 J/g · °C (or 0.2 cal/g · °C)? Assume that half of the heat is lost.

Solution: Basis: 1 g cement = 7.4 g mix. From the data in Fig. 13–2.5,

a)   (7.4 g/g cement)(0.8 J/g · °C)ΔT = (865 J/g cement)(0.5);

$$\Delta T = 73°C.$$

b) $\Delta T = 0.5(500 \text{ J})/(5.92 \text{ J}/°C) = 42°C.$

c) $\Delta T = 0.5(260 \text{ J})/(5.92 \text{ J}/°C) = 22°C.$

Comments. Not only do $Ca_3Al_2O_6$ and Ca_3SiO_5 release more heat than does Ca_2SiO_4, they react in a relatively short period of time. Therefore, a smaller fraction of the heat escapes from a large structure during the early stages of setting. The temperature rise may become critical unless only small fractions of $Ca_3Al_2O_6$ and Ca_3SiO_5 are present in the cement. ◀

• 13–3   WOOD

Although complex, wood is sufficiently familiar so that we can appreciate its structure and properties. Its macrostructure is evident in its grain; its microstructure is of biological origin; and its atomic coordinations are molecular.

Wood, of course, is a very important engineering material. It has a high strength-to-weight ratio. It is easily processed, even on the job. And finally, it is a replaceable resource in this day of diminishing raw materials. We are very much aware of its directional properties, which must be taken into consideration whenever it is used. To better understand wood, let's examine its structure.

Wood is a natural polymeric composite. The principal polymeric molecules are those of *cellulose*.

(13–3.1)

Since cellulose is isotactic and has no side branches, it develops a fair degree of crystallinity. In addition to its more than 50% cellulose composition, wood contains 10% to 35% *lignin*, a more complex 3-dimensional, cross-linked polymer.

Larger than polymeric molecules, wood's next-most-prevalent structural units are biological cells, of which the most extensive are the *tracheids*. These are hollow, spindle-shaped cells that are elongated in the longitudinal direction of the wood. The most visible structural unit is the *grain* of the wood, which is made up of *spring* and *summer* layers. The biological cells of the spring wood are larger and have thinner walls than those of the summer wood (Fig. 13–3.1). In this respect, biological cells are much more variable in structure than the unit cells of crystals.

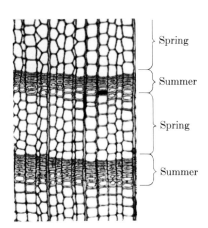

Fig. 13–3.1 Microstructure of wood. The units of structure are biological cells. The cells that form with early (spring) growth are larger but have thinner walls than those that form with late (summer) growth. Cellulose (Eq. 13–3.1) is the major molecular constituent.

The above description of wood is, of course, oversimplified.* However, it does indicate the source of the anisotropies in properties that are so characteristic of wood.

In materials that have simple structures, *density* is a structure-insensitive property. This is not the case in a complex material such as wood. First, the amount of early wood and late wood varies from species to species. Second, the ratio of cellulose to lignin varies ($\rho = \sim 1.55$ g/cm^3 and ~ 1.35 g/cm^3, respectively). As a result of these two factors, the density can range from about 0.15 g/cm^3 for *balsa* to 1.3 g/cm^3 for the dense *lignum vitae*. In addition to the above factors, wood is hygroscopic; hence it absorbs moisture as a function of humidity of the surrounding atmosphere. Because this is so, most woods, when they are wet, exhibit a net increase in density.

Anisotropy   It should come as no surprise that *dimensional changes* that accompany variations in temperature, moisture, and mechanical loading in wood are anisotropic. Wood technologists point out that *thermal expansion* is greater in the tangential, t, direction (Fig. 13–3.2) and in the radial, r, direction, than in the longitudinal, l, direction. Thermal expansion in the longitudinal direction is relatively independent of density, while in the other two directions it depends on density, ρ (in g/cm^3). Between $-50°$ and $+55°C$,

$$\alpha_t \simeq [6\rho + 3] \times 10^{-5}/°C;$$

$$\alpha_r \simeq [6\rho + 2] \times 10^{-5}/°C;$$

$$\alpha_l \simeq 0.4 \times 10^{-5}/°C$$

Thus, it is typical to have the cross-sectional thermal expansion to be more than ten times the lengthwise expansion.

As an order of magnitude, longitudinal values for *Young's modulus* are between 7,000 MPa and 14,000 MPa (1,000,000 psi–2,000,000 psi) when measured in tension.

---

* It does not take into account the wide variety of minor wood chemicals, nor the *vascular rays*, which are rows of single, almost equidimensional cells that radiate from the center of the tree. These become important during deformation.

Fig. 13-3.2 Macrostructure of wood. Properties are anisotropic because the growth produces layers of cells with different properties. Strength, elastic moduli, and expansion coefficients all vary with direction.

Tangential values are normally between 400 MPa and 700 MPa (60,000 psi–100,000 psi), while the radial values are commonly in the 500 MPa to 1,000 MPa (75,000 psi–150,000 psi) range. We would expect the longitudinal value to be the highest. However, we could also have predicted that the tangential modulus would be higher than the radial. But in Fig. 13-3.1 we did not take into account the vascular rays cited in the accompanying footnote. This feature provides an additional rigidity in the radial direction.

Shrinkage is also anisotropic; longitudinal changes are negligible, but tangential shrinkage is very high (~ 0.25 l/o* per 1 w/o moisture for Douglas fir). Radial shrinkage is intermediate (~ 0.15 l/o* per 1 w/o moisture for the same wood) because of the restraining effects of the vascular rays. Effects of shrinkage are summarized in Fig. 13-3.3. The consequences of dimensional distortion on *warpage* are readily apparent.

Fig. 13-3.3 Shrinkage of wood. Since the structure is anisotropic, shrinkage varies with orientation. This may cause warpage if the wood is cut before it has dried completely. (Shrinkage is exaggerated in this sketch.) Q = quartersawed; P = plainsawed.

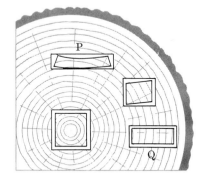

The longitudinal *tensile strength*, S_l, is in excess of 20 times the radial tensile strength, S_r, because any fracture that takes place must occur across the elongated tracheid cells. Increased densities (for a given moisture content) reflect an increase in the thickness of cell walls and, therefore, a proportional increase in the longitudinal strength. The transverse strength increases more accordingly (again for a given

---

* Based on the dimensions of green wood.

moisture content), because the denser the wood, the less the opportunity for failure parallel to the hollow tracheid cells.

Except in cases in which designs capitalize on the longitudinal strength, the anisotropies of wood are undesirable. Therefore much has been done to modify its structure. Examples include (1) *plywood*, in which the longitudinal strength is developed in *two* coordinates; (2) *fiberboard*, in which individual wood fibers are separated and randomly oriented in the plane of the panels; (3) *particleboard*, in which thin wood flakes or particles are randomly oriented in a plane and adhesively bonded to each other; and (4) *paper*, in which delignified fibers (pulped wood) are randomly oriented in a sheet and held together by hydrogen bonds. (5) Finally, the *cellulose* may be extracted from the wood (and cotton fibers) to serve as a raw material for certain polymeric products, such as cellulose acetate (rayon, etc.).

Example 13–3.1 A solid wood door was made by gluing together edges of quartersawed boards of Douglas fir. That is, the width of the board is in the radial direction, and the thickness is in the tangential direction (and, of course, the length is the longitudinal direction of the wood). The door was initially trimmed to 761 mm by 2035 mm (to fit a 765 × 2050 mm opening) in late fall when the moisture content of the wood was 9 percent. Will the door "stick" when the moisture content increases to 14 percent?

Solution

$$\text{Radial change} = 0.0015/\% \times [14 - 9\%] = +0.0075.$$

$$\Delta L = (761 \text{ mm})(+0.0075) = +5.7 \text{ mm}.$$

It will not fit, since 761 + 5.7 mm is greater than 765 mm. The longitudinal change is negligible.

Comment. Suggest a design feature that may be used to minimize this problem. ◀

Example 13–3.2 A plainsawed board of Douglas fir (width is tangential) is not permitted to expand its width as its moisture content increases from 5 to 10 percent. What pressure develops parallel to the grain?

Solution: If expansion were allowed,

$$\Delta L/L = 0.0025/\% \times (10 - 5\%) = +0.0125.$$

Assume $E = 550$ MPa, that is, the mid-value of those indicated in the text for the tangential modulus. Then compress $(-)$ the wood back to its original dimension.

$$s = (-0.0125)(550 \text{ MPa})$$
$$= 7 \text{ MPa compression} \quad \text{(or 1,000 psi).} \quad ◀$$

Example 13–3.3 Water absorption into wood from moist air is initially within the wood fibers (and not into the pores). The fiber saturation point for most wood is ~ 30 w/o (of oven-dry wood). What is the density increase of birch (oven-dry density $= \sim 0.56$ g/cm^3) when it is changed from 10% moisture to 20% moisture? (Assume the same shrinkage as Douglas fir.)

Answer: Basis: 1 cm³ dry wood

	With 10% moisture	With 20% moisture
	$l = 1.000$ cm	$l = 1.000$ cm
	$r = 1.000[1 + (0.0015/\%)(10\%)]$ $= 1.015$ cm	$r = 1.030$ cm
	$t = 1.000[1 + (0.0025/\%)(10\%)]$ $= 1.025$ cm	$t = 1.050$ cm
Volume $= 1.04$ cm³		$V = 1.08$ cm³
Density $= \dfrac{(0.56 \text{ g})(1.10)}{1.04 \text{ cm}^3}$		$\rho = \dfrac{(0.56 \text{ g})(1.20)}{1.08 \text{ cm}^3}$
$= 0.59$ g/cm³.		$= 0.62$ g/cm³.

$$\Delta\rho = (0.62 \text{ g/cm}^3 - 0.59 \text{ g/cm}^3)/(0.59 \text{ g/cm}^3) = 5\% \text{ increase.} \quad \blacktriangleleft$$

13–4 COMPOSITES

We have already considered certain composite materials—those which are reinforced on a microscopic scale. Bainite, tempered martensite, and precipitation-hardened (age-hardened) alloys are strengthened by a dispersion of fine particles of harder phases. For example, the tensile strength of the $(\alpha + \bar{C})$ in tempered martensite can be more than 1400 MPa ($> 200{,}000$ psi), whereas the tensile strength of ferrite, α, alone is less than 20% of those values. The strengthening occurs because the deformable phase cannot undergo strain independent of the rigid phase.

Composites of reinforced materials are well known on a macroscopic scale. Examples include reinforced concrete and fiber-reinforced plastics (FRP). Consider a "glass" fishing rod that consists of a bundle of parallel glass fibers bonded together by a polymeric resin (or conversely, a plastic rod that is longitudinally reinforced by glass fibers). If this composite is placed under tension, the plastic and glass must deform in concert, even though the two components may have markedly different elastic moduli ($E_{gl}/E_{pl} > 10$), or appreciably different individual strengths ($S_{gl}/S_{pl} > 20$).

Let us analyse the fiber-reinforced plastic just cited, when it is under tension. If the fibers are continuous, $e_{gl} = e_{pl}$, because the two are bonded together. We will find therefore that $s_{gl} \neq s_{pl}$ because the two have different elastic moduli, E:

$$s_1/E_1 = e_1 = e_2 = s_2/E_2, \tag{13–4.1}$$

where the subscripts are for the two components. Thus,

$$s_1/s_2 = E_1/E_2. \tag{13–4.2}$$

This was encountered previously in our discussion of elastic anisotropy (Fig. 6–3.4). It leads us to the first rule for reinforcement. The *reinforcing phase must be the higher-modulus phase* so that it can carry added load.

We can write mixture rules not only for density (Eq. 10–2.1), and specific heat (Eq. 10–4.2) and conductivities (Eqs. 10–4.3 and 10–4.4) but also for certain mechanical properties such as Young's modulus. Again give attention to the fiber-reinforced plastics of the past two paragraphs. If f_1 and f_2 represent the volume fraction of the two components, and we use Eq. (13–4.1), since $e_1 = e_2$, then

$$\frac{F_1/f_1 A}{E_1} = \frac{F/A}{\bar{E}} = \frac{F_2/f_2 A}{E_2}, \tag{13–4.3}*$$

where $\bar{E}$ is the modulus of the composite. Also, since the total load F is equal to the sum, $F = F_1 + F_2$;

$$F = f_1 E_1 F/\bar{E} + f_2 E_2 F/\bar{E}.$$

$$\bar{E} = f_1 E_1 + f_2 E_2. \tag{13–4.4}*$$

- **Interfacial stresses** The civil engineer is aware that shear stresses develop at the interface between a reinforcing rod and concrete. For this reason, he specifies "deformed" rods with merloned surfaces (ASTM, A305). Comparable shear stresses are encountered between fiber reinforcement and the surrounding plastic matrix. Here, however, the shear stresses are supported by a chemical rather than a mechanical bond.

    Interfacial shear stresses become particularly important if the reinforcement is not continuous. This is illustrated in Fig. 13–4.1. In this figure, $s_{f\,max}$ represents the stress that is carried by the fiber if there are no end effects (infinite length). This corresponds to the calculation in Example 13–4.1, which follows shortly, and depends on the volume fraction of the reinforcement, as well as the Young's moduli of the two

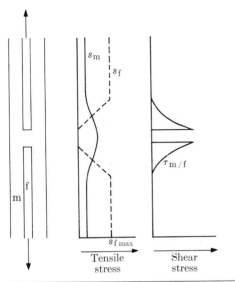

Fig. 13–4.1 Stress distribution (at a break in a reinforcing fiber). The fiber stress, s_f, drops from its maximum value to zero. The load must be transferred across the interface from the fiber to the matrix by shear stresses, $\tau_{m/f}$. The matrix has to carry a higher stress, s_m, in the vicinity of the break.

* These calculations assume that the Poisson ratios v of the two components are equal.

materials. If the fiber is broken, however, the fiber stress drops to zero at that point. When this occurs, the load must be transferred to the matrix by shear stresses. Two features stand out in this load transfer.

1. The bonding between the two materials must be sufficiently good to carry the shear stresses.

2. Reinforcement is most effective if it is continuous, or at least has a long aspect ratio (l/d). Local interruptions transfer the load onto the weaker matrix. Consequently, a deformable matrix has an advantage because the load can be distributed over a larger area for lower maximum stresses. (Of course, there is a limit, because an extremely weak matrix will fail completely.)

Example 13–4.1 A glass-reinforced polyvinylidene chloride rod contains 25 w/o borosilicate glass fibers. All the fibers are aligned longitudinally. What fraction of the load is carried by the glass?

Solution: Basis: 1 g. (Data from Appendix C. Use g/cm^3.)

$$v/o \ glass = \frac{(0.25 \ g)/(2.4 \ g/cm^3)}{(0.25/2.4)_{gl} + (0.75)/(1.7)_{pvc}}$$

$$= 19 \ v/o \ (= 19 \ area \ percent);$$

$$load_{gl}/A_{gl}E_{gl} = e_{gl} = e_{pvc} = load_{pvc}/A_{pvc}E_{pvc},$$

$$\frac{load_{gl}}{load_{pvc}} = \frac{(0.19)(70{,}000 \ MPa)}{(0.81)(350 \ MPa)} \simeq \frac{98\%}{2\%}. \ \blacktriangleleft$$

• **Example 13–4.2** Formulate a mixture rule for Young's modulus of a laminate, when the stress is applied perpendicular ($\perp$) to the "grain" of the laminate.

Derivation: With transverse loading, $s_1 = s = s_2$.

$$\bar{E}_\perp = s_\perp/e = s/(e_1 f_1 + e_2 f_2)$$
$$= 1/(f_1/E_1 + f_2/E_2),$$
$$1/\bar{E}_\perp = f_1/E_1 + f_2/E_2. \tag{13–4.5}$$

Comment. This derivation and the one for Eq. (13–4.4) assume that the Poisson ratios for the two components are comparable. Thus there would be no secondary stresses because of differences in lateral strains. ◀

Example 13–4.3 A 0.25-mm (0.10-in.) iron sheet that is to be used in a household oven is coated on *both* sides with a glassy enamel. The final processing occurs above the 500°C (930°F) strain-point (Section 8–8), to give a 0.5-mm (0.02-in.) coating. The glass has a Young's modulus of 70,000 MPa (10^7 psi) and a thermal expansion of $8.0 \times 10^{-6}/°C$.
a) What are the stresses in the glass at 20°C?
b) At 200°C? (Assume no plastic strain.)

Solution: Since $\Delta l/l =$ thermal expansion + elastic strain, and in this case $(\Delta l/l)_{gl} = (\Delta l/l)_{Fe}$, we may write

$$\alpha_{gl}\,\Delta T + s_{gl}/E_{gl} = \alpha_{Fe}\,\Delta T + s_{Fe}/E_{Fe}.$$

a)   By using data from above and from Appendix C,

$$s_{Fe}/205{,}000\ \text{MPa} - s_{gl}/70{,}000\ \text{MPa} = (8.00 - 11.75)(10^{-6}/^\circ\text{C})(-480^\circ\text{C}).$$

But $A_{Fe} = 2.5 A_{gl}$ and $F_{Fe} = -F_{gl}$, so $s_{gl} = -2.5 s_{Fe}$. Thus

$$s_{Fe}\left[\frac{1}{205{,}000} + \frac{2.5}{70{,}000}\right] = 0.0018.$$

Solving, we obtain

$$s_{Fe} = +44\ \text{MPa} \qquad (\text{or }6400\ \text{psi}) \qquad (+ : \text{tension});$$

$$s_{gl} = -110\ \text{MPa} \qquad (\text{or }16{,}000\ \text{psi}) \qquad (- : \text{compression}).$$

b)   By similar calculations for $\Delta T = (500 - 200)^\circ\text{C}$,

$$s_{gl} = -69\ \text{MPa (or }10{,}000\ \text{psi}), \quad \text{and} \quad s_{Fe} = +27.5\ \text{MPa (or }4000\ \text{psi}).$$

Comments. We assumed unidirectional strain. In reality, plane (i.e., two-dimensional) strain occurs. The necessary correction gives a higher stress by the factor of $(1 - v)^{-1}$, where v is Poisson's ratio (Section 6–3). ◀

REVIEW AND STUDY

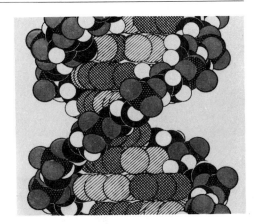

Fig. 13–5.1 Submicroscopic structure (DNA molecule: Watson–Crick model). When uncoiled by other enzymes, each molecule and new double helix will have the same structure and properties as the original molecule. (Courtesy of Dr. L. D. Hamilton, Brookhaven National Laboratory, and The Upjohn Company, Kalamazoo, Michigan. Not to be reproduced without their permission.)

CONCLUSION

This final chapter has selected four examples of materials that are widely used and are therefore familiar to the reader. Admittedly, none of these is a simple material. However, each of them has *its properties governed by its internal structure.* In short, this is the concept of Materials Science and Engineering.

This concept can be extended to all types of materials, whether they be submicroscopic like the DNA molecule with its double helix (Fig. 13–5.1), or they be macroscopic like a radial tire with its tread stock, bead wire, plies, liner rubber, etc.

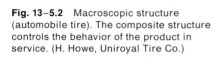

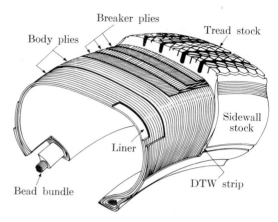

Fig. 13–5.2 Macroscopic structure (automobile tire). The composite structure controls the behavior of the product in service. (H. Howe, Uniroyal Tire Co.)

(Fig. 13–5.2). The scientist, who analyzes the structure, and the engineer, who either designs a material for a given application or selects a material to complete the design of a product, must continuously utilize this concept.

KEY TERMS AND CONCEPT

Absorption

Aggregate

Cast iron

  ferritic

  gray

  malleable

  nodular (ductile)

  pearlitic

  white

Cellulose

Cement

  portland

Composite

Concrete

Damping capacity

Entrained air

Fiber-reinforced plastics (FRP)

Grain (wood)

Graphite flakes

Graphitization

Hydration

Lignin

Malleabilization

Pores

  closed (impermeable voids)

  open (permeable voids)

Porosity

  apparent

  total

Sieve series

Sintering

Tracheids

Unit mixes

Volume

  apparent

  total (bulk)

  true

Warpage

Wood

  spring

  summer

FOR CLASS DISCUSSION

A_{13} Indicate several reasons why cast iron is one of the least expensive metallic materials.

B_{13} Outline the processing differences that lead to gray cast iron, to nodular cast iron, and to malleable cast iron.

C_{13} Outline the processing differences that lead to ferritic, to pearlitic, and to bainitic malleable iron.

D_{13} Can a pearlitic cast iron be changed to a ferritic cast iron without remelting the metal? If so, how? Can a ferritic cast iron be changed to a pearlitic cast iron without remelting the metal? If so, how?

E_{13} Can a gray cast iron be changed to a malleable cast iron without remelting the metal? If so, how? Can a malleable cast iron be changed to a white cast iron without remelting the metal? Why, or why not?

F_{13} Based on microstructures, compare the expected ductilities of ferritic gray cast iron, ferritic malleable cast iron, ferritic steel, ferritic nodular cast iron. Explain your answer.

G_{13} Explain why it is impossible to make large castings out of malleable cast iron.

H_{13} Steels that contain silicon are poor choices for applications that must withstand high temperatures for extended periods of time. Suggest why.

I_{13} Cast irons are usually brazed rather than welded. A brazing alloy is commonly a copper-base filler metal that bonds to the cast iron but doesn't melt it. Why not weld it as we do steels?

J_{13} Assume sand is composed of spheres. How do particles on successive Tyler screens compare in their maximum cross-sectional area? In their volumes?

K_{13} Distinguish among apparent, bulk, and true densities; between total and apparent porosities.

L_{13} If other factors such as shape are equivalent, a fine-mesh sand will have the same total porosity as a coarse-mesh sand. An exception to this generalization occurs if the particle size begins to approach the size of the container. Explain.

M_{13} Water is required to hydrate portland cement and therefore give it strength. However, excess water weakens concrete. Explain.

N_{13} Many consider that concrete dries as it hardens. If so, how does a bridge pier harden below tide level?

O_{13} Compare and contrast the composition, structure, and characteristics of a mortar for a brick wall and a concrete.

P_{13} In order to smooth out the top of a concrete patio, the amateur mason will commonly "pat" the surface of the wet concrete with a trowel. How does this change the structure and properties of the concrete? What kind of adjustment could be made to partially compensate for its effects?

Q_{13} Why does a mason sometimes sprinkle the brick with water before spreading the mortar?

R_{13} Concrete blocks are sometimes made with light-weight aggregate. Suggest how a light-weight aggregate may be manufactured. What properties of these blocks warrant the premium price that they command?

S_{13} A denser wood absorbs more moisture in humid weather than a lighter wood. Why?

T_{13} The low latitudes do not have spring and summer. Account for the grain in their woods. (Cf. Fig. 13–3.1.)

U_{13} Account for the warpage concavity of the upper cut of wood in Fig. 13–3.3.

V_{13} Why is the longitudinal shrinkage of wood negligible compared to the shrinkage in the other directions?

W_{13} Soft woods, such as pine, are cured before milling. Why?

X_{13} Refer to an encyclopedia and report on the difference between "soft" woods and "hard" woods.

Y_{13} Distinguish between particle board and fiber board. What are the anticipated differences in properties?

Z_{13} Plainsawed lumber is more subject to warping than quartersawed lumber. Why?

AA_{13} Within your field of engineering, select a materials application. What processing steps are required to assure the correct initial structure and properties? Will service conditions affect the structure (and therefore the properties)? Present your selection as a report to the class.

STUDY PROBLEMS

13–1.1 A micrographic analysis of a gray cast iron shows 12 v/o graphite and 88 v/o ferrite. Estimate the carbon content.

Answer: 3.7% C and 96.3% (Fe + Si)

13–1.2 Estimate the density of a ferritic gray cast iron (3.5% C).

13–1.3 A white iron contains 0.5% silicon and 3% carbon. (a) What is the approximate carbon content of γ if the metal has been malleabilized at 800°C? (b) How much graphite will form at that temperature?

Answer: (a) ~0.95% C   (b) ~2% graphite

13–1.4 Estimate the density of the pearlitic malleable iron from Study Problem 13–1.3. (Use necessary data from Example 13–1.3.)

13–1.5 What heat-treating sequence will produce a bainitic malleable cast iron? Give a basis for your choices.

13–1.6 What heat-treating sequence will produce a pearlitic malleable cast iron? Give a basis for your choices.

13–2.1 A sample of crushed basalt weighed 482 g after being completely dried. The water level changed from 452 cm³ to 630 cm³ when it was added to a large graduated cylinder. After the water was drained away, but the surface water remained, the sample weighed 487 g. (a) What is the apparent volume? The total volume of the particles? (b) The open porosity of the crushed particles? (c) The apparent density?

Answer: (a) 178 cm^3; 182 cm^3 (b) 2.7 v/o   (c) 2.65 g/cm^3 ($=2.65$ Mg/m^3)

13–2.2 In order to determine the volume of an irregularly shaped concrete building block, three measurements were made: weight (dry) $= 17.57$ kg; weight (saturated with water) $= 18.93$ kg; weight (suspended in water) $= 10.43$ kg (a) What is the bulk volume of the block? (b) The apparent volume of the block?

13–2.3 The bulk density of crushed limestone is 1.827 Mg/m^3 (114 lbs/ft^3) as loaded into a truck. What is its total porosity in the truck if the true density of the limestone is 2.7 Mg/m^3?

Answer: 32 v/o

13–2.4 A small construction truck holds 3500 kg (7700 lb) of gravel when level full. (Capacity $=$ 2 m^3 or 70.6 ft^3.) The apparent density of this gravel is measured by placing 5 kg (11 lb) of gravel into a bucket with a volume of 0.0125 m^3 (0.44 ft^3). This partially filled bucket is then filled with water. The water plus the gravel has a net weight of 15.6 kg (34.3 lb). (a) What is the apparent density of the gravel? (b) How much volume of sand should be obtained to add to this truckload of gravel to produce the greatest packing factor?

13–2.5 A mix of concrete comprises the following volumes (as-received basis): cement, 1; sand, 2.25; gravel, 2.85; water, 0.8. Using the data of Example 13–2.3, calculate the final volume of this mix.

Answer: 4.6 (m^3, cm^3, or ft^3 may be used)

13–2.6 A unit mix of concrete comprises 42.7 kg (94 lb) of cement, 125 kg (275 lb) of sand, 160 kg (350 lb) of gravel, and 21 kg (46 lb) of water. Using the data of Example 13–2.3, calculate the number of unit mixes required for a driveway of 20 m^3 (700 ft^3).

13–2.7 On a dry-weight basis, a unit mix of concrete was to contain 2.7 gravel, 2.2 sand, and 1 cement, plus 0.50 water. The gravel is that of Example 13–2.1, but it is not dry. Rather, the aggregate particles are saturated with water (capillaries and open pores filled). What adjustments are necessary in the unit mix?

Answer: 0.46 units of H$_2$O, rather than 0.50

13–2.8 The mix in Study Problem 13–2.5 is changed to cement, 1; sand, 2; gravel, 2.5; and water, 0.7. In addition, 4 v/o of entrained air is specified. What is the volume of the final mix?

13–2.9 A cement contains 10% Ca$_3$Al$_2$O$_6$, 25% Ca$_3$SiO$_5$, and 65% Ca$_2$SiO$_4$. The temperature rises 7°C during the initial stages of setting. What percent of the total heat of hydration is retained after this initial period if the concrete mix is comparable to that in Example 13–2.4?

Answer: 11%

13–3.1 The bulk density of some dry wood is ~ 0.54 g/cm^3. Estimate the percent porosity if the true density of the dry wood is 1.5 g/cm^3 ($=1.5$ Mg/m^3).

Answer: 64 v/o total porosity

13–3.2 A quartersawed board of Douglas fir was trimmed to 18 mm $\times$ 91 mm (nominal 1 $\times$ 4) when it contained 18 percent moisture. What cross-sectional dimensions will it have if it loses two thirds of its moisture?

13–3.3 The equilibrium moisture content for wood can vary from 6 percent in winter to 14 percent in the summer. Calculate the biannual dimensional change of a solid Douglas fir table top, 900 mm $\times$ 1500 mm $\times$ 20 mm, (a) when it is quartersawed (width is radial); (b) when it is plainsawed (width is tangential).

Answer: (a) $\Delta w = 11$ mm, $\Delta t = 0.4$ mm, $\Delta l = $ nil
          (b) $\Delta w = 18$ mm, $\Delta t = 0.2$ mm, $\Delta l = $ nil

13–3.4 The trimmed door in Example 13–3.1 was shut, then its moisture content increased from 9 to 14 percent. What force would develop if the door is 30 mm thick? Assume that the doorway casing is rigid.

13–3.5 Assume in case (a) of Study Problem 13–3.3 that the edges of the table were clamped in the summer. Would the top split in the winter if $S_l = 110$ MPa (16,000 psi)? Where ranges are given, state your assumed values.

Answer: Assume $S_l = 20S_r$, and use the more critical value of Young's modulus: $s = 12$ MPa > 5.5 MPa. It will split. (1750 psi > 800 psi.)

13–3.6 Birchwood veneer is impregnated with phenol-formaldehyde (Fig. 7–4.1b) to ensure resistance to water and to increase the hardness of the final product. Although dry birch weighs only 0.56 g/cm^3, the true specific gravity of the cellulose–lignin combination is 1.52. (a) How many grams of phenol-formaldehyde (PF) are required to impregnate 10,000 mm^3 (0.6 $in.^3$) of dry birchwood? (b) What is the final density?

Answer: (a) 8.2 g PF   (b) 1.38 g/cm^3 ($= 1.38$ Mg/m^3)

13–3.7 How much water will the table top of Study Problem 13–3.3 absorb going from winter to summer if it were made of (a) balsa (0.16 g/cm^3 dry)? (b) Birch (0.56 g/cm^3 dry)? (c) Explain the greater absorption of the birch.

Answer: (a) ~ 350 g   (b) $\sim 1{,}200$ g

13–4.1 What is the longitudinal elastic modulus for a rod of borosilicate glass-reinforced polystyrene in which all of the glass fibers are oriented lengthwise? There is 80 w/o glass.

Answer: 46,000 MPa (6,500,000 psi)

13–4.2 What are the elastic moduli in the two dissimilar directions of the laminate of Study Problem 10–4.7?

13–4.3 A glass-reinforced plastic rod (fishing pole) is made of 67 v/o borosilicate glass fibers in a polystyrene matrix. What is the thermal expansion coefficient?

Answer: $\sim 4 \times 10^{-6}/°C$

13–4.4 An AISI–SAE 1040 steel wire (cross section 1 mm^2) has an aluminum coating, so that the total cross-sectional area is 1.2 mm^2. (a) What fraction of a 450 N load (100-lb$_f$) will be carried by the steel? (b) What is the electrical resistance of this wire per unit length?

13–4.5 A 2.5-mm (0.10-in.) steel wire coated with 0.5 mm of copper [total dia. $= 3.5$ mm (or 0.14 in.)] is loaded with 4450 N (~ 1000 lb$_f$). (a) What is the elastic strain? (b) How much strain occurs if the total diameter is composed of 1040 steel? (c) If the total wire is composed of copper?

Answer: (a) 0.003

13–4.6 The copper-coated steel (1020) wire of the previous problem is stress-relieved at 400°C (750°F) and is cooled rapidly to 10°C (50°F). (a) Which metal is in tension? (b) What is the stress?

13–4.7 A 2.5-mm (0.1-in.) steel wire ($\rho = 200$ ohm · nm) is to be copper-coated so its resistance is 3.3×10^{-3} ohm/m (~ 0.001 ohm/ft). How thick should the coating be?

Answer: 0.5 mm (0.02 in.)

Appendixes

Appendix A CONSTANTS AND CONVERSIONS

Constants*

Acceleration of gravity, g	$9.80 \ldots$ m/s^2
Atomic mass unit, amu	$1.66 \ldots \times 10^{-24}$ g
Avogadro's number, N	$0.6022 \ldots \times 10^{24}$ mole^{-1}
Boltzmann's constant, k	$86.1 \ldots \times 10^{-6}$ eV/K
	$13.8 \ldots \times 10^{-24}$ J/K
	8.31 J/mole $\cdot$ K
Capacitivity (vacuum), ϵ	$8.85 \ldots \times 10^{-12}$ C/V$\cdot$m
Electron charge, q	$0.1602 \ldots \times 10^{-18}$ C
Electron moment, β	$9.27 \ldots \times 10^{-24}$ A$\cdot$m^2
Electron volt, eV	$0.160 \ldots \times 10^{-18}$ J
Faraday, $\mathscr{F}$	$96.5 \ldots \times 10^3$ C
Fe–Fe$_3$C eutectoid composition	0.77 w/o carbon
Fe–Fe$_3$C eutectoid temperature	727°C (1340°F)
Gas constant, R	$8.31 \ldots$ J/mole $\cdot$ K
	$1.987 \ldots$ cal/mole $\cdot$ K
Gas volume (STP)	$22.4 \ldots \times 10^{-3}$ m^3/mole
Planck's constant, h	$0.662 \ldots \times 10^{-33}$ J$\cdot$s
Velocity of light, c	$0.299 \ldots \times 10^9$ m/s

Conversions*

1 ampere	$= 1$ C/s
1 angstrom	$= 10^{-10}$ m
	$= 10^{-8}$ cm
	$= 0.1$ nm
	$= 3.937 \times 10^{-9}$ in.
1 amu	$= 1.66 \ldots \times 10^{-24}$ g
1 Btu	$= 1.055 \ldots \times 10^3$ J
1 Btu/°F	$= 1.899 \ldots \times 10^3$ J/°C
1 [Btu/(ft$^2 \cdot$s)]/[°F/in.]	$= 0.519 \ldots \times 10^3$ [J/(m$^2 \cdot$s)]/[°C/m]
	$= 0.519 \ldots \times 10^3$ (W/m^2)/(°C/m)
1 Btu$\cdot$ft^2	$= 11.3 \ldots \times 10^3$ J/m^2
1 calorie, gram	$= 4.18 \ldots$ J
1 centimeter	$= 10^{-2}$ m
	$= 0.3937$ in.
1 coulomb	$= 1$ A$\cdot$s
1 cubic centimeter	$= 0.0610 \ldots$ in^3.
1 cubic inch	$= 16.3 \ldots \times 10^{-6}$ m^3
1°C difference	$= 1.8$ °F
1 electron volt	$= 0.160 \ldots \times 10^{-18}$ J
1°F difference	$= 0.555 \ldots$ °C

---

* All irrational values are rounded downward.

Conversions (*continued*)

1 foot	$= 0.3048 \ldots$ m
1 foot·pound$_f$	$= 1.355 \ldots$ J
1 gallon (U S. liq.)	$= 3.78 \ldots \times 10^{-3}$ m^3
1 gram	$= 0.602 \ldots \times 10^{24}$ amu
	$= 2.20 \ldots \times 10^{-3}$ lb$_m$
1 gram/cm^3	$= 62.4 \ldots$ lb$_m$/ft^3
	$= 1000$ kg/m^3
	$= 1$ Mg/m^3
1 inch	$= 0.0254 \ldots$ m
1 joule	$= 0.947 \ldots \times 10^{-3}$ Btu
	$= 0.239 \ldots$ cal, gram
	$= 6.24 \ldots \times 10^{18}$ eV
	$= 0.737 \ldots$ ft·lb$_f$
	$= 1$ watt·sec
1 joule/meter^2	$= 8.80 \ldots \times 10^{-5}$ Btu/ft^2
1 [joule/(m^2·s)]/[°C/m]	$= 1.92 \ldots \times 10^{-3}$ [Btu/(ft^2·s)]/[°F/in.]
1 kilogram	$= 2.20 \ldots$ lb$_m$
1 megagram/meter^3	$= 1$ g/cm^3
	$= 10^6$ g/m^3
	$= 1000$ kg/m^3
1 meter	$= 10^{10}$ Å
	$= 10^9$ nm
	$= 3.28 \ldots$ ft
	$= 39.37$ in.
1 micrometer	$= 10^{-6}$ m
1 nanometer	$= 10^{-9}$ m
1 newton	$= 0.224 \ldots$ lb$_f$
1 ohm·inch	$= 0.0254 \ldots$ Ω·m
1 ohm·meter	$= 39.37$ Ω·in.
1 pascal	$= 0.145 \ldots \times 10^{-3}$ lb$_f$/in.^{2}
1 poise	$= 0.1$ Pa·s
1 pound (force)	$= 4.44 \ldots$ newtons
1 pound (mass)	$= 0.453 \ldots$ kg
1 pound/foot^3	$= 16.0 \ldots$ kg/m^3
1 pound/inch^2	$= 6.89 \ldots \times 10^{-3}$ MPa
1 watt	$= 1$ J/s
1 (watt/m^2)/(°C/m)	$= 1.92 \ldots \times 10^{-3}$ [Btu/(ft^2·s)]/[°F/in.]

SI prefixes

giga	G	10^9
mega	M	10^6
kilo	k	10^3
milli	m	10^{-3}
micro	μ	10^{-6}
nano	n	10^{-9}

Appendix B TABLE OF SELECTED ELEMENTS

Element	Symbol	Atomic number	Atomic mass, amu	Orbitals	Melting point, °C	Density (solid), Mg/m³ (= g/cm³)	Crystal structure, 20°C	Approx. atomic radius, nm†	Valence (most common)	Approx. ionic radius, nm‡
				1s						
Hydrogen	H	1	1.0078	1	−259.14	—	—	0.046	1+	Very small
Helium	He	2	4.003	2	−272.2	—	—	0.176	Inert	small
				2s 2p						
Lithium	Li	3	6.94	He + 1	180	0.534	bcc	0.1519	1+	0.068
Beryllium	Be	4	9.01	He + 2	1289	1.85	hcp	0.114	2+	0.035
Boron	B	5	10.81	He + 2 1	2103	2.34		0.046	3+	~0.025
Carbon	C	6	12.011	He + 2 2	>3500	2.25	hex	0.077		—
Nitrogen	N	7	14.007	He + 2 3	−210			0.071	3−	—
Oxygen	O	8	15.999	He + 2 4	−218.4			0.060	2−	0.140
Fluorine	F	9	19.00	He + 2 5	−220			0.06	1−	0.133
Neon	Ne	10	20.18	He + 2 6	−248.7		fcc	0.160	Inert	—
				3s 3p						
Sodium	Na	11	22.99	Ne + 1	97.8	0.97	bcc	0.1857	1+	0.097
Magnesium	Mg	12	24.31	Ne + 2	649	1.74	hcp	0.161	2+	0.066
Aluminum	Al	13	26.98	Ne + 2 1	660.4	2.70	fcc	0.14315	3+	0.051
Silicon	Si	14	28.09	Ne + 2 2	1414	2.33	*	0.1176	4+	0.042
Phosphorus	P	15	30.97	Ne + 2 3	44	1.8		0.11	5+	~0.035
Sulfur	S	16	32.06	Ne + 2 4	112.8	2.07		0.106	2−	0.184
Chlorine	Cl	17	35.45	Ne + 2 5	−101			0.0905	1−	0.181
Argon	Ar	18	39.95	Ne + 2 6	−189.2		fcc	0.192	Inert	—

Element				Config	3d	4s	4p	T	ρ	Structure	R†	Valence	Radius‡
Potassium	K	19	39.10	Ar +		1		63	0.86	bcc	0.2312	1+	0.133
Calcium	Ca	20	40.08	Ar +		2		840	1.54	fcc	0.1969	2+	0.099
Titanium	Ti	22	47.90	Ar +	2	2		1672	4.51	hcp	0.146	4+	0.068
Chromium	Cr	24	52.00	Ar +	5	1		1863	7.20	bcc	0.1249	3+	0.063
Manganese	Mn	25	54.94	Ar +	5	2		1246	7.2	—	0.112	2+	0.080
Iron	Fe	26	55.85	Ar +	6	2		1538	7.88	bcc	0.1241	2+	0.074
										fcc	0.1269	3+	0.064
Cobalt	Co	27	58.93	Ar +	7	2		1494	8.9	hcp	0.125	2+	0.072
Nickel	Ni	28	58.71	Ar +	8	2		1455	8.90	fcc	0.1246	2+	0.069
Copper	Cu	29	63.54	Ar +	10	1		1084.5	8.92	fcc	0.1278	1+	0.096
Zinc	Zn	30	65.37	Ar +	10	2		419.6	7.14	hcp	0.139	2+	0.074
Germanium	Ge	32	72.59	Ar +	10	2	2	937	5.35	*	0.1224	4+	—
Arsenic	As	33	74.92	Ar +	10	2	3	~809	5.73	—	0.125	3+	—
Krypton	Kr	36	83.80	Ar +	10	2	6	-157	—	fcc	0.201	Inert	—

Element				Config	4d	5s	5p	T	ρ	Structure	R†	Valence	Radius‡
Silver	Ag	47	107.87	Kr + 10	10	1		961.9	10.5	fcc	0.1444	1+	0.126
Tin	Sn	50	118.69	Kr + 10	10	2	2	232	7.3	bct	0.1509	4+	0.071
Antimony	Sb	51	121.75	Kr + 10	10	2	3	630.7	6.7	ortho	0.1452	5+	—
Iodine	I	53	126.9	Kr + 10	10	2	5	114	4.93	ortho	0.135	1−	0.220
Xenon	Xe	54	131.3	Kr + 10	10	2	6	-112	2.7	fcc	0.221	Inert	—

Element				Config	4f	5d	6s	T	ρ	Structure	R†	Valence	Radius‡
Cesium	Cs	55	132.9	Xe +			1	28.4	1.9	bcc	0.262	1+	0.167
Tungsten	W	74	183.9	Xe + 14	14	4	2	3387	19.4	bcc	0.1367	4+	0.070
Gold	Au	79	197.0	Xe + 14	14	10	1	1064.4	19.32	fcc	0.1441	1+	0.137
Mercury	Hg	80	200.6	Xe + 14	14	10	2	-38.86	—	—	0.155	2+	0.110
Lead	Pb	82	207.2	Hg + 6p²				327.5	11.34	fcc	0.1750	2+	0.120
Uranium	U	92	238.0	Rn + 5f³	6d	7s²		1133	19	—	0.138	4+	0.097

* Diamond cubic
† One half of closest approach of two atoms in the elemental solid. For noncubic structures, the average interatomic distance is given; e.g., in hcp, the atom is slightly ellipsoidal.
‡ Radii for CN = 6; otherwise, $0.97 R_{CN=8} \approx R_{CN=6} \approx 1.1 R_{CN=4}$. Patterned after Ahrens.

$$\Delta L / L = \alpha_L \, \Delta T$$

For Volume
$$\Delta V / V = \alpha_V \, \Delta T \quad \text{where} \quad \alpha_V = 3\alpha_L$$

Appendix C PROPERTIES OF SELECTED ENGINEERING MATERIALS (20°C)*

Material	Density Mg/m^3 $(=g/cm^3)$	Thermal conductivity, K $\left(\dfrac{watts}{mm^2}\right)\Big/\left(\dfrac{°C}{mm}\right)$ **	Linear expansion, α_L $°C^{-1}$ †	Electrical resistivity, ρ $ohm \cdot m$ ‡	Average modulus of elasticity, $\bar{E}$ MPa	psi
Metals						
Aluminum (99.9+)	2.7	0.22	22.5×10^{-6}	29×10^{-9}	70,000	10×10^6
Aluminum alloys	2.7(+)	0.16	22×10^{-6}	$\sim45 \times 10^{-9}$	70,000	10×10^6
Brass (70 Cu–30 Zn)	8.5	0.12	20×10^{-6}	62×10^{-9}	110,000	16×10^6
Bronze (95 Cu–5 Sn)	8.8	0.08	18×10^{-6}	$\sim100 \times 10^{-9}$	110,000	16×10^6
Cast iron (gray)	7.15	—	10×10^{-6}	—	140,000($\pm$)	$20 \times 10^6\pm$
Cast iron (white)	7.7	—	9×10^{-6}	660×10^{-9}	205,000	30×10^6
Copper (99.9+)	8.9	0.40	17×10^{-6}	17×10^{-9}	110,000	16×10^6
Iron (99.9+)	7.88	0.072	11.7×10^{-6}	98×10^{-9}	205,000	30×10^6
Lead (99+)	11.34	0.033	29×10^{-6}	206×10^{-9}	14,000	2×10^6
Magnesium (99+)	1.74	0.16	25×10^{-6}	45×10^{-9}	45,000	6.5×10^6
Monel (70 Ni–30 Cu)	8.8	0.025	15×10^{-6}	482×10^{-9}	180,000	26×10^6
Silver (sterling)	10.4	0.41	18×10^{-6}	18×10^{-9}	75,000	11×10^6
Steel (1020)	7.86	0.050	11.7×10^{-6}	169×10^{-9}	205,000	30×10^6
Steel (1040)	7.85	0.048	11.3×10^{-6}	171×10^{-9}	205,000	30×10^6
Steel (1080)	7.84	0.046	10.8×10^{-6}	180×10^{-9}	205,000	30×10^6
Steel (18 Cr–8 Ni stainless)	7.93	0.015	9×10^{-6}	700×10^{-9}	205,000	30×10^6
Ceramics						
Al_2O_3	3.8	0.029	9×10^{-6}	$>10^{12}$	350,000	50×10^6
Brick						
Building	2.3($\pm$)	0.0006	9×10^{-6}	—	—	—
Fireclay	2.1	0.0008	4.5×10^{-6}	1.4×10^6	—	—
Graphite	1.5	—	5×10^{-6}	—	—	—

Material						
Paving	2.5	—	4×10^{-6}	—	—	—
Silica	1.75	0.0008	—	1.2×10^{6}	—	—
Concrete	2.4($\pm$)	0.0010	13×10^{-6}	—	14,000	2×10^{6}
Glass						
Plate	2.5	0.00075	9×10^{-6}	10^{12}	70,000	10×10^{6}
Borosilicate	2.4	0.0010	2.7×10^{-6}	$>10^{15}$	70,000	10×10^{6}
Silica	2.2	0.0012	0.5×10^{-6}	10^{18}	70,000	10×10^{6}
Vycor	2.2	0.0012	0.6×10^{-6}	—	—	—
Wool	0.05	0.00025	—	—	—	—
Graphite (bulk)	1.9	—	5×10^{-6}	10^{-5}	7,000	1×10^{6}
MgO	3.6	—	9×10^{-6}	10^{3} (1100°C)	205,000	30×10^{6}
Quartz (SiO_2)	2.65	0.012	—	10^{12}	310,000	45×10^{6}
SiC	3.17	0.012	4.5×10^{-6}	0.025 (1100°C)	—	—
TiC	4.5	0.030	7×10^{-6}	50×10^{-8}	350,000	50×10^{6}
Polymers						
Melamine-formaldehyde	1.5	0.00030	27×10^{-6}	10^{11}	9,000	1.3×10^{6}
Phenol-formaldehyde	1.3	0.00016	72×10^{-6}	10^{10}	3,500	0.5×10^{6}
Urea-formaldehyde	1.5	0.00030	27×10^{-6}	10^{10}	10,300	1.5×10^{6}
Rubbers (synthetic)	1.5	0.00012	—	—	4–75	600–11,000
Rubber (vulcanized)	1.2	0.00012	81×10^{-6}	10^{12}	3,500	0.5×10^{6}
Polyethylene (L.D.)	0.92	0.00034	180×10^{-6}	10^{13}–10^{16}	100–350	14,000–50,000
Polyethylene (H.D.)	0.96	0.00052	120×10^{-6}	10^{12}–10^{16}	350–1,250	50,000–180,000
Polystyrene	1.05	0.00008	63×10^{-6}	10^{16}	2,800	0.4×10^{6}
Polyvinylidene chloride	1.7	0.00012	190×10^{-6}	10^{11}	350	0.05×10^{6}
Polytetrafluoroethylene	2.2	0.00020	100×10^{-6}	10^{14}	350–700	50,000–100,000
Polymethyl methacrylate	1.2	0.00020	90×10^{-6}	10^{14}	3,500	0.5×10^{6}
Nylon	1.15	0.00025	100×10^{-6}	10^{12}	2,800	0.4×10^{6}

* Data in this table were taken from numerous sources.

** Alternatively, $(W/mm^2)/(K/mm)$. Multiply by 1.92 to get $Btu/(ft^2 \cdot s)/(°F/in.)$.

† Or, K^{-1}; divide by 1.8 to get $°F^{-1}$.

‡ Multiply ohm·m by 39 to get ohm·in.

Appendix D GLOSSARY OF TERMS AS APPLIED TO MATERIALS

Abrasive   Hard, mechanically resistant material used for grinding or cutting; commonly made of a ceramic material.

Absorption   Volume assimilation (cf. *ad*sorption).

$A_mB_nX_p$ compounds   Ternary compounds. In this text X is generally oxygen; A and B are commonly metal atoms.

Acceptor levels   Energy levels of *p*-type (electron–hole) carriers.

Acceptor saturation   Filled acceptor levels in *p*-type semiconductors. As a result, additional thermal activation does not increase the number of extrinsic carriers.

Activation energy   Energy barrier that must be met prior to reaction.

Addition polymerization   Polymerization by sequential addition of monomers.

Additive properties   Properties of mixtures that depend on geometry only (and not on phase interactions).

Adsorption   Surface adhesion (cf. *ab*sorption).

Age-hardening   *See* Precipitation hardening.

Agglomerated materials   Small particles bonded together into an integrated mass.

Aggregate   Coarse particles used in concrete; for example, sand and gravel.

AISI–SAE steels   Standardized identification code for plain-carbon and low-alloy steels, which is based on composition. Last two numbers indicate the carbon content. (*See* Table 9–9.1.)

Allotropism   *See* Polymorphism.

Alloy   A metal containing two or more elements.

Alloy retardation   Decrease in the rate of austenite decomposition because of alloying elements.

Alloying elements   Elements added to form an alloy (sometimes referred to as alloys).

Alpha iron   Iron with a body-centered cubic structure that is stable at room temperature.

Amorphous   Noncrystalline and without long-range order.

Anion   Negative ion.

Anisotropic   Having different properties in different directions.

Annealing   Heating and cooling to produce softening.

Annealing point (glass)   Stress-relief heat treatment. The temperature should provide a viscosity of $\sim 10^{13.5}$ Pa·s.

Anode   The electrode that supplies electrons to an external circuit.

Anodized   Surface-coated with an oxide layer; achieved by making the component an anode in an electrolytic bath.

Arrhenius equation   Thermal activation relationship. (*See* comments with Example 4–6.2.)

Asbestos   A fibrous silicate material.

ASTM grain-size number   Standardized grain counts. *See* Eq. (4–4.2).

Atactic   Lack of long-range repetition in a polymer (as contrasted to isotactic).

Atomic mass unit (amu)   One twelfth of the mass of C^{12}; gram/$(0.602 \ldots \times 10^{24})$.

Atomic number   The number of electrons possessed by an uncharged atom.

Atomic packing factor   Fraction of volume occupied by "spherical" atoms (or ions).

Atomic radius (elements)   Half of interatomic distance.

Atomic weight   Atomic mass expressed in atomic mass units.

Austempering   Process of isothermal transformation to form bainite.

Austenite (γ)   Face-centered cubic iron, or iron alloy based on this structure.

Austenite decomposition   Eutectoid reaction which changes austenite to (α + carbide).

Austenite grain size   *See* ASTM grain size.

Austenization   Heat treatment to dissolve carbon into fcc iron, thereby forming austenite.

Avogadro's number (N)   Number of amu's per gram; hence, the number of molecules per mole.

AX compounds   Binary compounds with a 1-to-1 ratio of the two elements; commonly ionic.

AX structure (CsCl)   Structure of a binary compound with CN = 8.

AX structure (NaCl)   Structure of a binary compound with CN = 6.

AX structure (ZnS)   Structure of a binary compound with CN = 4.

A_mX_p structure   Binary compound with unequal component ratio. (*See* Section 8–3.)

Axis (crystals)   One of three principal crystal directions.

Bainite   Microstructure of carbide dispersed in ferrite, obtained by low-temperature isothermal transformation.

Base   The center zone of a transistor.

Bifunctional   Molecule with two reaction sites for joining with adjacent molecules.

Body-centered cubic (bcc)   A cubic unit cell with the center position equivalent to corner positions.

Bohr magneton (β)   Magnetic moment of individual electron ($9.27 \ldots \times 10^{-24}$ amp·m²).

Boltzmann's constant (k)   Thermal energy coefficient ($13.8 \ldots \times 10^{-24}$ J/K).

Bond angle   Angle between stereospecific bonds in molecules, or covalent solids.

Bond energy   Energy required to separate two chemically bonded atoms. Generally expressed as energy per mole of 0.6×10^{24} bonds.

Bond length   Interatomic distance of stereospecific bond.

Boundary (microstructures)   Surface between two grains or between two phases.

Bragg's law   Diffraction law for periodic structures (Eq. 3–8.2).

Branching   Bifurcation in addition polymerization.

Brass   An alloy of copper and zinc.

Bravais lattices   The 14 basic crystal lattices.

Brazing   Joining metals at temperatures above 425°C (800°F), but below the melting point of the joined metals.

Bridging oxygens   Oxygen atoms shared by two adjacent silica tetrahedra.

Brinell   A hardness test utilizing a spherical indenter. The hardness is determined from the diameter of the indentation.

Brittle   Opposite of tough; fractures with little energy absorption.

Bronze   An alloy of copper and tin (unless otherwise specified; e.g., an aluminum bronze is an alloy of copper and aluminum).

Burgers vector (b)   Displacement vector around a dislocation. *See* slip vector.

Butadiene-type compound   Prototype for several rubbers based on C=C—C=C. (*See* Table 7–2.2.)

Calcination   Solid dissociation to a gas and another solid; for example, $CaCO_3 \rightarrow CaO + CO_2$.

Carbide ($\overline{C}$)   Compound of metal and carbon. Unless specifically stated, it refers to iron-base carbides.

Carbon steel   Steel in which carbon is the chief variable alloying element (other alloying elements may be present only in nominal amounts).

Carburize   Introduction of carbon through the surface of the steel by diffusion, to change the surface properties.

Case   Subsurface zone (usually of a carburized steel).

Case-hardening   Hardening by forming a case of higher carbon content.

Cast iron   Fe–C alloy, sufficiently rich in carbon to produce a eutectic liquid during solidification. In practice, this generally means more carbon than can be dissolved in austenite ($>2\%$). Also see gray, malleable, nodular, and white cast iron.

Casting   The process of pouring a liquid or suspension into a mold, or the object produced by this process.

Catalyst   A reusable agent for activating a chemical reaction.

Cathode   The electrode that receives electrons from an external circuit.

Cathode reactions   Reduction (electron-consuming) reactions on a galvanic electrode.

Cation   Positive ion.

C-curve   Isothermal transformation curve.

Cell (galvanic)   A combination of two electrodes in an electrolyte.

Cell, composition   Galvanic cell between electrodes of different compositions.

Cell, concentration   Galvanic cell arising from nonequal electrolyte concentrations. (The more dilute solution produces the anode.)

Cell, oxidation   Galvanic cell arising from nonequal oxygen potentials. (The oxygen-deficient area becomes the anode.)

Cell, stress   Galvanic cell arising from a plastically deformed anode.

Cell, unit   *See* unit cell.

Cellulose   Natural polymer of $C_6H_{10}O_5$. (*See* Eq. 13–3.1.)

Cement   A material (usually ceramic) for bonding solids together.

Cement, hydraulic   Cement that bonds by a reaction with water.

Cement, portland   A hydraulic calcium silicate cement.

Cementite   Iron carbide (Fe_3C).

Center-of-gravity method   Calculation method for determining phases. The overall composition is at the center of gravity of the weighted components.

Ceramics   Materials consisting of compounds of metallic and nonmetallic elements.

Charge carriers   Electrons in the conduction band provide n-type (negative) carriers. Electron holes in the valence band provide p-type (positive) carriers.

Charge density ($\mathcal{D}$)   Coulombs per unit area.

Charpy   One of two standardized impact tests, utilizing a square notched bar.

Cis **(polymers)**   A prefix denoting unsaturated positions on the same side of the polymer chain.

Clay   Fine soil particles (<0.1 mm). In ceramics, clays are specifically sheetlike alumino-silicates.

Cleavage   Plane of easy splitting.

Coalescence   Change from many small particles to fewer, large particles.

Coercive force (electric, $\mathcal{E}_c$)   Electric field required to remove residual polarization.

Coercive force (magnetic, H_c)   Magnetic field required to remove residual magnetization.

Cold work, %   Plastic strain calculated from change in cross-sectional area, $100\,(A_o - A_f)/A_o$.

Cold-working   Deformation below the recrystallization temperature.

Collector   The zone of a transistor that receives the charge carriers across the base from the emitter.

Compact   Compressed shapes of powders prior to sintering.

Component (design)   The individual parts of a machine or similar engineering design.

Component (phases)   The basic chemical substances required to create a chemical mixture or solution.

Composite   Material containing two or more distinct materials.

Compound   A phase composed of two or more elements in a given ratio.

Concrete   Agglomerate of aggregate and a hydraulic cement.

Conduction band   Energy band of conduction electrons. Electrons must be in this band to be carriers.

Conduction electron   Electron raised above the energy gap to serve as negative charge carrier.

Conductivity   Transfer of thermal or electrical energy along a potential gradient.

Configuration   Arrangement of mers along a polymer chain. (Rearrangements require bond breaking.)

Conformation   Twisting and/or kinking of a polymer chain. (Changes require bond rotation only.)

Continuous cooling transformation   A thermal reaction during cooling, particularly austenite decomposition.

Cooling rate   Decrease in temperature per second; specifically, the rate of change at the transformation temperature.

Coordination number (CN)   Number of closest ionic or atomic neighbors.

Copolymer   Polymers with more than one type of mer.

Core (heat-treating)   Center of a bar, inside the case.

Corrosion   Deterioration and removal by chemical attack.

Coulombic forces   Forces between charged particles, particularly ions.

Covalent bond   Interatomic bond created when two adjacent atoms share a pair of electrons.

Creep   A slow deformation by stresses below the normal yield strength (commonly occurring at elevated temperatures).

Creep rate   Creep strain per unit of time.

Critical cooling rates (austenite decomposition) CR_M is the slowest cooling rate, which produces only *martensite*. CR_P is the fast cooling rate, which produces all *pearlite*.

Critical shear stress   Minimum resolved stress to produce shear.

Cross-linking   The tying together of adjacent polymer chains.

Crystal   A physically uniform solid, in three dimensions, with long-range repetitive order.

Crystal direction [uvw]   A ray from an arbitrary origin through a selected unit-cell location. The indices are the lattice coefficients of that location.

Crystal lattice   The spatial arrangement of equivalent sites within a crystal.

Crystal plane   Two-dimensional array of atoms. (*See* also Miller indices.)

Crystal system   Categorization of unit cells by axial and dimensional symmetry (Table 3–1.1).

Crystallinity (polymers)   Volume fraction of a solid that has a crystalline (as contrasted to an amorphous) structure.

Curie point (magnetic)   Transition temperature between ferromagnetism and paramagnetism.

Current (I)   Flow of positive charge (opposite to electron movement).

Current density (i)   Amperage per unit area.

Damping capacity   Attentuation of mechanical vibrations.

Defect structure   Compounds with noninteger ratios of atoms or ions. These compounds contain either vacancies or interstitials within the structure.

Deformation, elastic   Reversible deformation without permanent atomic (or molecular) displacements.

Deformation, plastic   Permanent deformation arising from the displacement of atoms (or molecules) to new surroundings.

Deformation crystallization   Crystallization occurring as polymers are unkinked into parallel linear orientations.

Degradation   Reduction of polymers to smaller molecules.

Degree of polymerization (n)   Mers per average molecular weight.

Delocalization   Multiple-atom orbitals.

Density, apparent   Mass divided by apparent volume (material + closed pores).

Density, bulk   Mass divided by total volume.

Density, true   Mass divided by true (pore-free) volume.

Dielectric An insulator. A material that can be placed between two electrodes without conduction.

Dielectric constant, relative (κ) Ratio of charge density arising from an electric field (1) with and (2) without the material present.

Dielectric strength Electrical breakdown potential of an insulator per unit thickness.

Diffraction (x-ray) Deviation of an x-ray beam by regularly spaced atoms.

Diffusion The movement of atoms or molecules in a material.

Diffusivity (D) Diffusion flux per unit concentration gradient.

Dipole An electrical couple with positively and negatively charged ends.

Dipole moment Product of electric charge and charge separation distance.

Directional solidification Heat extraction from one end of a mold to produce freezing.

Dislocation, edge ($\perp$) Linear defect at the edge of an extra crystal plane. The slip vector is perpendicular to the defect line.

Dislocation, screw ($\int$) Linear defect with slip vector parallel to the defect line.

Dispersed phases Microstructure of very fine particles within a matrix phase.

Domains Microstructural areas of coordinated magnetic alignments (or of electrical dipole alignments).

Donor exhaustion Depletion of donor electrons. Because of it, additional thermal activation does not increase the number of extrinsic carriers.

Donor levels Energy levels of n-type (electron) carriers.

Drawing Mechanical forming by tension through a die, e.g., wire drawing (Fig. 6–2.3) and sheet drawing; usually carried out at temperatures below the recrystallization temperature.

Drawing (glass) Shaping by drawing viscous glass through an orifice into sheet, rod or fiber.

Drift velocity ($\bar{v}$) Net velocity of electrons in an electric field.

Ductile fracture Fracture accompanied by plastic deformation, and therefore by energy absorption.

Ductility Permanent deformation before fracture; measured as elongation or reduction in area.

Elastic strain Nonpermanent deformation.

Elastomer Polymer with a large elastic strain. This strain arises from the unkinking of the polymer chain.

Electric field ($\mathscr{E}$) Voltage gradient, volts/cm.

Electrical conductivity (σ) Coefficient between charge flux and electric field. Reciprocal of electrical resistivity.

Electrical resistivity (ρ) Resistance of a material with unit dimensions. Reciprocal of electrical conductivity.

Electrode potential (ϕ) Voltage developed at an electrode (as compared with a standard reference electrode).

Electrolyte Conductive ionic solution (liquid or solid).

Electron charge (q) The charge of 0.16×10^{-18} coul (or 0.16×10^{-18} amp·sec) carried by each electron.

Electron hole (p) Electron vacancy in the valence band that serves as a positive charge carrier.

Electron–hole pair A conduction electron in the conduction band and an accompanying electron hole in the valence band, which result when an electron jumps the gap in an intrinsic semiconductor.

Electronegativity Measure of electron attraction for nonmetallic characteristics.

Electron repulsion Repelling force of too many electrons in the same vicinity. Counteracts the attractive bonding forces.

Electroplating Cathodic reduction process, opposite of corrosion. Electrons are supplied by external circuit.

Elongation Axial strain accompanying fracture. (A gage length must be stated.)

End-quench test Standardized test, by quenching from one end only, for determining hardenability.

Endurance limit The maximum stress allowable for unlimited cycling.

Energy band Permissible energy levels for valence electrons.

Energy distribution Spectrum of energy levels determined by thermal activation.

Energy gap (E_g) Unoccupied energies between the valence band and the conduction band.

Energy well Interatomic potential energy minimum.

Entrained air Bubbles (~ 1 mm) of air added to concrete to facilitate workability and to diminish freezing damage.

Equiaxial Comparable dimensions in the three principal directions.

Equicohesive temperature The temperature of equal strength for grains and grain boundaries.

Equilibrium (chemical) Reaction ceases (because the minimum free energy has been reached).

Equilibrium diagram *See* Phase diagram.

Equivalent sites Crystal lattice sites with fully identical surroundings.

Eutectic analysis (composition) Analysis of liquid-solution phase with the minimum melting temperature (at the intersection of two solubility curves).

Eutectic reaction $L_2 \underset{\text{Heating}}{\overset{\text{Cooling}}{\rightleftarrows}} S_1 + S_3$.

Eutectic temperature Temperature of the eutectic reaction at the intersection of two solubility curves.

Eutectoid analysis (composition) Analysis of solid-solution phase with the minimum decomposition temperature (at the intersection of two solid solubility curves).

Eutectoid ferrite *See* Ferrite, eutectoid.

Eutectoid reaction $S_2 \underset{\text{Heating}}{\overset{\text{Cooling}}{\rightleftarrows}} S_1 + S_3$.

Eutectoid shift The change in temperature and the carbon analysis of the eutectoid reaction arising from alloying element additions.

Eutectoid temperature Temperature of the eutectoid reaction at the intersection of two solid solubility curves.

Expansion coefficient *See* Thermal expansion.

Extrinsic semiconductors *See* Semiconductors.

Extrusion Shaping by pushing the material through a die (Fig. 6–2.2).

Face-centered cubic (fcc) A unit cell with face positions equivalent to corner positions.

Family of directions $\langle uvw \rangle$ Crystal directions that are identical except for our arbitrary choice of axes.

Family of planes $\{hkl\}$ Crystal planes that are identical except for our arbitrary choice of axes.

Faraday ($\mathscr{F}$) A unit charge per 0.6×10^{24} electrons (96,500 coul).

Fatigue Tendency to fracture under cyclic stresses.

Ferrite (ceramics) Compounds containing trivalent iron; commonly magnetic.

Ferrite (metals, α) Body-centered cubic iron, or an iron alloy based on this structure.

Ferrite, eutectoid Ferrite that forms (along with carbide) during austenite decomposition.

Ferrite, proeutectoid Ferrite that separates from austenite above the eutectoid temperature.

Ferroelectric Materials with spontaneous dipole alignment.

Ferromagnetic Materials with spontaneous magnetic alignment.

Fiber-reinforced plastics (FRP) Composite of glass fibers and plastics.

Fick's first law Proportionality between diffusion flux and concentration gradient.

Firing   *See* Sintering.

Flame-hardening   Hardening by means of surface heating by flames (followed by quenching)

Fluidity (f)   Coefficient of flowability; reciprocal of viscosity.

Fluorescence   Luminescence that occurs immediately after excitation.

Flux (diffusion, J)   Transport per unit area and time.

Forging   Mechanical shaping by compression (Fig. 6–2.2).

Form $\{hkl\}$ Crystal planes that are identical except for our arbitrary choice of axes.

Fracture, brittle   Failure by crack propagation and with the absence of significant ductility.

Fracture, ductile   Failure by crack propagation accompanied by plastic deformation.

Frenkel defect   Atom or ion displacement (combined vacancy and interstitial).

Functionality   Number of available reaction sites for polymerization.

Fusion (heat of)   Thermal energy required for the melting of crystalline solid.

Gage length   Initial dimensions, for determining elongation. (*See* Fig. 1–2.2.)

Galvanic cell   A cell containing two dissimilar metals and an electrolyte.

Galvanic protection   Protection given to a material by making it the cathode to a sacrificial anode.

Galvanization   The process of coating steel with zinc to give galvanic protection.

Gamma iron (γ)   *See* Austenite.

Glass   An amorphous solid below its transition temperature. A glass lacks long-range crystalline order but normally has short-range order.

Glass transition temperature (T_g)   Transition temperature between a supercooled liquid and its glass solid.

Grading   Size distribution of aggregate.

Grain (metals and ceramics)   Individual crystal of a microstructure.

Grain (wood)   Macrostructure from growth cycle.

Grain boundary   The zone of crystalline mismatch between adjacent grains.

Grain boundary area (S_V)   Area/unit volume; for example, in^2/in^3 or mm^2/mm^3.

Grain growth   Increase in average grain size by atoms diffusing across grain (or phase) boundaries.

Grain size   Statistical grain diameter in a random cross section. (Austenite grain size is reported as the number of former austenite grains within a standardized area. *See* Section 4–4.)

Graphitization   Decomposition of iron carbide into iron and carbon (Eq. 11–1.1).

Hard-drawn   Cold-worked to high hardnesses by drawing.

Hardenability   The ability to develop maximum hardness by avoiding the ($\gamma \rightarrow \alpha$ + carbide) reaction.

Hardenability curve   Hardness profile of end-quench test bar.

Hardness   Resistance to penetration.

Hardness traverse   Profile of hardness values.

Heat capacity (c)   Energy per unit temperature, dH/dT.

Hexagonal close-packed (hcp)   The lattice of a hexagonal crystal with lattice points at the corners *and* in offset positions at midheight. (*See* Fig. 3–3.2.)

High-speed steels   Steels with their carbides stabilized against overtempering by alloy additions.

Hot-working   Deformation that is performed above the recrystallization temperature, so that annealing occurs concurrently.

Hydration   Chemical reaction consuming water:

$$\text{Solid}_1 + H_2O \rightarrow \text{Solid}_2.$$

Hydrogen bridge   Van der Waals bond in which the hydrogen atom (proton) is attracted to electrons of neighboring atoms.

Hydrogen electrode   Standard reference electrode with the half-cell reaction:

$$H_2 \rightarrow 2\,H^+ + 2\,e^-.$$

Hydroplastic   Plastic when wet, e.g., clays.

Hypoeutectoid   *See* Steels, hypoeutectoid.

Impact strength   *See* Toughness.

Impressed voltage   Dc voltage applied to make a metal cathodic during service.

Induced dipole   Electric dipole produced by external electric field.

Induction hardening   Hardening by high-frequency induced currents for surface heating.

Ingot   A large casting that is to be subsequently rolled or forged.

Inhibitor   An additive to the electrolyte that promotes passivation.

Injection molding   Process of molding a material in a closed die. For thermoplasts, the die is appropriately cooled. For thermosets, the die is maintained at the curing temperature for the plastic.

Insulator   Nonconductor of (a) electrical or (b) thermal energy; in either case, the insulator has significant electronic resistivity. Material with filled valence bands and a large energy gap.

Interactive properties   Properties arising from the correlated behavior of two or more phases.

Intermetallic phase   Compound of two metals.

Internal structure   Arrangements of atoms, molecules, crystals, and grains within a material.

Interplanar spacing   Perpendicular distance between two adjacent crystal planes with the same index.

Interrupted quench   Two-stage quenching of steel that involves heating to form austenite and an initial quench to temperature above the start of martensite formation, followed by a second cooling to room temperature.

Interstice   Unoccupied space between atoms or ions.

Interstitial site (n-fold)   An interstice with n (4, 6, or 8) immediate atomic (or ionic) neighbors.

Ion   An atom that possesses a charge because it has had electrons added or removed.

Ion stuffing   Ion exchange that produces compressive forces because the new ions are larger than the original ion sites.

Ion vacancy   Unoccupied ion site within a crystal structure. The charge of the missing ion must be appropriately compensated.

Ionic bond   Atomic bonding by coulombic attraction of unlike ions.

Ionic radius   Semiarbitrary assigned radius to ions. Varies with coordination number. (*See* Appendix B.)

Ionization   Process of removing (or adding) electrons to neutral atoms.

Iso   Prefix indicating "the same."

Isomer   Molecules with the same composition but different structures.

Isotactic (polymers)   Long-range repetition in a polymer chain (in contrast to atactic).

Isostatic molding   Compression of powder into the desired shape, utilizing a rubber die and a hydraulic fluid (Fig. 8–8.1).

Isotherm   Line of constant temperature.

Isothermal precipitation   Precipitation from supersaturation at constant temperature.

Isothermal transformation   Transformation with time by holding at a specific temperature.

Jominy distance   Cooling rate as determined by the Jominy test.

Junction (semiconductor)   Interface between n-type and p-type semiconductors.

Lattice   The space arrangement of equivalent sites in a crystal.

Lattice constants   Dimensions of the unit cell.

Lever rule (inverse)   Calculation method for determining phases. The overall composition is at the fulcrum of the lever.

Light emitting diode, LED   A *p-n* junction device designed to produce photons by recombination.

Lignin   Important constituent of many woods; denser than cellulose.

Linear density   Items, e.g., atoms, per unit length.

Linear polymer   *See* Polymers.

Liquidus   The locus of temperatures above which only liquid is stable.

Lone pair   Electrons in a nonconnecting sp^3 orbital.

Long-range order   A repetitive pattern over many atomic distances.

Low-alloy steels   *See* Steels.

Luminescence   Light emitted by the energy released as conduction electrons recombine with electron holes.

Macromolecules   Molecules made up of hundreds to thousands of atoms.

Macrostructure   Structural features observed at very low (or without) magnification.

Magnet, permanent (hard)   Magnet with a large $(-BH)$ energy product, so that it maintains domain alignment.

Magnet, soft   Magnet that requires negligible energy for domain randomization.

Magnetic saturation   The maximum magnetization that can occur in a material.

Malleable cast iron   Cast iron that undergoes graphitization after solidification. Graphite is present as "clusters."

Martensite   Metastable body-centered phase of iron supersaturated with carbon; produced from austenite by a shear transformation during quenching.

Materials balance   Mathematical calculation of "the whole equals the sum of the parts."

Matrix   The enveloping phase in which another phase is embedded.

Mean ($\bar{X}$)   The average value (cf. Median).

Mean free path   Mean distance traveled by electrons or by elastic waves between deflections or reflections.

Mechanical properties   Characteristics of a material in response to externally applied forces.

Mechanical working   Shaping by the use of forces.

Median $\bar{M}$ The middle value (cf. Mean).

Mer   The smallest repetitive unit in a polymer.

Mesh   The screen size for particle measurement.

Metal   Materials consisting primarily of elements that release part of their valence electrons. (*See* Fig. 2–1.1.) Characterized by conductivities that decrease at elevated temperatures. Material with partially filled valence bands.

Metallic bond   Interatomic bonds in metals characterized by delocalized electrons in energy bands.

Metallic conductor   Material with a conductivity greater than 1 ohm$^{-1}\cdot$m^{-1}.

Metastability   Nonequilibrium condition.

Mho   Unit of conductance; reciprocal of ohm.

Microstructure   Structure of grains and phases. Generally requires magnification for observation.

Miller indices (*hkl*)   Index relating a plane to the reference axes of a crystal. Reciprocals of axial intercepts.

Mixed particles   Aggregate with a size distribution.

Mixture   Combination of two phases.

Mobility (μ)   The drift velocity $\bar{v}$ of an electric charge per unit electric field, $\mathscr{E}$, (m/sec)/(volt/m). Alternatively, the diffusion coefficient of a charge per volt, (m^2/sec)/volt.

Modulus of elasticity   Stress per unit strain. (*Young's modulus* is the most commonly encountered elastic modulus.)

Modulus, bulk   Hydrostatic pressure per unit volume strain.

Modulus, shear   Shear stress per unit shear strain.

Modulus, Young's   Axial stress per unit normal strain.

Molding (plastics)   Shaping by means of a contoured die.

Mole   Mass equal to the molecular weight of a material; 0.6×10^{24} molecules.

Molecular crystal   Crystals with molecules as basic units (as contrasted to atoms).

Molecular length   End-to-end root-mean-length.

Molecular size   Mass of one molecule (expressed in amu), or mass of 0.6×10^{24} molecules (expressed in grams).

Molecular size (mass-average, $\bar{M}_m$)   Average molecular size based on the mass fraction.

Molecular size (number-average, $\bar{M}_n$)   Average molecular size based on the number fraction.

Molecule   Finite groups of atoms bonded by strong attractive forces. Bonding between molecules is weak.

Molecules, linear   *See* Polymers.

Molecules, network   *See* Polymers.

Molecules, polar   Molecules containing a permanent electrical dipole.

Monel   An alloy of copper and nickel.

Monomer   A molecule with a single mer.

Multiphase microstructure   Microstructure containing two or more correlated phases.

n-type   Semiconductor having negative charge carriers, i.e., electrons.

NaCl-type structure   Fcc arrangement of ions with oppositely charged ions in the 6-fold sites.

Nernst equation   Electrode potential as a function of electrolyte concentration.

Noble   Nonreactive.

Nomenclature (AISI–SAE)   *See* Table 9–9.1.

Nonbridging oxygens   Oxygens attached to one SiO_4 tetrahedron only, thus not tying tetrahedra together.

Nonmetals   Materials consisting primarily of elements that accept or share electrons in their valence shells. (*See* Fig. 2–1.1.) They characteristically form molecules or anions.

Nonstoichiometric compounds   Compounds with noninteger atom (or ion) ratios.

Normal stresses   Stresses perpendicular to the surface.

Normalizing   Heating steel to $\sim 50°C$ ($100°F$) into the austenite range so that it will contain a uniform, fine-grained microstructure.

Notch sensitivity   A reduction in properties by the presence of stress concentrations.

Nucleation   The start of the growth of a new phase.

Octahedron   An eight-sided volume.

Orbital   Wave probabilities of atomic and molecular electrons.

Ordered (crystal)   Structure with a long-range repetitive pattern.

Orientation (polymers)   Strain process by which molecules are elongated into one preferred alignment.

Orthorhombic   A crystal with three unequal but perpendicular axes.

Over-aging   Continued aging until softening occurs.

Overtempering   Excessive heating in order to cause coalescence of carbides (and therefore softening) of tempered martensite.

Oxidation (general)   The raising of the valence level of an element.

Oxidation cell   *See* Cell, oxidation.

p-type   Semiconductor having positive charge carriers, i.e., electron holes.

p–n junction   Device having an interface between p-type and n-type semiconductors; a rectifying diode.

Packing factor   True volume per unit of bulk volume.

Passivation   The condition in which normal corrosion is impeded by an adsorbed surface film on the electrode.

Pearlite   A microstructure of ferrite and lamellar carbide of eutectoid analysis.

Periodic table   *See* Fig. 2–1.1.

Phase (material)   A physically homogeneous part of a materials system. (*See* Section 4–5.)

Phase, transition   Metastable phase that forms as an intermediate step in a reaction.

Phases, analysis of   Phase compositions expressed in terms of chemical components.

Phases, quantity fraction of   Material compositions expressed in terms of phase fractions.

Phase boundary   Compositional and/or structural discontinuity between two phases.

Phase diagram   Graph of phase stability areas with analysis and environment (usually temperature) as coordinates.

Phase diagram, isothermal cut   A constant temperature section of a phase diagram.

Phase diagram, one-phase area   Part of a phase diagram where both temperature *and* the chemical analysis of the phase may be varied.

Phase diagram, two-phase area   Part of a phase diagram beyond the solubility limit curves so a second phase is necessary. Temperature and the phase analyses *cannot* be varied independently.

Phase diagram, three-phase temperature   Invariant temperatures at which three phases can coexist.

Phenol-formaldehyde (PF)   Condensation or step-reaction polymer of phenol, C_6H_5OH, and formaldehyde, CH_2O. (*See* Fig. 7–4.1.)

Phosphorescence   Luminescence that is prolonged until a period of time elapses after excitation.

Photoconduction   Conduction arising from activation of electrons across the energy gap by means of light.

Photon   A quantum of light.

Piezoelectric   Dielectric materials with structures that are asymmetric, so that their centers of positive and negative charges are not coincident. As a result the polarity is sensitive to pressures that change the dipole distance, and the polarization.

Planar density   Items, e.g., atoms per unit area.

Plastic constraint   Prevention of plastic deformation in a ductile material by the presence of an adjacent nonductile material.

Plastic deformation   Permanent deformation arising from the displacement of atoms (or molecules) to new lattice sites.

Plasticizer   Micromolecules added among macromolecules to induce deformation and flexibility.

Plastics   Materials consisting predominantly of large molecules of nonmetallic elements or compounds. (*See* Polymers.) Moldable organic resins.

Point defect   Crystal imperfection involving one (or a very few) atoms.

Poisson's ratio (ν)   Ratio (negative) of lateral to axial strain.

Polar group   A local electric dipole within the polymer molecule.

Polarization (chemical)   Depletion of reactants at a cathode surface, thus reducing the corrosion current.

Polarization (electric, $\mathscr{P}$)   The dipole moment $\mu\,(=Qd)$ per unit volume.

Polarization (molecules)   Displacement of centers of positive and negative charges.

Polycrystalline   Materials with more than one crystal; therefore with grain boundaries.

Polyester   Polymer with $-\overset{\overset{\displaystyle O}{\|}}{C}-O-$ segments.

Polyethylene   Polymer of $(C_2H_4)_n$.

Polyfunctional Molecule with three or more sites at which there can be joining reactions with adjacent molecules.

Polymer Nonmetallic material consisting of (large) macromolecules composed of many repeating units; the technical term for plastics.

Polymer, addition Polymer formed by a double bond opening up into two single bonds as the monomer joins a reactive site of a growing chain (chain-reaction polymerization).

Polymer, condensation Polymer formed by a reaction that also produces a small by-product molecule.

Polymer, linear Polymer of bifunctional mers.

Polymer, network Polymers containing polyfunctional mers that form a 3-dimensional structure.

Polymorphism A composition with more than one crystal structure.

Pores, closed Pores without access to the surrounding environment.

Pores, open Interconnecting pores.

Porosity, apparent (Open-pore volume)/(Total volume).

Porosity, total (Open- + Closed-pore volume)/(Total volume).

Precipitation hardening Hardening by the formation of clusters prior to precipitation (also called age-hardening).

Precipitation reactions Exsolution from supersaturation.

Preferred orientation A nonrandom alignment of crystals or molecules.

Pressing (ceramic) Agglomeration of particulate materials by pressure.

Primary bond Strong (> 200 kJ/mole) interatomic bonds of the covalent, ionic, or metallic type.

Process anneal (steel) Annealing close to, but below, the eutectoid temperature.

Propagation Polymer growth step through the reaction: Reactive site + Monomer → New reactive site.

Properties Quantitative attributes of materials, e.g., density, strength, conductivity.

Quantity fraction (phase mixtures) Material composition expressed in terms of phases.

Quartz The most common phase of SiO_2.

Quench Cooling accelerated by immersion in agitated water or oil.

Radiation damage Structural defects arising from exposure to radiation.

Radius ratio (r/R) Ratio of ionic radii of coordinated cations, r, and anions, R.

Rate (solids) Transformation per unit time.

Reactive site (•) Open end of a free radical.

Recombination Annihilation of electron–hole pairs.

Recombination time (τ) *See* relaxation time.

Recovery Loss of resistivity, or related behaviors, originating during strain hardening, by annealing out point defects. (Cf. recrystallization.)

Recrystallization The formation of new annealed grains from previously strain-hardened grains.

Recrystallization temperature Temperature at which recrystallization is spontaneous.

Rectifier Electric "valve" that permits forward current and prevents reverse current.

Reduction Removal of oxygen from an oxide; the lowering of the valence level of an element.

Reduction of area (Red. of A) Measure of plastic deformation at the point of fracture.

Reinforcement Component of composites with high elastic modulus and high strength.

Relative dielectric constant (κ) *See* Dielectric constant.

Relaxation Decay of a property parameter.

Relaxation time (τ) Time required to decay an exponentially dependent value to 37%, that is, $1/e$, of the original value.

Residual stresses   Stresses induced as a result of differences in temperature or volume.

Resistivity (ρ)   Reciprocal of conductivity (usually expressed in ohm · m).

Resistivity coefficient, solid-solution (ρ_x)   Resistivity arising from additions of solute.

Resistivity coefficient, thermal (γ_T)   Electrical resistivity arising from thermal agitation.

Resolved shear stress   Stress vector in slip plane.

Rockwell hardness (R)   A test utilizing an indenter; the depth of indentation is a measure of the hardness.

Rolling   Mechanical working between two circular rolls (Fig. 6–2.2).

Root-mean-square length   Statistical end-to-end length of molecules.

Rubber   A polymeric material with a high elastic yield strain.

Rubbery plateau   For a plastic, the range of temperature between the glass temperature and melting temperature, which has a viscoelastic modulus that is relatively constant.

Sacrificial anode   Expendable metal that is anodic to the product it is to protect.

Schmid's law   Relates axial stress to resolved shear stress (Eq. 6–4.1).

Schottky defect   Ion-pair vacancies.

Scission   Degradation of polymers by radiation.

Secondary bond   Weak (< 40 kJ/mole) interatomic bonds arising from dipoles within the atoms or molecules.

Segregation   Heterogeneities in composition.

Self-diffusion   Diffusion of atoms within their own structure.

Semiconductor   A material with controllable conductivities, intermediate between insulators and conductors.

Semiconductor, compound   compounds of two or more elements with an average of four shared electrons per atom.

Semiconductor, defect   Nonstoichiometric transition metal oxides that gain their conductivity from multivalent ions.

Semiconductor, extrinsic   Semiconduction from impurity sources.

Semiconductor, intrinsic   Semiconduction of pure material. The electrons are excited across the energy gap.

Semiconductor, n-type   Impurities provide donor electrons to the conduction band. These electrons are the majority charge carriers.

Semiconductor, p-type   Impurities provide acceptor sites for electrons from the valence band. Electron holes are the majority charge carriers.

Shear strain (γ)   Tangent of shear angle α developed from shear stress.

Shear stress (τ)   Shear force per unit area.

Shear stress, critical   Minimum resolved stress to produce shear.

Shear stress, resolved   Stress vector in slip plane.

Short-range order   Specific first-neighbor arrangement of atoms, but random long-range arrangements.

Sieve series   One of several standardized sets of screens or meshes used to determine particle size distributions.

Silicate structures   Materials containing SiO_4 tetrahedra.

Silicates, chain   Polymerized silicates with two oxygens of each tetrahedron, jointly shared with two neighboring tetrahedra to provide a chainlike structure.

Silicates, network   Three-dimensional silicate structure with each tetrahedral oxygen shared.

Silicates, sheet   Polymerized silicate with three oxygens of each tetrahedron shared to provide a two-dimensional, sheetlike structure, e.g., mica.

Silicone   "Silicates" with organic side radicals; thus silicon-based polymeric molecules.

Simple cubic   A cubic unit cell with equivalent points at the corners only.

Single-phase materials   Materials containing only one basic structure.

Sintering   Agglomeration by thermal means.

Sintering, liquid-phase   Agglomeration by capillary action of a liquid, *and* diffusion through a liquid (which later solidifies).

Sintering, solid   Agglomeration by solid diffusion.

SiO$_4$ tetrahedra   Coordination unit of four oxygens surrounding a silicon atom.

Slip (ceramic processing)   A slurry, or suspension for casting into porous mold of the desired product shape. The mold absorbs the liquid.

Slip (deformation)   Shear deformation along a crystal plane.

Slip direction   Crystal direction of the displacement vector.

Slip plane   Crystal plane along which slip occurs.

Slip system   Combination of slip directions on slip planes that have low critical shear stresses.

Slip vector   Displacement distance of a dislocation. It is parallel to a screw dislocation and perpendicular to an edge dislocation.

Snell's law   Law of refraction at a boundary (Eq. 10–4.1).

Soda-lime glass   The most widely produced glass; it serves as a basis for window and container glass. Contains $Na_2O/CaO/SiO_2$ in approximately a 1/1/6 ratio.

Solder   Metals that melt below 425°C ($\sim$800°F) that are used for joining. Commonly, Pb–Sn alloys, but may also be other materials, even glass.

Solid-phase reactions   Reactions involving phase changes within solids.

Solid solution, interstitial   Crystals that contain a second component in their interstices. The basic structure is unaltered.

Solid solution, ordered   A substitutional solid solution with a preference by each of the components for specific lattice sites.

Solid solution, substitutional   Crystals with a second component substituted for solvent atoms in the basic structure.

Solidus   The locus of temperatures below which only solids are stable.

Solubility limit   Maximum solute addition without supersaturation.

Solute   The minor component of a solution.

Solution   A single phase containing more than one component.

Solution hardening   Increased strength arising from the creation of solid solutions (or from pinning of dislocations by solute atoms).

Solution treatment   Heating to induce solid solutions.

Solvent   The major component of a solution.

Specific gravity   Ratio of density of a material to the density of water.

Specific heat   Ratio of heat capacity of a material to the heat capacity of water.

Spheroidite   Microstructure of coarse spherical carbides in a ferrite matrix.

Spheroidization   Process of making spheroidite, generally by extensive overtempering.

Spin (magnetism)   The assumed rotational movement of an electron in its orbit within an atom.

Spinel   Cubic $[A_mB_nX_p]$ compounds in which A is divalent and B is trivalent. These compounds are commonly used in ceramic magnets and for refractory purposes.

Spinning   Mechanical working on a mandrel for sheet metal (Fig. 6–2.3).

Stamping   Mechanical working process in a die for sheet metal.

Standard deviation (SD)   Measure of variation of data.

Steel   Iron-base alloys, commonly containing carbon. In practice, the carbon can all be dissolved by heat treatment; hence, <2.0 w/o C.

Steel, eutectoid   Steel with a carbon content to give 100% pearlite on annealing.

Steel, hypereutectoid   Steel with *more* carbon than an eutectoid steel; hence it can contain proeutectoid *carbide*.

Steel, hypoeutectoid   Steel with *less* carbon than an eutectoid steel; hence it can contain pro-eutectoid *ferrite*.

Steel, low-alloy   Steel containing up to 5% alloying elements other than carbon. Phase equilibria are related to the Fe–C diagram.

Steel, plain-carbon   Basically Fe–C alloys with minimal alloy content.

Steel, stainless   High-alloy steel (usually containing Cr, or Cr + Ni) designed for resistance to corrosion and/or oxidation.

Steel, tool   Steel with high tempering temperatures, usually containing carbide stabilizers such as Cr, Mn, Mo, V, W.

Stereoisomers   Configuration isomers of polymers.

Stereospecific   Covalent bonding between specific atom pairs (in contrast to nondirectional coulombic attractions).

Sterling silver   An alloy of 92.5 Ag and 7.5 Cu. (This corresponds to nearly the maximum solubility of copper in silver.)

Strain (e)   Deformation from an applied stress.

Strain, elastic   Reversible deformation.

Strain, plastic   Permanent deformation following slip.

Strain, true (ϵ)   Plastic strain calculated on the basis of the concurrent cross-sectional area.

Strain hardening   Increased hardness (and strength) arising from plastic deformation.

Strain point (glass)   Temperature at which the viscosity, η, of a glass is $10^{13.5}$ Pa·s ($10^{14.5}$ poises).

Strength   Resistance to mechanical stresses.

Strength, breaking   Stress at fracture.

Strength, tensile (S_t)   Maximum load per unit original area. This is the "ultimate strength" used for design purposes.

Strength, yield (S_y)   Stress to give the initial significant plastic deformation.

Stress s   Force per unit area.

Stress concentration factor   Increase in stress at a notch.

Stress relaxation   Decay of stress at constant strain by molecular rearrangement.

Stress relief   Removal of residual stresses by heating.

Stress rupture   Time-dependent rupture resulting from constant stress (usually at elevated temperatures).

Structure   Geometric relationships of material components.

Supercooling   Cooling below the solubility limit without precipitation.

Surface   Boundary between a condensed phase and gas.

Symmetry   Structural correspondence of size, shape, and relative position.

System (phase diagram)   Compositions of equilibrated components.

Systéme International   Nearly worldwide standards for units and dimensions.

Temper (hardness)   Extent of strain-hardening.

Tempered glass   Glass with surface compressive stresses induced by heat treatment.

Tempered martensite   A microstructure of ferrite and carbide obtained by heating martensite.

Tempering   A toughening process in which martensite is heated to initiate a ferrite-plus-carbide microstructure.

Termination   Finalizing step of addition polymerization. A common reaction involves the joining of the reactive sites at the growing ends of two propagating molecules.

Tetragonal (crystal)   Two of three axes equal; all three at right angles.

Tetrahedron   A four-sided solid.

Texture Macroscopic structures.

Thermal agitation Thermally induced movements of atoms and molecules.

Thermal conductivity (k) Coefficient between thermal flux and thermal gradient.

Thermal diffusivity (h) Diffusion coefficient for thermal energy; $k/\rho c_p$.

Thermal expansion Expansion caused by increased atomic vibrations due to increased thermal energy.

Thermal expansion coefficient (α) (change in dimensions)/(change in temperature).

Thermistor Semiconductor device with a high resistance dependence on temperature. It may be calibrated as a thermometer.

Thermocouple A temperature-measuring device utilizing the thermoelectric effect of dissimilar wires.

Thermoplastic polymers Polymers that soften and are moldable due to the effect of heat. They are hardened by cooling, but soften again during subsequent heating cycles.

Thermosetting polymers Polymers that polymerize further on heating; therefore, heat causes them to take on an additional set. They do not soften with subsequent heating.

Toughness A measure of the energy required for mechanical failure.

Tracheid Biological cell structure of wood.

Trans- A prefix indicating "across" (cf. *Cis*).

Transducer A material or device that converts energy from one form to another, specifically electrical energy to or from mechanical energy.

Transformation temperature Temperature of an equilibrium phase change.

Transistor Semiconductor device for amplification of current.

Transition temperature (steels) Temperature (range) of change from ductile to nonductile fracture.

Translation Vector displacement between equivalent lattice sites.

Trifunctional Molecule with three reaction sites for joining with adjacent molecules.

Ultimate strength *See* Tensile strength.

Unit cell A small (commonly the smallest) repetitive volume that comprises the complete lattice pattern of a crystal.

Unit mix Ratio of gravel, sand, and water to portland cement ($=1$) in a concrete mix. May be based on either weight or volume.

Vacancy Unfilled lattice site.

Valence band Filled energy band below the energy gap. Conduction in this band requires holes.

Valence electrons Electrons from the outer shell of an electron.

Van der Waals forces Secondary bonds arising from structural polarization.

Vinyl-type compounds *See* Table 7–2.1.

Viscoelasticity Combination of viscous flow and elastic behavior.

Viscoelastic modulus (M_{ve}) Ratio of shear stress to the sum of elastic deformation, γ_e, and viscous flow, v_f.

Viscosity (η) The ratio of shear stress to velocity gradient.

Viscous forming Thermoplastic shaping of glass or polymeric products.

Vitreous Glassy or glasslike.

Volume, apparent True volume plus closed-pore volume (or total volume minus open-pore volume).

Volume, total (bulk) True volume plus pore volumes (both closed and open).

Volume, true Volume inherent to the material (exclusive of pore volumes).

Vulcanization Treatment of rubber with sulfur to cross-link the elastomer chains.

Warpage (wood) Nonuniform shrinkage.

Wiedemann–Franz ratio (k/σ) Ratio of thermal to electrical conductivity.

Wood, spring   Grain with larger, less dense tracheids (early wood).

Wood, summer   Grain with smaller, thicker-walled cells (late wood).

Working range (glass)   Temperature range lying between viscosities of $10^{2.5}$ Pa·s to 10^6 Pa·s.

X-ray diffraction   Method for determining interplanar spacings of crystals.

Yield point   The point on a stress–strain curve of sudden plastic yield at the start of plastic deformation (common only in low-carbon steels).

Young's modulus (E) *See* Modulus of elasticity.

Zener diode   A p–n junction with controlled breakdown voltage under reverse bias.

Zone melting   Purification process utilizing segregation by directional solidification.

Index

Italicized page numbers refer to glossary terms: F refers to Foreword.